MOBILE FADING CHANNELS

MOBILE FADING CHANNELS

Matthias Pätzold
Professor of Mobile Communications
Agder University College, Grimstad, Norway

JOHN WILEY & SONS, LTD

Originally published in the German language by Friedr. Vieweg & Sohn Verlagsgesellschaft mbH, D-65189 Wiesbaden, Germany, under the title "Matthias Pätzold: Mobilfunkkanäle. 1. Auflage (1st Edition)". Copyright Friedr. Vieweg & Sohn Verlagsgesellschaft mbH, Braunschweig/Wiesbaden, 1999.

Baffins Lane, Chichester,
West Sussex, PO19 1UD, England

National 01243 779777
International (+44) 1243 779777

e-mail (for orders and customer service enquiries): cs-books@wiley.co.uk
Visit our Home Page on http://www.wiley.co.uk or http://www.wiley.com

Other Wiley Editorial Offices

John Wiley & Sons, Inc., 605 Third Avenue,
New York, NY 10158-0012, USA

WILEY-VCH Verlag GmbH
Pappelallee 3, D-69469 Weinheim, Germany

John Wiley & Sons Australia Ltd, 33 Park Road, Milton,
Queensland 4064, Australia

John Wiley & Sons (Canada) Ltd, 22 Worcester Road
Rexdale, Ontario, M9W 1L1, Canada

John Wiley & Sons (Asia) Pte Ltd, 2 Clementi Loop #02-01,
Jin Xing Distripark, Singapore 129809

A catalogue record for this book is available from the British Library

ISBN 0471 49549 2

Produced from PostScript files supplied by the author.
Printed and bound in Great Britain by Biddles Ltd, Guildford and King's Lynn.
This book is printed on acid-free paper responsibly manufactured from sustainable forestry, in which at least t
trees are planted for each one used for paper production.

Preface

This book results from my teaching and research activities at the Technical University of Hamburg-Harburg (TUHH), Germany. It is based on my German book "Mobilfunkkanäle — Modellierung, Analyse und Simulation" published by Vieweg & Sohn, Braunschweig/Wiesbaden, Germany, in 1999. The German version served as a text for the lecture *Modern Methods for Modelling of Networks*, which I gave at the TUHH from 1996 to 2000 for students in electrical engineering at masters level.

The book mainly is addressed to engineers, computer scientists, and physicists, who work in the industry or in research institutes in the wireless communications field and therefore have a professional interest in subjects dealing with mobile fading channels. In addition to that, it is also suitable for scientists working on present problems of stochastic and deterministic channel modelling. Last, but not least, this book also is addressed to master students of electrical engineering who are specialising in mobile radio communications.

In order to be able to study this book, basic knowledge of probability theory and system theory is required, with which students at masters level are in general familiar. In order to simplify comprehension, the fundamental mathematical tools, which are relevant for the objectives of this book, are recapitulated at the beginning. Starting from this basic knowledge, nearly all statements made in this book are derived in detail, so that a high grade of mathematical unity is achieved. Thanks to sufficient advice and help, it is guaranteed that the interested reader can verify the results with reasonable effort. Longer derivations interrupting the flow of the content are found in the Appendices. There, the reader can also find a selection of MATLAB-programs, which should give practical help in the application of the methods described in the book. To illustrate the results, a large number of figures have been included, whose meanings are explained in the text. Use of abbreviations has generally been avoided, which in my experience simplifies the readability considerably. Furthermore, a large number of references is provided, so that the reader is led to further sources of the almost inexhaustible topic of mobile fading channel modelling.

My aim was to introduce the reader to the fundamentals of modelling, analysis, and simulation of mobile fading channels. One of the main focuses of this book is the treatment of deterministic processes. They form the basis for the development of efficient channel simulators. For the design of deterministic processes with given correlation properties, nearly all the methods known in the literature up to now are introduced, analysed, and assessed on their performance in this book. Further focus is put on the derivation and analysis of stochastic channel models as well as on the development of highly precise channel simulators for various classes of frequency-selective and frequency-nonselective mobile radio channels. Moreover, a primary topic is the fitting of the statistical properties of the designed channel models to the statistics of real-world channels.

At this point, I would like to thank those people, without whose help this book would never have been published in its present form. First, I would like to express my warmest thanks to Stephan Kraus and Can Karadogan, who assisted me with the English translation considerably. I would especially like to thank Frank Laue for performing the computer experiments in the book and for making the graphical plots, which decisively improved the vividness and simplified the comprehension of the text. Sincerely, I would like to thank Alberto Díaz Guerrero and Qi Yao for reviewing most parts of the manuscript and for giving me numerous suggestions that have helped me to shape the book into its present form. Finally, I am also grateful to Mark Hammond and Sarah Hinton my editors at John Wiley & Sons, Ltd.

Matthias Pätzold

Grimstad
January 2002

Contents

1

INTRODUCTION

1.1 THE EVOLUTION OF MOBILE RADIO SYSTEMS

For several years, the mobile communications sector has definitely been the fastest-growing market segment in telecommunications. Experts agree that today we are just at the beginning of a global development, which will increase considerably during the next years. Trying to find the factors responsible for this development, one immediately discovers a broad range of reasons. Certainly, the liberalization of the telecommunication services, the opening and deregulation of the European markets, the topping of frequency ranges around and over 1 GHz, improved modulation and coding techniques, as well as impressive progress in the semiconductor technology (e.g., large-scale integrated CMOS- and GaAs-technology), and, last but not least, a better knowledge of the propagation processes of electromagnetic waves in an extraordinary complex environment have made their contribution to this success.

The beginning of this turbulent development now can be traced to more than 40 years ago. The *first generation mobile radio systems* developed at that time were entirely based on analog technique. They were strictly limited in their capacity of subscribers and their accessibility. The first mobile radio network in Germany was in service between 1958 and 1977. It was randomly named A-net and was still based on manual switching. Direct dialling was at first possible with the B-net, introduced in 1972. Nevertheless, the calling party had to know where the called party was located and, moreover, the capacity limit of 27 000 subscribers was reached fairly quickly. The B-net was taken out of service on the 31st of December 1994. Automatic localization of the mobile subscriber and passing on to the next cell was at first possible with the cellular C-net introduced in 1986. It operates at a frequency range of 450 MHz and has a Germany-wide accessibility with a capacity of 750 000 subscribers.

Second generation mobile radio systems are characterized by digitalization of the networks. The GSM standard (GSM: Groupe Spécial Mobile)[1] developed in Europe is generally accepted as the most elaborated standard worldwide. The D-net, brought into service in 1992, is based on the GSM standard. It operates at a frequency range of 900 MHz and offers all subscribers a Europe-wide coverage. In addition to this, the E-net (Digital Cellular System, DCS 1800) has been running parallel to the

[1] By now GSM stands for "Global System for Mobile Communications".

D-net since 1994, operating at a frequency range of 1800 MHz. Mainly, these two networks only differ in their respective frequency range. In Great Britain, however, the DCS 1800 is known as PCN (Personal Communications Network). Estimates say the amount of subscribers using mobile telephones will in Europe alone grow from 92 million at present to 215 million at the end of 2005. In consequence, it is expected that in Europe the number of employees in this branch will grow from 115 000 at present to 1.89 million (source of information: Lehman Brothers Telecom Research estimates). The originally European GSM standard has in the meantime become a worldwide mobile communication standard that has been accepted by 129 (110) countries at the end of 1998 (1997). The network operators altogether ran 256 GSM networks with over 70.3 million subscribers at the end of 1997 worldwide. But only one year later (at the end of 1998), the amount of GSM networks had increased to 324 with 135 million subscribers. In addition to the GSM standard, a new standard for cordless telephones, the DECT standard (DECT: Digital European Cordless Telephone), was introduced by the European Telecommunications Standard Institute (ETSI). The DECT standard allows subscribers moving at a fair pace to use cordless telephones at a maximum range of about 300 m.

In Europe, *third generation mobile radio systems* is expected to be practically ready for use at the beginning of the twenty-first century with the introduction of the Universal Mobile Telecommunications System (UMTS) and the Mobile Broadband System (MBS). With UMTS, in Europe one is aiming at integrating the various services offered by second generation mobile radio systems into one universal system [Nie92]. An individual subscriber can then be called at any time, from any place (car, train, aircraft, etc.) and will be able to use all services via a universal terminal. With the same aim, the system IMT 2000 (International Mobile Telecommunications 2000)[2] is being worked on worldwide. Apart from that, UMTS/IMT 2000 will also provide multimedia services and other broadband services with maximum data rates up to 2 Mbit/s at a frequency range of 2 GHz. MBS plans mobile broadband services up to a data rate of 155 Mbit/s at a frequency range between 60 and 70 GHz. This concept is aimed to cover the whole area with mobile terminals, from fixed optical fibre networks over optical fibre connected base stations to the indoor area. For UMTS/IMT 2000 as for MBS, communication by satellites will be of vital importance.

From future satellite communication it will be expected — besides supplying areas with weak infrastructure — that mobile communication systems can be realized for global usage. The present INMARSAT-M system, based on four geostationary satellites (35 786 km altitude), will at the turn of the century be replaced by satellites flying on non-geostationary orbits at medium height (Medium Earth Orbit, MEO) and at low height (Low Earth Orbit, LEO). The MEO satellite system is represented by ICO with 12 satellites circling at an altitude of 10 354 km, and typical representatives of the LEO satellite systems are IRIDIUM (66 satellites, 780 km altitude), GLOBALSTAR (48 satellites, 1 414 km altitude), and TELEDESIC (288 satellites,

[2] IMT 2000 was formally known as FPLMTS (Future Public Land Mobile Telecommunications System).

1 400 km altitude)[3] [Pad95]. A coverage area at 1.6 GHz is intended for hand-portable terminals that have about the same size and weight as GSM mobile telephones today. On the 1st of November 1998, IRIDIUM took the first satellite telephone network into service. An Iridium satellite telephone cost 5 999 DM in April 1999, and the price for a call was, depending on the location, settled at 5 to 20 DM per minute. Despite the high prices for equipment and calls, it is estimated that in the next ten years about 60 million customers worldwide will buy satellite telephones.

At the end of this technical evolution from today's point of view is the development of the *fourth generation mobile radio systems*. The aim of this is integration of broadband mobile services, which will make it necessary to extend the mobile communication to frequency ranges up to 100 GHz.

Before the introduction of each newly developed mobile communication systems a large number of theoretical and experimental investigations have to be made. These help to answer open questions, e.g., how existing resources (energy, frequency range, labour, ground, capital) can be used economically with a growing number of subscribers and how reliable, secure data transmission can be provided for the user as cheaply and as simple to handle as possible. Also included are estimates of environmental and health risks that almost inevitably exist when mass-market technologies are introduced and that are only to a certain extent tolerated by a public becoming more and more critical. Another boundary condition growing in importance during the development of new transmission techniques is often the demand for compatibility with existing systems. To solve the technical problems related to these boundary conditions, it is necessary to have a firm knowledge of the specific characteristics of the mobile radio channel. The term mobile radio channel in this context is the physical medium that is used to send the signal from the transmitter to the receiver [Pro95]. However, when the channel is modelled, the characteristics of the transmitting and the receiving antenna are in general included in the channel model. The basic characteristics of mobile radio channels are explained later. The thermal noise is not taken into consideration in the following and has to be added separately to the output signal of the mobile radio channel, if necessary.

1.2 BASIC KNOWLEDGE OF MOBILE RADIO CHANNELS

In mobile radio communications, the emitted electromagnetic waves often do not reach the receiving antenna directly due to obstacles blocking the line-of-sight path. In fact, the received waves are a superposition of waves coming from all directions due to reflection, diffraction, and scattering caused by buildings, trees, and other obstacles. This effect is known as *multipath propagation*. A typical scenario for the terrestrial mobile radio channel is shown in Figure 1.1. Due to the multipath propagation, the received signal consists of an infinite sum of attenuated, delayed, and phase-shifted replicas of the transmitted signal, each influencing each other. Depending on the phase of each partial wave, the superposition can be constructive or destructive. Apart

[3] Originally TELEDESIC planned to operate 924 satellites circling at an altitude between 695 and 705 km.

from that, when transmitting digital signals, the form of the transmitted impulse can be distorted during transmission and often several individually distinguishable impulses occur at the receiver due to multipath propagation. This effect is called the *impulse dispersion*. The value of the impulse dispersion depends on the propagation delay differences and the amplitude relations of the partial waves. We will see later on that multipath propagation in a frequency domain expresses itself in the non-ideal frequency response of the transfer function of the mobile radio channel. As a consequence, the channel distorts the frequency response characteristic of the transmitted signal. The distortions caused by multipath propagation are linear and have to be compensated for on the receiver side, for example, by an equalizer.

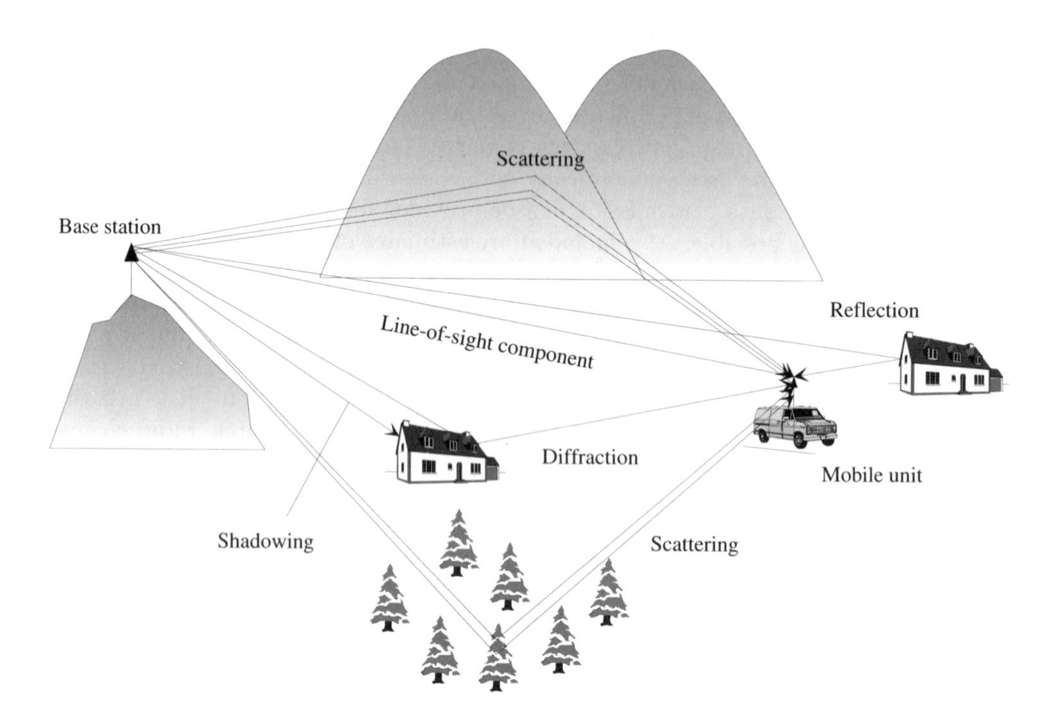

Figure 1.1: Typical mobile radio scenario illustrating multipath propagation in a terrestrial mobile radio environment.

Besides the multipath propagation, also the *Doppler effect* has a negative influence on the transmission characteristics of the mobile radio channel. Due to the movement of the mobile unit, the Doppler effect causes a frequency shift of each of the partial waves. The *angle of arrival* α_n, which is defined by the direction of arrival of the nth incident wave and the direction of motion of the mobile unit as shown in Figure 1.2, determines the *Doppler frequency* (*frequency shift*) of the nth incident wave according to the relation

$$f_n := f_{max} \cos \alpha_n \, . \tag{1.1}$$

In this case, f_{max} is the *maximum Doppler frequency* related to the speed of the mobile unit v, the speed of light c_0, and the carrier frequency f_0 by the equation

$$f_{max} = \frac{\mathrm{v}}{c_0} f_0 \,. \tag{1.2}$$

The maximum (minimum) Doppler frequency, i.e., $f_n = f_{max}$ ($f_n = -f_{max}$), is reached for $\alpha_n = 0$ ($\alpha_n = \pi$). In comparison, though, $f_n = 0$ for $\alpha_n = \pi/2$ and $\alpha_n = 3\pi/2$. Due to the Doppler effect, the spectrum of the transmitted signal undergoes a frequency expansion during transmission. This effect is called the *frequency dispersion.* The value of the frequency dispersion mainly depends on the maximum Doppler frequency and the amplitudes of the received partial waves. In the time domain, the Doppler effect implicates that the impulse response of the channel becomes time-variant. One can easily show that mobile radio channels fulfil the principle of superposition [Opp75, Lue90] and therefore are linear systems. Due to the time-variant behaviour of the impulse response, mobile radio channels therefore generally belong to the class of linear time-variant systems.

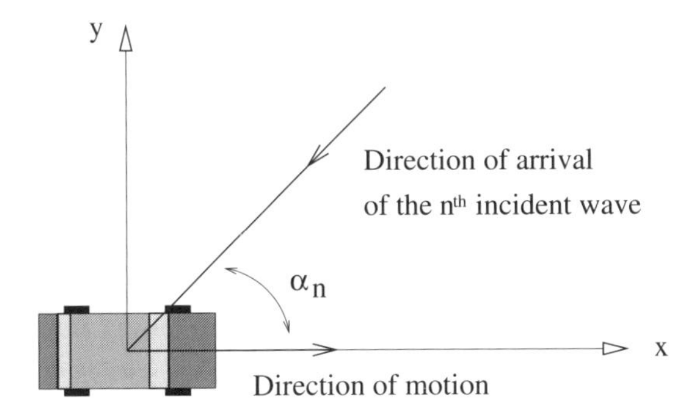

Figure 1.2: Angle of arrival α_n of the nth incident wave illustrating the Doppler effect.

Multipath propagation in connection with the movement of the receiver and/or the transmitter leads to drastic and random fluctuations of the received signal. Fades of 30 to 40 dB and more below the mean value of the received signal level can occur several times per second, depending on the speed of the mobile unit and the carrier frequency [Jak93]. A typical example of the behaviour of the received signal in mobile communications is shown in Figure 1.3. In this case, the speed of the mobile unit is $\mathrm{v} = 110\,\mathrm{km/h}$ and the carrier frequency is $f_0 = 900\,\mathrm{MHz}$. According to (1.2), this corresponds to a maximum Doppler frequency of $f_{max} = 91\,\mathrm{Hz}$. In the present example, the distance covered by the mobile unit during the chosen period of time from 0 to 0.327 s is equal to 10 m.

In digital data transmission, the momentary fading of the received signal causes *burst errors*, i.e., errors with strong statistical connections to each other [Bla84]. Therefore, a fading interval produces burst errors, where the burst length is determined by the

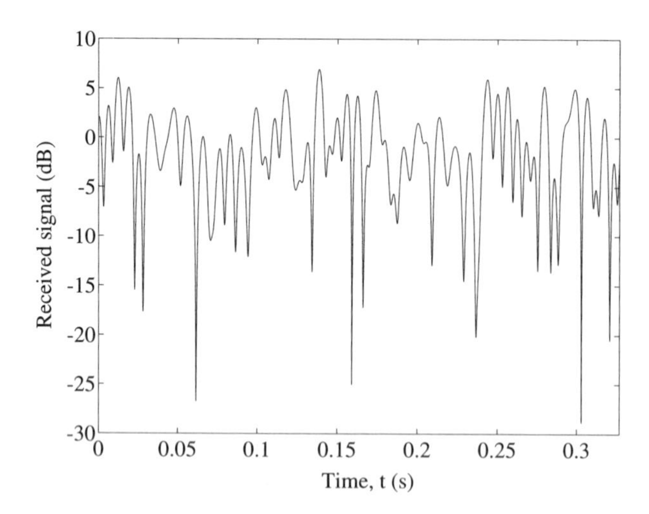

Figure 1.3: Typical behaviour of the received signal in mobile communications.

duration of the fading interval for which the expression *duration of fades* has been introduced in [Kuc82]. Corresponding to this, a connecting interval produces a bit sequence almost free of errors. Its length depends on the duration of the connecting interval for which the term *connecting time interval* has been established [Kuc82]. As suitable measures for error protection, high performance procedures for channel coding are called in to help. Developing and dimensioning of codes require knowledge of the statistical distribution of the duration of fades and of the connecting time intervals as exact as possible. The task of channel modelling now is to record and to model the main influences on signal transmission to create a basis for the development of transmission systems [Kit82].

Modern methods of modelling mobile radio channels are especially useful, for they not only can model the statistical properties of real-world (measured) channels regarding the probability density function (first order statistics) of the channel amplitude sufficiently enough, but also regarding the level-crossing rate (second order statistics) and the average duration of fades (second order statistics). Questions connected to this theme will be treated in detail in this book. Mainly, two goals are aimed at. The first one is to find stochastic processes especially suitable for modelling frequency-nonselective and frequency-selective mobile radio channels. In this context, we will establish a channel model described by ideal (not realizable) stochastic processes as the *reference model* or as the *analytical model.* The second goal is the derivation of efficient and flexible simulation models for various typical mobile radio scenarios. Following these aims, the relations shown in Figure 1.4, which demonstrates the connections between the physical channel, the stochastic reference model, and the therefrom derivable deterministic simulation model, will accompany us throughout the book. The usefulness and the quality of a reference model and the corresponding simulation model are ultimately judged on how well its individual statistics can be adapted to the statistical properties of measured or specified channels.

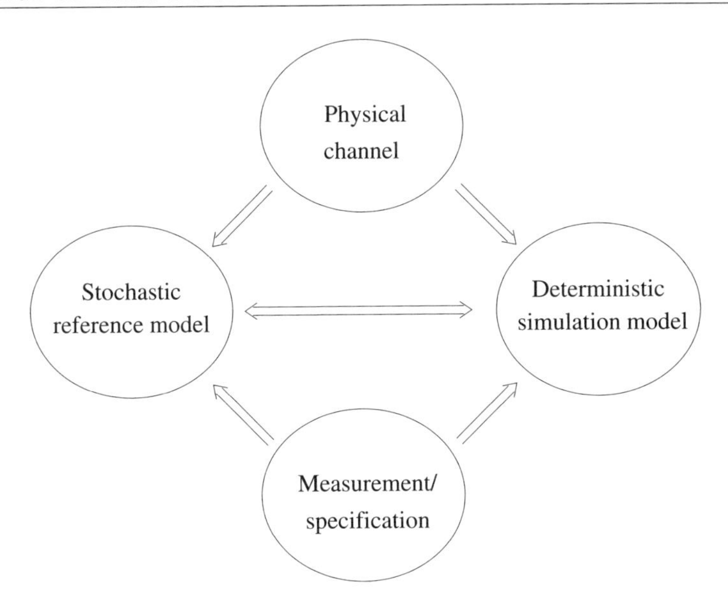

Figure 1.4: Relationships between the physical channel, the stochastic reference model, the deterministic simulation model, and the measurement or specification.

1.3 STRUCTURE OF THIS BOOK

A good knowledge of statistics and system theory are the necessary tools for engineers in practice as well as for scientists working in research areas, making the approach to a deeper understanding of channel modelling possible. Therefore, in **Chapter 2** some important terms, definitions, and formulae often referred to in following chapters will be recapitulated. Chapter 2 makes the reader familiar with the nomenclature used in the book.

Building on the terms introduced in Chapter 2, in **Chapter 3** Rayleigh and Rice processes are dealt with as reference models to describe frequency-nonselective mobile radio channels. These are at first described in general (Section 3.1). Then, a description of the most frequently used Doppler power spectral densities (Jakes or Clarke power spectral density and Gaussian power spectral density) and their characteristic quantities such as the Doppler shift and the Doppler spread are given (Section 3.2). After that, in Section 3.3 the statistical properties of the first kind (probability density of amplitude and phase) and of the second kind (level-crossing rate and average duration of fades) are investigated. Chapter 3 ends with an analysis of the statistics of the duration of fades of Rayleigh processes.

In **Chapter 4**, it is at first made clear that, from the developers of simulation models point of view, analytical models represent reference models to a certain extent. Their relevant statistical properties will be modelled sufficiently exactly with the smallest possible realization expenditure. To solve this problem, various deterministic and statistic methods have been proposed in the literature. The heart of many procedures for channel modelling is based on the principle that filtered Gaussian random processes can be approximated by a sum of weighted harmonic functions. This principle in itself is not at all new, but can historically be traced back to basic works of S. O. Rice [Ric44, Ric45]. In principle, all attempts to compute the parameters of a simulation model can be classified as either statistic, deterministic or as a combination of both. A fact though is that the resulting simulation model is definitely of pure deterministic nature, which is made clear in Section 4.1. The analysis of the elementary properties of deterministic simulation systems is therefore mainly performed by the system theory and signal theory (Section 4.2). Investigating the statistical properties of the first kind and of the second kind, however, we will again make use of the probability theory and statistics (Section 4.3).

Chapter 5 contains a comprehensive description of the most important procedures presently known for computing the model parameters of deterministic simulation models (Sections 5.1 and 5.2). The performance of each procedure will be assessed with the help of quality criteria. Often, also the individual methods are compared in their performance to allow the advantages and disadvantages stand out. Chapter 5 ends with an analysis of the duration of fades of deterministic Rayleigh processes (Section 5.3).

It is well known that the statistics (of the first kind and of the second kind) of Rayleigh and Rice processes can only be influenced by a small number of parameters. On the one hand, this makes the mathematical description of the model much easier, but on the other hand, however, it narrows the flexibility of these stochastic processes. A consequence of this is that the statistical properties of real-world channels can only be roughly modelled with Rayleigh and Rice processes. For a finer adaptation to reality, one therefore needs more sophisticated model processes. **Chapter 6** deals with the description of stochastic and deterministic processes for modelling frequency-nonselective mobile radio channels. The so-called extended Suzuki processes of Type I (Section 6.1) and of Type II (Section 6.2) as well as generalized Rice and Suzuki processes (Section 6.3) are derived and their statistical properties are analysed. Apart from that, in Section 6.4, a modified version of the Loo model is introduced, containing the classical Loo model as a special case. To demonstrate the usefulness of all channel models suggested in this chapter, the statistical properties (probability density of the channel amplitude, level-crossing rate and average duration of fades) of each model are fitted to measurement results in the literature and are compared with the corresponding simulation results.

Chapter 7 is dedicated to the description of frequency-selective stochastic and deterministic channel models beginning with the ellipses model introduced by Parsons and Bajwa, illustrating the path geometry for multipath fading channels (Section 7.1). In Section 7.2, a description of linear time-variant systems is given. With the help of system theory, four important system functions are introduced allowing us to describe the input-output behaviour of linear time-variant systems in different ways. Section 7.3 is devoted to the theory of linear time-variant stochastic systems going back to Bello [Bel63]. In connection with this, stochastic system functions and characteristic quantities derivable from these are defined. Also the reference to frequency-selective stochastic channel models is established and, moreover, the channel models for typical propagation areas specified in the European work group COST 207 [COS89] are given. Section 7.4 deals with the derivation and analysis of frequency-selective deterministic channel models. Chapter 7 ends with the design of deterministic simulation models for the channel models according to COST 207.

Chapter 8 deals with the derivation, analysis, and realization of fast channel simulators. For the derivation of fast channel simulators, the periodicity of harmonic functions is exploited. It is shown how alternative structures for the simulation of deterministic processes can be derived. In particular, for complex Gaussian random processes it is extraordinarily easy to derive simulation models merely based on adders, storage elements, and simple address generators. During the actual simulation of the complex-valued channel amplitude, time-consuming trigonometric operations as well as multiplications are then no longer required. This results in high-speed channel simulators, which are suitable for all frequency-selective and frequency-nonselective channel models dealt with in previous chapters. Since the proposed principle can be generalized easily, we will in Chapter 8 restrict our attention to the derivation of fast channel simulators for Rayleigh channels. Therefore, we will exclusively employ the discrete-time representation and will introduce so-called discrete-time deterministic processes in Section 8.1. With these processes there are new possibilities for indirect realization. The three most important of them are introduced in Section 8.2. In the following Section 8.3, the elementary and statistical properties of discrete deterministic processes are examined. Section 8.4 deals with the analysis of the required realization expenditure and with the measurement of the simulation speed of fast channel simulators. Chapter 8 ends with a comparison between the Rice method and the filter method (Section 8.5).

2

RANDOM VARIABLES, STOCHASTIC PROCESSES, AND DETERMINISTIC SIGNALS

Besides clarifying the used nomenclature, we will in this chapter introduce some important terms, which will later often be used in the context of describing stochastic and deterministic channel models. However, the primary aim is to familiarize the reader with some basic principles and definitions of probability, random signals, and systems theory, as far as it is necessary for the understanding of this book. A complete and detailed description of these subjects will not be presented here; instead, some relevant technical literature will be recommended for further studies. As technical literature for the subject of probability theory, random variables, and stochastic processes, the books by Papoulis [Pap91], Peebles [Pee93], Therrien [The92], Dupraz [Dup86], as well as Shanmugan and Breipohl [Sha88] are recommended. Also the classical works of Middleton [Mid60], Davenport [Dav70], and the book by Davenport and Root [Dav58] are even nowadays still worth reading. A modern German introduction to the basic principles of probability and stochastic processes can be found in [Boe98, Bei97, Hae97]. Finally, the excellent textbooks by Oppenheim and Schafer [Opp75], Papoulis [Pap77], Rabiner and Gold [Rab75], Kailath [Kai80], Unbehauen [Unb90], Schüßler [Sch91], and Fettweis [Fet96] provide a deep insight into systems theory as well as into the principles of digital signal processing.

2.1 RANDOM VARIABLES

In the context of this book, random variables are of central importance, not only to the statistical but also to the deterministic modelling of mobile radio channels. Therefore, we will at first review some basic definitions and terms which are frequently used in connection with random variables.

An experiment whose outcome is not known in advance is called a *random experiment.* We will call points representing the outcomes of a random experiment *sample points* s. A collection of possible outcomes of a random experiment is an *event* A. The event $A = \{s\}$ consisting of a single element s is an *elementary event.* The set of all possible outcomes of a given random experiment is called the *sample space* Q of that experiment. Hence, a sample point is an element of the event, i.e., $s \in A$, and the event itself is a subset of the sample space, i.e., $A \subset Q$. The sample space Q is called the *certain event*, and the *empty set* or *null set*, denoted by $\emptyset$, is the *impossible event*. Let $\mathcal{A}$ be a class (collection) of subsets of a sample space Q. In probability theory, $\mathcal{A}$ is often called σ*-field* (or σ*-algebra*), if and only if the following conditions are fulfilled:

(i) The empty set $\emptyset \in \mathcal{A}$.

(ii) If $A \in \mathcal{A}$, then also $Q - A \in \mathcal{A}$, i.e., if the event A is an element of the class $\mathcal{A}$, then so is its complement.

(iv) If $A_n \in \mathcal{A}$ $(n = 1, 2, \ldots)$, then also $\cup_{n=1}^{\infty} A_n \in \mathcal{A}$, i.e., if the events A_n are all elements of the class $\mathcal{A}$, then so is their countable union.

A pair $(Q, \mathcal{A})$ consisting of a sample space Q and a σ-field $\mathcal{A}$ is called a *measurable space*.

A mapping $P : \mathcal{A} \to \mathbb{R}$ is called the *probability measure* or briefly *probability*, if the following conditions are fulfilled:

(i) If $A \in \mathcal{A}$, then $0 \leq P(A) \leq 1$.

(ii) $P(Q) = 1$.

(iii) If $A_n \in \mathcal{A}$ $(n = 1, 2, \ldots)$ with $\cup_{n=1}^{\infty} A_n \in \mathcal{A}$ and $A_n \cap A_k = \emptyset$ for any $n \neq k$, then also $P(\cup_{n=1}^{\infty} A_n) = \sum_{n=1}^{\infty} P(A_n)$.

A *probability space* is the triple $(Q, \mathcal{A}, P)$.

A *random variable* $\mu \in Q$ is a mapping which assigns to every outcome s of a random experiment a number $\mu(s)$, i.e.,

$$\mu : Q \to \mathbb{R}, \quad s \longmapsto \mu(s) . \tag{2.1}$$

This mapping has the property that the set $\{s|\mu(s) \leq x\}$ is an event of the considered σ-algebra for all $x \in \mathbb{R}$, i.e., $\{s|\mu(s) \leq x\} \in \mathcal{A}$. Hence, a random variable is a function of the elements of a sample space Q.

For the probability that the random variable μ is less or equal to x, we use the simplified notation

$$P(\mu \leq x) := P(\{s|\mu(s) \leq x\}) \tag{2.2}$$

in the sequel.

Cumulative distribution function: The function F_μ, defined by

$$F_\mu : \mathbb{R} \to [0, 1] , \quad x \longmapsto F_\mu(x) = P(\mu \leq x) , \tag{2.3}$$

is called the *cumulative distribution function* of the random variable μ. The cumulative distribution function $F_\mu(x)$ satisfies the following properties: a) $F_\mu(-\infty) = 0$; b) $F_\mu(\infty) = 1$; and c) $F_\mu(x)$ is non-decreasing, i.e., $F_\mu(x_1) \leq F_\mu(x_2)$ if $x_1 \leq x_2$.

Probability density function: The function p_μ, defined by

$$p_\mu : \mathbb{R} \to \mathbb{R}, \quad x \longmapsto p_\mu(x) = \frac{dF_\mu(x)}{dx}, \tag{2.4}$$

is called the *probability density function* (or *probability density* or simply *density*) of the random variable μ, where it is assumed that the cumulative distribution function $F_\mu(x)$ is differentiable with respect to x. The probability density function $p_\mu(x)$ satisfies the following properties: a) $p_\mu(x) \geq 0$ for all x; b) $\int_{-\infty}^{\infty} p_\mu(x)\, dx = 1$; and c) $F_\mu(x) = \int_{-\infty}^{x} p_\mu(x)\, dx$.

Joint cumulative distribution function: The function $F_{\mu_1\mu_2}$, defined by

$$F_{\mu_1\mu_2} : \mathbb{R}^2 \to [0,1], \quad (x_1, x_2) \longmapsto F_{\mu_1\mu_2}(x_1, x_2) = P(\mu_1 \leq x_1, \mu_2 \leq x_2), \tag{2.5}$$

is called the *joint cumulative distribution function* (or *bivariate cumulative distribution function*) of the random variables μ_1 and μ_2.

Joint probability density function: The function $p_{\mu_1\mu_2}$, defined by

$$p_{\mu_1\mu_2} : \mathbb{R}^2 \to \mathbb{R}, \quad (x_1, x_2) \longmapsto p_{\mu_1\mu_2}(x_1, x_2) = \frac{\partial^2 F_{\mu_1\mu_2}(x_1, x_2)}{\partial x_1 \partial x_2}, \tag{2.6}$$

is called the *joint probability density function* (or *bivariate density function* or simply *bivariate density*) of the random variables μ_1 and μ_2, where it is assumed that the joint cumulative distribution function $F_{\mu_1\mu_2}(x_1, x_2)$ is partially differentiable with respect to x_1 and x_2.

The random variables μ_1 and μ_2 are said to be *statistically independent*, if the events $\{s|\mu_1(s) \leq x_1\}$ and $\{s|\mu_2(s) \leq x_2\}$ are independent for all $x_1, x_2 \in \mathbb{R}$. In this case, we can write $F_{\mu_1\mu_2}(x_1, x_2) = F_{\mu_1}(x_1) \cdot F_{\mu_2}(x_2)$ and $p_{\mu_1\mu_2}(x_1, x_2) = p_{\mu_1}(x_1) \cdot p_{\mu_2}(x_2)$.

The *marginal probability density functions* (or *marginal densities*) of the joint probability density function $p_{\mu_1\mu_2}(x_1, x_2)$ are obtained by

$$p_{\mu_1}(x_1) = \int_{-\infty}^{\infty} p_{\mu_1\mu_2}(x_1, x_2)\, dx_2, \tag{2.7a}$$

$$p_{\mu_2}(x_2) = \int_{-\infty}^{\infty} p_{\mu_1\mu_2}(x_1, x_2)\, dx_1. \tag{2.7b}$$

Expected value (mean value): The quantity

$$E\{\mu\} = \int_{-\infty}^{\infty} x\, p_\mu(x)\, dx \tag{2.8}$$

is called the *expected value* (or *mean value* or *statistical average*) of the random variable μ, where $E\{\cdot\}$ denotes the *expected value operator*. The expected value operator $E\{\cdot\}$ is linear, i.e., the relations $E\{\alpha\mu\} = \alpha E\{\mu\}$ ($\alpha \in \mathbb{R}$) and $E\{\mu_1 + \mu_2\} = E\{\mu_1\} + E\{\mu_2\}$ hold. Let $f(\mu)$ be a function of the random variable μ. Then, the expected value of $f(\mu)$ can be determined by applying the fundamental relationship

$$E\{f(\mu)\} = \int_{-\infty}^{\infty} f(x)\, p_\mu(x)\, dx\,. \tag{2.9}$$

The generalization to two random variables μ_1 and μ_2 leads to

$$E\{f(\mu_1, \mu_2)\} = \int_{-\infty}^{\infty}\int_{-\infty}^{\infty} f(x_1, x_2)\, p_{\mu_1\mu_2}(x_1, x_2)\, dx_1\, dx_2\,. \tag{2.10}$$

Variance: The value

$$\begin{aligned} \mathrm{Var}\,\{\mu\} &= E\left\{(\mu - E\{\mu\})^2\right\} \\ &= E\{\mu^2\} - (E\{\mu\})^2 \end{aligned} \tag{2.11}$$

is called the *variance* of the random variable μ, where $\mathrm{Var}\,\{\cdot\}$ denotes the *variance operator*. The variance of a random variable μ is a measure of the concentration of μ near its expected value.

Covariance: The *covariance* of two random variables μ_1 and μ_2 is defined by

$$\mathrm{Cov}\,\{\mu_1, \mu_2\} = E\{(\mu_1 - E\{\mu_1\})(\mu_2 - E\{\mu_2\})\} \tag{2.12a}$$

$$= E\{\mu_1\mu_2\} - E\{\mu_1\} \cdot E\{\mu_2\}\,. \tag{2.12b}$$

Moments: The *kth moment* of the random variable μ is defined by

$$E\{\mu^k\} = \int_{-\infty}^{\infty} x^k\, p_\mu(x)\, dx\,, \quad k = 0, 1, \ldots \tag{2.13}$$

Characteristic function: The characteristic function of a random variable μ is defined as the expected value

$$\Psi_\mu(\nu) = E\left\{e^{j2\pi\nu\mu}\right\} = \int_{-\infty}^{\infty} p_\mu(x)\, e^{j2\pi\nu x}\, dx\,, \tag{2.14}$$

where ν is a real-valued variable. It should be noted that $\Psi_\mu(-\nu)$ is the Fourier transform of the probability density function $p_\mu(x)$. The characteristic function often provides a simple technique for determining the probability density function of a sum of statistically independent random variables.

Chebyshev inequality: Let μ be an arbitrary random variable with a finite expected value and a finite variance. Then, the *Chebyshev inequality*

$$P(|\mu - E\{\mu\}| \geq \epsilon) \leq \frac{\mathrm{Var}\,\{\mu\}}{\epsilon^2} \tag{2.15}$$

holds for any $\epsilon > 0$. The Chebyshev inequality is often used to obtain bounds on the probability of finding μ outside of the interval $E\{\mu\} \pm \epsilon\sqrt{\mathrm{Var}\,\{\mu\}}$.

Central limit theorem: Let μ_n $(n = 1, 2, \ldots, N)$ be statistically independent random variables with $E\{\mu_n\} = m_{\mu_n}$ and $\mathrm{Var}\,\{\mu_n\} = \sigma^2_{\mu_n}$. Then, the random variable

$$\mu = \lim_{N\to\infty} \frac{1}{\sqrt{N}} \sum_{n=1}^{N} (\mu_n - m_{\mu_n}) \tag{2.16}$$

is asymptotically normally distributed with the expected value $E\{\mu\} = 0$ and the variance $\mathrm{Var}\,\{\mu\} = \sigma^2_\mu = \lim\limits_{N\to\infty} \frac{1}{N} \sum_{n=1}^{N} \sigma^2_{\mu_n}$.

The central limit theorem plays a fundamental role in statistical asymptotic theory. The density of the sum (2.16) of merely seven statistically independent random variables with almost identical variance often results in a good approximation of the normal distribution.

2.1.1 Important Probability Density Functions

In the following, a summary of some important probability density functions often used in connection with channel modelling will be presented. The corresponding statistical properties such as the expected value and the variance will be dealt with as well. At the end of this section, we will briefly present some rules of calculation, which are of importance to the addition, multiplication, and transformation of random variables.

Uniform distribution: Let θ be a real-valued random variable with the probability density function

$$p_\theta(x) = \begin{cases} \dfrac{1}{2\pi}, & x \in [-\pi, \pi)\,, \\ 0\,, & \text{else}\,. \end{cases} \tag{2.17}$$

Then, $p_\theta(x)$ is called the *uniform distribution* and θ is said to be *uniformly distributed* in the interval $[-\pi, \pi)$. The expected value and the variance of a uniformly distributed random variable θ are $E\{\theta\} = 0$ and $\mathrm{Var}\,\{\theta\} = \pi^2/3$, respectively.

Gaussian distribution (normal distribution): Let μ be a real-valued random variable with the probability density function

$$p_\mu(x) = \frac{1}{\sqrt{2\pi}\sigma_\mu}\, e^{-\frac{(x-m_\mu)^2}{2\sigma_\mu^2}}\,, \quad x \in \mathbb{R}\,. \tag{2.18}$$

Then, $p_\mu(x)$ is called the *Gaussian distribution* (or *normal distribution*) and μ is said to be *Gaussian distributed* (or *normally distributed*). In the equation above, the quantity $m_\mu \in \mathbb{R}$ denotes the expected value and $\sigma^2_\mu \in (0, \infty)$ is the variance of μ, i.e.,

$$E\{\mu\} = m_\mu \tag{2.19a}$$

and

$$\mathrm{Var}\,\{\mu\} = E\{\mu^2\} - m_\mu^2 = \sigma_\mu^2\,. \tag{2.19b}$$

To describe the distribution properties of Gaussian distributed random variables μ, we often use the short notation $\mu \sim N(m_\mu, \sigma_\mu^2)$ instead of giving the complete expression (2.18). Especially, for $m_\mu = 0$ and $\sigma_\mu^2 = 1$, $N(0,1)$ is called the *standard normal distribution.*

Multivariate Gaussian distribution: Let us consider n real-valued Gaussian distributed random variables $\mu_1, \mu_2, \ldots, \mu_n$ with the expected values m_{μ_i} $(i = 1, 2, \ldots, n)$ and the variances $\sigma_{\mu_i}^2$ $(i = 1, 2, \ldots, n)$. The *multivariate Gaussian distribution* (or *multivariate normal distribution*) of the Gaussian random variables $\mu_1, \mu_2, \ldots, \mu_n$ is defined by

$$p_{\mu_1\mu_2\ldots\mu_n}(x_1, x_2, \ldots, x_n) = \frac{1}{\left(\sqrt{2\pi}\right)^n \sqrt{\det \boldsymbol{C}_\mu}}\, e^{-\frac{1}{2}(\boldsymbol{x}-\boldsymbol{m}_\mu)^T \boldsymbol{C}_\mu^{-1}(\boldsymbol{x}-\boldsymbol{m}_\mu)}\,, \tag{2.20}$$

where T denotes the transpose of a vector (or a matrix). In the above expression, $\boldsymbol{x}$ and $\boldsymbol{m}_\mu$ are column vectors, which are given by

$$\boldsymbol{x} = \begin{pmatrix} x_1 \\ x_2 \\ \vdots \\ x_n \end{pmatrix} \in \mathbb{R}^{n\times 1} \tag{2.21a}$$

and

$$\boldsymbol{m}_\mu = \begin{pmatrix} E\{\mu_1\} \\ E\{\mu_2\} \\ \vdots \\ E\{\mu_n\} \end{pmatrix} = \begin{pmatrix} m_{\mu_1} \\ m_{\mu_2} \\ \vdots \\ m_{\mu_n} \end{pmatrix} \in \mathbb{R}^{n\times 1}, \tag{2.21b}$$

respectively, and $\det \boldsymbol{C}_\mu$ $(\boldsymbol{C}_\mu^{-1})$ denotes the determinant (inverse) of the *covariance matrix*

$$\boldsymbol{C}_\mu = \begin{pmatrix} C_{\mu_1\mu_1} & C_{\mu_1\mu_2} & \cdots & C_{\mu_1\mu_n} \\ C_{\mu_2\mu_1} & C_{\mu_2\mu_2} & \cdots & C_{\mu_2\mu_n} \\ \vdots & \vdots & \ddots & \vdots \\ C_{\mu_n\mu_1} & C_{\mu_n\mu_2} & \cdots & C_{\mu_n\mu_n} \end{pmatrix} \in \mathbb{R}^{n\times n}\,. \tag{2.22}$$

The elements of the covariance matrix $\boldsymbol{C}_\mu$ are given by

$$C_{\mu_i\mu_j} = \mathrm{Cov}\,\{\mu_i, \mu_j\} = E\{(\mu_i - m_{\mu_i})(\mu_j - m_{\mu_j})\}\,, \quad \forall\, i, j = 1, 2, \ldots, n\,. \tag{2.23}$$

If the n random variables μ_i are normally distributed and uncorrelated in pairs, then the covariance matrix $\boldsymbol{C}_\mu$ results in a diagonal matrix with diagonal entries $\sigma_{\mu_i}^2$. In this case, the joint probability density function (2.20) decomposes into a product of n

Gaussian distributions of the normally distributed random variables $\mu_i \sim N(m_{\mu_i}, \sigma_{\mu_i}^2)$. This implies that the random variables μ_i are statistically independent for all $i = 1, 2, \ldots, n$.

Rayleigh distribution: Let us consider two zero-mean statistically independent normally distributed random variables μ_1 and μ_2, each having a variance σ_0^2, i.e., $\mu_1, \mu_2 \sim N(0, \sigma_0^2)$. Furthermore, let us derive a new random variable from μ_1 and μ_2 according to $\zeta = \sqrt{\mu_1^2 + \mu_2^2}$. Then, ζ represents a *Rayleigh distributed* random variable. The probability density function $p_\zeta(x)$ of Rayleigh distributed random variables ζ is given by

$$p_\zeta(x) = \begin{cases} \dfrac{x}{\sigma_0^2} e^{-\frac{x^2}{2\sigma_0^2}}, & x \geq 0, \\ 0, & x < 0. \end{cases} \tag{2.24}$$

Rayleigh distributed random variables ζ have the expected value

$$E\{\zeta\} = \sigma_0 \sqrt{\frac{\pi}{2}} \tag{2.25a}$$

and the variance

$$\mathrm{Var}\,\{\zeta\} = \sigma_0^2 \left(2 - \frac{\pi}{2}\right). \tag{2.25b}$$

Rice distribution: Let $\mu_1, \mu_2 \sim N(0, \sigma_0^2)$ and $\rho \in \mathbb{R}$. Then, the random variable $\xi = \sqrt{(\mu_1 + \rho)^2 + \mu_2^2}$ is a so-called *Rice distributed* random variable. The probability density function $p_\xi(x)$ of Rice distributed random variables ξ is

$$p_\xi(x) = \begin{cases} \dfrac{x}{\sigma_0^2} e^{-\frac{x^2+\rho^2}{2\sigma_0^2}} I_0\left(\dfrac{x\rho}{\sigma_0^2}\right), & x \geq 0, \\ 0, & x < 0, \end{cases} \tag{2.26}$$

where $I_0(\cdot)$ denotes the modified Bessel function of 0th order. For $\rho = 0$, the Rice distribution $p_\xi(x)$ results in the Rayleigh distribution $p_\zeta(x)$ described above. The first and second moment of Rice distributed random variables ξ are [Wol83a]

$$E\{\xi\} = \sigma_0 \sqrt{\frac{\pi}{2}}\, e^{-\frac{\rho^2}{4\sigma_0^2}} \left\{\left(1 + \frac{\rho^2}{2\sigma_0^2}\right) I_0\left(\frac{\rho^2}{4\sigma_0^2}\right) + \frac{\rho^2}{2\sigma_0^2} I_1\left(\frac{\rho^2}{4\sigma_0^2}\right)\right\} \tag{2.27a}$$

and

$$E\{\xi^2\} = 2\sigma_0^2 + \rho^2, \tag{2.27b}$$

respectively, where $I_n(\cdot)$ denotes the modified Bessel function of nth order. From (2.27a), (2.27b), and by using (2.11), the variance of Rice distributed random variables ξ can easily be calculated.

Lognormal distribution: Let μ be a Gaussian distributed random variable with the expected value m_μ and the variance σ_μ^2, i.e., $\mu \sim N(m_\mu, \sigma_\mu^2)$. Then, the random variable $\lambda = e^{\mu}$ is said to be *lognormally distributed*. The probability density function $p_\lambda(x)$ of lognormally distributed random variables λ is given by

$$p_\lambda(x) = \begin{cases} \dfrac{1}{\sqrt{2\pi}\sigma_\mu x} e^{-\frac{(\ln x - m_\mu)^2}{2\sigma_\mu^2}}, & x \geq 0, \\ 0, & x < 0. \end{cases} \tag{2.28}$$

The expected value and the variance of lognormally distributed random variables λ are given by

$$E\{\lambda\} = e^{m_\mu + \frac{\sigma_\mu^2}{2}} \tag{2.29a}$$

and

$$\mathrm{Var}\{\lambda\} = e^{2m_\mu + \sigma_\mu^2}\left(e^{\sigma_\mu^2} - 1\right), \tag{2.29b}$$

respectively.

Suzuki distribution: Consider a Rayleigh distributed random variable ζ with the probability density function $p_\zeta(x)$, according to (2.24), and a lognormally distributed random variable λ with the probability density function $p_\lambda(x)$, according to (2.28). Let us assume that ζ and λ are statistically independent. Furthermore, let η be a random variable defined by the product $\eta = \zeta \cdot \lambda$. Then, the probability density function $p_\eta(z)$ of η, that is

$$p_\eta(z) = \begin{cases} \dfrac{z}{\sqrt{2\pi}\sigma_0^2\sigma_\mu} \displaystyle\int_0^\infty \frac{1}{y^3} \cdot e^{-\frac{z^2}{2y^2\sigma_0^2}} \cdot e^{-\frac{(\ln y - m_\mu)^2}{2\sigma_\mu^2}} \, dy, & z \geq 0, \\ 0, & z < 0, \end{cases} \tag{2.30}$$

is called the *Suzuki distribution* [Suz77]. Suzuki distributed random variables η have the expected value

$$E\{\eta\} = \sigma_0 \sqrt{\frac{\pi}{2}} e^{m_\mu + \frac{\sigma_\mu^2}{2}} \tag{2.31}$$

and the variance

$$\mathrm{Var}\{\eta\} = \sigma_0^2 \cdot e^{2m_\mu + \sigma_\mu^2} \cdot \left(2e^{\sigma_\mu^2} - \frac{\pi}{2}\right). \tag{2.32}$$

Nakagami distribution: Consider a random variable ω distributed according to the probability density function

$$p_\omega(x) = \begin{cases} \dfrac{2m^m x^{2m-1} e^{-(m/\Omega)x^2}}{\Gamma(m)\Omega^m}, & m \geq 1/2, \quad x \geq 0, \\ 0, & x < 0. \end{cases} \tag{2.33}$$

Then, ω denotes a *Nakagami distributed* random variable and the corresponding probability density function $p_\omega(x)$ is called the *Nakagami distribution* or *m-distribution*

[Nak60]. In (2.33), the symbol $\Gamma(\cdot)$ represents the Gamma function, the second moment of the random variable ω has been introduced by $\Omega = E\{\omega^2\}$, and the parameter m denotes the reciprocal value of the variance of ω^2 normalized to Ω^2, i.e., $m = \Omega^2/E\{(\omega^2 - \Omega)^2\}$. From the Nakagami distribution, we obtain the one-sided Gaussian distribution and the Rayleigh distribution as special cases if $m = 1/2$ and $m = 1$, respectively. In certain limits, the Nakagami distribution, moreover, approximates both the Rice distribution and the lognormal distribution [Nak60, Cha79].

2.1.2 Functions of Random Variables

In some parts of this book, we will deal with functions of two and more random variables. In particular, we will often make use of fundamental rules in connection with the addition, multiplication, and transformation of random variables. In the sequel, the mathematical principles necessary for this will briefly be reviewed.

Addition of two random variables: Let μ_1 and μ_2 be two random variables, which are statistically characterized by the joint probability density function $p_{\mu_1\mu_2}(x_1, x_2)$. Then, the probability density function of the sum $\mu = \mu_1 + \mu_2$ can be obtained as follows

$$\begin{aligned} p_\mu(y) &= \int_{-\infty}^{\infty} p_{\mu_1\mu_2}(x_1, y - x_1)\, dx_1 \\ &= \int_{-\infty}^{\infty} p_{\mu_1\mu_2}(y - x_2, x_2)\, dx_2 \,. \end{aligned} \tag{2.34}$$

If the two random variables μ_1 and μ_2 are statistically independent, then it follows that the probability density function of μ is given by the convolution of the probability densities of μ_1 and μ_2. Thus,

$$\begin{aligned} p_\mu(y) &= p_{\mu_1}(y) * p_{\mu_2}(y) \\ &= \int_{-\infty}^{\infty} p_{\mu_1}(x_1) p_{\mu_2}(y - x_1)\, dx_1 \\ &= \int_{-\infty}^{\infty} p_{\mu_1}(y - x_2) p_{\mu_2}(x_2)\, dx_2 \,, \end{aligned} \tag{2.35}$$

where $*$ denotes the convolution operator.

Multiplication of two random variables: Let ζ and λ be two random variables, which are statistically described by the joint probability density function $p_{\zeta\lambda}(x, y)$. Then, the probability density function of the random variable $\eta = \zeta \cdot \lambda$ is equal to

$$p_\eta(z) = \int_{-\infty}^{\infty} \frac{1}{|y|} p_{\zeta\lambda}\left(\frac{z}{y}, y\right)\, dy \,. \tag{2.36}$$

From this relation, we obtain the expression

$$p_\eta(z) = \int_{-\infty}^{\infty} \frac{1}{|y|} p_\zeta\left(\frac{z}{y}\right) \cdot p_\lambda(y)\, dy \tag{2.37}$$

for statistically independent random variables ζ, λ.

Functions of random variables: Let us assume that $\mu_1, \mu_2, \ldots, \mu_n$ are random variables, which are statistically described by the joint probability density function $p_{\mu_1\mu_2\ldots\mu_n}(x_1, x_2, \ldots, x_n)$. Furthermore, let us assume that the functions $f_1, f_2, \ldots, f_n$ are given. If the system of equations $f_i(x_1, x_2, \ldots, x_n) = y_i \quad (i = 1, 2, \ldots, n)$ has real-valued solutions $x_{1\nu}, x_{2\nu}, \ldots, x_{n\nu}$ $(\nu = 1, 2, \ldots, m)$, then the joint probability density function of the random variables $\xi_1 = f_1(\mu_1, \mu_2, \ldots, \mu_n)$, $\xi_2 = f_2(\mu_1, \mu_2, \ldots, \mu_n)$, $\ldots$, $\xi_n = f_n(\mu_1, \mu_2, \ldots, \mu_n)$ can be expressed by

$$p_{\xi_1\xi_2\ldots\xi_n}(y_1, y_2, \ldots, y_n) = \sum_{\nu=1}^{m} \frac{p_{\mu_1\mu_2\ldots\mu_n}(x_{1\nu}, x_{2\nu}, \ldots, x_{n\nu})}{|J(x_{1\nu}, x_{2\nu}, \ldots, x_{n\nu})|}, \tag{2.38}$$

where

$$J(x_1, x_2, \ldots, x_n) = \begin{vmatrix} \frac{\partial f_1}{\partial x_1} & \frac{\partial f_1}{\partial x_2} & \cdots & \frac{\partial f_1}{\partial x_n} \\ \frac{\partial f_2}{\partial x_1} & \frac{\partial f_2}{\partial x_2} & \cdots & \frac{\partial f_2}{\partial x_n} \\ \vdots & \vdots & \ddots & \vdots \\ \frac{\partial f_n}{\partial x_1} & \frac{\partial f_n}{\partial x_2} & \cdots & \frac{\partial f_n}{\partial x_n} \end{vmatrix} \tag{2.39}$$

denotes the *Jacobian determinant.*

Furthermore, we can compute the joint probability density function of the random variables $\xi_1, \xi_2, \ldots, \xi_k$ for $k < n$ by using (2.38) as follows

$$p_{\xi_1\xi_2\ldots\xi_k}(y_1, y_2, \ldots, y_k) = \int_{-\infty}^{\infty}\int_{-\infty}^{\infty} \cdots \int_{-\infty}^{\infty} p_{\xi_1\xi_2\ldots\xi_n}(y_1, y_2, \ldots, y_n)\, dy_{k+1}\, dy_{k+2} \ldots dy_n\,. \tag{2.40}$$

2.2 STOCHASTIC PROCESSES

Let $(Q, \mathcal{A}, P)$ be a probability space. Now let us assign to every particular outcome $s = s_i \in Q$ of a random experiment a particular function of time $\mu(t, s_i)$ according to a rule. Hence, for a particular $s_i \in Q$, the function $\mu(t, s_i)$ denotes a mapping from $\mathbb{R}$ to $\mathbb{R}$ (or $\mathbb{C}$) according to

$$\mu(\cdot, s_i) : \mathbb{R} \to \mathbb{R} \text{ (or } \mathbb{C}), \quad t \mapsto \mu(t, s_i)\,. \tag{2.41}$$

The individual functions $\mu(t, s_i)$ of time are called *realizations* or *sample functions.* A *stochastic process* $\mu(t, s)$ is a family (or an ensemble) of sample functions $\mu(t, s_i)$, i.e., $\mu(t, s) = \{\mu(t, s_i)|s_i \in Q\} = \{\mu(t, s_1), \mu(t, s_2), \ldots\}$.

On the other hand, at a particular time instant $t = t_0 \in \mathbb{R}$, the stochastic process $\mu(t_0, s)$ only depends on the outcome s and, thus, equals a random variable. Hence, for a particular $t_0 \in \mathbb{R}$, $\mu(t_0, s)$ denotes a mapping from Q to $\mathbb{R}$ (or C) according to

$$\mu(t_0, \cdot) : \; Q \to \mathbb{R} \text{ (or } \mathbb{C}) \,, \quad s \mapsto \mu(t_0, s) \,. \tag{2.42}$$

The probability density function of the random variable $\mu(t_0, s)$ is determined by the occurrence of the outcomes.

Therefore, a stochastic process is a function of two variables $t \in \mathbb{R}$ and $s \in Q$, so that the correct notation is $\mu(t, s)$. Henceforth, however, we will drop the second argument and simply write $\mu(t)$ as in common practice.

From the statements above, we can conclude that a stochastic process $\mu(t)$ can be interpreted as follows [Pap91]:

(i) If t is a variable and s is a random variable, then $\mu(t)$ represents a family or an ensemble of sample functions $\mu(t, s)$.

(ii) If t is a variable and $s = s_0$ is a constant, then $\mu(t) = \mu(t, s_0)$ is a realization or a sample function of the stochastic process.

(iii) If $t = t_0$ is a constant and s is a random variable, then $\mu(t_0)$ is a random variable as well.

(iv) If both $t = t_0$ and $s = s_0$ are constants, then $\mu(t_0)$ is a real-valued (complex-valued) number.

The relationships following from the statements (i)–(iv) made above are illustrated in Figure 2.1.

Complex-valued stochastic processes: Let $\mu'(t)$ and $\mu''(t)$ be two real-valued stochastic processes, then a *(complex-valued) stochastic process* is defined by $\mu(t) = \mu'(t) + j\mu''(t)$.

We have stated above that a stochastic process $\mu(t)$ can be interpreted as a random variable for fixed values of $t \in \mathbb{R}$. This random variable can again be described by a distribution function $F_\mu(x; t) = P(\mu(t) \leq x)$ or a probability density function $p_\mu(x; t) = dF_\mu(x; t)/dx$. The extension of the concept of the expected value, which was introduced for random variables, to stochastic processes leads to the *expected value function*

$$m_\mu(t) = E\{\mu(t)\} \,. \tag{2.43}$$

Let us consider the random variables $\mu(t_1)$ and $\mu(t_2)$, which are assigned to the stochastic process $\mu(t)$ at the time instants t_1 and t_2, then

$$r_{\mu\mu}(t_1, t_2) = E\{\mu^*(t_1)\mu(t_2)\} \tag{2.44}$$

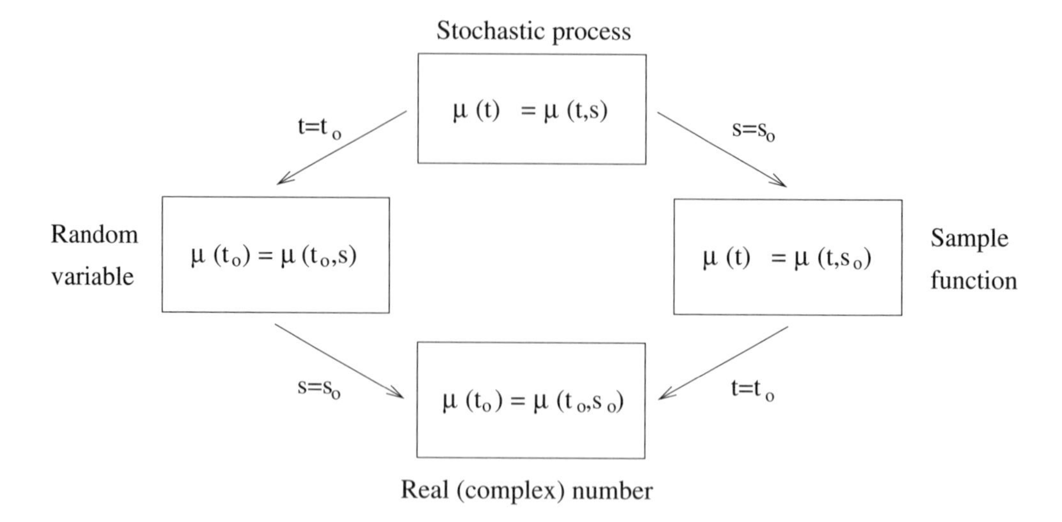

Figure 2.1: Relationships between stochastic processes, random variables, sample functions, and real-valued (complex-valued) numbers.

is called the *autocorrelation function* of $\mu(t)$, where the superscripted asterisk $*$ denotes the complex conjugation. Note that the complex conjugation is associated with the first independent variable in $r_{\mu\mu}(t_1, t_2)$.[1] The so-called *variance function* of a complex-valued stochastic process $\mu(t)$ is defined as

$$\begin{aligned} \sigma_\mu^2(t) &= \text{Var}\{\mu(t)\} = E\{|\mu(t) - E\{\mu(t)\}|^2\} \\ &= E\{\mu^*(t)\mu(t)\} - E\{\mu^*(t)\}E\{\mu(t)\} \\ &= r_{\mu\mu}(t,t) - |m_\mu(t)|^2 \,, \end{aligned} \tag{2.45}$$

where $r_{\mu\mu}(t,t)$ denotes the autocorrelation function (2.44) at the time instant $t_1 = t_2 = t$, and $m_\mu(t)$ represents the expected value function according to (2.43). Finally, the expression

$$r_{\mu_1\mu_2}(t_1, t_2) = E\{\mu_1^*(t_1)\mu_2(t_2)\} \tag{2.46}$$

introduces the *cross-correlation function* of the stochastic processes $\mu_1(t)$ and $\mu_2(t)$ at the time instants t_1 and t_2.

2.2.1 Stationary Processes

Stationary processes are of crucial importance to the modelling of mobile radio channels and will therefore be dealt with briefly here. One often distinguishes between strict-sense stationary processes and wide-sense stationary processes.

---

[1] It should be noted that in the literature, the complex conjugation is often also associated with the second independent variable of the autocorrelation function $r_{\mu\mu}(t_1, t_2)$, i.e., $r_{\mu\mu}(t_1, t_2) = E\{\mu(t_1)\mu^*(t_2)\}$.

A stochastic process $\mu(t)$ is said to be *strict-sense stationary*, if its statistical properties are invariant to a shift of the origin, i.e., $\mu(t_1)$ and $\mu(t_1 + t_2)$ have the same statistics for all $t_1, t_2 \in \mathbb{R}$. This leads to the following conclusions:

$$\text{(i)} \quad p_\mu(x; t) = p_\mu(x)\,, \tag{2.47a}$$

$$\text{(ii)} \quad E\{\mu(t)\} = m_\mu = const.\,, \tag{2.47b}$$

$$\text{(iii)} \quad r_{\mu\mu}(t_1, t_2) = r_{\mu\mu}(|t_1 - t_2|)\,. \tag{2.47c}$$

A stochastic process $\mu(t)$ is said to be *wide-sense stationary* if (2.47b) and (2.47c) are fulfilled. In this case, the expected value function $E\{\mu(t)\}$ is independent of t and, thus, simplifies to the expected value m_μ introduced for random variables. Furthermore, the autocorrelation function $r_{\mu\mu}(t_1, t_2)$ merely depends on the time difference $t_1 - t_2$. From (2.44) and (2.47c), with $t_1 = t$ and $t_2 = t + \tau$, it then follows for $\tau > 0$

$$r_{\mu\mu}(\tau) = r_{\mu\mu}(t, t+\tau) = E\{\mu^*(t)\mu(t+\tau)\}\,, \tag{2.48}$$

where $r_{\mu\mu}(0)$ represents the *mean power* of $\mu(t)$. Analogously, for the cross-correlation function (2.46) of two wide-sense stationary processes $\mu_1(t)$ and $\mu_2(t)$, we obtain

$$r_{\mu_1\mu_2}(\tau) = E\{\mu_1^*(t)\mu_2(t+\tau)\} = r^*_{\mu_2\mu_1}(-\tau)\,. \tag{2.49}$$

Let $\mu_1(t)$, $\mu_2(t)$, and $\mu(t)$ be three wide-sense stationary stochastic processes. The Fourier transform of the autocorrelation function $r_{\mu\mu}(\tau)$, defined by

$$S_{\mu\mu}(f) = \int_{-\infty}^{\infty} r_{\mu\mu}(\tau)\, e^{-j2\pi f\tau}\, d\tau\,, \tag{2.50}$$

is called the *power spectral density* (*power density spectrum*). The general relation given above between the power spectral density and the autocorrelation function is also known as the *Wiener-Khinchine relationship*. The Fourier transform of the cross-correlation function $r_{\mu_1\mu_2}(\tau)$, defined by

$$S_{\mu_1\mu_2}(f) = \int_{-\infty}^{\infty} r_{\mu_1\mu_2}(\tau)\, e^{-j2\pi f\tau}\, d\tau\,, \tag{2.51}$$

is called the *cross-power spectral density* (*cross-power density spectrum*). Taking (2.49) into account, we immediately realize that $S_{\mu_1\mu_2}(f) = S^*_{\mu_2\mu_1}(f)$ holds.

Let $\nu(t)$ be the input process and $\mu(t)$ the output process of a linear time-invariant stable system with the impulse response $h(t)$. Furthermore, let us assume that the system is deterministic, meaning that it only operates on the time variable t. Then, the output process $\mu(t)$ is the convolution of the input process $\nu(t)$ and the impulse response $h(t)$, i.e., $\mu(t) = \nu(t) * h(t)$. It is well known that the transfer function $H(f)$ of the system is the Fourier transform of the impulse response $h(t)$. Moreover, the following relations hold:

$$r_{\nu\mu}(\tau) = r_{\nu\nu}(\tau) * h(\tau) \quad \circ\!\!-\!\!\bullet \quad S_{\nu\mu}(f) = S_{\nu\nu}(f) \cdot H(f)\,, \tag{2.52a, b}$$

$$r_{\mu\nu}(\tau) = r_{\nu\nu}(\tau) * h^*(-\tau) \quad \circ\!\!-\!\!\bullet \quad S_{\mu\nu}(f) = S_{\nu\nu}(f) \cdot H^*(f)\,, \tag{2.52c, d}$$

$$r_{\mu\mu}(\tau) = r_{\nu\nu}(\tau) * h(\tau) * h^*(-\tau) \quad \circ\!\!-\!\!\bullet \quad S_{\mu\mu}(f) = S_{\nu\nu}(f) \cdot |H(f)|^2\,, \tag{2.52e, f}$$

where the symbol $\circ\!\!-\!\!\bullet$ denotes the Fourier transform. We will assume in the sequel that all systems under consideration are linear, time-invariant, and stable.

It should be noted that, strictly speaking, no stationary processes can exist. Stationary processes are merely used as mathematical models for processes, which hold their statistical properties over a relatively long time. From now on, a stochastic process will be assumed as a strict-sense stationary stochastic process, as long as nothing else is said.

A system with the transfer function

$$\check{H}(f) = -j\,\mathrm{sgn}\,(f) \tag{2.53}$$

is called the *Hilbert transformer*. We observe that this system causes a phase shift of $-\pi/2$ for $f > 0$ and a phase shift of $+\pi/2$ for $f < 0$. It should also be observed that $H(f) = 1$ holds. The inverse Fourier transform of the transfer function $\check{H}(f)$ results in the impulse response

$$\check{h}(t) = \frac{1}{\pi t}\,. \tag{2.54}$$

Since $\check{h}(t) \neq 0$ holds for $t < 0$, it follows that the Hilbert transformer is not causal. Let $\nu(t)$ with $E\{\nu(t)\} = 0$ be a real-valued input process of the Hilbert transformer, then the output process

$$\check{\nu}(t) = \nu(t) * \check{h}(t) = \frac{1}{\pi} \int_{-\infty}^{\infty} \frac{\nu(t')}{t - t'}\, dt' \tag{2.55}$$

is said to be the *Hilbert transform* of $\nu(t)$. One should note that the computation of the integral in (2.55) must be performed according to Cauchy's principal value.

With (2.52) and (2.54), the following relations hold:

$$r_{\nu\check{\nu}}(\tau) = \check{r}_{\nu\nu}(\tau) \quad \circ\!\!-\!\!\bullet \quad S_{\nu\check{\nu}}(f) = -j\,\mathrm{sgn}\,(f) \cdot S_{\nu\nu}(f)\,, \tag{2.56a, b}$$

$$r_{\nu\check{\nu}}(\tau) = -r_{\check{\nu}\nu}(\tau) \quad \circ\!\!-\!\!\bullet \quad S_{\nu\check{\nu}}(f) = -S_{\check{\nu}\nu}(f)\,, \tag{2.56c, d}$$

$$r_{\check{\nu}\check{\nu}}(\tau) = r_{\nu\nu}(\tau) \quad \circ\!\!-\!\!\bullet \quad S_{\check{\nu}\check{\nu}}(f) = S_{\nu\nu}(f)\,. \tag{2.56e, f}$$

2.2.2 Ergodic Processes

The description of the statistical properties of stochastic processes, like the expected value or the autocorrelation function, is based on ensemble means (statistical means), which takes all possible sample functions of the stochastic process into account. In practice, however, one almost always observes and records only a finite number of sample functions (mostly even only one single sample function). Nevertheless, in order to make statements on the statistical properties of stochastic process, one refers to the ergodicity hypothesis.

The ergodicity hypothesis deals with the question, whether it is possible to evaluate only a single sample function of a stationary stochastic process instead of averaging over the whole ensemble of sample functions at one or more specific time instants. Of particular importance is the question whether the expected value and the autocorrelation function of a stochastic process $\mu(t)$ equal the temporal means taken over any arbitrarily sample function $\mu(t, s_i)$. According to the *ergodic theorem*, the expected value $E\{\mu(t)\} = m_\mu$ equals the temporal average of $\mu(t, s_i)$, i.e.,

$$m_\mu = \tilde{m}_\mu := \lim_{T\to\infty} \frac{1}{2T} \int_{-T}^{+T} \mu(t, s_i)\, dt\,, \tag{2.57}$$

and the autocorrelation function $r_{\mu\mu}(\tau) = E\{\mu^*(t)\mu(t+\tau)\}$ equals the temporal autocorrelation function of $\mu(t, s_i)$, i.e.,

$$r_{\mu\mu}(\tau) = \tilde{r}_{\mu\mu}(\tau) := \lim_{T\to\infty} \frac{1}{2T} \int_{-T}^{+T} \mu^*(t, s_i)\, \mu(t+\tau, s_i)\, dt\,. \tag{2.58}$$

A stationary stochastic process $\mu(t)$ is said to be *strict-sense ergodic*, if all expected values, which take all possible sample functions into account, are identical to the respective temporal averages taken over an arbitrary sample function. If this condition is only fulfilled for the expected value and the autocorrelation function, i.e., if only (2.57) and (2.58) are fulfilled, then the stochastic process $\mu(t)$ is said to be *wide-sense ergodic*. A strict-sense ergodic process is always stationary. The inverse statement is not always true, although commonly assumed.

2.2.3 Level-Crossing Rate and Average Duration of Fades

Apart from the probability density function and the autocorrelation function, other characteristic quantities describing the statistics of mobile fading channels are of importance. These quantities are the *level-crossing rate* and the *average duration of fades*.

As we know, the received signal in mobile radio communications often undergoes heavy statistical fluctuations, which can reach as high as 30 dB and more. In digital communications, a heavy decline of the received signal directly leads to a drastic increase of the bit error rate. For the optimization of coding systems, which are required for error correction, it is not only important to know how often the received signal crosses a given signal level per time unit, but also for how long on average the signal is below a certain level. Suitable measures for this are the level-crossing rate and the average duration of fades.

Level-crossing rate: The *level-crossing rate*, denoted by $N_\zeta(r)$, describes how often a stochastic process $\zeta(t)$ crosses a given level r from up to down (or from down to up) within one second. According to [Ric44, Ric45], the level-crossing rate $N_\zeta(r)$ can be calculated by

$$N_\zeta(r) = \int_0^\infty \dot{x}\, p_{\zeta\dot{\zeta}}(r, \dot{x})\, d\dot{x}\,, \quad r \geq 0\,, \tag{2.59}$$

where $p_{\zeta\dot{\zeta}}(x, \dot{x})$ denotes the joint probability density function of the process $\zeta(t)$ and its time derivative $\dot{\zeta}(t) = d\zeta(t)/dt$ at the same time instant. Analytical expressions for the level-crossing rate of Rayleigh and Rice processes can be calculated easily.

Consider two uncorrelated real-valued zero-mean Gaussian random processes $\mu_1(t)$ and $\mu_2(t)$ with identical autocorrelation functions, i.e., $r_{\mu_1\mu_1}(\tau) = r_{\mu_2\mu_2}(\tau)$. Then, for the level-crossing rate of the resulting Rayleigh processes $\zeta(t) = \sqrt{\mu_1^2(t) + \mu_2^2(t)}$, we obtain the following expression [Jak93]

$$\begin{aligned} N_\zeta(r) &= \sqrt{\frac{\beta}{2\pi}} \cdot \frac{r}{\sigma_0^2} e^{-\frac{r^2}{2\sigma_0^2}} \\ &= \sqrt{\frac{\beta}{2\pi}} \cdot p_\zeta(r)\,, \quad r \geq 0\,, \end{aligned} \tag{2.60}$$

where $\sigma_0^2 = r_{\mu_i\mu_i}(0)$ denotes the mean power of the underlying Gaussian random processes $\mu_i(t)$ $(i = 1, 2)$. Here, β is a short notation for the negative curvature of the autocorrelation functions $r_{\mu_1\mu_1}(\tau)$ and $r_{\mu_2\mu_2}(\tau)$ at the origin $\tau = 0$, i.e.,

$$\beta = -\left.\frac{d^2}{d\tau^2} r_{\mu_i\mu_i}(\tau)\right|_{\tau=0} = -\ddot{r}_{\mu_i\mu_i}(0)\,, \quad i = 1, 2\,. \tag{2.61}$$

For the Rice process $\xi(t) = \sqrt{(\mu_1(t) + \rho)^2 + \mu_2^2(t)}$, we obtain the following expression for the level-crossing rate [Ric48]

$$\begin{aligned} N_\xi(r) &= \sqrt{\frac{\beta}{2\pi}} \cdot \frac{r}{\sigma_0^2} e^{-\frac{r^2+\rho^2}{2\sigma_0^2}} I_0\left(\frac{r\rho}{\sigma_0^2}\right) \\ &= \sqrt{\frac{\beta}{2\pi}} \cdot p_\xi(r)\,, \quad r \geq 0\,. \end{aligned} \tag{2.62}$$

Average duration of fades: The *average duration of fades*, denoted by $T_{\zeta_-}(r)$, is the expected value for the length of the time intervals in which the stochastic process

$\zeta(t)$ is below a given level r. The average duration of fades $T_{\zeta_-}(r)$ can be calculated by means of [Jak93]

$$T_{\zeta_-}(r) = \frac{F_{\zeta_-}(r)}{N_\zeta(r)}, \tag{2.63}$$

where $F_{\zeta_-}(r)$ denotes the cumulative distribution function of the stochastic process $\zeta(t)$ being the probability that $\zeta(t)$ is less or equal to the level r, i.e.,

$$F_{\zeta_-}(r) = P(\zeta(t) \leq r) = \int_0^r p_\zeta(x)\, dx\,. \tag{2.64}$$

For the Rayleigh processes $\zeta(t)$, the average duration of fades is given by

$$T_{\zeta_-}(r) = \sqrt{\frac{2\pi}{\beta}} \cdot \frac{\sigma_0^2}{r} \left(e^{\frac{r^2}{2\sigma_0^2}} - 1 \right), \quad r \geq 0\,, \tag{2.65}$$

where the quantity β is again given by (2.61).

For Rice processes $\xi(t)$, however, we find by substituting (2.26), (2.64), and (2.62) in (2.63) the following integral expression

$$T_{\xi_-}(r) = \sqrt{\frac{2\pi}{\beta}} \cdot \frac{e^{\frac{r^2}{2\sigma_0^2}}}{r\, I_0\left(\frac{r\rho}{\sigma_0^2}\right)} \int_0^r x\, e^{-\frac{x^2}{2\sigma_0^2}} I_0\left(\frac{x\rho}{\sigma_0^2}\right) dx\,, \quad r \geq 0\,, \tag{2.66}$$

which has to be evaluated numerically.

Analogously, the *average connecting time interval* $T_{\zeta_+}(r)$ can be introduced. This quantity describes the expected value for the length of the time intervals, in which the stochastic process $\zeta(t)$ is above a given level r. Thus,

$$T_{\zeta_+}(r) = \frac{F_{\zeta_+}(r)}{N_\zeta(r)}, \tag{2.67}$$

where $F_{\zeta_+}(r)$ is called the *complementary cumulative distribution function* of $\zeta(t)$. This function describes the probability that $\zeta(t)$ is larger than r, i.e., $F_{\zeta_+}(r) = P(\zeta(t) > r)$. The complementary cumulative distribution function F_{ζ_+} and the cumulative distribution function $F_{\zeta_-}(r)$ are related by $F_{\zeta_+}(r) = 1 - F_{\zeta_-}(r)$.

2.3 DETERMINISTIC CONTINUOUS-TIME SIGNALS

In principle, one distinguishes between continuous-time and discrete-time signals. For deterministic signals, we will in what follows use the continuous-time representation wherever it is possible. Only in those sections where the numerical simulations of channel models play a significant role, is the discrete-time representation of signals chosen.

A *deterministic (continuous-time) signal* is usually defined over $\mathbb{R}$. The set $\mathbb{R}$ is considered as the time space in which the variable t takes its values, i.e., $t \in \mathbb{R}$.

A deterministic signal is described by a function (mapping) in which each value of t is definitely assigned to a real-valued (or complex-valued) number. Furthermore, in order to distinguish deterministic signals from stochastic processes better, we will put the tilde-sign onto the symbols chosen for deterministic signals. Thus, under a deterministic signal $\tilde{\mu}(t)$, we will understand a mapping of the kind

$$\tilde{\mu} : \mathbb{R} \to \mathbb{R} \quad (\text{or } \mathbb{C}) \,, \quad t \longmapsto \tilde{\mu}(t) \,. \tag{2.68}$$

In connection with deterministic signals, the following terms are of importance.

Mean value: The *mean value* of a deterministic signal $\tilde{\mu}(t)$ is defined by

$$\tilde{m}_\mu := \lim_{T\to\infty} \frac{1}{2T} \int_{-T}^{T} \tilde{\mu}(t) dt \,. \tag{2.69}$$

Mean power: The *mean power* of a deterministic signal $\tilde{\mu}(t)$ is defined by

$$\tilde{\sigma}_\mu^2 := \lim_{T\to\infty} \frac{1}{2T} \int_{-T}^{T} |\tilde{\mu}(t)|^2 dt \,. \tag{2.70}$$

From now on, we will always assume that the power of a deterministic signal is finite.

Autocorrelation function: Let $\tilde{\mu}(t)$ be a deterministic signal. Then, the *autocorrelation function* of $\tilde{\mu}(t)$ is defined by

$$\tilde{r}_{\mu\mu}(\tau) := \lim_{T\to\infty} \frac{1}{2T} \int_{-T}^{T} \tilde{\mu}^*(t)\, \tilde{\mu}(t+\tau) dt \,, \quad \tau \in \mathbb{R} \,. \tag{2.71}$$

Comparing (2.70) with (2.71), we realize that the value of $\tilde{r}_{\mu\mu}(\tau)$ at $\tau = 0$ is identical to the mean power of $\tilde{\mu}(t)$, i.e., the relation $\tilde{r}_{\mu\mu}(0) = \tilde{\sigma}_\mu^2$ holds.

Cross-correlation function: Let $\tilde{\mu}_1(t)$ and $\tilde{\mu}_2(t)$ be two deterministic signals. Then, the *cross-correlation function* of $\tilde{\mu}_1(t)$ and $\tilde{\mu}_2(t)$ is defined by

$$\tilde{r}_{\mu_1\mu_2}(\tau) := \lim_{T\to\infty} \frac{1}{2T} \int_{-T}^{T} \tilde{\mu}_1^*(t)\, \tilde{\mu}_2(t+\tau)\, dt \,, \quad \tau \in \mathbb{R} \,. \tag{2.72}$$

Here, $\tilde{r}_{\mu_1\mu_2}(\tau) = \tilde{r}^*_{\mu_2\mu_1}(-\tau)$ holds.

Power spectral density: Let $\tilde{\mu}(t)$ be a deterministic signal. Then, the Fourier transform of the autocorrelation function $\tilde{r}_{\mu\mu}(\tau)$, defined by

$$\tilde{S}_{\mu\mu}(f) := \int_{-\infty}^{\infty} \tilde{r}_{\mu\mu}(\tau) e^{-j2\pi f\tau} d\tau \,, \quad f \in \mathbb{R} \,, \tag{2.73}$$

is called the *power spectral density* (or *power density spectrum*) of $\tilde{\mu}(t)$.

Cross-power spectral density: Let $\tilde{\mu}_1(t)$ and $\tilde{\mu}_2(t)$ be two deterministic signals. Then, the Fourier transform of the cross-correlation function $\tilde{r}_{\mu_1\mu_2}(\tau)$

$$\tilde{S}_{\mu_1\mu_2}(f) := \int_{-\infty}^{\infty} \tilde{r}_{\mu_1\mu_2}(\tau) e^{-j2\pi f\tau} d\tau \,, \quad f \in \mathbb{R} \,, \tag{2.74}$$

is called the *cross-power spectral density* (or *cross-power density spectrum*). From (2.74) and the relation $\tilde{r}_{\mu_1\mu_2}(\tau) = \tilde{r}^*_{\mu_2\mu_1}(-\tau)$ it follows that $\tilde{S}_{\mu_1\mu_2}(f) = \tilde{S}^*_{\mu_2\mu_1}(f)$ holds.

Let $\tilde{\nu}(t)$ and $\tilde{\mu}(t)$ be the deterministic input signal and the deterministic output signal, respectively, of a linear time-invariant stable system with the transfer function $H(f)$. Then, the relationship

$$\tilde{S}_{\mu\mu}(f) = |H(f)|^2 \tilde{S}_{\nu\nu}(f) \tag{2.75}$$

holds.

2.4 DETERMINISTIC DISCRETE-TIME SIGNALS

By equidistant sampling of a continuous-time signal $\tilde{\mu}(t)$ at the discrete time instants $t = t_k = kT_s$, where $k \in \mathbb{Z}$ and T_s symbolizes the *sampling interval*, we obtain the sequence of numbers $\{\tilde{\mu}(kT_s)\} = \{\ldots, \tilde{\mu}(-T_s), \tilde{\mu}(0), \tilde{\mu}(T_s), \ldots\}$. In specific questions of many engineering fields, it is occasionally strictly distinguished between the sequence $\{\tilde{\mu}(kT_s)\}$ itself, which is then called a *discrete-time signal*, and the kth element $\tilde{\mu}(kT_s)$ of it. For our purposes, however, this differentiation is not connected to any advantage worth mentioning. In what follows, we will therefore simply write $\tilde{\mu}(kT_s)$ for discrete-time signals or sequences, and we will make use of the notation $\bar{\mu}[k] := \tilde{\mu}(kT_s) = \tilde{\mu}(t)|_{t=kT_s}$.

It is clear that by sampling a deterministic continuous-time signal $\tilde{\mu}(t)$, we obtain a discrete-time signal $\bar{\mu}[k]$, which is deterministic as well. Under a deterministic discrete-time signal $\bar{\mu}[k]$, we understand a mapping of the kind

$$\bar{\mu} : \mathbb{Z} \to \mathbb{R} \quad (\text{or } \mathbb{C}) \,, \quad k \longmapsto \bar{\mu}[k] \,. \tag{2.76}$$

The terms such as mean value, autocorrelation function, and power spectral density, which were previously introduced for deterministic continuous-time signals, can also be applied to deterministic discrete-time signals. The most important definitions and relationships will only be introduced here, as far as they are actually used, especially in Chapter 8. The reader can find a detailed presentation of the relationships, e.g., in [Opp75, Kam98, Unb90].

Mean value: The *mean value* of a deterministic sequence $\bar{\mu}[k]$ is defined by

$$\bar{m}_\mu := \lim_{K\to\infty} \frac{1}{2K+1} \sum_{k=-K}^{K} \bar{\mu}[k] \,. \tag{2.77}$$

Mean power: The *mean power* of a deterministic sequence $\bar{\mu}[k]$ is defined by

$$\bar{\sigma}_\mu^2 := \lim_{K\to\infty} \frac{1}{2K+1} \sum_{k=-K}^{K} |\bar{\mu}[k]|^2 \,. \tag{2.78}$$

Autocorrelation sequence: Let $\bar{\mu}[k]$ be a deterministic sequence, then the corresponding *autocorrelation sequence* is defined by

$$\bar{r}_{\mu\mu}[\kappa] := \lim_{K\to\infty} \frac{1}{2K+1} \sum_{k=-K}^{K} \bar{\mu}^*[k]\, \bar{\mu}[k+\kappa]\,, \quad \kappa \in \mathbb{Z}\,. \tag{2.79}$$

Thus, in connection with (2.78), it follows $\bar{\sigma}_\mu^2 = \bar{r}_{\mu\mu}[0]$.

Cross-correlation sequence: Let $\bar{\mu}_1[k]$ and $\bar{\mu}_2[k]$ be two deterministic sequences, then the *cross-correlation sequence* is defined by

$$\bar{r}_{\mu_1\mu_2}[\kappa] := \lim_{K\to\infty} \frac{1}{2K+1} \sum_{k=-K}^{K} \bar{\mu}_1^*[k]\, \bar{\mu}_2[k+\kappa]\,, \quad \kappa \in \mathbb{Z}\,. \tag{2.80}$$

Here, the relation $\bar{r}_{\mu_1\mu_2}[\kappa] = \bar{r}^*_{\mu_2\mu_1}[-\kappa]$ holds.

Power spectral density: Let $\bar{\mu}[k]$ be a deterministic sequence, then the *discrete Fourier transform* of the autocorrelation sequence $\bar{r}_{\mu\mu}[\kappa]$, defined by

$$\bar{S}_{\mu\mu}(f) := \sum_{\kappa=-\infty}^{\infty} \bar{r}_{\mu\mu}[\kappa]\, e^{-j2\pi f T_s \kappa}\,, \quad f \in \mathbb{R}\,, \tag{2.81}$$

is called the *power spectral density* or *power density spectrum* of $\bar{\mu}[k]$.

Between (2.81) and (2.73), the relation

$$\bar{S}_{\mu\mu}(f) := \frac{1}{T_s} \sum_{m=-\infty}^{\infty} \tilde{S}_{\mu\mu}(f - mf_s) \tag{2.82}$$

holds, where $f_s = 1/T_s$ is called the *sampling frequency* or the *sampling rate*. Obviously, the power spectral density $\bar{S}_{\mu\mu}(f)$ is periodic with the period f_s, since $\bar{S}_{\mu\mu}(f) = \bar{S}_{\mu\mu}(f - mf_s)$ holds for all $m \in \mathbb{Z}$. The relation (2.82) states that the power spectral density $\bar{S}_{\mu\mu}(f)$ of $\bar{\mu}[k]$ follows from the power spectral density $\tilde{S}_{\mu\mu}(f)$ of $\tilde{\mu}(t)$, if the latter one is weighted by $1/T_s$ and periodically continued at instants mf_s, where $m \in \mathbb{Z}$.

The *inverse discrete Fourier transform* of the power spectral density $\bar{S}_{\mu\mu}(f)$ again results in the autocorrelation sequence $\bar{r}_{\mu\mu}[\kappa]$ of $\bar{\mu}[k]$, i.e.,

$$\bar{r}_{\mu\mu}[\kappa] := \frac{1}{f_s} \int_{-f_s/2}^{f_s/2} \bar{S}_{\mu\mu}(f)\, e^{j2\pi f T_s \kappa}\, df\,, \quad \kappa \in \mathbb{Z}\,. \tag{2.83}$$

Cross-power spectral density: Let $\bar{\mu}_1[k]$ and $\bar{\mu}_2[k]$ be two deterministic sequences. Then, the discrete Fourier transform of the cross-correlation sequence $\bar{r}_{\mu_1\mu_2}[\kappa]$, defined by

$$\bar{S}_{\mu_1\mu_2}(f) := \sum_{\kappa=-\infty}^{\infty} \bar{r}_{\mu_1\mu_2}[\kappa]\, e^{-j2\pi f T_s \kappa}, \quad f \in \mathbb{R}, \tag{2.84}$$

is called the *cross-power spectral density* or the *cross-power density spectrum.* From the above equation and $\bar{r}_{\mu_1\mu_2}[\kappa] = \bar{r}^*_{\mu_2\mu_1}[-\kappa]$ it follows that $\bar{S}_{\mu_1\mu_2}(f) = \bar{S}^*_{\mu_2\mu_1}(f)$ holds.

Sampling theorem: Let $\tilde{\mu}(t)$ be a band-limited continuous-time signal with the cut-off frequency f_c. If this signal is sampled with a sampling frequency f_s greater than the double of its cut-off frequency f_c, i.e.,

$$f_s > 2f_c\,, \tag{2.85}$$

then $\tilde{\mu}(t)$ is completely determined by the corresponding sampling values $\bar{\mu}[k] = \tilde{\mu}(kT_s)$. In particular, the continuous-time signal $\tilde{\mu}(t)$ can be reconstructed from the sequence $\bar{\mu}[k]$ by means of the relation

$$\tilde{\mu}(t) = \sum_{k=-\infty}^{\infty} \bar{\mu}[k]\ \mathrm{sinc}\left(\pi \frac{t - kT_s}{T_s}\right), \tag{2.86}$$

where $\mathrm{sinc}\,(\cdot)$ denotes the sinc function, which is defined by $\mathrm{sinc}\,(x) = \sin(x)/x$.

It should be added that the sampling condition (2.85) can be replaced by the less restrictive condition $f_s \geq 2f_c$, if the power spectral density $\tilde{S}_{\mu\mu}(f)$ has no δ-components at the limits $f = \pm f_c$ [Fet96]. In this case, even on condition that $f_s \geq 2f_c$ holds, the validity of the sampling theorem is absolutely guaranteed.

3

RAYLEIGH AND RICE PROCESSES AS REFERENCE MODELS

From now on, we assume that the transmitter is stationary. The transmitted electromagnetic waves mostly do not, at least in urban areas, arrive at the vehicle antenna of the receiver over the direct path. On the other hand, due to reflections from buildings, from the ground, and from other obstacles with vast surfaces, as well as scatters from trees and other scatter-objects, a multitude of partial waves arrive at the receiver antenna from different directions. This effect is known as *multipath propagation*. Due to multipath propagation, the received partial waves increase or weaken each other, depending on the phase relations of the waves. Consequently, the received electromagnetic field strength and, thus, also the received signal are both strongly fluctuating functions of the receiver's position [Lor85] or, in case of a moving receiver, strongly fluctuating functions of time. Besides, as a result of the Doppler effect, the motion of the receiver leads to a *frequency shift* (*Doppler shift*)[1] of the partial waves hitting the antenna. Depending on the direction of arrival of these partial waves, different Doppler shifts occur, so that for the sum of all scattered (and reflected) components, we finally obtain a continuous spectrum of Doppler frequencies, which is called the *Doppler power spectral density.*

If the propagation delay differences among the scattered signal components at the receiver are negligible compared to the symbol interval, what we will assume in the following, then the channel is said to be *frequency-nonselective*. In this case, the fluctuations of the received signal can be modelled by multiplying the transmitted signal with an appropriate stochastic model process. After extensive measurements of the envelope of the received signal [You52, Nyl68, Oku68] in urban and suburban areas, i.e., in regions where the line-of-sight component is often blocked by obstacles, the Rayleigh process was suggested as suitable stochastic model process. In rural regions, however, the line-of-sight component is often a part of the received signal, so

---

[1] In the two-dimensional horizontal plane, the Doppler shift (Doppler frequency) of an elementary wave is equal to $f = f_{max} \cos \alpha$, where α is the angle of arrival as illustrated in Figure 1.2 and $f_{max} = \mathrm{v} f_0 / c_0$ denotes the maximum Doppler frequency (v: velocity of the vehicle, f_0: carrier frequency, c_0: speed of light) [Jak93].

that the Rice process is the more suitable stochastic model for these channels.

However, the validity of these models is limited to relatively small areas with dimensions in the order of about some few tens of wavelengths, where the local mean of the envelope is approximately constant [Jak93]. In larger areas, however, the local mean fluctuates due to shadowing effects and is approximately lognormally distributed [Oku68, Par92].

The knowledge of the statistical properties of the received signal envelope is necessary for the development of digital communication systems and for planning mobile radio networks. Usually, Rayleigh and Rice processes are preferred for modelling *fast-term fading*, whereas *slow-term fading* is modelled with a lognormal process [Par92]. Slow-term fading not only has a strong influence on channel availability, selection of the carrier frequency, handover, etc., but is also important in the planning of mobile radio networks. For the choice of the transmission technique and the design of digital receivers, however, the properties of the fast-term statistics, on which we will concentrate in this chapter, are of vital importance [Fec93b].

In order to better assess the performance of deterministic processes and deterministic simulation models derivable from these, we will often refer to stochastic reference models. As reference models — depending on the objective — the respective reference models for Gaussian, Rayleigh or Rice processes will be used. The aim of this chapter is to describe these reference models. At first, an introductory description of the reference models is given in Section 3.1. After some elementary properties of these models have been examined closer in Section 3.2, we will finally analyse in Section 3.3 the statistical properties of the first order (Subsection 3.3.1) and of the second order (Subsection 3.3.2), as far as it is necessary for the further aims of this book. Chapter 3 ends with an analysis of the fading intervals of Rayleigh processes (Subsection 3.3.3).

3.1 GENERAL DESCRIPTION OF RICE AND RAYLEIGH PROCESSES

The sum of all scattered components of the received signal is — when transmitting an unmodulated carrier over a frequency-nonselective mobile radio channel — in the equivalent complex baseband often described by a zero-mean complex Gaussian random process

$$\mu(t) = \mu_1(t) + j\mu_2(t) \,. \tag{3.1}$$

Usually, it is assumed that the real-valued Gaussian random processes $\mu_1(t)$ and $\mu_2(t)$ are statistically uncorrelated. Let the variance of the processes $\mu_i(t)$ be equal to $\mathrm{Var}\{\mu_i(t)\} = \sigma_0^2$ for $i = 1, 2$, then the variance of $\mu(t)$ is given by $\mathrm{Var}\{\mu(t)\} = 2\sigma_0^2$.

The line-of-sight component of the received signal will in the following be described by a general time-variant part

$$m(t) = m_1(t) + jm_2(t) = \rho e^{j(2\pi f_\rho t + \theta_\rho)} \,, \tag{3.2}$$

where ρ, f_ρ, and θ_ρ denote the amplitude, the Doppler frequency, and the phase of the line-of-sight component, respectively. One should note about this that, due to the

Doppler effect, the relation $f_\rho = 0$ only holds if the direction of arrival of the incident wave is orthogonal to the direction of motion of the mobile user. Consequently, (3.2) then becomes a time-invariant component, i.e.,

$$m = m_1 + jm_2 = \rho e^{j\theta_\rho} \,. \tag{3.3}$$

At the receiver antenna, we have the superposition of the sum of the scattered components with the line-of-sight component. In the model chosen here, this superposition is equal to the addition of (3.1) and (3.2). For this reason, we introduce a further complex Gaussian random process

$$\mu_\rho(t) = \mu_{\rho_1}(t) + j\mu_{\rho_2}(t) = \mu(t) + m(t) \tag{3.4}$$

with time-variant mean value $m(t)$.

As we know, forming the absolute values of (3.1) and (3.4) leads to Rayleigh and Rice processes [Ric48], respectively. In order to distinguish these processes clearly from each other, we will in the following denote Rayleigh processes by

$$\zeta(t) = |\mu(t)| = |\mu_1(t) + j\mu_2(t)| \tag{3.5}$$

and Rice processes by

$$\xi(t) = |\mu_\rho(t)| = |\mu(t) + m(t)| \,. \tag{3.6}$$

3.2 ELEMENTARY PROPERTIES OF RICE AND RAYLEIGH PROCESSES

The shape of the power spectral density of the complex Gaussian random process (3.4) is identical to the Doppler power spectral density, which is obtained from both the power of all electromagnetic waves arriving at the receiver antenna and the distribution of the angles of arrival. In addition to that, the antenna radiation pattern of the receiving antenna has a decisive influence on the shape of the Doppler power spectral density.

By modelling mobile radio channels, one frequently simplifies matters by assuming that the propagation of electromagnetic waves occurs in the two-dimensional plane, hence, horizontally. Furthermore, mostly the idealized assumption is made that the angles of incidence of the waves arriving at the antenna of the mobile participant (receiver) are uniformly distributed from 0 to 2π. For omnidirectional antennas, we can then easily calculate the (Doppler) power spectral density $S_{\mu\mu}(f)$ of the scattered components $\mu(t) = \mu_1(t) + j\mu_2(t)$. For $S_{\mu\mu}(f)$, one finds the following expression [Cla68, Jak93]

$$S_{\mu\mu}(f) = S_{\mu_1\mu_1}(f) + S_{\mu_2\mu_2}(f) \,, \tag{3.7}$$

where

$$S_{\mu_i\mu_i}(f) = \begin{cases} \dfrac{\sigma_0^2}{\pi f_{max}\sqrt{1-(f/f_{max})^2}} \,, & |f| \leq f_{max} \,, \\ 0 \,, & |f| > f_{max} \,, \end{cases} \tag{3.8}$$

holds for $i = 1, 2$ and f_{max} denotes the maximum Doppler frequency. In the literature, (3.8) is often called *Jakes power spectral density* (*Jakes PSD*), although it was originally derived by Clarke [Cla68]. The reader can find a full derivation of the Jakes power spectral density in Appendix A.

In principle, the electromagnetic waves arriving at the receiver have besides the vertical also a horizontal component. The latter is considered in the three-dimensional propagation model derived in [Aul79]. The only difference between the resulting power spectral density and (3.8) is that there are no poles at $f = \pm f_{max}$. Apart from that, the course of the curve is similar to that of (3.8).

A stochastic model for a land mobile radio channel with communication between two moving vehicles (mobile-to-mobile communication) was introduced in [Akk86]. It was shown there that the channel can again be represented by a narrow-band complex Gaussian random process with symmetrical Doppler power spectral density, which has poles though at the points $f = \pm(f_{max_1} - f_{max_2})$. Here, f_{max_1} (f_{max_2}) denotes the maximum Doppler frequency due to the motion of the receiver (transmitter). The shape of the curve differs considerably from the Jakes power spectral density (3.8), but contains it as a special case for $f_{max_1} = 0$ or $f_{max_2} = 0$. The statistical properties (of second order) for this channel model were analysed in a further paper [Akk94].

Considering (3.7) and (3.8), we see that $S_{\mu\mu}(f)$ is an even function. This property no longer exists, however, as soon as either a spatially limited shadowing prevents an isotropic distribution of the received waves or sector antennas with a formative directional antenna radiation pattern are used at the receiver [Cla68, Gan72]. The electromagnetic reflecting power of the environment can also be in such a condition that waves from certain directions are reflected with different intensities. In this case, the Doppler power spectral density $S_{\mu\mu}(f)$ of the complex Gaussian random process (3.1) is also unsymmetrical [Kra90b]. We will return to this subject in Chapter 5.

The inverse Fourier transform of $S_{\mu\mu}(f)$ results for the Jakes power spectral density (3.8) in the autocorrelation function derived in Appendix A

$$r_{\mu\mu}(\tau) = r_{\mu_1\mu_1}(\tau) + r_{\mu_2\mu_2}(\tau)\,, \tag{3.9}$$

where

$$r_{\mu_i\mu_i}(\tau) = \sigma_0^2 J_0(2\pi f_{max}\tau)\,, \quad i = 1, 2\,, \tag{3.10}$$

holds, and $J_0(\cdot)$ denotes the 0th-order Bessel function of the first kind.

By way of illustration, the Jakes power spectral density (3.8) is presented together with the corresponding autocorrelation function (3.10) in Figures 3.1(a) and 3.1(b), respectively.

Besides the Jakes power spectral density (3.8), the so-called *Gaussian power spectral density* (*Gaussian PSD*)

$$S_{\mu_i\mu_i}(f) = \frac{\sigma_0^2}{f_c}\sqrt{\frac{\ln 2}{\pi}}\, e^{-\ln 2\left(\frac{f}{f_c}\right)^2}\,, \quad i = 1, 2\,, \tag{3.11}$$

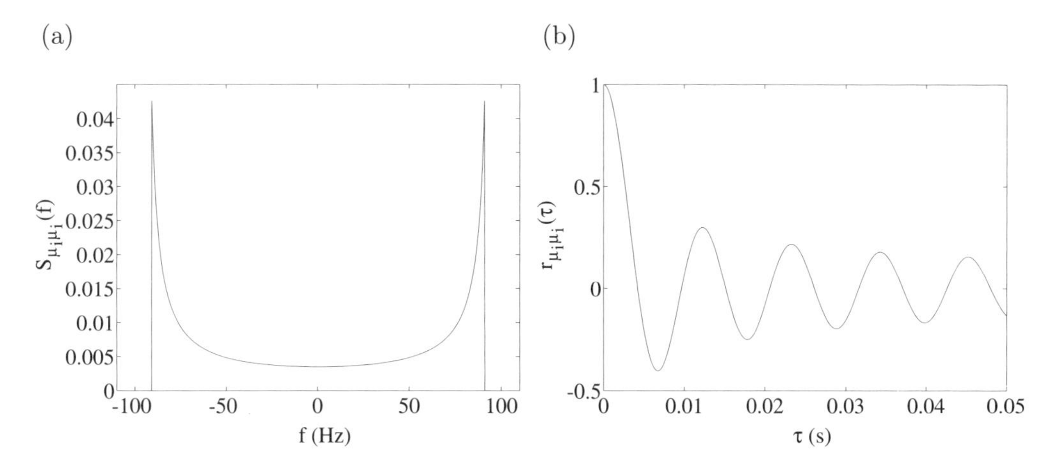

Figure 3.1: (a) Jakes power spectral density $S_{\mu_i\mu_i}(f)$ and (b) the corresponding autocorrelation function $r_{\mu_i\mu_i}(\tau)$ ($f_{max} = 91\,\text{Hz}$, $\sigma_0^2 = 1$).

will play an important role in the following, where f_c denotes the 3-dB-cut-off frequency.

Theoretical investigations in [Bel73] have shown that the Doppler power spectral density of aeronautical channels has a Gaussian shape. Further information on the measurements concerning the propagation characteristics of aeronautical satellite channels can be found, for example, in [Neu87]. Although no absolute correspondence to the obtained measurements could be proved, (3.11) can in most cases very well be used as a sufficiently good approximation [Neu89]. For signal bandwidths up to some 10 kHz, the aeronautical satellite channel belongs to the class of frequency-nonselective mobile radio channels [Neu89].

Especially for frequency-selective mobile radio channels, it has been shown [Cox73] that the Doppler power spectral density of the far echoes deviates strongly from the shape of the Jakes power spectral density. Hence, the Doppler power spectral density is approximately Gaussian shaped and is generally shifted from the origin of the frequency plane, because the far echoes mostly dominate from a certain direction of preference. Specifications for frequency-shifted Gaussian power spectral densities for the pan-European, terrestrial, cellular GSM system can be found in [COS86].

The inverse Fourier transform results for the Gaussian power spectral density (3.11) in the autocorrelation function

$$r_{\mu_i\mu_i}(\tau) = \sigma_0^2\, e^{-\left(\pi \frac{f_c}{\sqrt{\ln 2}}\tau\right)^2} . \tag{3.12}$$

In Figure 3.2, the Gaussian power spectral density (3.11) is illustrated with the corresponding autocorrelation function (3.12).

Characteristic quantities for the Doppler power spectral density $S_{\mu_i\mu_i}(f)$ are the *average Doppler shift* $B^{(1)}_{\mu_i\mu_i}$ and the *Doppler spread* $B^{(2)}_{\mu_i\mu_i}$ [Bel63]. The average Doppler shift (Doppler spread) describes the average frequency shift (frequency spread) that

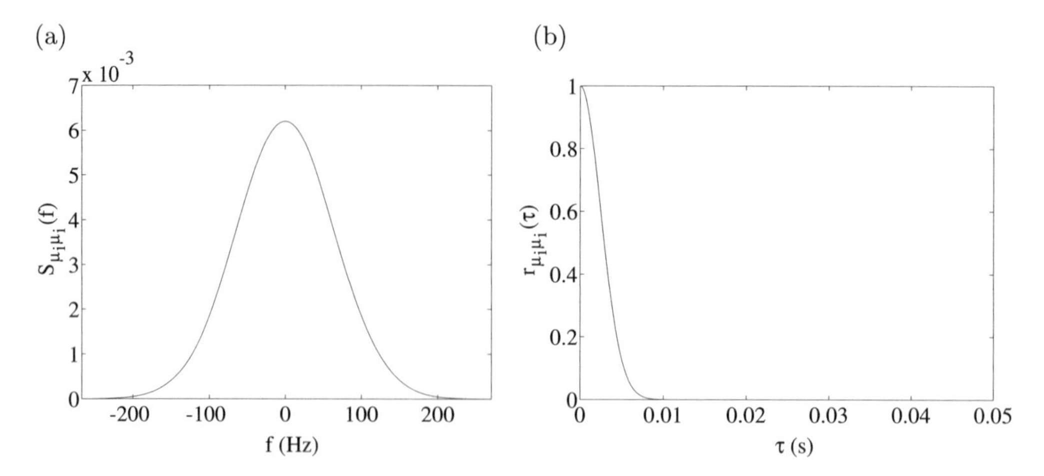

Figure 3.2: (a) Gaussian power spectral density $S_{\mu_i\mu_i}(f)$ and (b) corresponding autocorrelation function $r_{\mu_i\mu_i}(\tau)$ ($f_c = \sqrt{\ln 2} f_{max}$, $f_{max} = 91\,\text{Hz}$, $\sigma_0^2 = 1$).

a carrier signal experiences during transmission. The average Doppler shift $B^{(1)}_{\mu_i\mu_i}$ is the first moment of $S_{\mu_i\mu_i}(f)$ and the Doppler spread $B^{(2)}_{\mu_i\mu_i}$ is the square root of the second central moment of $S_{\mu_i\mu_i}(f)$. Consequently, $B^{(1)}_{\mu_i\mu_i}$ and $B^{(2)}_{\mu_i\mu_i}$ are defined by

$$B^{(1)}_{\mu_i\mu_i} := \frac{\int_{-\infty}^{\infty} f S_{\mu_i\mu_i}(f) df}{\int_{-\infty}^{\infty} S_{\mu_i\mu_i}(f) df} \tag{3.13a}$$

and

$$B^{(2)}_{\mu_i\mu_i} := \sqrt{\frac{\int_{-\infty}^{\infty} (f - B^{(1)}_{\mu_i\mu_i})^2 S_{\mu_i\mu_i}(f) df}{\int_{-\infty}^{\infty} S_{\mu_i\mu_i}(f) df}}, \tag{3.13b}$$

for $i = 1, 2$, respectively. Equivalent — but often easier to calculate — expressions for (3.13a) and (3.13b) can be obtained by using the autocorrelation function $r_{\mu_i\mu_i}(\tau)$ as well as its first and second time derivative at the origin, i.e.,

$$B^{(1)}_{\mu_i\mu_i} := \frac{1}{2\pi j} \cdot \frac{\dot{r}_{\mu_i\mu_i}(0)}{r_{\mu_i\mu_i}(0)} \quad \text{and} \quad B^{(2)}_{\mu_i\mu_i} = \frac{1}{2\pi} \sqrt{\left(\frac{\dot{r}_{\mu_i\mu_i}(0)}{r_{\mu_i\mu_i}(0)}\right)^2 - \frac{\ddot{r}_{\mu_i\mu_i}(0)}{r_{\mu_i\mu_i}(0)}} \tag{3.14a, b}$$

for $i = 1, 2$, respectively.

For the important special case where the Doppler power spectral densities $S_{\mu_1\mu_1}(f)$ and $S_{\mu_2\mu_2}(f)$ are identical and symmetrical, $\dot{r}_{\mu_i\mu_i}(0) = 0$ ($i = 1, 2$) holds. Hence, by using (3.7), we obtain the following expressions for the corresponding characteristic quantities of the Doppler power spectral density $S_{\mu\mu}(f)$

$$B^{(1)}_{\mu\mu} = B^{(1)}_{\mu_i\mu_i} = 0 \quad \text{and} \quad B^{(2)}_{\mu\mu} = B^{(2)}_{\mu_i\mu_i} = \frac{\sqrt{\beta}}{2\pi\sigma_0}, \tag{3.15a, b}$$

where $\sigma_0^2 = r_{\mu_i\mu_i}(0) \geq 0$ and $\beta = -\ddot{r}_{\mu_i\mu_i}(0) \geq 0$.

By making use of (3.15a, b), especially for the Jakes power spectral density [see (3.8)] and the Gaussian power spectral density [see (3.11)], the expressions

$$B_{\mu_i\mu_i}^{(1)} = B_{\mu\mu}^{(1)} = 0 \quad \text{and} \quad B_{\mu_i\mu_i}^{(2)} = B_{\mu\mu}^{(2)} = \begin{cases} \dfrac{f_{max}}{\sqrt{2}}, & \text{Jakes PSD}, \\ \dfrac{f_c}{\sqrt{2\ln 2}}, & \text{Gaussian PSD}, \end{cases} \tag{3.16a, b}$$

for $i = 1, 2$ follow for the average Doppler shift $B_{\mu_i\mu_i}^{(1)}$ and the Doppler spread $B_{\mu_i\mu_i}^{(2)}$, respectively. From (3.16b), it follows that the Doppler spread of the Jakes power spectral density is identical to the Doppler spread of the Gaussian power spectral density, if the 3-dB-cut-off frequency f_c and the maximum Doppler frequency f_{max} are related by $f_c = \sqrt{\ln 2} f_{max}$.

3.3 STATISTICAL PROPERTIES OF RICE AND RAYLEIGH PROCESSES

Besides the probability density of the amplitude and the phase, we will in this section also analyse the level-crossing rate as well as the average duration of fades of Rice processes $\xi(t) = |\mu(t) + m(t)|$ [see (3.6)] with time-variant line-of-sight components $m(t)$. Analysing the influence of the power spectral density $S_{\mu\mu}(f)$ of the complex Gaussian random process $\mu(t)$ on the statistical properties of $\xi(t)$, we will restrict ourselves to the Jakes and Gaussian power spectral densities introduced above.

3.3.1 Probability Density Function of the Amplitude and the Phase

The probability density function of the Rice process $\xi(t)$, $p_\xi(x)$, is described by the so called Rice distribution [Ric48]

$$p_\xi(x) = \begin{cases} \dfrac{x}{\sigma_0^2} e^{-\frac{x^2+\rho^2}{2\sigma_0^2}} I_0\left(\dfrac{x\rho}{\sigma_0^2}\right), & x \geq 0, \\ 0, & x < 0, \end{cases} \tag{3.17}$$

where $I_0(\cdot)$ is the 0th-order modified Bessel function of the first kind and $\sigma_0^2 = r_{\mu_i\mu_i}(0) = r_{\mu\mu}(0)/2$ again denotes the power of the real-valued Gaussian random process $\mu_i(t)$ ($i = 1, 2$). Obviously, neither the time variance of the mean (3.2) caused by the Doppler frequency of the line-of-sight component nor the exact shape of the Doppler power spectral density $S_{\mu\mu}(f)$ influences the probability density function $p_\xi(x)$. Merely the amplitude of the line-of-sight component ρ and the power σ_0^2 of the real part or the imaginary part of the scattered component determine the behaviour of $p_\xi(x)$.

Of particular interest is in this context the *Rice factor*, denoted by c_R, which describes the ratio of the power of the line-of-sight component to the sum of the power of all scattered components. Thus, the Rice factor is defined by

$$c_R := \frac{\rho^2}{2\sigma_0^2} \,. \tag{3.18}$$

From the limit $\rho \to 0$, i.e., $c_R \to 0$, the Rice process $\xi(t)$ results in the Rayleigh process $\zeta(t)$, whose statistical amplitude variations are described by the Rayleigh distribution [Pap91]

$$p_\zeta(x) = \begin{cases} \frac{x}{\sigma_0^2} e^{-\frac{x^2}{2\sigma_0^2}}, & x \geq 0\,, \\ 0\,, & x < 0\,. \end{cases} \tag{3.19}$$

The probability density functions $p_\xi(x)$ and $p_\zeta(x)$ according to (3.17) and (3.19) are shown in the Figures 3.3(a) and 3.3(b), respectively.

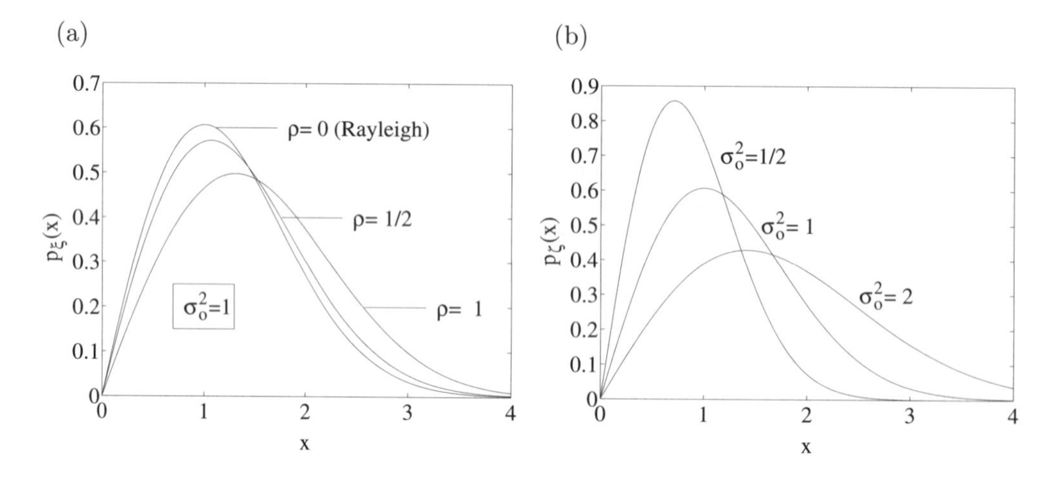

Figure 3.3: The probability density function of (a) Rice and (b) Rayleigh processes.

As mentioned before, the exact shape of the Doppler power spectral density $S_{\mu\mu}(f)$ has no effect on the probability density of the absolute value of the complex Gaussian random process, i.e., $\xi(t) = |\mu_\rho(t)|$. Analogously, this statement is also valid for the probability density function of the phase $\vartheta(t) = \arg\{\mu_\rho(t)\}$, where $\vartheta(t)$ can be expressed with (3.1), (3.2), and (3.4) as follows

$$\vartheta(t) = \arctan\left\{\frac{\mu_2(t) + \rho\sin(2\pi f_\rho t + \theta_\rho)}{\mu_1(t) + \rho\cos(2\pi f_\rho t + \theta_\rho)}\right\}. \tag{3.20}$$

In order to confirm this statement, we study the probability density function $p_\vartheta(\theta; t)$ of the phase $\vartheta(t)$ given by the following relation [Pae98d]

$$p_\vartheta(\theta; t) = \frac{e^{-\frac{\rho^2}{2\sigma_0^2}}}{2\pi} \left\{ 1 + \frac{\rho}{\sigma_0} \sqrt{\frac{\pi}{2}} \cos(\theta - 2\pi f_\rho t - \theta_\rho) e^{\frac{\rho^2 \cos^2(\theta - 2\pi f_\rho t - \theta_\rho)}{2\sigma_0^2}} \left[1 + \operatorname{erf}\left(\frac{\rho \cos(\theta - 2\pi f_\rho t - \theta_\rho)}{\sigma_0 \sqrt{2}} \right) \right] \right\}, \quad -\pi < \theta \leq \pi, \tag{3.21}$$

where erf$(\cdot)$ is called the *error function.*[2] The dependence of the probability density function $p_\vartheta(\theta; t)$ on the time t is due to the Doppler frequency f_ρ of the line-of-sight component $m(t)$. According to Subsection 2.2.1, the stochastic process $\vartheta(t)$ is not stationary in the strict sense, because the condition (2.47a) is not fulfilled. Only for the special case that $f_\rho = 0$ $(\rho \neq 0)$, the phase $\vartheta(t)$ is a strict-sense stationary process which is then described by the probability density function shown in [Par92]

$$p_\vartheta(\theta) = \frac{e^{-\frac{\rho^2}{2\sigma_0^2}}}{2\pi} \left\{ 1 + \frac{\rho}{\sigma_0} \sqrt{\frac{\pi}{2}} \cos(\theta - \theta_\rho) e^{\frac{\rho^2 \cos^2(\theta - \theta_\rho)}{2\sigma_0^2}} \left[1 + \operatorname{erf}\left(\frac{\rho \cos(\theta - \theta_\rho)}{\sigma_0 \sqrt{2}} \right) \right] \right\}, \quad -\pi < \theta \leq \pi. \tag{3.22}$$

As $\rho \to 0$, it follows $\mu_\rho(t) \to \mu(t)$ and, thus, $\xi(t) \to \zeta(t)$, and from (3.22), we obtain the uniform distribution

$$p_\vartheta(\theta) = \frac{1}{2\pi}, \quad -\pi < \theta \leq \pi. \tag{3.23}$$

Therefore, the phase of zero-mean complex Gaussian random processes with uncorrelated real and imaginary parts is always uniformly distributed. Finally, it should be mentioned that in the limit $\rho \to \infty$, (3.22) tends to $p_\vartheta(\theta) = \delta(\theta - \theta_\rho)$.

By way of illustration, the probability density function $p_\vartheta(\theta)$ is depicted in Figure 3.4 for several values of ρ.

3.3.2 Level-Crossing Rate and Average Duration of Fades

As further statistical quantities, we will in this subsection study the level-crossing rate and the average duration of fades. Therefore, we at first turn to the Rice process $\xi(t)$ introduced by (3.6), and we impose on our reference model that the real-valued zero-mean Gaussian random processes $\mu_1(t)$ and $\mu_2(t)$ are uncorrelated and both have identical autocorrelation functions, i.e., $r_{\mu_1\mu_2}(\tau) = 0$ and $r_{\mu_1\mu_1}(\tau) = r_{\mu_2\mu_2}(\tau)$. When calculating the level-crossing rate $N_\xi(r)$ of the Rice process $\xi(t) = |\mu_\rho(t)|$, however, it must be taken into consideration that a correlation exists between the real and imaginary part of the complex Gaussian random process $\mu_\rho(t)$ [see (3.4)] due to the time-variant line-of-sight component (3.2).

---

[2] The error function is defined as $\operatorname{erf}(x) = \frac{2}{\sqrt{\pi}} \int_0^x e^{-t^2} dt$.

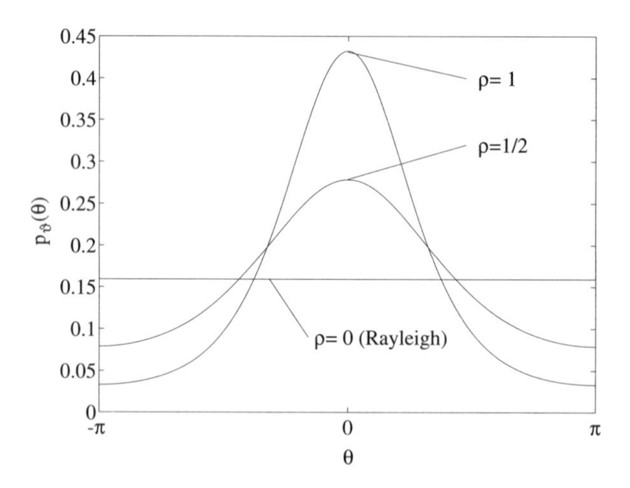

Figure 3.4: The probability density function of the phase $p_\vartheta(\theta)$ ($f_\rho = 0, \theta_\rho = 0, \sigma_0^2 = 1$).

For the level-crossing rate $N_\xi(r)$ it then holds [Pae98d]

$$N_\xi(r) = \frac{r\sqrt{2\beta}}{\pi^{3/2}\sigma_0^2} \, e^{-\frac{r^2+\rho^2}{2\sigma_0^2}} \int_0^{\pi/2} \cosh\left(\frac{r\rho}{\sigma_0^2}\cos\theta\right) \left\{e^{-(\alpha\rho\sin\theta)^2} + \sqrt{\pi}\alpha\rho\sin(\theta)\cdot \operatorname{erf}(\alpha\rho\sin\theta)\right\} d\theta \,, \quad r \geq 0 \,, \tag{3.24}$$

where the quantities α and β are given by

$$\alpha = 2\pi f_\rho / \sqrt{2\beta} \tag{3.25}$$

and

$$\beta = \beta_i = -\ddot{r}_{\mu_i\mu_i}(0) \,, \quad i = 1, 2 \,, \tag{3.26}$$

respectively. Considering (3.25), we notice that the Doppler frequency f_ρ of the line-of-sight component $m(t)$ has an influence on the level-crossing rate $N_\xi(r)$. However, if $f_\rho = 0$, and, thus, $\alpha = 0$, it follows from (3.24) the relation (2.62), which will at this point be given again for completeness, i.e.,

$$N_\xi(r) = \sqrt{\frac{\beta}{2\pi}} \cdot p_\xi(r) \,, \quad r \geq 0 \,. \tag{3.27}$$

Therefore, (3.27) describes the level-crossing rate of Rice processes with a time-invariant line-of-sight component. For $\rho \to 0$, it follows $p_\xi(r) \to p_\zeta(r)$, and for the level-crossing rate $N_\zeta(r)$ of Rayleigh processes $\zeta(t)$, we obtain the relation

$$N_\zeta(r) = \sqrt{\frac{\beta}{2\pi}} \cdot p_\zeta(r) \,, \quad r \geq 0 \,. \tag{3.28}$$

For Rice and Rayleigh processes, the expressions (3.27) and (3.28), respectively, clearly show the proportional relation between the level-crossing rate and the corresponding probability density function of the amplitude. The value of the proportional constant $\sqrt{\beta/(2\pi)}$ is due to (3.26) only depending on the negative curvature of the autocorrelation function of the real-valued Gaussian random processes at the origin. Especially for the Jakes and the Gaussian power spectral density, we obtain by using (3.10), (3.12), and (3.26), the following result for the quantity β:

$$\beta = \begin{cases} 2(\pi f_{max}\sigma_0)^2 \,, & \text{Jakes PSD}\,, \\ 2(\pi f_c \sigma_0)^2 / \ln 2 \,, & \text{Gaussian PSD}\,. \end{cases} \tag{3.29}$$

Despite the large differences existing between the shape of the Jakes and the Gaussian power spectral density, both Doppler power spectral densities enable the modelling of Rice or Rayleigh processes with identical level-crossing rates, as long as the relation $f_c = \sqrt{\ln 2} f_{max}$ between f_{max} and f_c holds.

The influence of the parameters f_ρ and ρ on the normalized level-crossing rate $N_\xi(r)/f_{max}$ is illustrated in Figures 3.5(a) and 3.5(b), respectively. Thereby, Figure 3.5(a) points out that an increase of $|f_\rho|$ leads to an increase of the level-crossing rate $N_\xi(r)$.

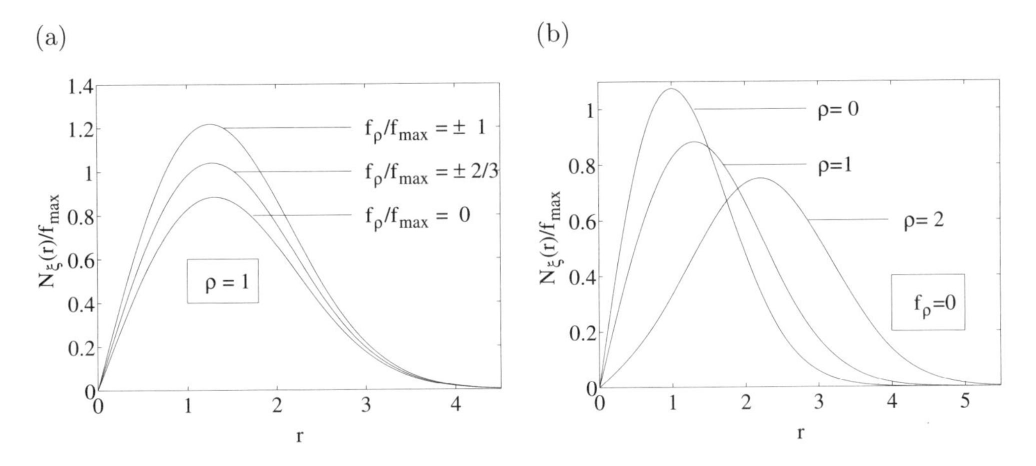

Figure 3.5: Normalized level-crossing rate $N_\xi(r)/f_{max}$ of Rice processes dependent on (a) f_ρ and (b) ρ (Jakes PSD, $f_{max} = 91\,\text{Hz}$, $\sigma_0^2 = 1$).

In some passages of this book, the case $r_{\mu_1\mu_1}(0) = r_{\mu_2\mu_2}(0)$ but $\beta_1 = -\ddot{r}_{\mu_1\mu_1}(0) \neq -\ddot{r}_{\mu_2\mu_2}(0) = \beta_2$ will be relevant for us. For the level-crossing rate $N_\xi(r)$ of the Rice process $\xi(t)$, we on this condition obtain the expression (B.13) derived in Appendix B

$$N_\xi(r) = \sqrt{\frac{\beta_1}{2\pi}} \frac{r}{\sigma_0^2} e^{-\frac{r^2+\rho^2}{2\sigma_0^2}}$$

$$\cdot \frac{1}{\pi} \int_0^{\pi} \cosh\left[\frac{r\rho}{\sigma_0^2}\cos(\theta - \theta_\rho)\right] \sqrt{1 - k^2 \sin^2\theta}\, d\theta\,, \quad r \geq 0\,, \tag{3.30}$$

where $k = \sqrt{(\beta_1 - \beta_2)/\beta_1}$, $\beta_1 \geq \beta_2$. In this case, the level-crossing rate is in general no longer proportional to the probability density function of the Rice distribution.

On the other hand, we again obtain the usual relations for Rayleigh processes $\zeta(t)$, whose level-crossing rate $N_\zeta(r)$ is obtained from (3.30) by taking the limit $\rho \to 0$, i.e.,

$$N_\zeta(r) = \sqrt{\frac{\beta_1}{2\pi}} \cdot \frac{r}{\sigma_0^2} e^{-\frac{r^2}{2\sigma_0^2}} \cdot \frac{1}{\pi} \int_0^{\pi} \sqrt{1 - k^2 \sin^2\theta}\, d\theta\,, \quad r \geq 0\,. \tag{3.31}$$

In the literature (see [Gra81, vol. II, eq. (8.111.3)]), the above integral with the form

$$E(\varphi, k) = \int_0^{\varphi} \sqrt{1 - k^2 \sin^2\theta}\, d\theta \tag{3.32}$$

is known as *elliptic integral of the second kind.* The parameter k denotes the modulus of the integral. For $\varphi = \pi/2$ these integrals are also called the *complete elliptic integrals of the second kind* and we write $\boldsymbol{E}(k) = E(\frac{\pi}{2}, k)$.

Let us use (3.19). Then, the level-crossing rate for Rayleigh processes can now be put in the following form

$$N_\zeta(r) = \sqrt{\frac{\beta_1}{2\pi}}\, p_\zeta(r) \cdot \frac{2}{\pi} \boldsymbol{E}(k)\,, \quad r \geq 0\,, \tag{3.33}$$

where for the modulus k again $k = \sqrt{(\beta_1 - \beta_2)/\beta_1}$, $\beta_1 \geq \beta_2$, holds. Thus, for Rayleigh processes the level-crossing rate is proportional to the probability density function of the amplitude even for the case $\beta_1 \neq \beta_2$. The proportional constant is here not only determined by β_1, but also by the difference $\beta_1 - \beta_2$.

Furthermore, we are interested in the level-crossing rate $N_\zeta(r)$ for the case that the relative deviation between β_1 and β_2 is very small. Therefore, let us assume that a positive number $\varepsilon = \beta_1 - \beta_2 \geq 0$ with $\varepsilon/\beta_1 << 1$ exists, so that

$$k = \sqrt{\frac{\beta_1 - \beta_2}{\beta_1}} = \sqrt{\frac{\varepsilon}{\beta_1}} << 1 \tag{3.34}$$

holds. Next, we make use of the relation (see [Gra81, vol. II, eq. (8.114.1)])

$$\begin{aligned} \boldsymbol{E}(k) &= \frac{\pi}{2} F\left(-\frac{1}{2}, \frac{1}{2}; 1; k^2\right) \\ &= \frac{\pi}{2}\left\{1 - \sum_{n=1}^{\infty}\left[\frac{1 \cdot 3 \cdot 5 \cdot \ldots \cdot (2n-1)}{2^n \quad n!}\right]^2 \frac{k^{2n}}{2n-1}\right\}, \end{aligned} \tag{3.35}$$

where $F(.,.;.;.)$ is called the *hypergeometric function.* By using the first two terms of the series for $\boldsymbol{E}(k)$, we obtain the following approximation formula

$$\boldsymbol{E}(k) \approx \frac{\pi}{2}\left(1-\frac{k^2}{4}\right) \approx \frac{\pi}{2}\sqrt{1-\frac{k^2}{2}}\,, \quad k << 1\,. \tag{3.36}$$

Now, substituting (3.34) into (3.36) and taking (3.33) into account, leads for the level-crossing rate $N_\zeta(r)$ to the approximation

$$N_\zeta(r) \approx \sqrt{\frac{\beta}{2\pi}} \cdot p_\zeta(r)\,, \quad r \geq 0\,, \tag{3.37}$$

which is valid for the case $(\beta_1 - \beta_2)/\beta_1 << 1$, where in (3.37) the quantity β is given by $\beta = (\beta_1 + \beta_2)/2$. Hence, (3.28) approximately keeps its validity if the relative deviations between β_1 and β_2 are small and if $\beta = \beta_1 = \beta_2$ is in (3.28) replaced by the arithmetical mean $\beta = (\beta_1 + \beta_2)/2$.

The average duration of fades, i.e., the average length of the duration while the channel amplitude is below a level r, is defined by the quotient of the distribution function of the channel amplitude over the level-crossing rate, according to (2.63). The probability density function and the level-crossing rate of Rice and Rayleigh processes considered here have already been analysed in detail, so that the analysis of the corresponding average duration of fades can easily be carried out. For completeness, however, the resulting relations will again be given here. For Rice processes with $f_\rho = 0$ and Rayleigh processes, we obtain for the average duration of fades [see also (2.66) and (2.65), respectively]:

$$T_{\xi_-}(r) = \frac{F_{\xi_-}(r)}{N_\xi(r)} = \sqrt{\frac{2\pi}{\beta}} \cdot \frac{e^{\frac{r^2}{2\sigma_0^2}}}{r\, I_0\left(\frac{r\rho}{\sigma_0^2}\right)} \int_0^r x\, e^{-\frac{x^2}{2\sigma_0^2}}\, I_0\left(\frac{x\rho}{\sigma_0^2}\right) dx\,, \quad r \geq 0\,, \tag{3.38a}$$

and

$$T_{\zeta_-}(r) = \frac{F_{\zeta_-}(r)}{N_\zeta(r)} = \sqrt{\frac{2\pi}{\beta}} \cdot \frac{\sigma_0^2}{r}\left(e^{\frac{r^2}{2\sigma_0^2}} - 1\right)\,, \quad r \geq 0\,, \tag{3.38b}$$

respectively, where $F_{\xi_-}(r) = P(\xi(t) \leq r)$ and $F_{\zeta_-}(r) = P(\zeta(t) \leq r)$ denote the cumulative distribution function of the Rice and Rayleigh process, respectively.

In channel modelling, we are especially interested in the behaviour of the average duration of fades at low levels r. We therefore wish to analyse this case separately. For this purpose, let $r << 1$, so that for moderate Rice factors, we may write $r\rho/\sigma_0^2 << 1$ and, consequently, both $I_0(r\rho/\sigma_0^2)$ and $I_0(x\rho/\sigma_0^2)$ can be approximated by one in (3.38a), since the independent variable x is within the relevant interval $[0, r]$. After a series expansion of the integrand in (3.38a), $T_{\xi_-}(r)$ can be given in a closed form. By this means, it quickly turns out that for low levels r, $T_{\xi_-}(r)$ converges to $T_{\zeta_-}(r)$ given by (3.38b). Furthermore, the relation(3.38b) can be simplified by using $e^x \approx 1 + x \quad (x << 1)$, so that we finally obtain the approximations

$$T_{\xi_-}(r) \approx T_{\zeta_-}(r) \approx r\sqrt{\frac{\pi}{2\beta}}\,, \quad r << 1\,, \tag{3.39}$$

where $r\rho/\sigma_0^2 << 1$ is assumed. The above result shows that the average duration of fades of Rice and Rayleigh processes are at low levels r approximately proportional to r.

An illustration of the results is shown in Figure 3.6. In Figure 3.6(a) it can be seen that an increase of $|f_\rho|$ leads to a decrease of the average duration of fades $T_{\xi_-}(r)$.

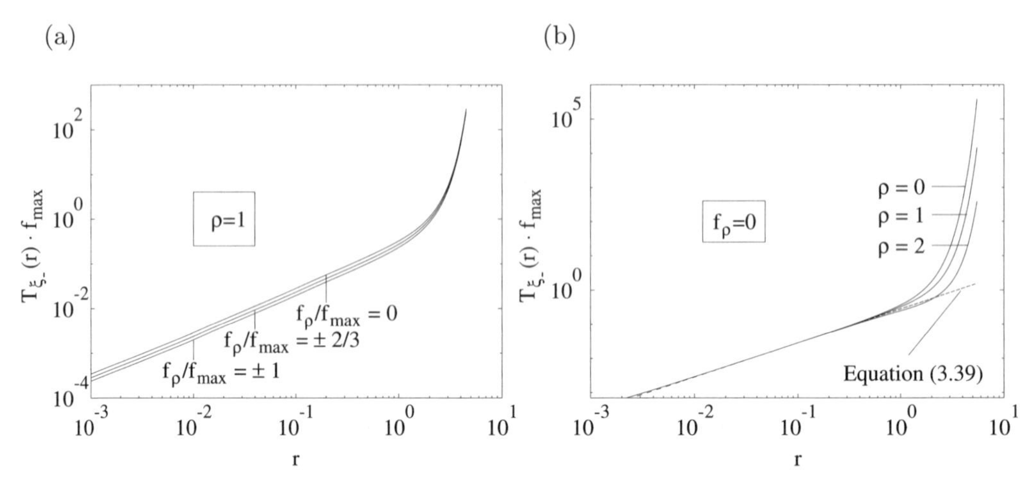

Figure 3.6: Normalized average duration of fades $T_{\xi_-}(r) \cdot f_{max}$ of Rice processes depending on (a) f_ρ and (b) ρ (Jakes PSD, $f_{max} = 91\,\text{Hz}$, $\sigma_0^2 = 1$).

3.3.3 The Statistics of the Fading Intervals of Rayleigh Processes

The statistical properties of Rayleigh and Rice processes analysed so far are independent of the behaviour of the autocorrelation function $r_{\mu_i\mu_i}(\tau)$ $(i = 1, 2)$ of the underlying Gaussian random processes for values $\tau > 0$. For example, we have seen that the probability density function of the amplitude $\zeta(t) = |\mu(t)|$ is totally determined by the behaviour of the autocorrelation function $r_{\mu_i\mu_i}(\tau)$ at the origin, i.e., by the variance $\sigma_0^2 = r_{\mu_i\mu_i}(0)$. The behaviour of $r_{\mu_i\mu_i}(\tau)$ at the origin determines the level-crossing rate $N_\zeta(r)$ and the average duration of fades $T_{\zeta_-}(r)$. These quantities are, besides on the variance $\sigma_0^2 = r_{\mu_i\mu_i}(0)$, also dependent on the negative curvature of the autocorrelation function at the origin $\beta = -\ddot{r}_{\mu_i\mu_i}(0)$. If we now ask ourselves which relevant statistical properties are at all affected by the behaviour of the autocorrelation function $r_{\mu_i\mu_i}(\tau)$ $(i = 1, 2)$ for $\tau > 0$, then this leads to the statistical distribution of the fading intervals.

The conditional probability density function for the event that a Rayleigh process $\zeta(t)$ crosses a given level r in an infinitesimal time interval $(t + \tau_-, t + \tau_- + d\tau_-)$ upwards for the first time on condition that the last down-crossing occurred within the time interval $(t, t+dt)$ is denoted as $p_{0_-}(\tau_-; r)$. An exact theoretical derivation for $p_{0_-}(\tau_-; r)$ is still today even for Rayleigh processes an unsolved problem. In [Ric58], Rice, however, managed to derive the probability density $p_{1-}(\tau_-; r)$ for the case that the Rayleigh process $\zeta(t)$ crosses the level r in the order mentioned, where, however, no information on the behaviour of $\zeta(t)$ between t and $t + \tau_-$ is given. For small τ_--values, at which the probability that further level-crossings occur between t and $t + \tau_-$ is very low, $p_{1_-}(\tau_-; r)$ can be considered a very good approximation for the desired probability density function $p_{0_-}(\tau_-; r)$. On the other hand, for large τ_--values, $p_{1_-}(\tau_-; r)$ cannot be used any further as a suitable approximation for $p_{0_-}(\tau_-; r)$.

The determination of $p_{1_-}(\tau_-; r)$ requires the numerical calculation of the threefold integral [Ric58]

$$p_{1_-}(\tau_-; r) = \frac{r M_{22}\, e^{\frac{r^2}{2}}}{\sqrt{2\pi\beta}(1 - r^2_{\mu_i\mu_i}(\tau_-))^2} \int_0^{2\pi} J(a, b)\, e^{-r^2 \frac{1 - r_{\mu_i\mu_i}(\tau_-)\cdot\cos\varphi}{1 - r^2_{\mu_i\mu_i}(\tau_-)}}\, d\varphi\,, \tag{3.40}$$

where

$$J(a, b) = \frac{1}{2\pi\sqrt{1 - a^2}} \int_b^\infty \int_b^\infty (x - b)(y - b)\, e^{-\frac{x^2 + y^2 - 2axy}{2(1 - a^2)}}\, dx\, dy\,, \tag{3.41}$$

$$a = \cos\varphi \cdot \frac{M_{23}}{M_{22}}\,, \tag{3.42}$$

$$b = \frac{r\, \dot{r}_{\mu_i\mu_i}(\tau_-) \cdot (r_{\mu_i\mu_i}(\tau_-) - \cos\varphi)}{1 - r^2_{\mu_i\mu_i}(\tau_-)} \cdot \sqrt{\frac{1 - r^2_{\mu_i\mu_i}(\tau_-)}{M_{22}}}\,, \tag{3.43}$$

$$M_{22} = \beta(1 - r^2_{\mu_i\mu_i}(\tau_-)) - \dot{r}^2_{\mu_i\mu_i}(\tau_-)\,, \tag{3.44}$$

$$M_{23} = \ddot{r}_{\mu_i\mu_i}(\tau_-)(1 - r^2_{\mu_i\mu_i}(\tau_-)) + r_{\mu_i\mu_i}(\tau_-)\dot{r}^2_{\mu_i\mu_i}(\tau_-)\,, \tag{3.45}$$

and β is again the quantity defined by (3.26).

Figures 3.7 and 3.8 show the evaluation of the probability density function $p_{1_-}(\tau_-; r)$ by using Jakes and Gaussian power spectral densities, respectively. For the 3-dB-cut-off frequency of the Gaussian power spectral density, the value $f_c = \sqrt{\ln 2} f_{max}$ was chosen. For the quantity β, we hereby obtain identical values for the Jakes and Gaussian power spectral density due to (3.29). Observing Figures 3.7(a) and 3.8(a), we see that at low levels ($r = 0.1$) the courses of the probability density functions $p_{1_-}(\tau_-; r)$ are identical. With increasing levels r, however, these courses differ more and more from each other (cf. Figures 3.7(b) and 3.8(b) for medium levels ($r = 1$) as well as Figures 3.7(c) and 3.8(c) for high levels ($r = 2.5$)).

In these figures, it should be observed that $p_{1_-}(\tau_-; r)$ does not converge to zero at medium and large values for the level r. Obviously, $p_{1_-}(\tau_-; r) \geq 0$ holds if $\tau_- \to \infty$ which extremely jeopardizes the accuracy of (3.40) — at least for the range of medium and high levels of r in connection with long fading intervals τ_-.

The validity of the approximate solution (3.40) can ultimately only be determined by simulating the level-crossing behaviour. Therefore, simulation models are needed, which reproduce the Gaussian random processes $\mu_i(t)$ of the reference model with

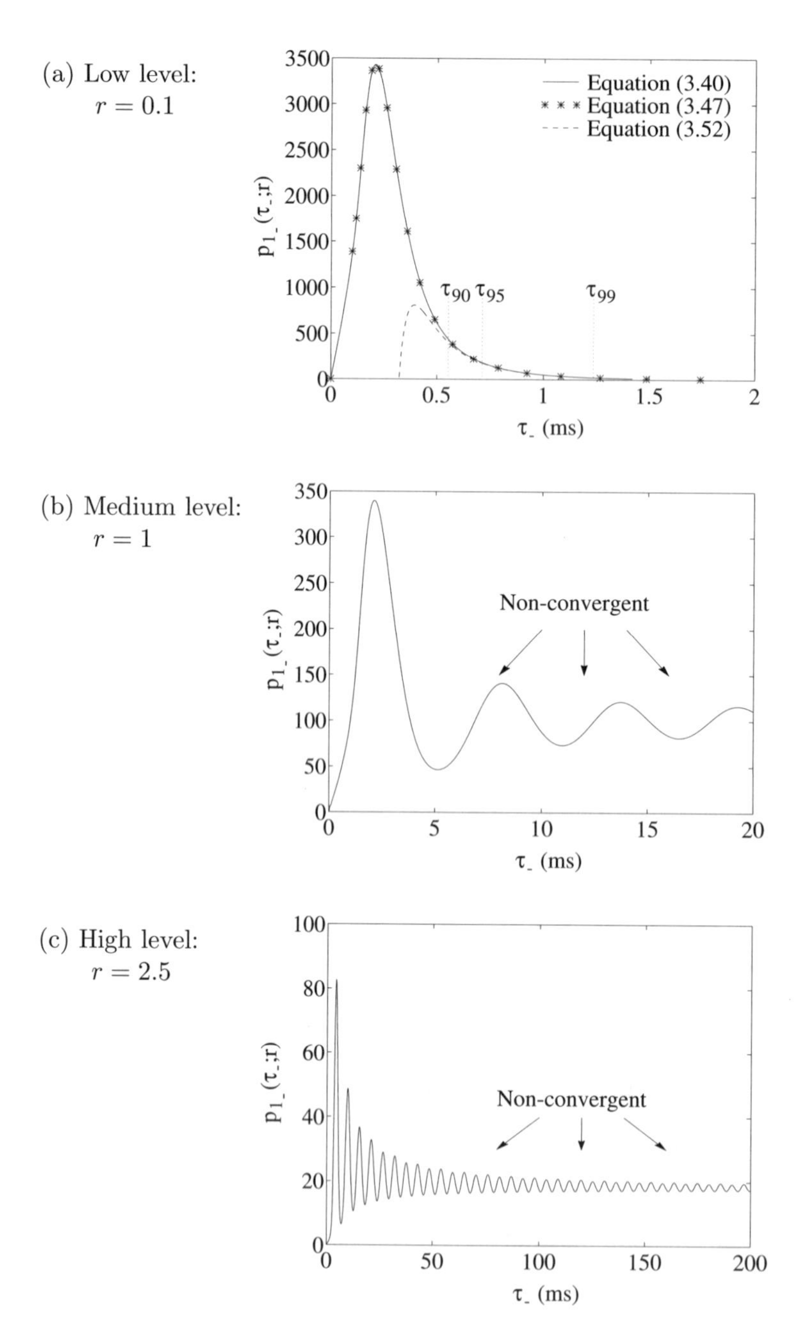

Figure 3.7: The probability density function $p_{1\_}(\tau_-; r)$ when using the Jakes power spectral density ($f_{max} = 91\,\text{Hz}$, $\sigma_0^2 = 1$).

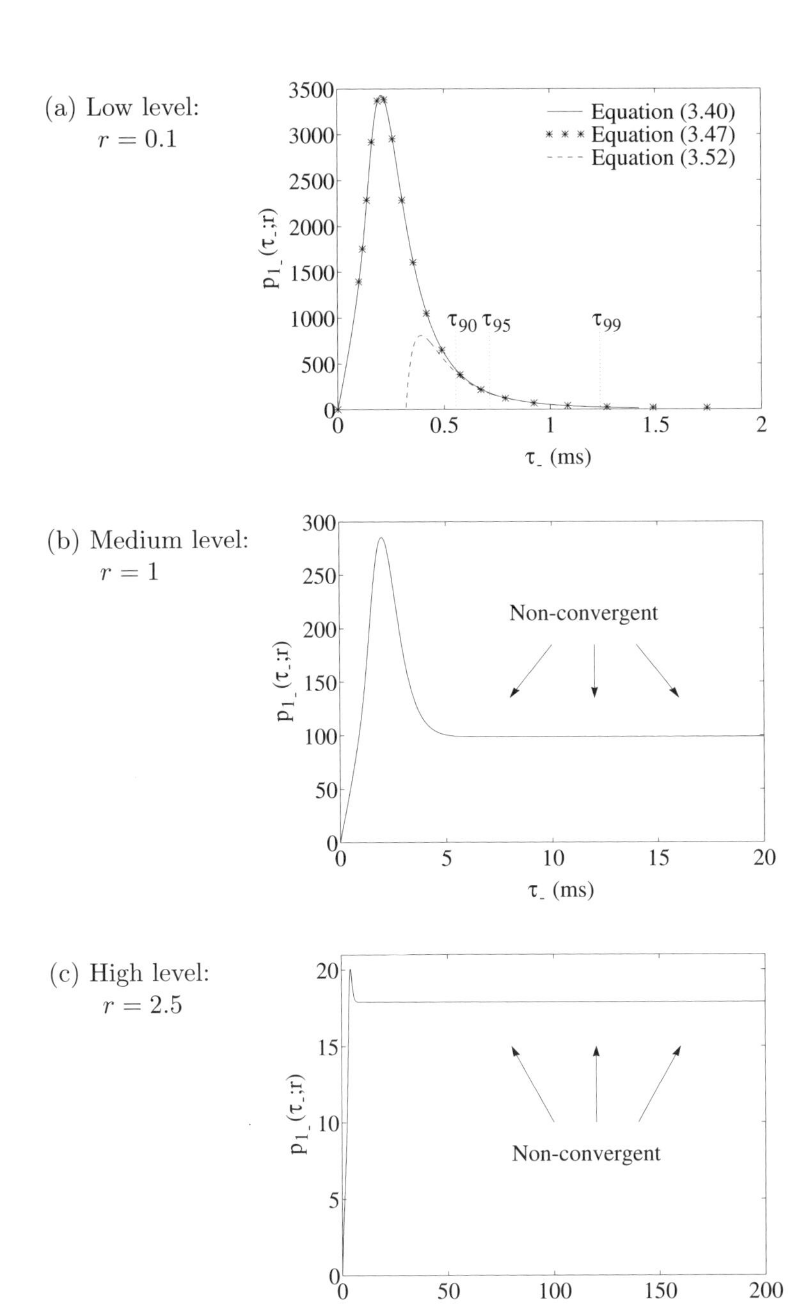

Figure 3.8: The probability density function $p_{1\_}(\tau_-; r)$ when using the Gaussian power spectral density ($f_c = \sqrt{\ln 2} f_{max}$, $f_{max} = 91\,\text{Hz}$, $\sigma_0^2 = 1$).

respect to the probability density function $p_{\mu_i}(x)$ and the autocorrelation function $r_{\mu_i\mu_i}(\tau)$ extremely accurate. We will return to this subject in Section 5.3. For our purposes, at first only the discovery that the probability density function of the fading intervals of Rayleigh channels at medium and high levels r decisively depends on the behaviour of the autocorrelation function $r_{\mu_i\mu_i}(\tau)$ for $\tau \geq 0$ is of importance.

In the following, we will analyse the statistics of the deep fades. The knowledge of the statistics of the deep fades is of great importance in mobile radio communications, since the bit and symbol error probability are mainly determined by the occurrence of deep fades. Hence, let $r << 1$. In this case, the duration of fades τ_- are short. Thus, the probability that further level-crossings occur between t and $t+\tau_-$ is very low, and the approximation $p_{0_-}(\tau_-;r) \approx p_{1_-}(\tau_-;r)$ is very good. In [Ric58] it is shown that the probability density function (3.40) converges to

$$p_{1_-}(\tau_-;r) = -\frac{1}{T_{\zeta_-}(r)} \frac{d}{du} \left[\frac{2}{u} \, I_1(z) \, e^{-z} \right] \tag{3.46}$$

as $r \to 0$, where $z = 2/(\pi u^2)$ and $u = \tau_-/T_{\zeta_-}(r)$ hold. After some algebraic manipulations, we find the following expression for this

$$p_{1_-}(\tau_-;r) = \frac{2\pi z^2 e^{-z}}{T_{\zeta_-}(r)} \left[I_0(z) - \left(1 + \frac{1}{2z}\right) I_1(z) \right], \quad r \to 0, \tag{3.47}$$

where $z = 2\big[T_{\zeta_-}(r)/\tau_-\big]^2/\pi$. Considering (3.46) or (3.47), we see that, besides on the level r, $p_{1-}(\tau_-;r)$ only depends on the average duration of fades $T_{\zeta_-}(r)$ and, hence, on $\sigma_0^2 = r_{\mu_i\mu_i}(0)$ and $\beta = -\ddot{r}_{\mu_i\mu_i}(0)$. Consequently, the probability density of the fading intervals at low levels ($r << 1$) is independent of the shape of the autocorrelation function $r_{\mu_i\mu_i}(\tau)$ for $\tau > 0$. The numerical evaluation of the probability density function (3.47) for the level $r = 0.1$ is also depicted in Figures 3.7(a) and 3.8(a). These figures clearly show that the deviations between (3.40) and (3.47) are negligible for low levels r.

In the limits $\tau_- \to 0$ and $\tau_- \to \infty$, (3.47) converges to $p_{1_-}(0;r) = p_{1_-}(\infty;r) = 0$. Finally, it should be mentioned that by using (3.47) one finds – after a short side-calculation – the following result for the expected value of the fading intervals τ_-

$$E\{\tau_-\} = \int_0^\infty \tau_- \, p_{1_-}(\tau_-;r) \, d\tau_- = T_{\zeta_-}(r) \, . \tag{3.48}$$

With τ_q, we will in the following denote the time interval of the duration of fades which includes q per cent of all fading intervals. Thus, by τ_q the lower bound of the integral

$$\int_{\tau_q}^\infty p_{0_-}(\tau_-;r) \, d\tau_- = 1 - \frac{q}{100} \tag{3.49}$$

is determined. The knowledge of the quantities τ_{90}, τ_{95}, and τ_{99} is of great importance to the (optimal) design of the interleaver/deinterleaver as well as for the channel coder/decoder. With the approximation $p_{0_-}(\tau_-;r) \approx p_{1_-}(\tau_-;r)$, we are now able to derive an approximate solution for τ_q in an explicit form. We at first proceed by developing (3.47) into a power series, where we make use of the series expansions [Abr72, eq. (4.2.1)]

$$e^{-z} = \sum_{n=0}^{\infty} \frac{(-z)^n}{n!} \tag{3.50}$$

and [Abr72, eq. (9.6.10)]

$$I_\nu(z) = \left(\frac{z}{2}\right)^\nu \sum_{n=0}^{\infty} \frac{(z^2/4)^n}{n!\,\Gamma(\nu+n+1)}, \quad \nu = 0, 1, 2, \ldots \tag{3.51}$$

In the latter expression $\Gamma(\cdot)$ denotes the gamma function.[3] Terminating the resulting series after the second term, leads for the right-sided tail of the distribution $p_{1_-}(\tau_-;r)$ to the approximation usable for our purposes

$$p_{1_-}(\tau_-;r) \approx \frac{\pi z^2}{2}\,(3-5z)/T_{\zeta_-}(r), \tag{3.52}$$

where z again represents $z = 2\big[T_{\zeta_-}(r)/\tau_-\big]^2/\pi$. If we now replace the probability density $p_{0_-}(\tau_-;r)$ in (3.49) by (3.52), then an explicit expression for the quantity $\tau_q = \tau_q(r)$ can be derived from the result of the integration.

Finally, we obtain the approximation valid for $75 \leq q \leq 100$ [Pae96e]

$$\tau_q(r) \approx \frac{T_{\zeta_-}(r)}{\{\frac{\pi}{4}[1-\sqrt{1-4(1-\frac{q}{100})}]\}^{\frac{1}{3}}}, \quad r << 1. \tag{3.53}$$

This equation clearly shows that the quantity $\tau_q(r)$ is at deep fades proportional to the average duration of fades. Especially for $\tau_{90}(r)$, $\tau_{95}(r)$, and $\tau_{99}(r)$, we obtain from (3.53):

$$\tau_{90}(r) \approx 1.78 \cdot T_{\zeta_-}(r), \tag{3.54}$$
$$\tau_{95}(r) \approx 2.29 \cdot T_{\zeta_-}(r), \tag{3.55}$$
$$\tau_{99}(r) \approx 3.98 \cdot T_{\zeta_-}(r). \tag{3.56}$$

Further simplifications are possible if we approximate the average duration of fades $T_{\zeta_-}(r)$ for $r << 1$ by $T_{\zeta_-}(r) \approx r\sqrt{\pi/(2\beta)}$ [cf. (3.39)]. If β is in this relation now replaced by the formula (3.29) found for the Jakes and Gaussian power spectral density, then we obtain, e.g., for the quantity $\tau_{90}(r)$ the approximation

$$\tau_{90}(r) \approx \begin{cases} \dfrac{r}{2\,\sigma_0\,f_{max}}, & \text{Jakes PSD}, \\ \dfrac{r\sqrt{\ln 2}}{2\,\sigma_0\,f_c}, & \text{Gaussian PSD}, \end{cases} \tag{3.57}$$

[3] According to Euler the gamma function $\Gamma(x)$ is defined for real numbers $x > 0$ by $\Gamma(x) := \int_0^\infty e^{-t}\,t^{x-1}\,dt$. If x is a natural number, then $\Gamma(x) = (x-1)!$ holds.

which is valid for all $r << 1$. By means of this result, we see that the quantity $\tau_{90}(r)$ and, hence, the general quantity $\tau_q(r)$ $(75 \leq q \leq 100)$ are proportional to r and reciprocally proportional to f_{max} or f_c for low levels r. Hereby, it is of major importance that the exact form of the power spectral density of the complex Gaussian random process, which generates the Rayleigh process, does not have any influence on the behaviour of $\tau_q(r)$. Hence, for the Jakes and the Gaussian power spectral density, we again obtain identical values for $\tau_q(r)$ by choosing $f_c = \sqrt{\ln 2}\, f_{max}$. Therefore, one may also compare Figures 3.7(a) and 3.8(a), where the approximation (3.52) and the quantities $\tau_{90}(r)$, $\tau_{95}(r)$, and $\tau_{99}(r)$ [see (3.54)–(3.56)] derived from that are illustrated. It should be noted that at the level $r = 0.1$, the relative deviations of the approximations (3.54)–(3.56) from the corresponding quantities $\tau_q(r)$, which were calculated over (3.49) in a numerical way, are less than one per thousand. The validity of all these approximate solutions for $\tau_q(r)$ can again ultimately only be judged by simulating the level-crossing behaviour. In Section 5.3, we will see that the approximations introduced in this section match the simulation results obtained there quite well.

In [Wol83a], computer simulations of probability density functions $p_{0_-}(\tau_-; r)$ were also performed for Rice processes. Thereby, it has turned out that a Rice process has practically the same probability density function of the fading intervals as the corresponding Rayleigh process. These results are at least for low levels not surprising any more, because from (3.47) it follows that $p_{1_-}(\tau_-; r)$ merely depends on $T_{\zeta_-}(r)$, and on the other hand, we have seen in Subsection 3.3.2 that $T_{\xi_-}(r) \approx T_{\zeta_-}(r)$ holds if $r << 1$ and $r\rho/\sigma_0^2 << 1$. The analytical approximations obtained for Rayleigh processes for $p_{0_-}(\tau_-; r)$ and $\tau_q(r)$ can in such cases be directly taken over for Rice processes.

At this point it should be mentioned that the calculation of the probability density function of fading intervals carried out by Rice [Ric58] caused various further research activities in this field (e.g., [McF56, McF58, Lon62, Rai65, Bre70]). They followed the goal of deriving new and more precise approximations than the approximate solutions (3.40) given by Rice. The mathematical treatment of the so-called level-crossing problem is even for Rayleigh channels connected with considerable difficulties and an exact general solution is still to be found. Special attention in this field should be paid to the works [Bre78, Mun82, Mun83, Wol83a, Wol83b, Mun86, Tez87] carried out at the Institute for Applied Physics of Frankfurt University, led by Prof. Wolf. In [Mun82], data have been reported about a 4-state model which gives a valid approximation for the probability density $p_{0_-}(\tau_-; r)$ over a much greater region than (3.40) [Wol83a]. The obtained approximate solutions could again be noticeably improved by extending the 4-state model to 6- and 8-state models [Mun83, Wol83b, Mun86, Tez87]. However, investigations on generalized Gaussian random processes, the so-called spherical invariant stochastic processes [Bre78], have shown [Bre89] that the 4- and 6-state models in this process class — especially for negative levels — often do not achieve satisfying results, whereas the approximation suggested in [Bre70] does quite well. In spite of all the progress in this field, the expenditure of mathematical and numerical calculation is considerable. Moreover, the reliability of all theoretically obtained approximations is not guaranteed from the start, so that we cannot get by without an experimental verification of the results.

From this point of view, it seems more sensible to give up the lavish numerical calculations and instead only carry out simulations on (however precisely) generated sample processes [Bre89]. This background will be taken into consideration in the following two chapters, where we will introduce and analyse methods for the efficient realization of highly precise simulation models for the generation of sample processes.

4

INTRODUCTION TO THE THEORY OF DETERMINISTIC PROCESSES

All channel models studied in the sequel of this book are based on the use of at least two real-valued coloured Gaussian random processes. In the previous chapter, for example, we have seen that the modelling of the classical Rayleigh or Rice processes requires the realization of two real-valued coloured Gaussian random processes. Whereas for a Suzuki process [Suz77], which is defined by the product process of a Rayleigh process and a lognormal process, three real-valued coloured Gaussian random processes are needed. In connection with digital data transmission over land mobile radio channels, we often refer to such processes (Rayleigh, Rice, Suzuki) as appropriate stochastic models in order to describe the random amplitude fluctuations of the received signal in the equivalent complex baseband. Mobile radio channels, whose statistical amplitude behaviour can be described by Rayleigh, Rice or Suzuki processes, consequently will be denoted as Rayleigh, Rice or Suzuki channels. These models can be classed into the group of frequency-nonselective channels [Pro95]. A further example can be given by the modelling of frequency-selective channels [Pro95] using finite impulse response (FIR) filters with $\mathcal{L}$ time-variant complex-valued coefficients. This requires the realization of $2\mathcal{L}$ real-valued coloured Gaussian random processes. With the help of these few examples, it already becomes clear that the development of efficient methods for the realization of coloured Gaussian random processes is of the utmost importance in the modelling of both frequency-nonselective and frequency-selective mobile radio channels.

For the solution of this problem, we will introduce in this chapter a fundamental method which is based on a superposition of a finite number of harmonic functions. The principle of this procedure is based on an approach of Rice [Ric44, Ric45]. In Section 4.1, we will first explain the principle of deterministic channel modelling. The following Section 4.2 deals with elementary properties of deterministic processes such as the autocorrelation function, power spectral density, Doppler spread, etc. The statistical properties of these processes are the subject of the discussions in Section 4.3. In this connection, we will also introduce suitable quality criteria, on the basis of which a fair assessment of the performance for all design methods, which are introduced later in Chapter 5, can be carried out. The application of these criteria allows us to state some rules for certain design methods. On the other hand, also the problems occurring when less suitable methods are used can be made clear.

4.1 PRINCIPLE OF DETERMINISTIC CHANNEL MODELLING

In the literature, one essentially finds two fundamental methods for the modelling of coloured Gaussian random processes: the *filter method* and the *Rice method.*

When using the filter method, as shown in Figure 4.1(a), white Gaussian noise (WGN) $\nu_i(t)$ is given to the input of a linear time-invariant filter, whose transfer function is denoted by $H_i(f)$. In the following, we assume that the filter is ideal, i.e., the transfer function $H_i(f)$ can be fitted to any given frequency response with arbitrary precision. If $\nu_i(t) \sim N(0,1)$, then we have a zero-mean stochastic Gaussian random process $\mu_i(t)$ at the filter output, where according (2.52e, f) the power spectral density $S_{\mu_i\mu_i}(f)$ of $\mu_i(t)$ matches the square of the absolute value of the transfer function, i.e., $S_{\mu_i\mu_i}(f) = |H_i(f)|^2$. Hence, by filtering of white Gaussian noise $\nu_i(t)$, we obtain a coloured Gaussian random process $\mu_i(t)$.

The principle of the Rice method [Ric44, Ric45] is illustrated in Figure 4.1(b). It is based on a superposition of an infinite number of weighted harmonic functions with equidistant frequencies and random phases. According to this principle, a stochastic Gaussian process $\mu_i(t)$ can be described mathematically as

$$\mu_i(t) = \lim_{N_i \to \infty} \sum_{n=1}^{N_i} c_{i,n} \cos(2\pi f_{i,n} t + \theta_{i,n}) \, , \tag{4.1}$$

where

$$c_{i,n} = 2\sqrt{\Delta f_i S_{\mu_i\mu_i}(f_{i,n})} \, , \tag{4.2a}$$

$$f_{i,n} = n \cdot \Delta f_i \, . \tag{4.2b}$$

The phases $\theta_{i,n}$ $(n = 1, 2, \ldots, N_i)$ are random variables, which are uniformly distributed in the interval $(0, 2\pi]$, and the quantity Δf_i is here chosen in such a way that (4.2b) covers the whole relevant frequency range, where it is furthermore assumed that the following property holds: $\Delta f_i \to 0$ as $N_i \to \infty$.

As we know, a Gaussian random process is completely characterized by its mean value and its colour, which can be described either by the power spectral density or, alternatively, by the autocorrelation function. According to Rice [Ric44, Ric45], the expression (4.1) represents a zero-mean Gaussian random process with the power spectral density $S_{\mu_i\mu_i}(f)$. Consequently, the analytical models shown in Figures 4.1(a) and 4.1(b) are equivalent, i.e., the two introduced methods — the filter method and the Rice method — result in identical stochastic processes. For both methods, however, one should take into account that these processes are not exactly realizable. When using the filter method, an exact realization is prevented by the assumption that the filter should be ideal. Strictly speaking, the input signal of the filter — the white Gaussian noise — can also not be realized exactly. When using the Rice method, a realization is

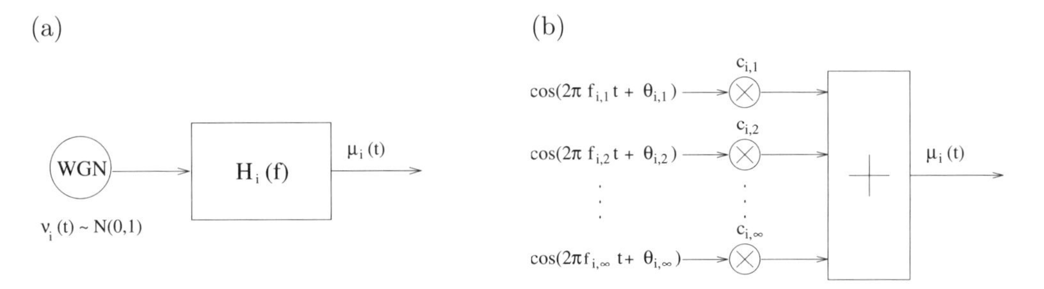

Figure 4.1: Stochastic reference models for coloured Gaussian random processes $\mu_i(t)$: (a) filter method and (b) Rice method.

impossible because an infinite number of harmonic functions N_i is not implementable on a computer or on a hardware platform. Hence, for a coloured Gaussian random process, the filter method and the Rice method only result in a stochastic analytical (ideal) model, which will be considered as reference model throughout the book.

As is well known, when using the filter method, the use of non-ideal but therefore realizable filters makes the realization of stochastic simulation models possible. Here, depending on the extent of the realization expenditure, one should take into consideration that the statistics of the filter output signal deviates more or less from that of the desired ideal Gaussian random process. In numerous publications (e.g., [Bre86a, Sch89, Fec93a, Mar94b, Lau94]), this method has been applied in order to design simulation models for mobile radio channels. In Section 8.5, we will return to the filter method once more. In the following sections, however, we will first present a detailed analysis of the Rice method. It should be noted that many of the results found for the Rice method can be applied directly to the filter method.

If the Rice method is applied by using only a finite number of harmonic functions N_i, then we obtain a further stochastic process denoted by

$$\hat{\mu}_i(t) = \sum_{n=1}^{N_i} c_{i,n} \cos(2\pi f_{i,n} t + \theta_{i,n}) \,, \tag{4.3}$$

where we assume for the moment that the parameters $c_{i,n}$ and $f_{i,n}$ are still given by (4.2a) and (4.2b), respectively, and $\theta_{i,n}$ are again uniformly distributed random variables. Now, this method can be applied to the realization of a simulation model whose general structure is shown in Figure 4.2(a). It is obvious that $\hat{\mu}_i(t) \to \mu_i(t)$ holds as $N_i \to \infty$. At this point, it should be emphasized that the simulation model is still of stochastic nature, since the phases $\theta_{i,n}$ are uniformly distributed random variables for all $n = 1, 2, \ldots, N_i$.

Only after the phases $\theta_{i,n}$ ($n = 1, 2, \ldots, N_i$) are taken out of a random generator with a uniform distribution in the interval $(0, 2\pi]$, the phases $\theta_{i,n}$ no longer represent random variables but constant quantities, since they are now realizations (outcomes) of a random variable. Thus, in connection with (4.2a), (4.2b), and (4.3), it becomes obvious that

$$\tilde{\mu}_i(t) = \sum_{n=1}^{N_i} c_{i,n} \cos(2\pi f_{i,n} t + \theta_{i,n}) \tag{4.4}$$

is a *deterministic process* or a *deterministic function.* Hence, from the stochastic simulation model shown in Figure 4.2(a), a deterministic simulation model follows, whose structure is presented in Figure 4.2(b) in its continuous-time form of representation. Note that in the limit $N_i \to \infty$, the deterministic process $\tilde{\mu}_i(t)$ tends to a sample function of the stochastic process $\mu_i(t)$.

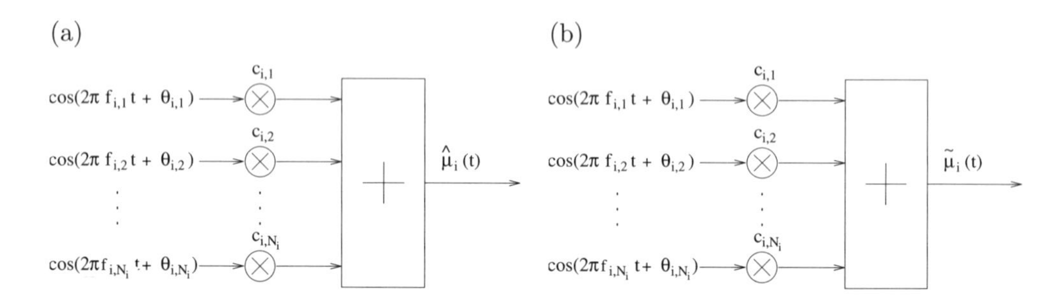

Figure 4.2: Simulation models for coloured Gaussian random processes: (a) stochastic simulation model (random phases $\theta_{i,n}$) and (b) deterministic simulation model (constant phases $\theta_{i,n}$).

In Section 4.3 and Chapter 5, it will be shown that by choosing the parameters describing the deterministic process (4.4) appropriately, a very good approximation can be achieved in such a way that the statistical properties of $\tilde{\mu}_i(t)$ are very close to those of the underlying zero-mean coloured Gaussian random process $\mu_i(t)$. For this reason, $\tilde{\mu}_i(t)$ will be called *real deterministic Gaussian process* and

$$\tilde{\mu}(t) = \tilde{\mu}_1(t) + j\tilde{\mu}_2(t) \tag{4.5}$$

will be named *complex deterministic Gaussian process.* With reference to (3.5), a so-called *deterministic Rayleigh process*

$$\tilde{\zeta}(t) = |\tilde{\mu}(t)| = |\tilde{\mu}_1(t) + j\tilde{\mu}_2(t)|\,. \tag{4.6}$$

follows from the absolute value of (4.5). Logically, by taking the absolute value of $\tilde{\mu}_\rho(t) = \tilde{\mu}(t) + m(t)$, a *deterministic Rice process*

$$\tilde{\xi}(t) = |\tilde{\mu}_\rho(t)| = |\tilde{\mu}(t) + m(t)| \tag{4.7}$$

can be introduced, where $m(t)$ again describes the line-of-sight component of the received signal, as defined by (3.2). The resulting structure of the simulation model for deterministic Rice processes is shown in Figure 4.3.

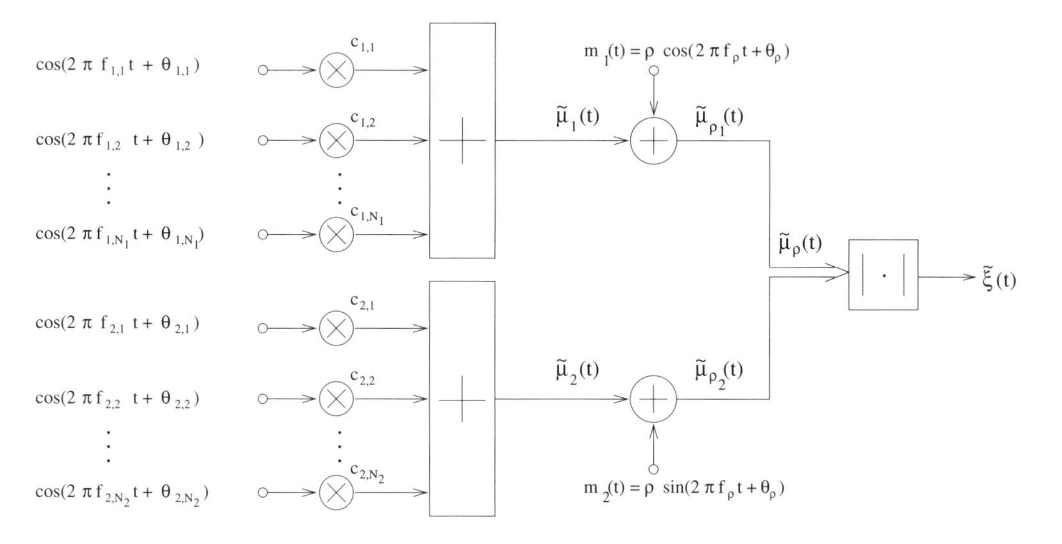

Figure 4.3: A deterministic simulation model for Rice processes.

The discrete-time simulation model, which is required for computer simulations, can directly be obtained by substituting the time variable t with $t = kT_s$, where T_s denotes the sampling interval and k is an integer. To carry out computer simulations, one generally proceeds by determining the parameters of the simulation model $c_{i,n}$, $f_{i,n}$, and $\theta_{i,n}$ for $n = 1, 2, \ldots, N_i$ during the simulation set-up phase. During the simulation run phase following this, all parameters are kept constant for the whole duration of the simulation.

Since for our purpose the deterministic processes are exclusively used for the modelling of the time-variant fading behaviour caused by the Doppler effect, we will in the following call the parameters $c_{i,n}$, $f_{i,n}$, and $\theta_{i,n}$ describing the deterministic process (4.4) the *Doppler coefficients*, *discrete Doppler frequencies*, and *Doppler phases*, respectively.

One aim of this book is to present methods by which the model parameters $(c_{i,n}, f_{i,n}, \theta_{i,n})$ can be determined in such a way that the statistical properties of the deterministic process $\tilde{\mu}_i(t)$ or $\tilde{\mu}_i(kT)$ match those of the (ideal) stochastic process $\mu_i(t)$ as closely as possible. Of course, this aim is pursued under the boundary condition that the realization expenditure should be kept as low as possible. Here, the realization expenditure is mainly determined by the number of harmonic functions N_i. However, before turning to this topic, we will first of all present some fundamental properties of deterministic processes.

4.2 ELEMENTARY PROPERTIES OF DETERMINISTIC PROCESSES

The interpretation of $\tilde{\mu}_i(t)$ as a deterministic process, i.e., as a mapping of the form

$$\tilde{\mu}_i : \mathbb{R} \to \mathbb{R}, \quad t \mapsto \tilde{\mu}_i(t), \tag{4.8}$$

enables us to derive for these kind of processes simple analytical closed-form solutions for most of the fundamental characteristic quantities like autocorrelation function, power spectral density, Doppler spread, etc.

First, a discussion of the terms introduced in Section 2.3 follows, where the definitions of these terms are now applied to the deterministic processes $\tilde{\mu}_i(t)$ $(i = 1, 2)$ defined by (4.4).

Mean value: Let $\tilde{\mu}_i(t)$ be a deterministic process with $f_{i,n} \neq 0$ $(n = 1, 2, \ldots, N_i)$. Then, it follows from (2.69) that the mean value of $\tilde{\mu}_i(t)$ is given by

$$\tilde{m}_{\mu_i} = 0\,. \tag{4.9}$$

In the following, it is always assumed that $f_{i,n} \neq 0$ holds for all $n = 1, 2, \ldots, N_i$ and $i = 1, 2$.

Mean power: Let $\tilde{\mu}_i(t)$ be a deterministic process. Then, it follows from (2.70) that the mean power of $\tilde{\mu}_i(t)$ is given by

$$\tilde{\sigma}^2_{\mu_i} = \sum_{n=1}^{N_i} \frac{c^2_{i,n}}{2}\,. \tag{4.10}$$

Obviously, the mean power $\tilde{\sigma}^2_{\mu_i}$ depends on the Doppler coefficients $c_{i,n}$ but not on the discrete Doppler frequencies $f_{i,n}$ and the Doppler phases $\theta_{i,n}$.

Autocorrelation function: For the autocorrelation function of deterministic processes $\tilde{\mu}_i(t)$, the following closed-form expression follows from (2.71)

$$\tilde{r}_{\mu_i\mu_i}(\tau) = \sum_{n=1}^{N_i} \frac{c^2_{i,n}}{2} \cos(2\pi f_{i,n}\tau)\,. \tag{4.11}$$

One should note that $\tilde{r}_{\mu_i\mu_i}(\tau)$ depends on the Doppler coefficients $c_{i,n}$ and the discrete Doppler frequencies $f_{i,n}$, but not on the Doppler phases $\theta_{i,n}$. Note also that the mean power $\tilde{\sigma}^2_{\mu_i}$ is identical to the autocorrelation function $\tilde{r}_{\mu_i\mu_i}(\tau)$ at $\tau = 0$, i.e., $\tilde{\sigma}^2_{\mu_i} = \tilde{r}_{\mu_i\mu_i}(0)$.

Cross-correlation function: Let $\tilde{\mu}_1(t)$ and $\tilde{\mu}_2(t)$ be two deterministic processes. Then, it follows from (2.72) that the cross-correlation function of $\tilde{\mu}_1(t)$ and $\tilde{\mu}_2(t)$ can be written as

$$\tilde{r}_{\mu_1\mu_2}(\tau) = 0\,, \quad \text{if } f_{1,n} \neq \pm f_{2,m}\,, \tag{4.12}$$

holds for all $n = 1, 2, \ldots, N_1$ and $m = 1, 2, \ldots, N_2$. This result shows that the deterministic processes $\tilde{\mu}_1(t)$ and $\tilde{\mu}_2(t)$ are uncorrelated if the absolute values of the respective discrete Doppler frequencies are different from each other. However, if $f_{1,n} = \pm f_{2,m}$ holds for some or all pairs of (n, m), then $\tilde{\mu}_1(t)$ and $\tilde{\mu}_2(t)$ are correlated, and we obtain the following expression for the cross-correlation function

$$\tilde{r}_{\mu_1\mu_2}(\tau) = \sum_{\substack{n=1 \\ f_{1,n}=\pm f_{2,m}}}^{N} \frac{c_{1,n}c_{2,m}}{2} \cos(2\pi f_{1,n}\tau - \theta_{1,n} \pm \theta_{2,m})\,, \tag{4.13}$$

where N denotes the largest number of N_1 and N_2, i.e., $N = \max\{N_1, N_2\}$. One should note that in this case $\tilde{r}_{\mu_1\mu_2}(\tau)$ also depends on the Doppler phases $\theta_{i,n}$. The cross-correlation function $\tilde{r}_{\mu_2\mu_1}(\tau)$ can be obtained from the relation $\tilde{r}_{\mu_2\mu_1}(\tau) = \tilde{r}^*_{\mu_1\mu_2}(-\tau) = \tilde{r}_{\mu_1\mu_2}(-\tau)$.

Power spectral density: Let $\tilde{\mu}_i(t)$ be a deterministic process. Then, it follows from (2.73) in connection with (4.11) that the power spectral density of $\tilde{\mu}_i(t)$ can be expressed as

$$\tilde{S}_{\mu_i\mu_i}(f) = \sum_{n=1}^{N_i} \frac{c_{i,n}^2}{4} [\delta(f - f_{i,n}) + \delta(f + f_{i,n})] . \tag{4.14}$$

Hence, the power spectral density of $\tilde{\mu}_i(t)$ is a symmetrical line spectrum, i.e., $\tilde{S}_{\mu_i\mu_i}(f) = \tilde{S}_{\mu_i\mu_i}(-f)$. The spectral lines are located at the discrete points $f = \pm f_{i,n}$ and weighted by the factor $c_{i,n}^2/4$.

Cross-power spectral density: Let $\tilde{\mu}_1(t)$ and $\tilde{\mu}_2(t)$ be two deterministic processes. Then, it follows from (2.74) with (4.12) and (4.13) that the cross-power spectral density of $\tilde{\mu}_1(t)$ and $\tilde{\mu}_2(t)$ is given by

$$\tilde{S}_{\mu_1\mu_2}(f) = 0 , \quad \text{if } f_{1,n} \neq \pm f_{2,m} , \tag{4.15}$$

and

$$\begin{aligned} \tilde{S}_{\mu_1\mu_2}(f) &= \sum_{\substack{n=1 \\ f_{1,n}=\pm f_{2,m}}}^{N} \frac{c_{1,n}c_{2,m}}{4} \left[\delta(f - f_{1,n}) \cdot e^{-j(\theta_{1,n} \mp \theta_{2,m})} \right. \\ &\qquad \left. + \, \delta(f + f_{1,n}) \cdot e^{j(\theta_{1,n} \mp \theta_{2,m})}\right] \end{aligned} \tag{4.16}$$

for all $n = 1, 2, \ldots, N_1$ and $m = 1, 2, \ldots, N_2$, where $N = \max\{N_1, N_2\}$. The cross-power spectral density $\tilde{S}_{\mu_2\mu_1}(f)$ can directly be obtained from the relation $\tilde{S}_{\mu_2\mu_1}(f) = \tilde{S}^*_{\mu_1\mu_2}(f)$.

Average Doppler shift: Let $\tilde{\mu}_i(t)$ be a deterministic process with the power spectral density $\tilde{S}_{\mu_i\mu_i}(f)$. Then, by analogy to (3.13a), the average Doppler shift $\tilde{B}^{(1)}_{\mu_i\mu_i}$ of $\tilde{\mu}_i(t)$ is defined by

$$\begin{aligned} \tilde{B}^{(1)}_{\mu_i\mu_i} &:= \frac{\int_{-\infty}^{\infty} f \, \tilde{S}_{\mu_i\mu_i}(f) \, df}{\int_{-\infty}^{\infty} \tilde{S}_{\mu_i\mu_i}(f) \, df} \\ &= \frac{1}{2\pi j} \cdot \frac{\dot{\tilde{r}}_{\mu_i\mu_i}(0)}{\tilde{r}_{\mu_i\mu_i}(0)} . \end{aligned} \tag{4.17}$$

Due to the symmetry property $\tilde{S}_{\mu_i\mu_i}(f) = \tilde{S}_{\mu_i\mu_i}(-f)$, it follows that

$$\tilde{B}^{(1)}_{\mu_i\mu_i} = 0 . \tag{4.18}$$

On condition that the real and imaginary parts are uncorrelated, one analogously obtains for complex deterministic processes $\tilde{\mu}(t) = \tilde{\mu}_1(t) + j\tilde{\mu}_2(t)$, the following relation between the average Doppler shifts of $\tilde{\mu}(t)$ and $\tilde{\mu}_i(t)$

$$\tilde{B}^{(1)}_{\mu\mu} = \tilde{B}^{(1)}_{\mu_i\mu_i} = 0\,, \quad i = 1, 2\,. \tag{4.19}$$

By considering the relations (4.19) and (3.15a), it turns out that the average Doppler shift of the simulation model is identical to that of the reference model.

Doppler spread: Let $\tilde{\mu}_i(t)$ be a deterministic process with the power spectral density $\tilde{S}_{\mu_i\mu_i}(f)$. Then, analogous to (3.13b), the Doppler spread $\tilde{B}^{(2)}_{\mu_i\mu_i}$ of $\tilde{\mu}_i(t)$ is defined by

$$\begin{aligned} \tilde{B}^{(2)}_{\mu_i\mu_i} &:= \sqrt{\frac{\int_{-\infty}^{\infty} (f - \tilde{B}^{(1)}_{\mu_i\mu_i})^2\, \tilde{S}_{\mu_i\mu_i}(f)\, df}{\int_{-\infty}^{\infty} \tilde{S}_{\mu_i\mu_i}(f)\, df}} \\ &= \frac{1}{2\pi}\sqrt{\left(\frac{\dot{\tilde{r}}_{\mu_i\mu_i}(0)}{\tilde{r}_{\mu_i\mu_i}(0)}\right)^2 - \frac{\ddot{\tilde{r}}_{\mu_i\mu_i}(0)}{\tilde{r}_{\mu_i\mu_i}(0)}}\,. \end{aligned} \tag{4.20}$$

Using (4.10) and (4.11), the last equation can be written as

$$\tilde{B}^{(2)}_{\mu_i\mu_i} = \frac{\sqrt{\tilde{\beta}_i}}{2\pi\tilde{\sigma}_{\mu_i}}\,, \tag{4.21}$$

where

$$\tilde{\beta}_i = -\ddot{\tilde{r}}_{\mu_i\mu_i}(0) = 2\pi^2 \sum_{n=1}^{N_i} (c_{i,n} f_{i,n})^2\,. \tag{4.22}$$

The comparison of the equation (3.15b) with (4.21) shows that the Doppler spreads $B^{(2)}_{\mu_i\mu_i}$ and $\tilde{B}^{(2)}_{\mu_i\mu_i}$ are always identical, if the Doppler coefficients $c_{i,n}$ and the discrete Doppler frequencies $f_{i,n}$ are determined in such a way that $\tilde{\sigma}^2_{\mu_i} = \sigma_0^2$ and $\tilde{\beta}_i = \beta$ hold. (In particular, it is sufficient that the condition $\tilde{\beta}_i/\tilde{\sigma}_0^2 = \beta/\sigma_0^2$ is fulfilled.)

Analogously, the Doppler spread $\tilde{B}^{(2)}_{\mu\mu}$ corresponding to the power spectral density $\tilde{S}_{\mu\mu}(f)$ of the complex deterministic process $\tilde{\mu}(t) = \tilde{\mu}_1(t) + j\tilde{\mu}_2(t)$ can be determined. On condition that $\tilde{\mu}_1(t)$ and $\tilde{\mu}_2(t)$ are uncorrelated, the Doppler spread $\tilde{B}^{(2)}_{\mu\mu}$ can be expressed as

$$\tilde{B}^{(2)}_{\mu\mu} = \frac{\sqrt{\tilde{\beta}}}{2\pi\tilde{\sigma}_{\mu}}\,, \tag{4.23}$$

where $\tilde{\sigma}^2_{\mu} = \tilde{\sigma}^2_{\mu_1} + \tilde{\sigma}^2_{\mu_2} > 0$ and $\tilde{\beta} = \tilde{\beta}_1 + \tilde{\beta}_2 > 0$ hold. In Chapter 5, we will get acquainted with methods for the design of deterministic processes $\tilde{\mu}_1(t)$ and $\tilde{\mu}_2(t)$

having the properties $\tilde{\sigma}_{\mu_1}^2 = \tilde{\sigma}_{\mu_2}^2$ and $\tilde{\beta}_1 \neq \tilde{\beta}_2$. Especially in this case, the Doppler spread $\tilde{B}_{\mu\mu}^{(2)}$ can be calculated from the quadratic mean of $\tilde{B}_{\mu_1\mu_1}^{(2)}$ and $\tilde{B}_{\mu_2\mu_2}^{(2)}$, i.e.,

$$\tilde{B}_{\mu\mu}^{(2)} = \sqrt{\frac{\left(\tilde{B}_{\mu_1\mu_1}^{(2)}\right)^2 + \left(\tilde{B}_{\mu_2\mu_2}^{(2)}\right)^2}{2}} \,. \tag{4.24}$$

Finally, it should be mentioned that if $\tilde{\sigma}_0^2 = \tilde{\sigma}_{\mu_1}^2 = \tilde{\sigma}_{\mu_2}^2$ and $\tilde{\beta} = \tilde{\beta}_1 = \tilde{\beta}_2$ hold, then we obtain the result

$$\tilde{B}_{\mu\mu}^{(2)} = \tilde{B}_{\mu_i\mu_i}^{(2)} = \frac{\sqrt{\tilde{\beta}}}{2\pi\tilde{\sigma}_0}, \quad i = 1, 2 \tag{4.25}$$

which is closely related to (3.15b). However, if the deviations between $\tilde{\beta}_1$ and $\tilde{\beta}_2$ are small, which is often the case, then the above expression is a very good approximation for $\tilde{B}_{\mu\mu}^{(2)}$ if $\tilde{\beta}$ is replaced by $\tilde{\beta} = \tilde{\beta}_1 \approx \tilde{\beta}_2$ there.

Periodicity: Let $\tilde{\mu}_i(t)$ be a deterministic process with arbitrary but nonzero parameters $c_{i,n}$, $f_{i,n}$ (and $\theta_{i,n}$). If the greatest common divisor of the discrete Doppler frequencies

$$F_i = \gcd\{f_{i,1}, f_{i,2}, \ldots, f_{i,N_i}\} \neq 0 \tag{4.26}$$

exists, then $\tilde{\mu}_i(t)$ is periodic with the period $T_i = 1/F_i$, i.e., it holds $\tilde{\mu}_i(t+T_i) = \tilde{\mu}_i(t)$ and $\tilde{r}_{\mu_i\mu_i}(\tau + T_i) = \tilde{r}_{\mu_i\mu_i}(\tau)$.

The proof of this theorem is relatively simple and will therefore be presented here only briefly. Since F_i is the greatest common divisor of $f_{i,1}, f_{i,2}, \ldots, f_{i,N_i}$, there are integers $q_{i,n} \in \mathbb{Z}$, so that $f_{i,n} = q_{i,n} \cdot F_i$ holds for all $n = 1, 2, \ldots, N_i$ and $i = 1, 2$. By putting $f_{i,n} = q_{i,n} \cdot F_i = q_{i,n}/T_i$ into (4.4) and (4.11), the validity of $\tilde{\mu}_i(t + T_i) = \tilde{\mu}_i(t)$ and $\tilde{r}_{\mu_i\mu_i}(\tau + T_i) = \tilde{r}_{\mu_i\mu_i}(\tau)$, respectively, can be proved directly.

4.3 STATISTICAL PROPERTIES OF DETERMINISTIC PROCESSES

Even though a discussion of the elementary properties of deterministic processes could be performed in the previous section without any problems, an analysis of the statistical properties first seems to be absurd, since statistical methods can meaningfully be applied only on random variables and stochastic processes. But, on the other hand, their application to deterministic processes (4.4) makes no sense. In order in this case to gain access to statistical quantities like the probability density function, the level-crossing rate, and the average duration of fades, we will study the behaviour of deterministic processes $\tilde{\mu}_i(t)$ at random time instants t. If nothing else is explicitly mentioned, we will assume throughout this section that the time variable t is a random variable uniformly distributed within the interval $\mathbb{R}$. It should also be noted that both the time variable t and the Doppler phases $\theta_{i,n}$ are in the argument of the cosine functions of (4.4). Therefore, we could alternatively assume that the time $t = t_0$ is a constant and the phases $\theta_{i,n}$ are uniformly distributed random variables. In both cases, however, we would obtain exactly the same results for the following computations.

4.3.1 Probability Density Function of the Amplitude and the Phase

In this subsection, we will analyse the probability density function of the amplitude and phase of complex deterministic processes $\tilde{\mu}(t) = \tilde{\mu}_1(t) + j\tilde{\mu}_2(t)$. It will be shown that these probability density functions are completely determined by the number of harmonic functions N_i and the choice of the Doppler coefficients $c_{i,n}$.

Therefore, we first consider a single weighted harmonic elementary function of the form

$$\tilde{\mu}_{i,n}(t) = c_{i,n} \cos(2\pi f_{i,n} t + \theta_{i,n}) \, , \tag{4.27}$$

where $c_{i,n}$, $f_{i,n}$, and $\theta_{i,n}$ are arbitrary but constant parameters different from zero and t is a uniformly distributed random variable. Due to the periodicity of $\tilde{\mu}_{i,n}(t)$, it is sufficient to restrict t on the open interval $(0, f_{i,n}^{-1})$ with $f_{i,n} \neq 0$. Since t was assumed to be a uniformly distributed random variable, $\tilde{\mu}_{i,n}(t)$ is no longer a deterministic function but a random variable as well, whose probability density function is given by [Pap91, p. 98]

$$\tilde{p}_{\mu_{i,n}}(x) = \begin{cases} \dfrac{1}{\pi \, c_{i,n}\sqrt{1-(x/c_{i,n})^2}} \, , & |x| < c_{i,n} \, , \\ 0 \, , & |x| \geq c_{i,n} \, . \end{cases} \tag{4.28}$$

The expected value and the variance of $\tilde{\mu}_{i,n}(t)$ are equal to 0 and $c_{i,n}^2/2$, respectively. If the random variables $\tilde{\mu}_{i,n}(t)$ are statistically independent, then the probability density function $\tilde{p}_{\mu_i}(x)$ of the sum

$$\tilde{\mu}_i(t) = \tilde{\mu}_{i,1}(t) + \tilde{\mu}_{i,2}(t) + \ldots + \tilde{\mu}_{i,N_i}(t) \tag{4.29}$$

can be obtained from the convolution of the individual probability density functions $\tilde{p}_{\mu_{i,n}}(x)$, i.e.,

$$\tilde{p}_{\mu_i}(x) = \tilde{p}_{\mu_{i,1}}(x) \; * \; \tilde{p}_{\mu_{i,2}}(x) \; * \; \ldots \; * \; \tilde{p}_{\mu_{i,N_i}}(x). \tag{4.30}$$

The expected value $\tilde{m}_{\mu_i}$ and the variance $\tilde{\sigma}_{\mu_i}^2$ of $\tilde{\mu}_i(t)$ are then given by

$$\tilde{m}_{\mu_i} = 0 \tag{4.31a}$$

and

$$\tilde{\sigma}_{\mu_i}^2 = \sum_{n=1}^{N_i} \frac{c_{i,n}^2}{2} \, , \tag{4.31b}$$

respectively. In principle, a rule for the computation of $\tilde{p}_{\mu_i}(x)$ is given by (4.30). But with regard to the following procedure, it is more advantageous to apply the concept of the characteristic function [see (2.14)]. After substituting (4.28) into (2.14), we find the following expression for the characteristic function $\tilde{\Psi}_{\mu_{i,n}}(\nu)$ of the random variables $\tilde{\mu}_{i,n}(t)$

$$\tilde{\Psi}_{\mu_{i,n}}(\nu) = J_0(2\pi c_{i,n} \nu) \, . \tag{4.32}$$

The N_i-fold convolution (4.30) of the probability density functions $\tilde{p}_{\mu_{i,n}}(x)$ can now be formulated as the N_i-fold product of the corresponding characteristic functions $\tilde{\Psi}_{\mu_{i,n}}(\nu)$

$$\begin{aligned}\tilde{\Psi}_{\mu_i}(\nu) &= \tilde{\Psi}_{\mu_{i,1}}(\nu) \cdot \tilde{\Psi}_{\mu_{i,2}}(\nu) \cdot \ \ldots \ \cdot \tilde{\Psi}_{\mu_{i,N_i}}(\nu) \\ &= \prod_{n=1}^{N_i} J_0(2\pi c_{i,n}\nu) \, . \end{aligned} \tag{4.33}$$

Concerning (4.30), an alternative expression for the probability density function $\tilde{p}_{\mu_i}(x)$ is then given by the inverse Fourier transform of $\tilde{\Psi}_{\mu_i}(-\nu) = \tilde{\Psi}_{\mu_i}(\nu)$ [Ben48]

$$\begin{aligned}\tilde{p}_{\mu_i}(x) &= \int_{-\infty}^{\infty} \tilde{\Psi}_{\mu_i}(\nu)\, e^{j2\pi\nu x}\, d\nu \\ &= 2\int_{0}^{\infty} \left[\prod_{n=1}^{N_i} J_0(2\pi c_{i,n}\nu)\right] \cos(2\pi\nu x)\, d\nu \, , \quad i = 1, 2 \, . \end{aligned} \tag{4.34}$$

It is important to realize that the probability density function $\tilde{p}_{\mu_i}(x)$ of $\tilde{\mu}_i(t)$ is completely determined by the number of harmonic functions N_i and by the Doppler coefficients $c_{i,n}$, whereas the discrete Doppler frequencies $f_{i,n}$ and the Doppler phases $\theta_{i,n}$ have no influence on $\tilde{p}_{\mu_i}(x)$.

In the following, let $c_{i,n} = \sigma_0\sqrt{2/N_i}$ and $f_{i,n} \neq 0$ for all $n = 1, 2, \ldots, N_i$ and $i = 1, 2$. Then, due to (4.31a) and (4.31b), the sum $\tilde{\mu}_i(t)$ introduced by (4.29) is a random variable with the expected value 0 and the variance

$$\tilde{\sigma}_0^2 = \tilde{\sigma}_{\mu_1}^2 = \tilde{\sigma}_{\mu_2}^2 = \sigma_0^2 \, . \tag{4.35}$$

Regarding the central limit theorem [see (2.16)], it turns out that in the limit $N_i \to \infty$, the sum $\tilde{\mu}_i(t)$ tends to a normally distributed random variable having the expected value 0 and the variance σ_0^2, i.e.,

$$\lim_{N_i\to\infty} \tilde{p}_{\mu_i}(x) = p_{\mu_i}(x) = \frac{1}{\sqrt{2\pi}\sigma_0}\, e^{-\frac{x^2}{2\sigma_0^2}} \, . \tag{4.36}$$

Hence, after computing the Fourier transform of this equation, one obtains the following relation for the corresponding characteristic functions

$$\lim_{N_i\to\infty} \tilde{\Psi}_{\mu_i}(\nu) = \Psi_{\mu_i}(\nu) = e^{-2(\pi\sigma_0\nu)^2} \, , \tag{4.37}$$

from which — by using (4.33) — the remarkable property

$$\lim_{N_i\to\infty} \left[J_0\left(2\pi\sigma_0\sqrt{\frac{2}{N_i}}\nu\right)\right]^{N_i} = e^{-2(\pi\sigma_0\nu)^2} \tag{4.38}$$

finally follows.

Of course, for a finite number of harmonic functions N_i, we have to write: $\tilde{p}_{\mu_i}(x) \approx p_{\mu_i}(x)$ and $\tilde{\Psi}_{\mu_i}(\nu) \approx \Psi_{\mu_i}(\nu)$. From Figure 4.4(a), illustrating $\tilde{p}_{\mu_i}(x)$ according to (4.34) with $c_{i,n} = \sigma_0\sqrt{2/N_i}$ for $N_i \in \{3, 5, 7, \infty\}$, it follows that in fact for $N_i \geq 7$, the approximation $\tilde{p}_{\mu_i}(x) \approx p_{\mu_i}(x)$ is astonishingly good. An appropriate measure of the approximation error is the mean-square error of the probability density function $\tilde{p}_{\mu_i}(x)$ defined by

$$E_{p_{\mu_i}} := \int_{-\infty}^{\infty} \left(p_{\mu_i}(x) - \tilde{p}_{\mu_i}(x)\right)^2 dx\,. \tag{4.39}$$

The behaviour of the mean-square error $E_{p_{\mu_i}}$ as a function of the number of harmonic functions N_i is shown in Figure 4.4(b). This figure gives us an impression of how fast $\tilde{p}_{\mu_i}(x)$ converges to $p_{\mu_i}(x)$ if N_i increases.

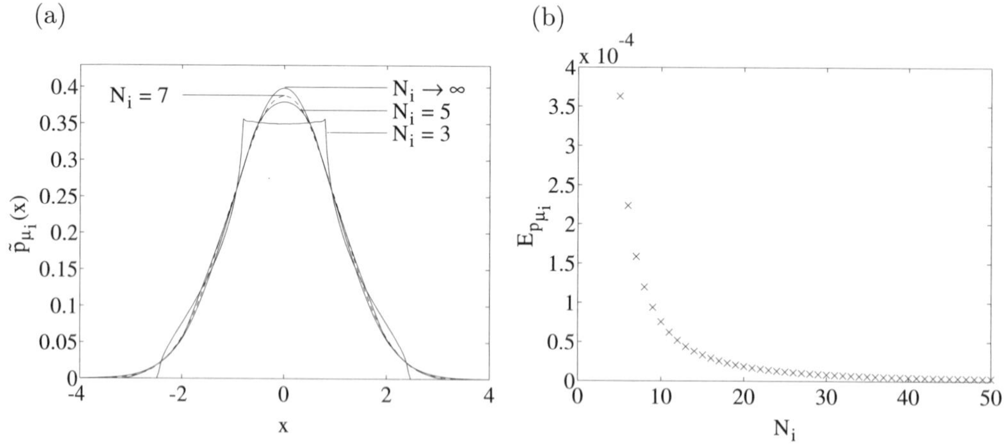

Figure 4.4: (a) Probability density function $\tilde{p}_{\mu_i}(x)$ for $N_i \in \{3, 5, 7, \infty\}$, (b) mean-square error $E_{p_{\mu_i}}$ as a function of N_i. (Analytical results obtained with $c_{i,n} = \sigma_0\sqrt{2/N_i}$, $\sigma_0^2 = 1$.)

Due to the good convergence behaviour that $\tilde{p}_{\mu_i}(x)$ exhibits in conjunction with $c_{i,n} = \sigma_0\sqrt{2/N_i}$, we will occasionally assume (without causing too large an error) that the identity

$$\tilde{p}_{\mu_i}(x) = p_{\mu_i}(x)\,, \quad \text{if } N_i \geq 7\,, \tag{4.40}$$

holds. In this case, many analytical problems, which are otherwise difficult to overcome, can then be solved in a relatively easy way.

Next, we will derive the probability density function of the absolute value and the phase of the complex-valued random variable

$$\tilde{\mu}_\rho(t) = \tilde{\mu}_{\rho_1}(t) + j\tilde{\mu}_{\rho_2}(t)\,, \tag{4.41}$$

where $\tilde{\mu}_{\rho_i}(t) = \tilde{\mu}_i(t) + m_i(t)$ $(i = 1, 2)$. Here, concerning $m_i(t)$ according to (3.2), we have to discuss the cases $f_\rho = 0$ and $f_\rho \neq 0$ separately.

At first, we consider the case $f_\rho = 0$. By doing this, m_i becomes independent of the random variable t. Consequently, m_i is a constant whose probability density function is described by $p_{m_i}(x) = \delta(x - m_i)$. Furthermore, it follows that the probability density function $\tilde{p}_{\mu_{\rho_i}}(x)$ of $\tilde{\mu}_{\rho_i}(t)$ can now be expressed directly by using $\tilde{p}_{\mu_i}(x)$ as

$$\begin{aligned} \tilde{p}_{\mu_{\rho_i}}(x) &= \tilde{p}_{\mu_i}(x) * p_{m_i}(x) \\ &= \tilde{p}_{\mu_i}(x - m_i) \,. \end{aligned} \tag{4.42}$$

On the assumption that $\tilde{\mu}_{\rho_1}(t)$ and $\tilde{\mu}_{\rho_2}(t)$ are statistically independent, and, thus, $f_{1,n} \neq \pm f_{2,m}$ holds for all $n = 1, 2, \ldots, N_1$ and $m = 1, 2, \ldots, N_2$, the joint probability density function of the random variables $\tilde{\mu}_{\rho_1}(t)$ and $\tilde{\mu}_{\rho_2}(t)$, denoted by $\tilde{p}_{\mu_{\rho_1}\mu_{\rho_2}}(x_1, x_2)$, can be expressed by

$$\tilde{p}_{\mu_{\rho_1}\mu_{\rho_2}}(x_1, x_2) = \tilde{p}_{\mu_{\rho_1}}(x_1) \cdot \tilde{p}_{\mu_{\rho_2}}(x_2) \,. \tag{4.43}$$

The transform of the Cartesian coordinates (x_1, x_2) into polar coordinates (z, θ) by means of

$$x_1 = z\cos\theta \,, \quad x_2 = z\sin\theta \tag{4.44a, b}$$

allows us to calculate the joint probability density function $\tilde{p}_{\xi\vartheta}(z, \theta)$ of the amplitude $\tilde{\xi}(t) = |\tilde{\mu}_\rho(t)|$ and the phase $\tilde{\vartheta}(t) = \arg\{\tilde{\mu}_\rho(t)\}$ as follows:

$$\begin{aligned} \tilde{p}_{\xi\vartheta}(z, \theta) &= z\, \tilde{p}_{\mu_{\rho_1}\mu_{\rho_2}}(z\cos\theta, z\sin\theta) && (4.45a) \\ &= z\, \tilde{p}_{\mu_{\rho_1}}(z\cos\theta) \cdot \tilde{p}_{\mu_{\rho_2}}(z\sin\theta) && (4.45b) \\ &= z\, \tilde{p}_{\mu_1}(z\cos\theta - \rho\cos\theta_\rho) \cdot \tilde{p}_{\mu_2}(z\sin\theta - \rho\sin\theta_\rho) \,. && (4.45c) \end{aligned}$$

By using (2.40), we obtain the probability density functions of the amplitude $\tilde{p}_\xi(z)$ and phase $\tilde{p}_\vartheta(\theta)$ from the preceding equation in the form:

$$\tilde{p}_\xi(z) = z \int_{-\pi}^{\pi} \tilde{p}_{\mu_1}(z\cos\theta - \rho\cos\theta_\rho) \cdot \tilde{p}_{\mu_2}(z\sin\theta - \rho\sin\theta_\rho)\, d\theta \,, \tag{4.46a}$$

$$\tilde{p}_\vartheta(\theta) = \int_0^\infty z\, \tilde{p}_{\mu_1}(z\cos\theta - \rho\cos\theta_\rho) \cdot \tilde{p}_{\mu_2}(z\sin\theta - \rho\sin\theta_\rho)\, dz \,. \tag{4.46b}$$

Putting (4.34) into the last two expressions gives us the following threefold integrals for the desired probability density functions:

$$\begin{aligned} \tilde{p}_\xi(z) = 4z \int_{-\pi}^{\pi} &\left\{ \int_0^\infty \left[\prod_{n=1}^{N_1} J_0(2\pi c_{1,n}\nu_1) \right] g_1(z, \theta, \nu_1)\, d\nu_1 \right\} \\ &\cdot \left\{ \int_0^\infty \left[\prod_{m=1}^{N_2} J_0(2\pi c_{2,m}\nu_2) \right] g_2(z, \theta, \nu_2)\, d\nu_2 \right\} d\theta \,, \end{aligned} \tag{4.47a}$$

$$\tilde{p}_\vartheta(\theta) = 4\int_0^\infty z \left\{ \int_0^\infty \left[\prod_{n=1}^{N_1} J_0(2\pi c_{1,n}\nu_1) \right] g_1(z,\theta,\nu_1)\, d\nu_1 \right\} \cdot \left\{ \int_0^\infty \left[\prod_{m=1}^{N_2} J_0(2\pi c_{2,m}\nu_2) \right] g_2(z,\theta,\nu_2)\, d\nu_2 \right\} dz\,, \tag{4.47b}$$

where

$$g_1(z,\theta,\nu_1) = \cos[2\pi\nu_1(z\cos\theta - \rho\cos\theta_\rho)]\,, \tag{4.48a}$$

$$g_2(z,\theta,\nu_2) = \cos[2\pi\nu_2(z\sin\theta - \rho\sin\theta_\rho)]\,. \tag{4.48b}$$

Up to now, there are no further simplifications known for (4.47b), so that the remaining three integrals must be solved numerically. In comparison with that, it is possible to reduce the threefold integral on the right-hand side of (4.47a) to a double integral by making use of the expression [Gra81, eq. (3.876.7)]

$$\int_0^1 \frac{\cos\left(2\pi\nu_2 z\sqrt{1-x^2}\right)}{\sqrt{1-x^2}} \cos(2\pi\nu_1 zx)\, dx = \frac{\pi}{2} J_0\left(2\pi z\sqrt{\nu_1^2+\nu_2^2}\right)\,. \tag{4.49}$$

After some algebraic manipulations, we finally come to the result

$$\tilde{p}_\xi(z) = 4\pi z \int_0^\pi \int_0^\infty \left[\prod_{n=1}^{N_1} J_0(2\pi c_{1,n} y\cos\theta) \right] \left[\prod_{m=1}^{N_2} J_0(2\pi c_{2,m} y\sin\theta) \right] \cdot J_0(2\pi zy)\, \cos\left[2\pi\rho y\cos(\theta-\theta_\rho)\right] y\, dy\, d\theta\,. \tag{4.50}$$

The results of the numerical evaluations of $\tilde{p}_\xi(z)$ and $\tilde{p}_\vartheta(\theta)$ for the special case $c_{i,n} = \sigma_0\sqrt{2/N_i}$ ($\sigma_0^2 = 1$) are illustrated in Figures 4.5(a) and 4.5(b), respectively. These illustrations again make clear that the approximation error can in general be ignored if $N_i \geq 7$.

We also want to show that if the Doppler coefficients $c_{i,n}$ are given by $c_{i,n} = \sigma_0\sqrt{2/N_i}$, then it follows in the limit $N_i \to \infty$ from (4.47a) and (4.47b) the expected result: $\tilde{p}_\xi(z) \to p_\xi(z)$ and $\tilde{p}_\vartheta(\theta) \to p_\vartheta(\theta)$, respectively. To show this, we apply the property (4.38) enabling us to express (4.47a) and (4.47b) as

$$\lim_{N_i\to\infty} \tilde{p}_\xi(z) = 4z \int_{-\pi}^\pi \left[\int_0^\infty e^{-2(\pi\sigma_0\nu_1)^2}\, g_1(z,\theta,\nu_1)\, d\nu_1 \right] \cdot \left[\int_0^\infty e^{-2(\pi\sigma_0\nu_2)^2}\, g_2(z,\theta,\nu_2)\, d\nu_2 \right] d\theta \tag{4.51}$$

and

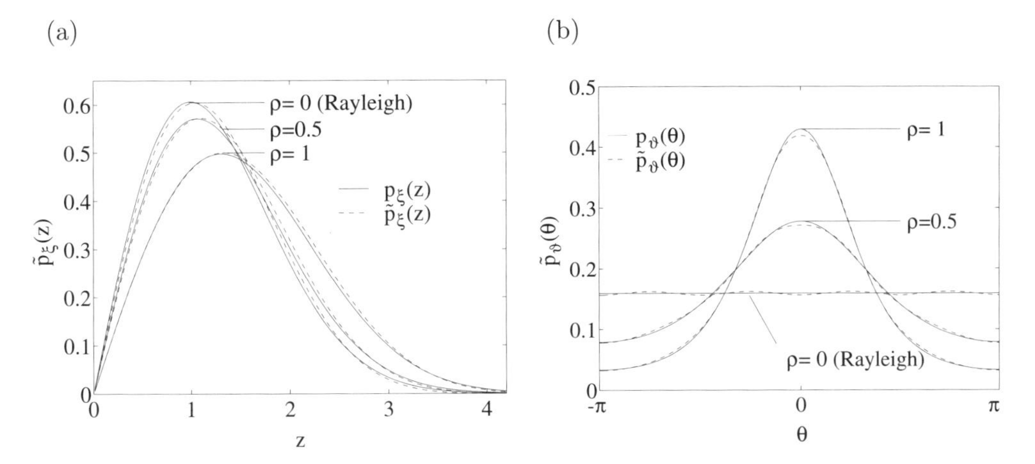

Figure 4.5: (a) Probability density function $\tilde{p}_\xi(z)$ of the amplitude $\tilde{\xi}(t) = |\tilde{\mu}_\rho(t)|$ and (b) probability density function $\tilde{p}_\vartheta(\theta)$ of the phase $\tilde{\vartheta}(t) = \arg\{\tilde{\mu}_\rho(t)\}$ for $N_1 = N_2 = 7$. (Analytical results obtained with $c_{i,n} = \sigma_0\sqrt{2/N_i}$ and $\sigma_0^2 = 1$.)

$$\lim_{N_i \to \infty} \tilde{p}_\vartheta(\theta) = 4 \int_0^z z \left[\int_0^\infty e^{-2(\pi\sigma_0\nu_1)^2} \, g_1(z, \theta, \nu_1) \, d\nu_1 \right] \cdot \left[\int_0^\infty e^{-2(\pi\sigma_0\nu_2)^2} \, g_2(z, \theta, \nu_2) \, d\nu_2 \right] dz \,, \tag{4.52}$$

respectively. The use of the integral [Gra81, eq. (3.896.2)]

$$\int_{-\infty}^{\infty} e^{-q^2x^2} \cos(px) \, dx = \frac{\sqrt{\pi}}{q} \, e^{-\frac{p^2}{4q^2}} \tag{4.53}$$

allows us to present the expressions (4.51) and (4.52) in the form

$$\lim_{N_i \to \infty} \tilde{p}_\xi(z) = \frac{z}{\sigma_0^2} e^{-\frac{z^2+\rho^2}{2\sigma_0^2}} \cdot \frac{1}{\pi} \int_0^\pi e^{\frac{z\rho}{\sigma_0^2}\cos(\theta - \theta_\rho)} d\theta \tag{4.54}$$

and

$$\lim_{N_i \to \infty} \tilde{p}_\vartheta(\theta) = \frac{1}{2\pi\sigma_0^2} e^{-\frac{\rho^2}{2\sigma_0^2}} \int_0^\infty z \, e^{-\frac{z^2}{2\sigma_0^2} + \frac{z\rho}{\sigma_0^2}\cos(\theta - \theta_\rho)} \, dz \,, \tag{4.55}$$

respectively. With the integral representation of the modified Bessel function of zeroth order [Abr72, eq. (9.6.16)]

$$I_0(z) = \frac{1}{\pi} \int_0^\pi e^{\pm z\cos\theta} \, d\theta \,, \tag{4.56}$$

we can immediately identify (4.54) with the Rice distribution $p_\xi(z)$ defined by (3.17), and from (4.55), using [Gra81, eq. (3.462.5)]

$$\int_0^\infty z\, e^{-qz^2-2pz} dz = \frac{1}{2q} - \frac{p}{2q}\sqrt{\frac{\pi}{q}}\, e^{\frac{p^2}{q}} \left[1 - \text{erf}\left(\frac{p}{\sqrt{q}}\right)\right],$$

$$|\arg\{p\}| < \frac{\pi}{2}, \quad \text{Re}\,\{q\} > 0\,, \tag{4.57}$$

we obtain, after elementary calculations, the probability density function $p_\vartheta(\theta)$ given by (3.22).

Furthermore, we will pay attention to the general case, where $f_\rho \neq 0$. The line-of-sight component $m(t) = m_1(t) + jm_2(t)$ [see (3.2)] will now be considered as time-variant mean value, whose real and imaginary parts can consequently be described by the probability density functions

$$p_{m_1}(x_1\,;\,t) = \delta(x_1 - m_1(t)) = \delta(x_1 - \rho\cos(2\pi f_\rho t + \theta_\rho)) \tag{4.58}$$

and

$$p_{m_2}(x_2\,;\,t) = \delta(x_2 - m_2(t)) = \delta(x_2 - \rho\sin(2\pi f_\rho t + \theta_\rho))\,, \tag{4.59}$$

respectively. The derivation of the probability density functions $\tilde{p}_\xi(z;t)$ and $\tilde{p}_\vartheta(\theta;t)$ can be performed analogously to the case $f_\rho = 0$. For these functions, one obtains expressions, which coincide with the right-hand side of (4.47a) and (4.47b), respectively, if there the functions $g_i(z, \theta, \nu_i)$ for $i = 1, 2$ are substituted by

$$g_1(z, \theta, \nu_1) = \cos\{2\pi\nu_1[z\cos\theta - \rho\cos(2\pi f_\rho t + \theta_\rho)]\} \tag{4.60a}$$

and

$$g_2(z, \theta, \nu_2) = \cos\{2\pi\nu_2[z\sin\theta - \rho\sin(2\pi f_\rho t + \theta_\rho)]\}\,. \tag{4.60b}$$

Concerning the convergence behaviour, it can be shown that for $N_i \to \infty$ with $c_{i,n} = \sigma_0\sqrt{2/N_i}$, it follows $\tilde{p}_\xi(z;t) \to p_\xi(z)$ and $\tilde{p}_\vartheta(\theta;t) \to p_\vartheta(\theta;t)$ as expected, where $p_\xi(z)$ and $p_\vartheta(\theta;t)$ are the probability density functions described by (3.17) and (3.21), respectively.

In order to complete this topic, we will verify the derived analytical expressions for the probability density functions $\tilde{p}_\xi(z)$ and $\tilde{p}_\vartheta(\theta)$ by simulation. In principle, we can proceed here by making use of the simulation model shown in Figure 4.3, where we have to substitute the time variable t by a uniformly distributed random variable, which already helped us to achieve our aim in deriving the analytical expressions. In the following, however, we will keep the conventional approach, i.e., we replace t by $t = kT_s$, where T_s denotes the sampling interval and $k = 1, 2, \ldots, K$. One should note here that the sampling interval T_s is chosen sufficiently small enough to assure that the statistical analysis and the evaluation of the deterministic sequences $\tilde{\xi}(kT_s) = |\tilde{\mu}_\rho(kT_s)|$ and $\tilde{\vartheta}(kT_s) = \arg\{\tilde{\mu}_\rho(kT_s)\}$ can be performed as precisely as possible. It will not be sufficient in this context if the sampling frequency $f_s = 1/T_s$ is merely given by the value $f_s = 2 \cdot \max\{f_{i,n}\}_{n=1}^{N_i}$; although this value would be completely sufficient to fulfil the sampling theorem. For our purposes, the inequality $f_s \gg \max\{f_{i,n}\}_{n=1}^{N_i}$ should rather hold. Experience shows that a good compromise between computational expenditure and attainable precision is achieved, if — depending on the case of application — $f_s \approx 20 \cdot \max\{f_{i,n}\}_{n=1}^{N_i}$ up to $f_s \approx 100 \cdot \max\{f_{i,n}\}_{n=1}^{N_i}$ holds, and for the number of iterations K, the value $K = 10^6$ is chosen. From the simulation of the discrete-time signals $\tilde{\xi}(kT_s)$ and $\tilde{\vartheta}(kT_s)$, the probability density functions $\tilde{p}_\xi(z)$ and $\tilde{p}_\vartheta(\theta)$ can then be determined by means of the histograms of the simulated signals. Here, the choice of the discrete Doppler frequencies $f_{i,n}$ is not decisive. On these parameters, we only impose that they should all be unequal and different from zero. Moreover, due to the periodical behaviour of $\tilde{\mu}_i(t)$, the discrete Doppler frequencies $f_{i,n}$ have to be determined in such a way that the period $T_i = 1/\gcd\{f_{i,n}\}_{n=1}^{N_i}$ is greater or equal to the simulation time T_{sim}, i.e., $T_i \geq T_{sim} = KT_s$.

Exemplary simulation results for the probability density functions $\tilde{p}_\xi(z)$ and $\tilde{p}_\vartheta(\theta)$ are depicted in Figure 4.6 for the case $c_{i,n} = \sigma_0\sqrt{2/N_i}$, where $N_i = 7$ $(i = 1, 2)$ and $\sigma_0^2 = 1$.

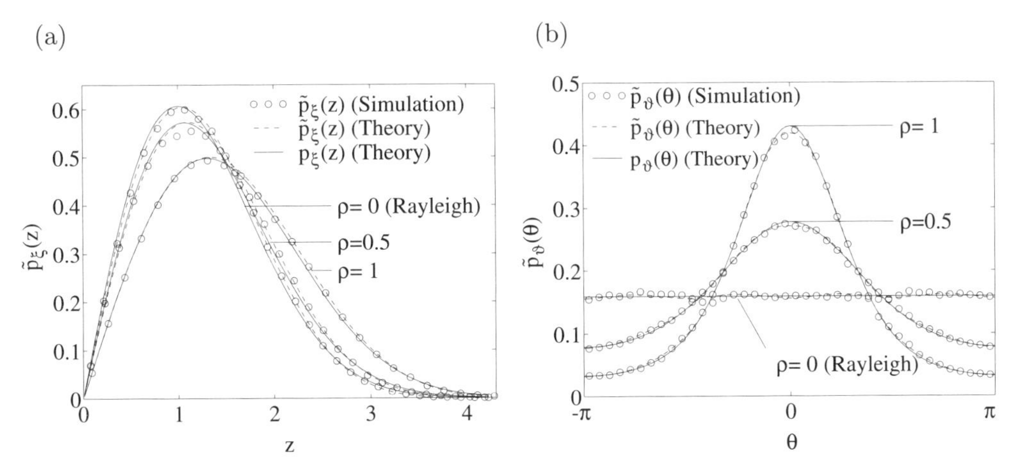

Figure 4.6: (a) Probability density function $\tilde{p}_\xi(z)$ of the amplitude and (b) probability density function $\tilde{p}_\vartheta(\theta)$ of the phase for $N_1 = N_2 = 7$ ($c_{i,n} = \sigma_0\sqrt{2/N_i}$, $\sigma_0^2 = 1$, $f_\rho = 0$, $\theta_\rho = 0$).

Figures 4.6(a) and 4.6(b) confirm that the probability density functions, which have been obtained from the simulation of deterministic processes, are identical to the analytical expressions describing the statistics of the underlying random variables. In the remainder of this book, we will therefore call $\tilde{p}_{\mu_i}(x)$ [see (4.34)] the probability density function of the deterministic process $\tilde{\mu}_i(t)$. Consequently, $\tilde{p}_\xi(z)$ [see (4.47a)] and $\tilde{p}_\vartheta(\theta)$ [see (4.47b)] describe the probability density functions of the amplitude $\tilde{\xi}(t)$ and the phase $\tilde{\vartheta}(t)$ of the complex deterministic process $\tilde{\mu}_\rho(t) = \tilde{\mu}_{\rho_1}(t) + j\tilde{\mu}_{\rho_2}(t)$, respectively.

4.3.2 Level-Crossing Rate and Average Duration of Fades

In this subsection, we will derive general analytical expressions for the level-crossing rate $\tilde{N}_\xi(r)$ and the average duration of fades $\tilde{T}_{\xi_-}(r)$ of the deterministic simulation model for Rice processes shown in Figure 4.3. The knowledge of analytical solutions makes the determination of $\tilde{N}_\xi(r)$ and $\tilde{T}_{\xi_-}(r)$ from time-consuming simulations superfluous. Moreover, they enable a deeper insight into the cause and the effect of statistical degradations, which can be attributed to the finite number of used harmonic functions N_i, on the one hand, but also to the applied method for the determination of the model parameters, on the other.

In the preceding Subsection 4.3.1, we have seen that the probability density function $\tilde{p}_{\mu_i}(x)$ of the deterministic process $\tilde{\mu}_i(t)$ is almost identical to the probability density function $p_{\mu_i}(x)$ of the (ideal) stochastic process $\mu_i(t)$, provided that the number of harmonic functions N_i is sufficiently high, let us say $N_i \geq 7$. On condition that the relations

$$\text{(i)} \qquad \tilde{p}_{\mu_i}(x) = p_{\mu_i}(x)\,, \tag{4.61a}$$

$$\text{(ii)} \qquad \tilde{\beta} = \tilde{\beta}_1 = \tilde{\beta}_2 \tag{4.61b}$$

hold, then the level-crossing rate $\tilde{N}_\xi(r)$ is still given by (3.24), if there the quantities α and β of the stochastic reference model are replaced by the corresponding quantities $\tilde{\alpha}$ and $\tilde{\beta}$ of the deterministic simulation model. Thus, one obtains the following expression for deterministic Rice processes $\tilde{\xi}(t)$ with $f_\rho \neq 0$:

$$\tilde{N}_\xi(r) = \frac{r\sqrt{2\tilde{\beta}}}{\pi^{3/2}\sigma_0^2} e^{-\frac{r^2+\rho^2}{2\sigma_0^2}} \int_0^{\pi/2} \cosh\left(\frac{r\rho}{\sigma_0^2}\cos\theta\right) \cdot \left\{ e^{-(\tilde{\alpha}\rho\sin\theta)^2} + \sqrt{\pi}\tilde{\alpha}\rho\sin(\theta)\,\mathrm{erf}\,(\tilde{\alpha}\rho\sin\theta)\right\} d\theta\,, \quad r \geq 0\,, \tag{4.62}$$

where

$$\tilde{\alpha} = 2\pi f_\rho \Big/ \sqrt{2\tilde{\beta}}\,, \tag{4.63a}$$

$$\tilde{\beta} = \tilde{\beta}_i = -\ddot{\tilde{r}}_{\mu_i\mu_i}(0) = 2\pi^2 \sum_{n=1}^{N_i} (c_{i,n} f_{i,n})^2\,. \tag{4.63b}$$

For the special case $f_\rho = 0$, it follows from (4.63a) that $\tilde{\alpha} = 0$ holds, so that (4.62) simplifies to the following expression

$$\tilde{N}_\xi(r) = \sqrt{\frac{\tilde{\beta}}{2\pi}} \cdot p_\xi(r)\,, \quad r \geq 0\,, \tag{4.64}$$

which is identical to (3.27) after replacing β by $\tilde{\beta}$ there. Obviously, the quality of the approximation $\tilde{\beta} \approx \beta$ quite decisively determines the deviation of the level-crossing rate of the deterministic simulation model from that of the underlying stochastic reference model.

For further analyses, we write

$$\tilde{\beta} = \beta + \Delta\beta\,, \tag{4.65}$$

where $\Delta\beta$ describes the *true error of* $\tilde{\beta}$ caused by the chosen method for the computation of the model parameters $c_{i,n}$ and $f_{i,n}$. In the following, we will call $\Delta\beta$ the *model error* for short. Let us assume that the *true relative error of* $\tilde{\beta}$, which is the *relative model error* $\Delta\beta/\beta$, is small, then, with the approximation $\sqrt{\beta + \Delta\beta} \approx \sqrt{\beta}(1 + \frac{\Delta\beta}{2\beta})$, the level-crossing rate (4.64) can be approximated by

$$\begin{aligned}\tilde{N}_\xi(r) &\approx N_\xi(r)\left(1 + \frac{\Delta\beta}{2\beta}\right) \\ &= N_\xi(r) + \Delta N_\xi(r)\,, \quad r \geq 0\,,\end{aligned} \tag{4.66}$$

where

$$\Delta N_\xi(r) = \frac{\Delta\beta}{2\beta} N_\xi(r) = \frac{\Delta\beta}{2\sqrt{2\pi\beta}}\, p_\xi(r) \tag{4.67}$$

describes the *true error of* $\tilde{N}_\xi(r)$. In this case, $\Delta N_\xi(r)$ behaves proportionally to $\Delta\beta$, or in other words: for any given level r, the relation $\Delta N_\xi(r)/\Delta\beta$ will be constant, and, thus, independent of the model error $\Delta\beta$ of the simulation model.

For $\rho \to 0$, it follows $\tilde{\xi}(t) \to \tilde{\zeta}(t)$, and, thus $\tilde{p}_\xi(r) \to \tilde{p}_\zeta(r)$. Consequently, with reference to the assumptions (4.61a) and (4.61b), we obtain the following relation for the level-crossing rate $\tilde{N}_\zeta(r)$ of deterministic Rayleigh processes $\tilde{\zeta}(t)$

$$\tilde{N}_\zeta(r) = \sqrt{\frac{\tilde{\beta}}{2\pi}} \cdot p_\zeta(r)\,, \quad r \geq 0\,. \tag{4.68}$$

Now, it is obvious that the approximation (4.66) also holds for $\tilde{N}_\zeta(r)$ in connection with (4.67) if the index ξ is replaced by ζ in both equations.

For reasons of illustration, the analytical expression for $\tilde{N}_\xi(r)$ given by (4.64) is shown in Figure 4.7 for a relative model error $\Delta\beta/\beta$ in the range of ± 10 per cent. This figure also shows the ideal conditions, i.e., $\tilde{\beta} = \beta$, which we have already seen in Figure 3.5(b).

We now want to concentrate on the analysis of the average duration of fades of deterministic Rice processes, where we again hold on to the assumptions (4.61a) and (4.61b). In particular, it follows from (4.61a) that the cumulative distribution function $\tilde{F}_{\xi_-}(r)$ of deterministic Rice processes is identical to that of stochastic Rice processes, i.e., $\tilde{F}_{\xi_-}(r) = F_{\xi_-}(r)$. Hence, by taking the definition (2.63) into consideration, it turns out that in this case the average duration of fades $\tilde{T}_{\xi_-}(r)$ of deterministic Rice processes $\tilde{\xi}(t)$ can be expressed by

$$\tilde{T}_{\xi_-}(r) = \frac{F_{\xi_-}(r)}{\tilde{N}_\xi(r)}\,, \quad r \geq 0\,, \tag{4.69}$$

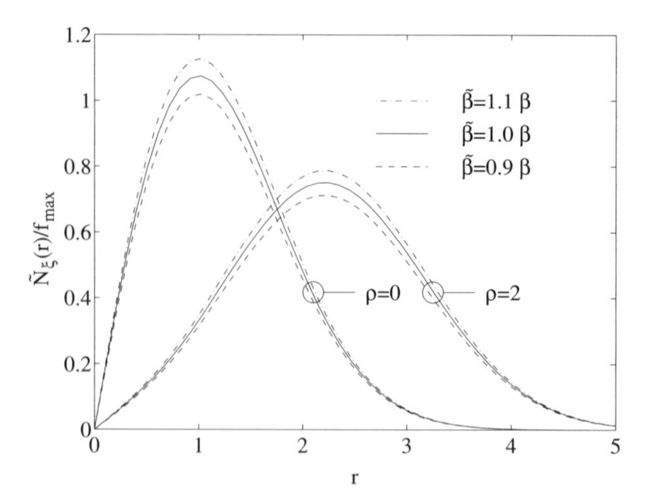

Figure 4.7: Normalized level-crossing rate $\tilde{N}_\xi(r)/f_{max}$ of deterministic Rice processes for various values of $\tilde{\beta} = \beta + \Delta\beta$ (Jakes PSD, $f_{max} = 91\,\text{Hz}$, $f_\rho = 0$).

where $\tilde{N}_\xi(r)$ is given by (4.62).

For the special case $f_\rho = 0$, simple approximate solutions can again be given if the relative model error $\Delta\beta/\beta$ is small. Hence, after substituting (4.66) into (4.69) and using the approximation formula $1 \Big/ \left(1 + \frac{\Delta\beta}{2\beta}\right) \approx 1 - \frac{\Delta\beta}{2\beta}$, we obtain the expression

$$\begin{aligned} \tilde{T}_{\xi_-}(r) &\approx T_{\xi_-}(r)\left(1 - \frac{\Delta\beta}{2\beta}\right) \\ &= T_{\xi_-}(r) + \Delta T_{\xi_-}(r)\,, \quad r \geq 0\,, \end{aligned} \tag{4.70}$$

where

$$\Delta T_{\xi_-}(r) = -\frac{\Delta\beta}{2\beta} T_{\xi_-}(r) \tag{4.71}$$

denotes the *true error of* $\tilde{T}_{\xi_-}(r)$. Hence, the approximation (4.70) states that the average duration of fades of the deterministic Rice process decreases (increases) approximately linearly with an increasing (decreasing) model error $\Delta\beta$.

For low levels r and moderate Rice factors $c_R = \rho^2/(2\sigma_0^2)$, we obtain the following approximation after a short calculation from (4.70) in connection with (3.39)

$$\tilde{T}_{\xi_-}(r) \approx \tilde{T}_{\zeta_-}(r) \approx r\sqrt{\frac{\pi}{2\beta}}\left(1 - \frac{\Delta\beta}{2\beta}\right), \tag{4.72}$$

where this result is valid for $r << 1$ and $r\rho/\sigma_0^2 << 1$. Thus, it appears that the fading intervals of deterministic Rice and Rayleigh processes are approximately identical at low signal levels, if the influence of the line-of-sight component is of no consequence.

By comparing (3.39) with (4.72) it also becomes clearer that the relative model error again determines the deviation from the average duration of fades of the corresponding reference model.

An interesting statement can also be made on the product $\tilde{N}_\xi(r) \cdot \tilde{T}_{\xi_-}(r)$. Namely, from (2.63) and (4.69), the *model error law* of deterministic channel modelling

$$\tilde{N}_\xi(r) \cdot \tilde{T}_{\xi_-}(r) = N_\xi(r) \cdot T_{\xi_-}(r) \tag{4.73}$$

follows, which means that the product of the level-crossing rate and the average duration of fades of deterministic Rice processes is independent of the model error $\Delta\beta$. With an increasing model error $\Delta\beta$, the level-crossing rate $\tilde{N}_\xi(r)$ may rise, but the average duration of fades $\tilde{T}_{\xi_-}(r)$ decreases by the same extent, so that the product $\tilde{N}_\xi(r) \cdot \tilde{T}_{\xi_-}(r)$ remains constant at any given level r = const. This result is also approximately obtained from the product of the approximations (4.66) and (4.70) if we ignore in the resulting product the quadratic term $[\Delta\beta/(2\beta)]^2$.

Since Rayleigh processes can naturally be considered as Rice processes for the special case $\rho = 0$, the relations (4.69)–(4.71) and (4.73) in principle hold for Rayleigh processes as well. Only the indices ξ and ξ_- have to be replaced by ζ and ζ_-, respectively.

The evaluation of the analytical expression for $\tilde{T}_{\xi_-}(r)$ [see (4.69)] is shown in Figure 4.8 for $\Delta\beta/\beta \in \{-0.1, 0, +0.1\}$.

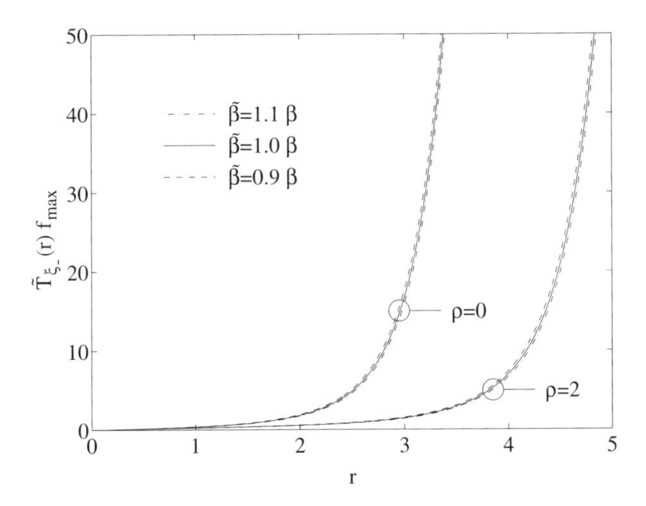

Figure 4.8: Normalized average duration of fades $\tilde{T}_{\xi_-}(r) \cdot f_{max}$ of deterministic Rice processes for various values of $\tilde{\beta} = \beta + \Delta\beta$ (Jakes PSD, f_{max} = 91 Hz, $f_\rho = 0$).

In Chapter 5, where the individual methods for the determination of the simulation model parameters are analysed, we will see that the condition (4.61b) can often not be fulfilled exactly. In most cases, however, the relative deviation between $\tilde{\beta}_1$ and $\tilde{\beta}_2$ is very small. Due to the analysis of the level-crossing rate and the average duration of fades of Rice and Rayleigh processes, which was dealt with in Subsection 3.3.2, we already know that for small relative deviations between β_1 and β_2, the ideal relations derived on condition that $\beta = \beta_1 = \beta_2$ holds will still keep their validity in a very good approximation, if we replace the quantity $\beta = \beta_1 = \beta_2$ by the arithmetical mean $\beta = (\beta_1 + \beta_2)/2$ [see (3.37)] in the corresponding expressions or if we directly identify β with β_1, i.e., $\beta = \beta_1 \approx \beta_2$ [see also Appendix B, eq. (B.17)]. Analogous results can be found for the deterministic model as well. For further simplification, we will therefore set $\tilde{\beta} = \tilde{\beta}_1 \approx \tilde{\beta}_2$ in the following, in case the relative deviation between $\tilde{\beta}_1$ and $\tilde{\beta}_2$ is small.

Finally, concerning this subject, it should be noted that even without the stated conditions (4.61a) and (4.61b), the level-crossing rate and, thus, also the average duration of fades of deterministic Rice processes can be calculated exactly. However, the numerical expenditure for the solution of the obtained integral equations is considerably high. Apart from that the achievable improvements are often only low, even for a small number of harmonic functions N_i, so that the comparatively high numerical expenditure does not seem to be justified. Not only against this background, it turns out that especially the condition (4.61a) is meaningful, even though — strictly speaking — this condition is only fulfilled exactly as $N_i \to \infty$.

For completeness, the exact calculation of both the level-crossing rate and the average duration of fades of deterministic Rice processes for any number of harmonic functions N_i is presented in Appendix C, where both conditions (4.61a) and (4.61b) have been dropped. In Appendix C, one finds the following analytical closed-form expression for the level-crossing rate $\tilde{N}_\xi(r)$

$$\tilde{N}_\xi(r) = 2r \int_0^\infty \int_{-\pi}^{\pi} w_1(r,\theta)\, w_2(r,\theta) \int_0^\infty j_1(z,\theta)\, j_2(z,\theta)\, \dot{z} \cos(2\pi z \dot{z})\, dz\, d\theta\, d\dot{z}\,, \quad (4.74)$$

where

$$\begin{aligned} w_1(r,\theta) &= \tilde{p}_{\mu_1}(r\cos\theta - \rho\cos\theta_\rho)\,, && (4.75a)\\ w_2(r,\theta) &= \tilde{p}_{\mu_2}(r\sin\theta - \rho\sin\theta_\rho)\,, && (4.75b)\\ j_1(z,\theta) &= \prod_{n=1}^{N_1} J_0(4\pi^2 c_{1,n} f_{1,n} z \cos\theta)\,, && (4.75c)\\ j_2(z,\theta) &= \prod_{n=1}^{N_2} J_0(4\pi^2 c_{2,n} f_{2,n} z \sin\theta)\,. && (4.75d) \end{aligned}$$

Now, if we substitute (4.74) into (4.69) and replace the cumulative distribution function $F_{\xi_-}(r)$ by $\tilde{F}_{\xi_-}(r)$ in the latter equation and using for $\tilde{F}_{\xi_-}(r)$ the expression (C.40) derived in Appendix C, then we also find an exact analytical expression for the average duration of fades $\tilde{T}_{\xi_-}(r)$ of deterministic Rice processes $\tilde{\xi}(t)$.

4.3.3 Statistics of the Fading Intervals at Low Levels

In this subsection, we will discuss the statistical properties of the fading intervals of deterministic Rayleigh processes. We will restrict ourselves to low levels here, because in this case very precise approximate solutions can be derived by analytical means. At medium and high levels, however, we have to rely on simulations to which we will come back in Section 5.3.

At first, we will study the probability density function of the fading intervals of deterministic Rayleigh processes $\tilde{p}_{0_-}(\tau_-; r)$. This density characterizes the conditional probability density function for the case that a deterministic Rayleigh process $\tilde{\zeta}(t)$ crosses a level r for the first time at the time instant $t_2 = t_1 + \tau_-$, provided that the last down-crossing occurred at the time instant t_1. If no further statements are made about the level-crossing behaviour of $\tilde{\zeta}(t)$ between t_1 and t_2, then the corresponding probability density function is denoted by $\tilde{p}_{1_-}(\tau_-; r)$. During the analysis proceeded for the stochastic reference model in Subsection 3.3.3, it was pointed out that, according to (3.47), $p_{1_-}(\tau_-; r)$ can be regarded as a very good approximation for $p_{0_-}(\tau_-; r)$ if the level r is low. Consequently, it also holds for the deterministic model: $\tilde{p}_{0_-}(\tau_-; r) \to \tilde{p}_{1_-}(\tau_-; r)$ if $r \to 0$, where $\tilde{p}_{1_-}(\tau_-; r)$ follows directly from (3.47) if there $T_{\zeta_-}(r)$ is substituted by $\tilde{T}_{\zeta_-}(r)$ [Pae96e], i.e.,

$$\tilde{p}_{1_-}(\tau_-; r) = \frac{2\pi \tilde{z}^2 \, e^{-\tilde{z}}}{\tilde{T}_{\zeta_-}(r)} \left[I_0(\tilde{z}) - \left(1 + \frac{1}{2\tilde{z}}\right) I_1(\tilde{z}) \right], \quad 0 \le r << 1, \tag{4.76}$$

where $\tilde{z} = 2\left[\tilde{T}_{\zeta_-}(r)/\tau_-\right]^2/\pi$. The use of the result (4.70) now gives us the opportunity to investigate the influence of the model error $\Delta\beta$ on the probability density function of the fading intervals of deterministic Rayleigh processes at deep signal levels analytically. The evaluation of $\tilde{p}_{1_-}(\tau_-; r)$ according to (4.76) for various values of $\tilde{\beta} = \beta + \Delta\beta$ is shown in Figure 4.9 for a level r of $r = 0.1$. In this figure, one recognizes that a positive model error $\Delta\beta > 0$ is always connected with a distinct decrease (increase) of the probability density function $\tilde{p}_{1_-}(\tau_-; r)$ in the range of relatively large (small) fading intervals τ_-. With a negative model error $\Delta\beta < 0$, logically, the inverse behaviour occurs.

In a similar way, we obtain an expression for $\tilde{\tau}_q$ describing the length of the time interval of those fading intervals of deterministic Rayleigh processes $\tilde{\zeta}(t)$ which include q per cent of all fading intervals. The quantity $\tilde{\tau}_q = \tilde{\tau}_q(r)$ follows directly from (3.53) if we again replace $T_{\zeta_-}(r)$ by $\tilde{T}_{\zeta_-}(r)$, i.e., for $75 \le q \le 100$ it follows that

$$\tilde{\tau}_q(r) \approx \frac{\tilde{T}_{\zeta_-}(r)}{\left\{ \frac{\pi}{4}\left[1 - \sqrt{1 - 4\left(1 - \frac{q}{100}\right)}\right]\right\}^{1/3}}, \quad r << 1. \tag{4.77}$$

In order to make the influence of the model error $\Delta\beta$ easily identifiable, we use $\tilde{T}_{\zeta_-}(r) = T_{\zeta_-}(r)[1 - \Delta\beta/(2\beta)]$ and in connection with (3.53), we write the relation above in the form

$$\tilde{\tau}_q(r) \approx \tau_q(r)\left(1 - \frac{\Delta\beta}{2\beta}\right), \quad r << 1. \tag{4.78}$$

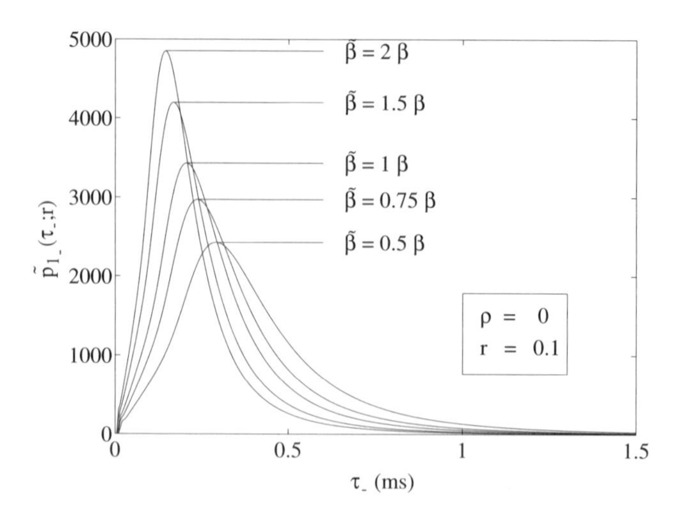

Figure 4.9: The influence of $\tilde{\beta} = \beta + \Delta\beta$ on the probability density function $\tilde{p}_{1_-}(\tau_-; r)$ of the fading intervals τ_- at low levels (Jakes PSD, $f_{max} = 91\,\text{Hz}$, $\sigma_0^2 = 1$).

This relation makes clear that a relative error of β in the order of $\pm\varepsilon$ approximately causes a relative error of $\tau_q(r)$ in the order of $\mp\varepsilon/2$.

In particular, after putting (3.54)–(3.56) into (4.78), the quantities $\tilde{\tau}_{90}(r)$, $\tilde{\tau}_{95}(r)$, and $\tilde{\tau}_{99}(\tau)$ can now approximately be expressed by:

$$\tilde{\tau}_{90}(r) \approx 1.78 \cdot T_{\zeta_-}(r)[1 - \Delta\beta/(2\beta)]\,, \tag{4.79a}$$

$$\tilde{\tau}_{95}(r) \approx 2.29 \cdot T_{\zeta_-}(r)[1 - \Delta\beta/(2\beta)]\,, \tag{4.79b}$$

$$\tilde{\tau}_{99}(r) \approx 3.98 \cdot T_{\zeta_-}(r)[1 - \Delta\beta/(2\beta)]\,. \tag{4.79c}$$

At this point, it should be explicitly emphasized that all approximations for $\tilde{p}_{0_-}(\tau_-; r)$ and $\tilde{\tau}_q(r)$, which were specially derived in this subsection for deterministic Rayleigh processes, are also valid for deterministic Rice processes with moderate Rice factors. This statement becomes clear immediately if we take into consideration that $\tilde{p}_{1_-}(\tau_-; r)$ [see (4.76)] and $\tilde{\tau}_q(r)$ [see (4.77)] only depend on the average duration of fades $\tilde{T}_{\zeta_-}(r)$, and, due to (4.72), $\tilde{T}_{\zeta_-}(r)$ can approximately be replaced at low signal levels r by the average duration of fades $\tilde{T}_{\xi_-}(r)$ of Rice processes.

4.3.4 Ergodicity and Criteria for the Performance Evaluation

As already mentioned, a deterministic process $\tilde{\mu}_i(t)$, defined by (4.4), is a sample function of the corresponding stochastic process $\hat{\mu}_i(t)$. In this subsection, we will discuss the ergodic properties of the stochastic process $\hat{\mu}_i(t)$. Hence, we first distinguish between the ergodicity with respect to the mean value and the ergodicity with respect to the autocorrelation function [Pap91]. After that, we will look at some criteria for the assessment of the performance, which will play a significant role in the following chapter.

Ergodicity with respect to the mean value: A stochastic process $\hat{\mu}_i(t)$ is said to be *ergodic with respect to the mean value* if the temporal mean value of $\tilde{\mu}_i(t)$ computed over the interval [-T,T] converges in the limit $T \to \infty$ to the statistical mean value $\hat{m}_{\mu_i} := E\{\hat{\mu}_i(t)\}$, i.e.,

$$\hat{m}_{\mu_i} = \tilde{m}_{\mu_i} := \lim_{T\to\infty} \frac{1}{2T} \int_{-T}^{T} \tilde{\mu}_i(t)\, dt \,. \tag{4.80}$$

Since the Doppler phases $\theta_{i,n}$ of the stochastic process $\hat{\mu}_i(t)$ are random variables uniformly distributed in the interval $(0, 2\pi]$, the left-hand side of the equation above is equal to zero. The right-hand side is also equal to zero if all discrete Doppler frequencies $f_{i,n}$ are unequal to zero, i.e., if $f_{i,n} \neq 0$ for all $n = 1, 2, \ldots, N_i$ and $i = 1, 2$. That requirement ($f_{i,n} \neq 0$) can be fulfilled without any difficulty by all parameter computation methods introduced in the next chapter. Hence, $\hat{m}_{\mu_i} = \tilde{m}_{\mu_i} = 0$ holds, and, thus, the stochastic process $\hat{\mu}_i(t)$ is ergodic with respect to the mean value.

Ergodicity with respect to the autocorrelation function: A stochastic process $\hat{\mu}_i(t)$ is said to be *ergodic with respect to the autocorrelation function* if the temporal mean of $\tilde{\mu}_i(t)\tilde{\mu}_i(t+\tau)$ computed over the interval $[-T, T]$ converges in the limit $T \to \infty$ to the statistical mean $\hat{r}_{\mu_i\mu_i}(\tau) := E\{\hat{\mu}_i(t)\hat{\mu}_i(t+\tau)\}$, i.e.,

$$\hat{r}_{\mu_i\mu_i}(\tau) = \tilde{r}_{\mu_i\mu_i}(\tau) := \lim_{T\to\infty} \frac{1}{2T} \int_{-T}^{T} \tilde{\mu}_i(t)\, \tilde{\mu}_i(t+\tau)\, dt \,. \tag{4.81}$$

If the discrete Doppler frequencies $f_{i,n}$ and the Doppler coefficients $c_{i,n}$ of the stochastic process $\hat{\mu}_i(t)$ are constant quantities and merely the Doppler phases $\theta_{i,n} \in (0, 2\pi]$ are uniformly distributed random variables, then the left-hand side of the equation above leads to

$$\hat{r}_{\mu_i\mu_i}(\tau) = \sum_{n=1}^{N_i} \frac{c_{i,n}^2}{2} \cos(2\pi f_{i,n}\tau) \,. \tag{4.82}$$

Due to (4.11), we are already familiar with the solution of the right-hand side of (4.81). The comparison of (4.11) with (4.82) shows us that $\hat{r}_{\mu_i\mu_i}(\tau) = \tilde{r}_{\mu_i\mu_i}(\tau)$ holds, and, thus, the stochastic process $\hat{\mu}_i(t)$ is ergodic with respect to the autocorrelation function.

Without wanting to jump ahead, it should be noted here that for the Monte Carlo method described in Subsection 5.1.4, the discrete Doppler frequencies $f_{i,n}$ of the stochastic process $\hat{\mu}_i(t)$ are not constants but random variables. We will see that in this case $\hat{r}_{\mu_i\mu_i}(\tau) \neq \tilde{r}_{\mu_i\mu_i}(\tau)$ holds, and, thus, the stochastic process $\hat{\mu}_i(t)$ is not ergodic with respect to the autocorrelation function.

Indeed, for channel modelling it is not the crucial factor whether the stochastic process $\hat{\mu}_i(t)$ is ergodic with respect to the mean value or the autocorrelation function. The deviations of statistical properties of the deterministic process $\tilde{\mu}_i(t)$ from the statistical properties of the underlying ideal stochastic process $\mu_i(t)$ are decisive. From these deviations, criteria can be gained for the performance evaluation of the methods for the computation of the model parameters presented in the next chapter.

Since the process $\mu_i(t)$ was introduced here as a zero-mean normally distributed stochastic process, i.e., $\mu_i(t) \sim N(0, \sigma_0^2)$, the mean-square error of the probability density function $\tilde{p}_{\mu_i}(x)$ [cf. (4.39)]

$$E_{p_{\mu_i}} := \int_{-\infty}^{\infty} \left(p_{\mu_i}(x) - \tilde{p}_{\mu_i}(x)\right)^2 dx \tag{4.83}$$

defines the first important criterion for the performance evaluation [Pae98b].

As is well known, real-valued Gaussian random processes are described completely by their probability density function and their autocorrelation function. A further important criterion for the performance evaluation is therefore the mean-square error of the autocorrelation function $\tilde{r}_{\mu_i\mu_i}(\tau)$ defined by

$$E_{r_{\mu_i\mu_i}} := \frac{1}{\tau_{\max}} \int_{0}^{\tau_{\max}} \left(r_{\mu_i\mu_i}(\tau) - \tilde{r}_{\mu_i\mu_i}(\tau)\right)^2 d\tau\,. \tag{4.84}$$

For the parameter $\tau_{\max}$, the value $\tau_{\max} = N_i/(2f_{max})$ has turned out to be suitable, especially for the Jakes power spectral density, as we will see in the next chapter.

Meanwhile, it has already been mentioned several times that the statistical properties of deterministic simulation models can deviate considerably from those of the underlying ideal stochastic reference model. We have seen that for many important statistical quantities, the model error $\Delta\beta$ could be held responsible for this. Therefore, a good parameter computation procedure should only cause a small relative model error $\Delta\beta/\beta$, even when the realization expenditure is low, i.e., for a small number of harmonic functions N_i. Hence, the model error $\Delta\beta$ and its convergence property $\Delta\beta \to 0$ or $\tilde{\beta} \to \beta$ for $N_i \to \infty$ will be paid attention to in the subsections of the following chapter as well.

5

METHODS FOR THE COMPUTATION OF THE MODEL PARAMETERS OF DETERMINISTIC PROCESSES

By now there is a multitude of various methods for the computation of the primary parameters of the simulation model (Doppler coefficients $c_{i,n}$ and discrete Doppler frequencies $f_{i,n}$). Exactly like the original Rice method [Ric44, Ric45], the method of equal distances [Pae94b, Pae96d] as well as the mean-square-error method [Pae96d] are characterized by the fact that the distances between two neighbouring discrete Doppler frequencies are equidistant. These three methods merely differ in the specific way of how the Doppler coefficients are adapted to the desired Doppler power spectral density. Due to the equidistant property of discrete Doppler frequencies, which are in neighbouring pairs, all three procedures have one decisive disadvantage in common, namely the comparatively small period of the designed deterministic Gaussian processes, and, thus, of the resulting simulation model. This disadvantage can be avoided, e.g., by using the method of equal areas [Pae94b, Pae96d], which has an acceptable performance when applied to the Jakes power spectral density. However, this method fails or leads to a comparatively high realization expenditure, if the procedure is used in connection with Gaussian shaped power spectral densities. In German-speaking countries, the Monte Carlo method [Schu89, Hoe92] has become quite popular. In comparison with other methods, however, the performance of this method is poor [Pae96d, Pae96e] if the approximation precision of the autocorrelation function of the resulting deterministic Gaussian processes is used as a criterion for the evaluation of the performance. The principle of the Monte Carlo method is that the discrete Doppler frequencies of the stochastic simulation system are obtained from the mapping of a uniformly distributed random variable into a random variable with a distribution proportional to the desired Doppler power spectral density. Consequently, the discrete Doppler frequencies themselves are random variables. The realization of a set $\{f_{i,n}\}$ of discrete Doppler frequencies can thus result in a deterministic Gaussian process $\tilde{\mu}_i(t)$, whose statistical properties may largely deviate from the desired properties of the (ideal) stochastic Gaussian random process $\mu_i(t)$. This even holds if the number of harmonic functions N_i chosen is very large, let us say $N_i = 100$ [Pae96e]. A quasi-optimal procedure is the method of exact Doppler spread [Pae98b, Pae96c].

This method is almost ideally suitable for Jakes shaped power spectral densities. The performance of the method of exact Doppler spread can only be outperformed by the L_p-norm method [Pae98b, Pae96c]. Unfortunately, the arising numerical complexity of this method is comparatively high, so that an application, especially in connection with the Jakes and Gaussian power spectral density, is often not worth the effort. The L_p-norm method only unfolds its full performance when the statistical properties of the deterministic simulation model have to be adapted to snapshot measurements of real-world mobile radio channels. A further design method is the Jakes method [Jak93], which, however, does not fulfil the often imposed requirement that the real and imaginary part of the complex Gaussian random processes describing the Rayleigh (Rice) process should be uncorrelated.

For an infinite number of harmonic functions, all these methods result in deterministic processes with identical statistical properties, which even match the ones of the reference model exactly. However, as soon as only a finite number of harmonic functions is used, we obtain deterministic processes with completely different statistical properties, which in particular cases can considerably deviate from those of the reference model. The discussion of these properties will be one objective of the following Section 5.1. Thus, in order to compute the model parameters $c_{i,n}$ and $f_{i,n}$, we will proceed by deriving the seven design procedures mentioned just in such a way that they are generally applicable. Afterwards, the methods are respectively applied to both of the often used Jakes and Gaussian power spectral densities. In general, we obtain simple equations allowing us to quickly determine the desired model parameters for the most important practical application cases. For each method, the characteristic properties as well as the advantages and disadvantages will be discussed. For a fair judgement of the performance, the criteria of assessment introduced in the preceding Subsection 4.3.4 will be made use of. At places in the text, where relations among the individual methods occur, these connections will be pointed out.

The computation of the Doppler phases $\{\theta_{i,n}\}$ can be carried out independently of these methods. Without restriction of generality, we at first assume that the elements of the set $\{\theta_{i,n}\}$ are generated from N_i statistically independent realizations of a random variable uniformly distributed in the interval $[0, 2\pi)$. Afterwards, in Section 5.2, we will discover a deterministic design method for the computation of the set $\{\theta_{i,n}\}$. At that stage, the relevance of the Doppler phases $\{\theta_{i,n}\}$ with regard to the statistical properties of $\tilde{\mu}_i(t)$ will also be analysed more precisely.

Finally, we will again deal with the analysis of the probability density function of the fading intervals of deterministic Rayleigh processes in Section 5.3.

5.1 METHODS FOR THE COMPUTATION OF THE DISCRETE DOPPLER FREQUENCIES AND DOPPLER COEFFICIENTS

5.1.1 Method of Equal Distances (MED)

One of the main characteristics of the method of equal distances (MED) [Pae94b, Pae96d] is that discrete Doppler frequencies, which are found in neighbouring pairs, have the same distance. This property is achieved by defining the discrete Doppler frequencies $f_{i,n}$ as

$$f_{i,n} := \frac{\Delta f_i}{2}(2n-1)\,, \quad n = 1, 2, \ldots, N_i\,, \tag{5.1}$$

where

$$\Delta f_i = f_{i,n} - f_{i,n-1}\,, \quad n = 2, 3, \ldots, N_i\,, \tag{5.2}$$

denotes the distance between two neighbouring discrete Doppler frequencies of the ith deterministic process $\tilde{\mu}_i(t)$ $(i = 1, 2)$.

In order to compute the Doppler coefficients $c_{i,n}$, we take a look at the frequency interval

$$I_{i,n} := \left[f_{i,n} - \frac{\Delta f_i}{2}, f_{i,n} + \frac{\Delta f_i}{2} \right)\,, \quad n = 1, 2, \ldots, N_i\,, \tag{5.3}$$

and demand that within this interval, the mean power of the power spectral density $S_{\mu_i\mu_i}(f)$ of the stochastic reference model is identical to that of the power spectral density $\tilde{S}_{\mu_i\mu_i}(f)$ of the deterministic simulation model, i.e.,

$$\int_{f \in I_{i,n}} S_{\mu_i\mu_i}(f)\, df = \int_{f \in I_{i,n}} \tilde{S}_{\mu_i\mu_i}(f)\, df \tag{5.4}$$

for all $n = 1, 2, \ldots, N_i$ and $i = 1, 2$. Thus, after substituting (4.14) into the above equation, the Doppler coefficients $c_{i,n}$ are determined by the expression

$$c_{i,n} = 2\sqrt{\int_{f \in I_{i,n}} S_{\mu_i\mu_i}(f)\, df}\,. \tag{5.5}$$

After substituting (5.5) into (4.11), one can easily prove that $\tilde{r}_{\mu_i\mu_i}(\tau) \to r_{\mu_i\mu_i}(\tau)$ holds as $N_i \to \infty$. Referring to the central limit theorem, it can furthermore be shown that the convergence property $\tilde{p}_{\mu_i}(x) \to p_{\mu_i}(x)$ holds as $N_i \to \infty$. Thus, for an infinite number of harmonic functions, the deterministic processes designed according to the method of equal distances can be interpreted as sample functions of the underlying ideal Gaussian random process.

The major disadvantage of this method is the resulting poor periodicity property of $\tilde{\mu}_i(t)$. To make this clear, we start from (4.26), and in connection with (5.1) it follows that the greatest common divisor of the discrete Doppler frequencies is

$$F_i = \gcd\{f_{i,n}\}_{n=1}^{N_i} = \frac{\Delta f_i}{2}\,. \tag{5.6}$$

Consequently, $\tilde{\mu}_i(t)$ is periodical with the period $T_i = 1/F_i = 2/\Delta f_i$. To obtain a large value for T_i, a small value for Δf_i is required, which is in general involved with a high realization amount as we will see below.

Jakes power spectral density: The frequency range of the Jakes power spectral density [see (3.8)] is limited to the range $|f| \leq f_{max}$, so that for a given number of harmonic functions N_i, a reasonable value for the difference between two neighbouring discrete Doppler frequencies Δf_i is given by $\Delta f_i = f_{max}/N_i$. Consequently, from (5.1) we obtain the following relation for the discrete Doppler frequencies $f_{i,n}$

$$f_{i,n} = \frac{f_{max}}{2N_i}(2n-1) \tag{5.7}$$

for all $n = 1, 2, \ldots, N_i$ and $i = 1, 2$. The corresponding Doppler coefficients $c_{i,n}$ can now easily be computed with (3.8), (5.3), (5.5), and (5.7). After an elementary computation, we find the expression

$$c_{i,n} = \frac{2\sigma_0}{\sqrt{\pi}} \left[\arcsin\left(\frac{n}{N_i}\right) - \arcsin\left(\frac{n-1}{N_i}\right) \right]^{1/2} \tag{5.8}$$

for all $n = 1, 2, \ldots, N_i$ and $i = 1, 2$.

The deterministic processes $\tilde{\mu}_i(t)$ designed with (5.7) and (5.8) obviously have the mean value $\tilde{m}_{\mu_i} = 0$ and the mean power

$$\tilde{\sigma}^2_{\mu_i} = \tilde{r}_{\mu_i\mu_i}(0) = \sum_{n=1}^{N_i} \frac{c^2_{i,n}}{2} = \sigma_0^2 \,. \tag{5.9}$$

Hence, both the mean value and the mean power of the deterministic process $\tilde{\mu}_i(t)$ exactly match the corresponding quantities of the stochastic process $\mu_i(t)$, i.e., the expected value and the variance.

Designing the complex deterministic processes $\tilde{\mu}(t) = \tilde{\mu}_1(t) + j\tilde{\mu}_2(t)$, the uncorrelatedness of $\tilde{\mu}_1(t)$ and $\tilde{\mu}_2(t)$ must be guaranteed. This can be ensured without difficulty by choosing N_2 in accordance with $N_2 = N_1 + 1$, so that due to (5.7) it follows: $f_{1,n} \neq f_{2,m}$ for $n = 1, 2, \ldots, N_1$ and $m = 1, 2, \ldots, N_2$. This again leads to the desired property that $\tilde{\mu}_1(t)$ and $\tilde{\mu}_2(t)$ are uncorrelated [cf. (4.12)].

As an example, the power spectral density $\tilde{S}_{\mu_i\mu_i}(f)$ and the corresponding autocorrelation function $\tilde{r}_{\mu_i\mu_i}(\tau)$ are depicted in Figure 5.1, where the value 25 has been chosen for the number of harmonic functions N_i.

For comparison, the autocorrelation function $r_{\mu_i\mu_i}(\tau)$ of the reference model is also presented in Figure 5.1(b). The shape of $\tilde{r}_{\mu_i\mu_i}(\tau)$ shown in this figure makes the periodical behaviour clearly recognizable. In general, the following relation holds

$$\tilde{r}_{\mu_i\mu_i}(\tau + mT_i/2) = \begin{cases} \tilde{r}_{\mu_i\mu_i}(\tau)\,, & m \text{ even}\,, \\ -\tilde{r}_{\mu_i\mu_i}(\tau)\,, & m \text{ odd}\,, \end{cases} \tag{5.10}$$

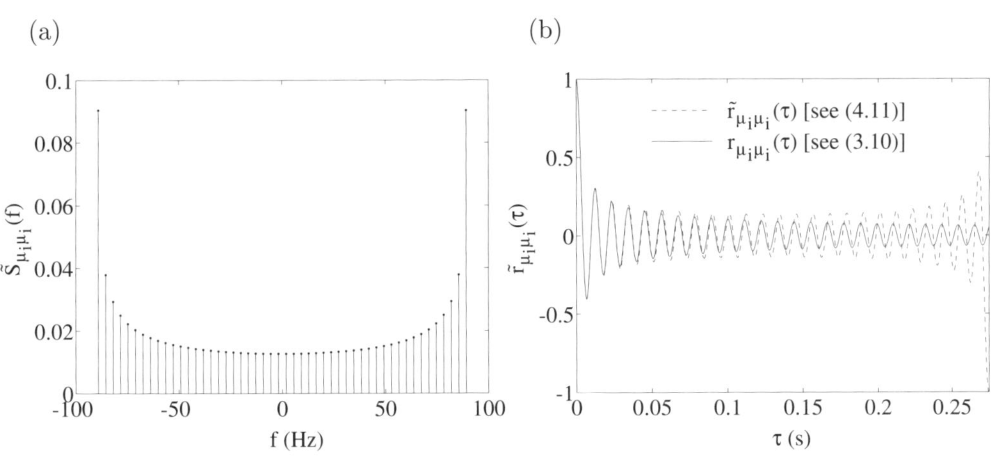

Figure 5.1: (a) Power spectral density $\tilde{S}_{\mu_i\mu_i}(f)$ and (b) autocorrelation function $\tilde{r}_{\mu_i\mu_i}(\tau)$ for $N_i = 25$ (MED, Jakes PSD, $f_{max} = 91$ Hz, $\sigma_0^2 = 1$).

where $T_i = 1/F_i = 2/\Delta f_i = 2N_i/f_{max}$. If we now choose the value $\tau_{max} = T_i/4 = N_i/(2f_{max})$ for the upper limit of the integral (4.84), the mean-square error $E_{r_{\mu_i\mu_i}}$ [see (4.84)] can thus be usefully evaluated, particularly since $E_{r_{\mu_i\mu_i}}$ represents a measure of the performance of the method of equal distances as a function of the realization complexity determined by N_i. The evaluation of the performance criteria $E_{r_{\mu_i\mu_i}}$ and $E_{p_{\mu_i}}$ according to (4.84) and (4.83), respectively, were performed on the basis of the method of equal distances. The obtained results are presented in Figures 5.2(a) and 5.2(b) showing the influence of the used number of harmonic functions N_i. For a better classification of the performance of this method, the results obtained for $c_{i,n} = \sigma_0\sqrt{2/N_i}$ are also shown in Figure 5.2(b). Hence, one realizes that the approximation of the Gaussian distribution using the Doppler coefficients $c_{i,n}$ according to (5.8) is worse in comparison with the results obtained by using $c_{i,n} = \sigma_0\sqrt{2/N_i}$.

Finally, we will also study the model error $\Delta\beta_i = \tilde{\beta}_i - \beta$. With (5.7), (5.8), (3.29), and (4.22), we find the closed-form expression

$$\Delta\beta_i = \beta\left[1 + \frac{1-4N_i}{2N_i^2} - \frac{8}{\pi N_i^2}\sum_{n=1}^{N_i-1} n \cdot \arcsin\left(\frac{n}{N_i}\right)\right], \tag{5.11}$$

whose right-hand side tends to 0 as $N_i \to \infty$, i.e., it holds $\lim_{N_i\to\infty}\Delta\beta_i = 0$. Figure 5.3 depicts the relative model error $\Delta\beta_i/\beta$. It should be observed that the ratio $\Delta\beta_i/\beta$ merely depends on N_i.

Gaussian power spectral density: The frequency range of the Gaussian power spectral density (3.11) must first be limited to the relevant range. Therefore, we introduce the quantity κ_c which is chosen in such a way that the mean power of the Gaussian power spectral density obtained within the frequency range $|f| \leq \kappa_c f_c$ makes up at least 99.99 per cent of its total mean power. This demand is fulfilled with

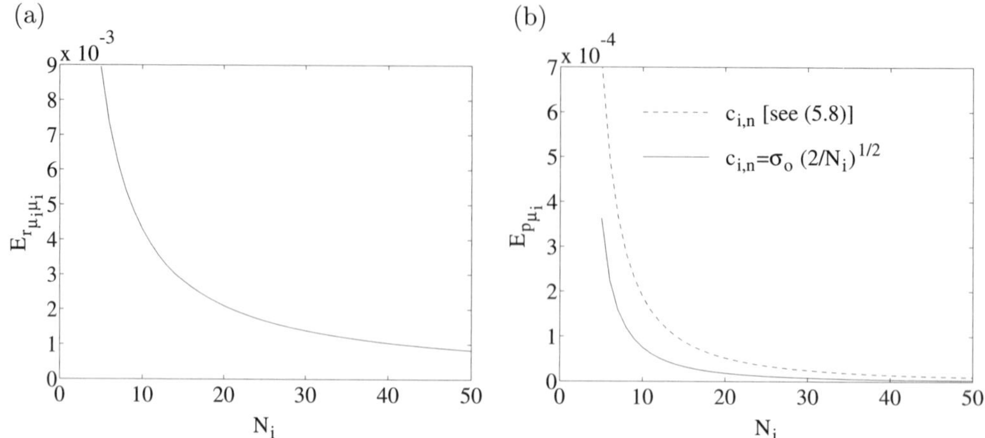

Figure 5.2: Mean-square errors: (a) $E_{r_{\mu_i\mu_i}}$ and (b) $E_{p_{\mu_i}}$ (MED, Jakes PSD, $f_{max} = 91\,\text{Hz}$, $\sigma_0^2 = 1$, $\tau_{max} = N_i/(2f_{max})$).

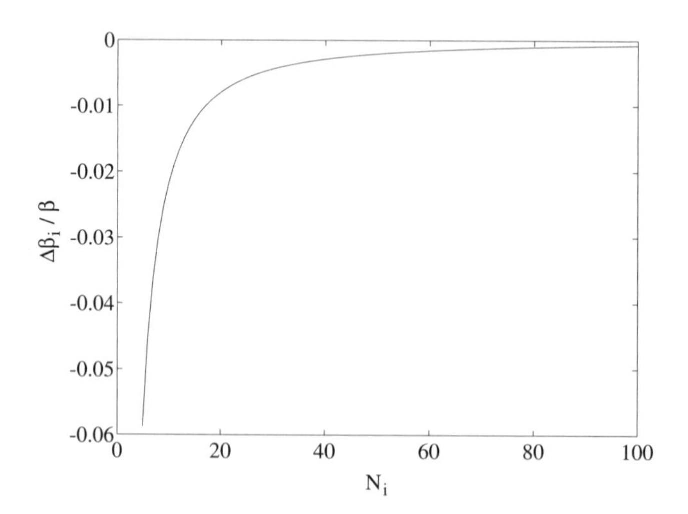

Figure 5.3: Relative model error $\Delta\beta_i/\beta$ (MED, Jakes PSD).

$\kappa_c = 2\sqrt{2/\ln 2}$. Depending on the number of harmonic functions N_i, the difference between two neighbouring discrete Doppler frequencies Δf_i can then be described by $\Delta f_i = \kappa_c f_c / N_i$. Thus, with (5.1), we obtain the following expression for the discrete Doppler frequencies $f_{i,n}$

$$f_{i,n} = \frac{\kappa_c f_c}{2N_i}(2n-1) \tag{5.12}$$

for all $n = 1, 2, \ldots, N_i$ and $i = 1, 2$. Now, using (3.11), (5.3), and (5.5), this enables the computation of the Doppler coefficients $c_{i,n}$. As a result, we find the formula

$$c_{i,n} = \sigma_0 \sqrt{2} \left[\operatorname{erf}\left(\frac{n\kappa_c \sqrt{\ln 2}}{N_i} \right) - \operatorname{erf}\left(\frac{(n-1)\kappa_c \sqrt{\ln 2}}{N_i} \right) \right]^{\frac{1}{2}} \tag{5.13}$$

for all $n = 1, 2, \ldots, N_i$ and $i = 1, 2$. Deterministic processes $\tilde{\mu}_i(t)$ designed with (5.12) and (5.13) have a mean value of zero and a mean power of

$$\begin{aligned} \tilde{\sigma}_{\mu_i}^2 &= \tilde{r}_{\mu_i \mu_i}(0) = \sum_{n=1}^{N_i} \frac{c_{i,n}^2}{2} \\ &= \sigma_0^2 \operatorname{erf}\left(\kappa_c \sqrt{\ln 2}\right) \\ &= 0.9999366 \cdot \sigma_0^2 \approx \sigma_0^2 \,, \end{aligned} \tag{5.14}$$

provided that κ_c is chosen as suggested, i.e., $\kappa_c = 2\sqrt{2/\ln 2}$. In the present case, the period of $\tilde{\mu}_i(t)$ is given by $T_i = 2/\Delta f_i = 2N_i/(\kappa_c f_c)$.

Figure 5.4(a) shows the power spectral density $\tilde{S}_{\mu_i \mu_i}(f)$ for $N_i = 25$ and Figure 5.4(b) illustrates the corresponding behaviour of the autocorrelation function $\tilde{r}_{\mu_i \mu_i}(\tau)$ in comparison with the autocorrelation function $r_{\mu_i \mu_i}(\tau)$ of the reference model in the range $0 \le \tau \le T_i/2$.

A suitable value for the upper limit of the integral (4.84) is also in this case a quarter of the period T_i, i.e., $\tau_{max} = T_i/4 = N_i/(2\kappa_c f_c)$. If the mean-square error $E_{r_{\mu_i \mu_i}}$ [see (4.84)] is evaluated with respect to the upper limit τ_{max} prescribed in this way, then we obtain the graph presented in Figure 5.5(a) showing the influence of the number of harmonic functions N_i. Figure 5.5(b) presents the results of the evaluation of the performance criterion $E_{p_{\mu_i}}$ according to (4.83). For comparison, the results obtained by using $c_{i,n} = \sigma_0 \sqrt{2/N_i}$ are shown in this figure as well.

Finally, we will also analyse the model error $\Delta\beta_i$. Using (4.22), (5.12), (5.13), and (3.29), we find the following closed-form solution for $\Delta\beta_i = \tilde{\beta}_i - \beta$

$$\Delta\beta_i = \beta \left\{ 2 \ln 2 \kappa_c^2 \left[\left(1 - \frac{1}{2N_i}\right)^2 \operatorname{erf}\left(\kappa_c \sqrt{\ln 2}\right) - \frac{2}{N_i^2} \sum_{n=1}^{N_i - 1} n \operatorname{erf}\left(\frac{n\kappa_c \sqrt{\ln 2}}{N_i} \right) \right] - 1 \right\} . \tag{5.15}$$

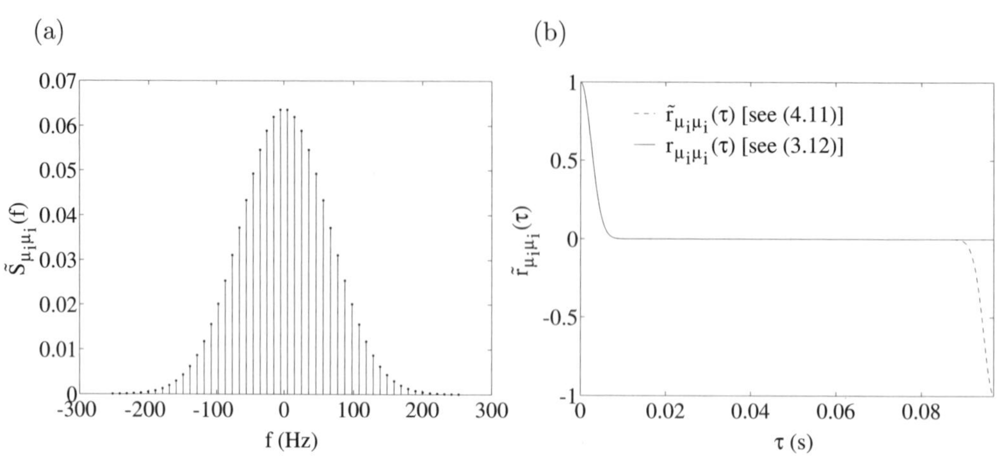

Figure 5.4: (a) Power spectral density $\tilde{S}_{\mu_i\mu_i}(f)$ and (b) autocorrelation function $\tilde{r}_{\mu_i\mu_i}(\tau)$ for $N_i = 25$ (MED, Gaussian PSD, $f_c = \sqrt{\ln 2} f_{max}$, $f_{max} = 91\,\text{Hz}$, $\sigma_0^2 = 1$, $\kappa_c = 2\sqrt{2/\ln 2}$).

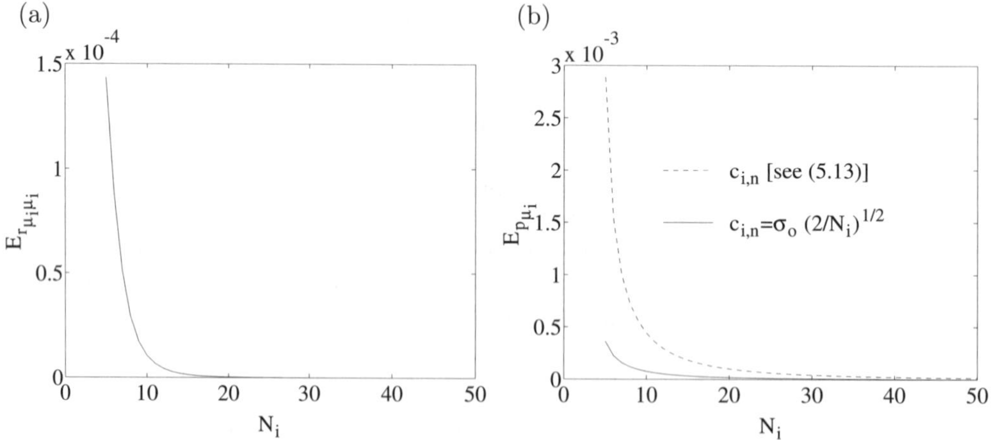

Figure 5.5: Mean-square errors: (a) $E_{r_{\mu_i\mu_i}}$ and (b) $E_{p_{\mu_i}}$ (MED, Gaussian PSD, $f_c = \sqrt{\ln 2}\, f_{max}$, $f_{max} = 91\,\text{Hz}$, $\sigma_0^2 = 1$, $\tau_{max} = N_i/(2\kappa_c f_c)$, $\kappa_c = 2\sqrt{2/\ln 2}$).

Let us choose $\kappa_c = 2\sqrt{2/\ln 2}$ again. From the above equation, it then follows the following expression for the relative model error $\Delta\beta_i/\beta$

$$\frac{\Delta\beta_i}{\beta} = 16\left[\left(1 - \frac{1}{2N_i}\right)^2 \operatorname{erf}\left(2\sqrt{2}\right) - \frac{2}{N_i^2}\sum_{n=1}^{N_i-1} n \cdot \operatorname{erf}\left(\frac{n2\sqrt{2}}{N_i}\right)\right] - 1\,, \quad (5.16)$$

whose behaviour is depicted in Figure 5.6 as a function of N_i. In addition to the rather small values for $\Delta\beta_i/\beta$, the fast convergence behaviour is to be assessed positively. When considering the limit $N_i \to \infty$, it turns out that the model error $\Delta\beta_i$ is very small but still larger than 0, because, due to the finite value for κ_c, the frequency range of the Gaussian power spectral density (3.11) is not covered completely by the discrete Doppler frequencies.

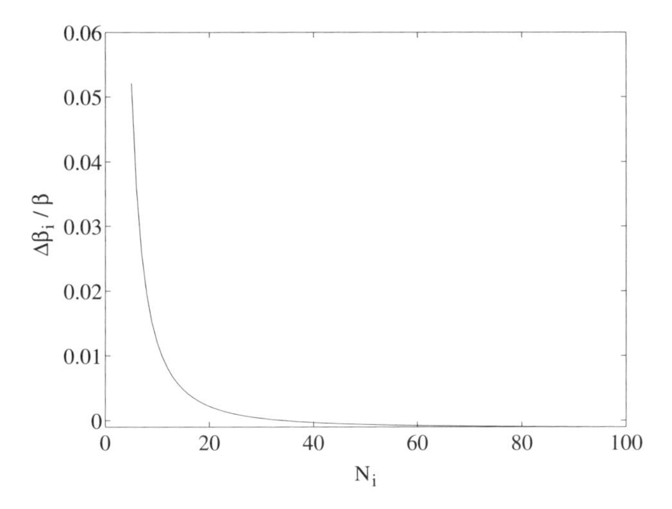

Figure 5.6: Relative model error $\Delta\beta_i/\beta$ (MED, Gaussian PSD, $\kappa_c = 2\sqrt{2/\ln 2}$).

In order to avoid correlations between $\tilde{\mu}_1(t)$ and $\tilde{\mu}_2(t)$, N_2 is again defined by $N_2 := N_1 + 1$. Hence, $\Delta\beta = \Delta\beta_1 \approx \Delta\beta_2$ holds, and we can easily analyse the characteristic quantities $\tilde{N}_\xi(r)$, $\tilde{T}_{\xi_-}(r)$, and $\tilde{\tau}_q(r)$ of deterministic Rice processes $\tilde{\xi}(t)$ by making use of (4.66), (4.70), and (4.78), respectively. Concerning the simulation of $\tilde{\xi}(t)$, it must be taken into account that the simulation time T_{sim} does not exceed the period T_i, i.e., $T_{sim} \leq T_i = 2N_i/f_{max}$ (Jakes PSD). As an example, we consider $N_i = 25$ and $f_{max} = 91\,\text{Hz}$ (v $= 110\,\text{km/h}$, $f_0 = 900\,\text{MHz}$). This results in a maximum simulation time of $T_{sim} = 0.549\,\text{s}$. Within this time, the vehicle covers a distance of $16.775\,\text{m}$, so that the model of the underlying mobile radio channel can be regarded as wide-sense stationary.[1] Nevertheless, this simulation time is not by a long way sufficient to determine typical characteristic quantities such as $\tilde{N}_\xi(r)$, $\tilde{T}_{\xi_-}(r)$, and $\tilde{\tau}_q(r)$ with acceptable precision. A more exact measurement of these quantities for the same parameter sets $\{f_{i,n}\}$ and $\{c_{i,n}\}$ can be achieved by ensemble averaging (statistical averaging). Therefore, various realizations of $\tilde{\xi}(t)$ are required, which can be generated by means of various sets for the Doppler phases $\{\theta_{i,n}\}$. Due to the relatively small period T_i, which only increases linearly with N_i, the method of equal distances is not recommendable for long-time simulations. For this reason, the properties of this method will not be investigated here in any detail. Further results of this approach can be found in [Pae96d].

5.1.2 Mean-Square-Error Method (MSEM)

The mean-square-error method (MSEM) is based on the idea that the model parameter sets $\{c_{i,n}\}$ and $\{f_{i,n}\}$ are computed in such a way that the mean-square error (4.84)

$$E_{r_{\mu_i\mu_i}} = \frac{1}{\tau_{\max}} \int_0^{\tau_{max}} (r_{\mu_i\mu_i}(\tau) - \tilde{r}_{\mu_i\mu_i}(\tau))^2 \, d\tau \tag{5.17}$$

becomes minimal [Pae96d]. Here, $r_{\mu_i\mu_i}(\tau)$ can be any autocorrelation function of the process $\mu_i(t)$ describing a theoretical reference model. Alternatively, $r_{\mu_i\mu_i}(\tau)$ can also be obtained from the measurement data of a real-world channel. The autocorrelation function $\tilde{r}_{\mu_i\mu_i}(\tau)$ of the deterministic model is again given by (4.11). In the equation given above, τ_{max} describes an appropriate time interval over which the approximation of the autocorrelation function $r_{\mu_i\mu_i}(\tau)$ is of interest. Unfortunately, a simple and closed-form solution for this problem only exists, if the discrete Doppler frequencies $f_{i,n}$ are again defined by (5.1) and, consequently, they are equidistant.

After substituting (4.11) into (5.17) and setting the partial derivatives of $E_{r_{\mu_i\mu_i}}$ with respect to the Doppler coefficients $c_{i,n}$ equal to zero, i.e., $\partial E_{r_{\mu_i\mu_i}}/\partial c_{i,n} = 0$, we obtain, in connection with (5.1), the following formula for $c_{i,n}$ [Pae96d]:

$$c_{i,n} = 2\sqrt{\frac{1}{\tau_{max}} \int_0^{\tau_{max}} r_{\mu_i\mu_i}(\tau) \cos(2\pi f_{i,n}\tau) \, d\tau} \tag{5.18}$$

for all $n = 1, 2, \ldots, N_i$ $(i = 1, 2)$, where τ_{max} shall again be given by $\tau_{max} = T_i/4 = 1/(2\Delta f_i)$.

In case of the limit $\Delta f_i \to 0$, one can show that from (5.18) the expression

$$c_{i,n} = \lim_{\Delta f_i \to 0} 2\sqrt{\Delta f_i S_{\mu_i\mu_i}(f_{i,n})} \tag{5.19}$$

follows, which is identical to the relation (4.2a) given by Rice [Ric44, Ric45]. Numerical analysis have shown that for $\Delta f_i > 0$, the formula

$$c_{i,n} = 2\sqrt{\Delta f_i S_{\mu_i\mu_i}(f_{i,n})}\,, \tag{5.20}$$

[1] Measurements have shown [Cox73] that in urban areas mobile radio channels can appropriately be modelled for signal bandwidths up to 10 MHz and covered distances up to 30 m by so-called GWSSUS channels ("Gaussian wide-sense stationary uncorrelated scattering").

which can easily be evaluated, even then shows a quite usable approximation of the exact solution (5.18), if the number of used harmonic functions N_i is moderate.

We also want to show that $\tilde{r}_{\mu_i\mu_i}(\tau) \to r_{\mu_i\mu_i}(\tau)$ follows as $N_i \to \infty$ ($\Delta f_i \to 0$). Putting (5.1) and (5.18) into (4.11) and taking $\tau_{max} = 1/(2\Delta f_i)$ into account, we may write

$$\begin{aligned}
\lim_{N_i\to\infty} \tilde{r}_{\mu_i\mu_i}(\tau) &= \lim_{N_i\to\infty} \sum_{n=1}^{N_i} \frac{c_{i,n}^2}{2} \cos(2\pi f_{i,n}\tau) \\
&= \lim_{N_i\to\infty} 4 \sum_{n=1}^{N_i} \int_0^{\frac{1}{2\Delta f_i}} r_{\mu_i\mu_i}(\tau') \cos(2\pi f_{i,n}\tau') \cos(2\pi f_{i,n}\tau)\, d\tau' \Delta f_i \\
&= 4 \int_0^{\infty} \int_0^{\infty} r_{\mu_i\mu_i}(\tau') \cos(2\pi f\tau') \cos(2\pi f\tau)\, d\tau'\, df \\
&= 2 \int_0^{\infty} S_{\mu_i\mu_i}(f) \cos(2\pi f\tau)\, df \\
&= r_{\mu_i\mu_i}(\tau)\,.
\end{aligned} \tag{5.21}$$

Next, we will study the application of the mean-square-error method (MSEM) on the Jakes and the Gaussian power spectral densities.

Jakes power spectral density: When using the MSEM, the formula for the computation of the discrete Doppler frequencies $f_{i,n}$ is identical to the relation (5.7), which has been obtained by applying the MED. For the corresponding Doppler coefficients $c_{i,n}$, however, we obtain quite different expressions. After substituting (3.10) into (5.18), we find

$$c_{i,n} = 2\sigma_0 \sqrt{\frac{1}{\tau_{max}} \int_0^{\tau_{max}} J_0(2\pi f_{max}\tau) \cos(2\pi f_{i,n}\tau)\, d\tau}\,, \tag{5.22}$$

where $\tau_{max} = 1/(2\Delta f_i) = N_i/(2f_{max})$. There is no closed-form solution for the definite integral appearing in (5.22), so that in this case a numerical integration technique has to be applied in order to calculate the Doppler coefficients $c_{i,n}$.

As an example, we consider Figure 5.7, where the power spectral density $\tilde{S}_{\mu_i\mu_i}(f)$ and the corresponding autocorrelation function $\tilde{r}_{\mu_i\mu_i}(\tau)$ for $N_i = 25$ are depicted. For reasons of comparison, the autocorrelation function $r_{\mu_i\mu_i}(\tau)$ of the reference model [see (3.10)] is also shown in Figure 5.7(b). The unwanted periodical behaviour of $\tilde{r}_{\mu_i\mu_i}(\tau)$, as a consequence of the equidistant discrete Doppler frequencies, is again clearly visible.

The evaluation of the performance criteria $E_{r_{\mu_i\mu_i}}$ and $E_{p_{\mu_i}}$ [see (4.84) and (4.83), respectively] has been performed for the MSEM. The obtained results, pointing out the influence of the number of harmonic functions N_i, are shown in Figures 5.8(a) and 5.8(b). For a better classification of the performance of the MSEM, the results found before by applying the MED as well as the results obtained by using the approximate solution (5.20) are likewise included in these figures.

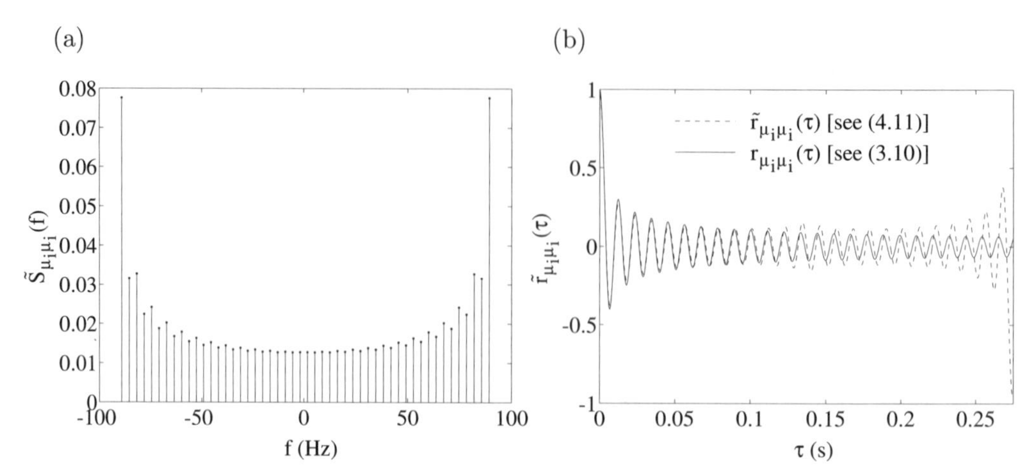

Figure 5.7: (a) Power spectral density $\tilde{S}_{\mu_i\mu_i}(f)$ and (b) autocorrelation function $\tilde{r}_{\mu_i\mu_i}(\tau)$ for $N_i = 25$ (MSEM, Jakes PSD, $f_{max} = 91\,\text{Hz}$, $\sigma_0^2 = 1$).

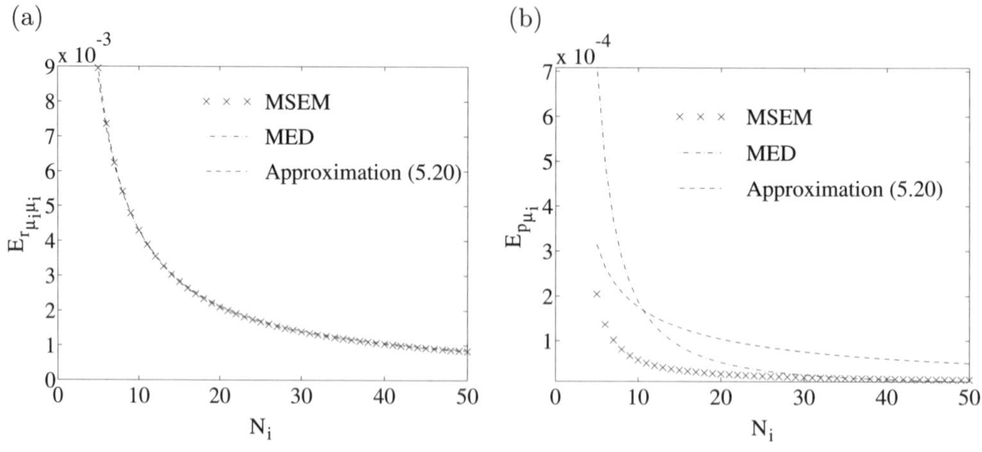

Figure 5.8: Mean-square errors: (a) $E_{r_{\mu_i\mu_i}}$ and (b) $E_{p_{\mu_i}}$ (MSEM, Jakes PSD, $f_{max} = 91\,\text{Hz}$, $\sigma_0^2 = 1$, $\tau_{max} = N_i/(2f_{max})$).

In case of the MSEM, a simple solution for the model error $\Delta\beta_i$ does not exist. By means of (5.7), (5.22), (3.29), and (4.22), the following formula for $\tilde{\beta}_i$ is obtained after a short computation

$$\tilde{\beta}_i = \beta \frac{1}{N_i} \sum_{n=1}^{N_i} (2n-1)^2 \int_0^1 J_0(\pi N_i u) \cos\left[\frac{\pi}{2}(2n-1)u\right] du \,. \tag{5.23}$$

With this expression and by making use of $\beta = 2(\pi f_{max}\sigma_0)^2$, the model error $\Delta\beta_i = \tilde{\beta}_i - \beta$ can be calculated. Figure 5.9 depicts the resulting relative model error $\Delta\beta_i/\beta$ in terms of N_i. This figure also shows the results which can be found when the approximate solution (5.20) is used. For reasons of comparison, the graph of $\Delta\beta_i/\beta$ obtained by applying the MED is also presented here once again.

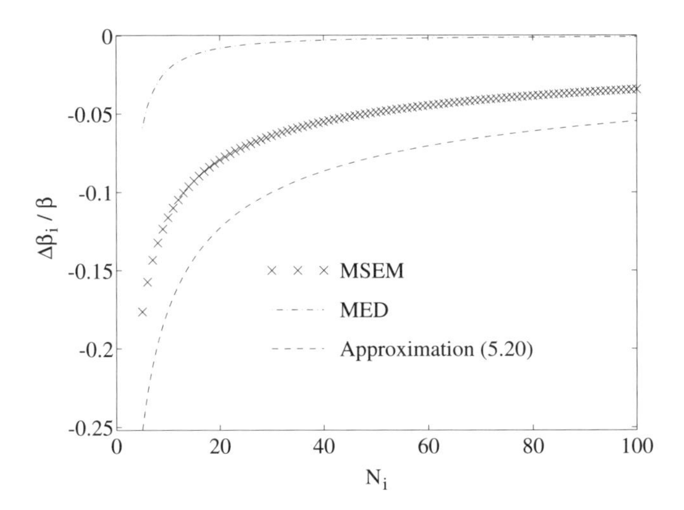

Figure 5.9: Relative model error $\Delta\beta_i/\beta$ (MSEM, Jakes PSD).

Gaussian power spectral density: The discrete Doppler frequencies $f_{i,n}$ are given by (5.12). For the Doppler coefficients $c_{i,n}$, we now obtain, after substituting (3.12) into (5.18), the expression

$$c_{i,n} = 2\sigma_0 \sqrt{\frac{1}{\tau_{max}} \int_0^{\tau_{max}} e^{-(\pi f_c \tau)^2/\ln 2} \cos(2\pi f_{i,n}\tau)\, d\tau} \tag{5.24}$$

for all $n = 1, 2, \ldots, N_i$ $(i = 1, 2)$, where $\tau_{max} = 1/(2\Delta f_i) = N_i/(2\kappa_c f_c)$. Let the quantity κ_c again be defined by $\kappa_c = 2\sqrt{2/\ln 2}$, so that the period T_i is given by $T_i = N_i/(\sqrt{2/\ln 2} f_c)$. The definite integral under the square root of (5.24) has to be solved numerically.

As an example, the power spectral density $\tilde{S}_{\mu_i\mu_i}(f)$ for $N_i = 25$ is shown in Figure 5.10(a). Figure 5.10(b) presents the corresponding autocorrelation function

$\tilde{r}_{\mu_i\mu_i}(\tau)$ in comparison with the autocorrelation function $r_{\mu_i\mu_i}(\tau)$ of the reference model in the range $0 \leq \tau \leq T_i/2$.

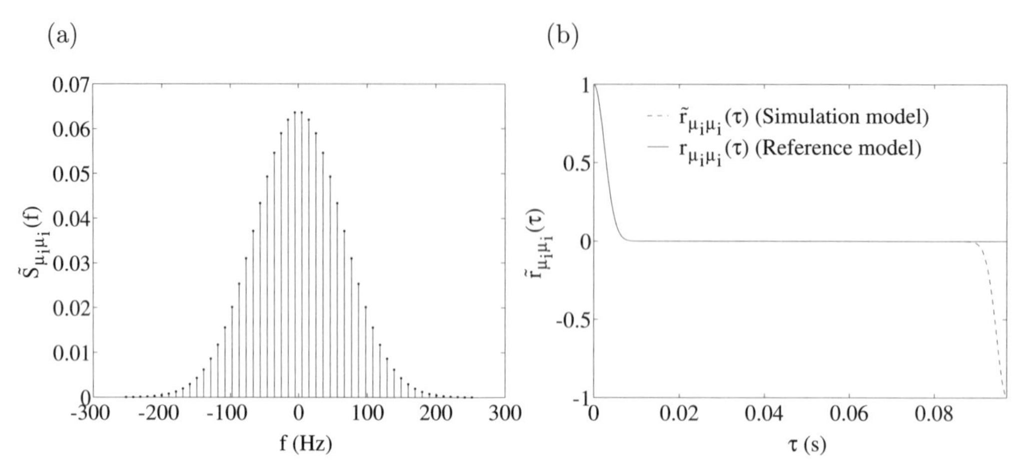

Figure 5.10: (a) Power spectral density $\tilde{S}_{\mu_i\mu_i}(f)$ and (b) autocorrelation function $\tilde{r}_{\mu_i\mu_i}(\tau)$ for $N_i = 25$ (MSEM, Gaussian PSD, $f_c = \sqrt{\ln 2} f_{max}$, $f_{max} = 91\,\text{Hz}$, $\sigma_0^2 = 1$, $\kappa_c = 2\sqrt{2/\ln 2}$).

The mean-square errors $E_{r_{\mu_i\mu_i}}$ and $E_{p_{\mu_i}}$ [see (4.84) and (4.83)], occurring when the MSEM is applied, are depicted in the Figures 5.11(a) and 5.11(b), respectively. For comparison, the results found for the MED before and the results by using the approximation (5.20) are also shown in these figures.

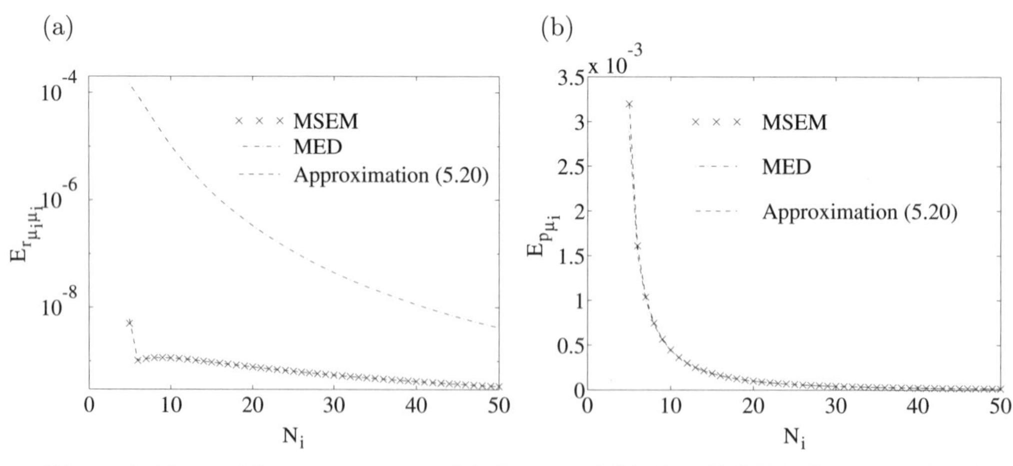

Figure 5.11: Mean-square errors: (a) $E_{r_{\mu_i\mu_i}}$ and (b) $E_{p_{\mu_i}}$ (MSEM, Gaussian PSD, $f_c = \sqrt{\ln 2} f_{max}$, $f_{max} = 91\,\text{Hz}$, $\sigma_0^2 = 1$, $\tau_{max} = N_i/(2\kappa_c f_c)$, $\kappa_c = 2\sqrt{2/\ln 2}$).

We briefly turn to the model error $\Delta\beta_i$. Putting (5.12) and (5.24) into the formula for $\tilde{\beta}_i$ [see (4.22)] and making use of (3.29), the expression

$$\tilde{\beta}_i = \beta \frac{\kappa_c^2 \ln 2}{N_i^2} \sum_{n=1}^{N_i} (2n-1)^2 \int_0^1 e^{-\left(\frac{\pi N_i}{2\kappa_c \sqrt{\ln 2}} u\right)^2} \cos\left[\frac{\pi}{2}(2n-1)u\right] du \tag{5.25}$$

follows, making the computation of the model error $\Delta\beta_i = \tilde{\beta}_i - \beta$ possible. Figure 5.12 displays the resulting relative model error $\Delta\beta_i/\beta$ as a function of N_i. This figure also presents the results which can be found by using the approximate solution (5.20) derived for the Doppler coefficients $c_{i,n}$. For comparison, this figure also shows the graph previously obtained for the MED.

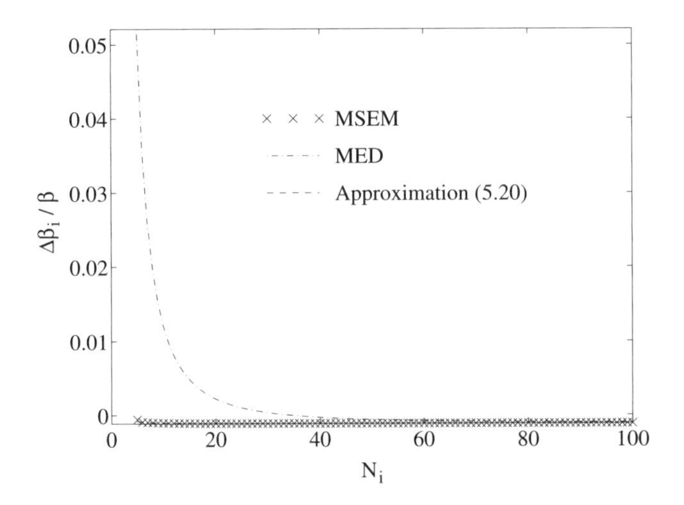

Figure 5.12: Relative model error $\Delta\beta_i/\beta$ (MSEM, Gaussian PSD, $\kappa_c = 2\sqrt{2/\ln 2}$).

5.1.3 Method of Equal Areas (MEA)

The method of equal areas (MEA) [Pae94b] is characterized by the fact that the discrete Doppler frequencies $f_{i,n}$ are determined in such a way that the area under the Doppler power spectral density $S_{\mu_i\mu_i}(f)$ is equal to $\sigma_0^2/(2N_i)$ within the frequency range $f_{i,n-1} < f \leq f_{i,n}$, i.e.,

$$\int_{f_{i,n-1}}^{f_{i,n}} S_{\mu_i\mu_i}(f)\, df = \frac{\sigma_0^2}{2N_i} \tag{5.26}$$

for all $n = 1, 2, \ldots, N_i$ and $i = 1, 2$, where $f_{i,0} := 0$. For an explicit computation of the discrete Doppler frequencies $f_{i,n}$, the introduction of the auxiliary function

$$G_{\mu_i}(f_{i,n}) := \int_{-\infty}^{f_{i,n}} S_{\mu_i\mu_i}(f)\, df \tag{5.27}$$

turns out to be helpful. In case of symmetrical Doppler power spectral densities, i.e., $S_{\mu_i\mu_i}(f) = S_{\mu_i\mu_i}(-f)$, and by using (5.26), we may express $G_{\mu_i}(f_{i,n})$ in the form

$$\begin{aligned} G_{\mu_i}(f_{i,n}) &= \frac{\sigma_0^2}{2} + \sum_{\nu=1}^{n} \int_{f_{i,\nu-1}}^{f_{i,\nu}} S_{\mu_i\mu_i}(f)\, df \\ &= \frac{\sigma_0^2}{2}\left(1 + \frac{n}{N_i}\right) . \end{aligned} \tag{5.28}$$

If the inverse function of G_{μ_i}, denoted by $G_{\mu_i}^{-1}$, exists, then the discrete Doppler frequencies $f_{i,n}$ are given by

$$f_{i,n} = G_{\mu_i}^{-1}\left[\frac{\sigma_0^2}{2}\left(1 + \frac{n}{N_i}\right)\right] \tag{5.29}$$

for all $n = 1, 2, \ldots, N_i$ and $i = 1, 2$.

The Doppler coefficients $c_{i,n}$ are now determined by imposing on both the reference model and the simulation model that within the frequency interval $I_{i,n} := (f_{i,n-1}, f_{i,n}]$, the mean power of the stochastic process $\mu_i(t)$ is identical to that of the deterministic process $\tilde{\mu}_i(t)$, i.e.,

$$\int_{f \in I_{i,n}} S_{\mu_i\mu_i}(f)\, df = \int_{f \in I_{i,n}} \tilde{S}_{\mu_i\mu_i}(f)\, df . \tag{5.30}$$

From the equation above and by using the relations (4.14) and (5.26), it now follows the following simple formula for the Doppler coefficients

$$c_{i,n} = \sigma_0 \sqrt{\frac{2}{N_i}} , \tag{5.31}$$

where $n = 1, 2, \ldots, N_i$ and $i = 1, 2$. Just as with the previous methods, we will also apply this procedure to the Jakes and Gaussian power spectral density.

Jakes power spectral density: With the Jakes power spectral density (3.8), we obtain the following expression for (5.27)

$$G_{\mu_i}(f_{i,n}) = \frac{\sigma_0^2}{2}\left[1 + \frac{2}{\pi} \arcsin\left(\frac{f_{i,n}}{f_{max}}\right)\right] , \tag{5.32}$$

where $0 < f_{i,n} \leq f_{max}$, $\forall\, n = 1, 2, \ldots, N_i$ and $i = 1, 2$. If we set up a relation between the right-hand side of (5.32) and (5.28), then the discrete Doppler frequencies $f_{i,n}$ can be computed explicitly. As a result, we find the equation

$$f_{i,n} = f_{max} \sin\left(\frac{\pi n}{2N_i}\right) , \tag{5.33}$$

which is valid for all $n = 1, 2, \ldots, N_i$ and $i = 1, 2$. The corresponding Doppler coefficients $c_{i,n}$ are furthermore given by (5.31). Theoretically, for all relevant values of N_i, say $N_i \geq 5$, the greatest common divisor $F_i := \gcd\{f_{i,n}\}_{n=1}^{N_i}$ is equal to zero, and, thus, the period $T_i = 1/F_i$ is infinite. Hence, in this idealized case, the deterministic process $\tilde{\mu}_i(t)$ is nonperiodic. In practical cases of application, however, the discrete Doppler frequencies $f_{i,n}$ can only be calculated with a finite precision. Let us assume that the discrete Doppler frequencies $f_{i,n}$, according to (5.33), are representable up to the lth decimal place after the comma, then the greatest common divisor is equal to $F_i = \gcd\{f_{i,n}\}_{n=1}^{N_i} = 10^{-l}\ \mathrm{s}^{-1}$. Consequently, the period T_i of the deterministic process $\tilde{\mu}_i(t)$ is $T_i = 1/F_i = 10^l$ s, so that $\tilde{\mu}_i(t)$ can be considered as *quasi-nonperiodic* if $l \geq 10$.

Deterministic processes $\tilde{\mu}_i(t)$ designed with (5.31) and (5.33) are characterized by the mean value $\tilde{m}_{\mu_i} = 0$ and the mean power

$$\tilde{\sigma}^2_{\mu_i} = \tilde{r}_{\mu_i\mu_i}(0) = \sum_{n=1}^{N_i} \frac{c^2_{i,n}}{2} = \sigma_0^2 \,. \tag{5.34}$$

When designing the complex deterministic processes $\tilde{\mu}(t) = \tilde{\mu}_1(t) + j\tilde{\mu}_2(t)$, the demand for uncorrelatedness of the real part and the imaginary part can be fulfilled sufficiently, if the number of harmonic functions N_2 is defined by $N_2 := N_1 + 1$. However, the fact that $f_{1,N_1} = f_{2,N_2} = f_{max}$ always holds for any chosen values of N_1 and N_2 has the consequence that $\tilde{\mu}_1(t)$ and $\tilde{\mu}_2(t)$ are not completely uncorrelated. But even for moderate values of N_i, the resulting correlation is very small, so that this effect will be ignored in order to simplify matters.

Let us choose $N_i = 25$, for example, then we obtain the results shown in Figures 5.13(a) for the power spectral density $\tilde{S}_{\mu_i\mu_i}(f)$. The corresponding autocorrelation function $\tilde{r}_{\mu_i\mu_i}(\tau)$ is presented in 5.13(b).

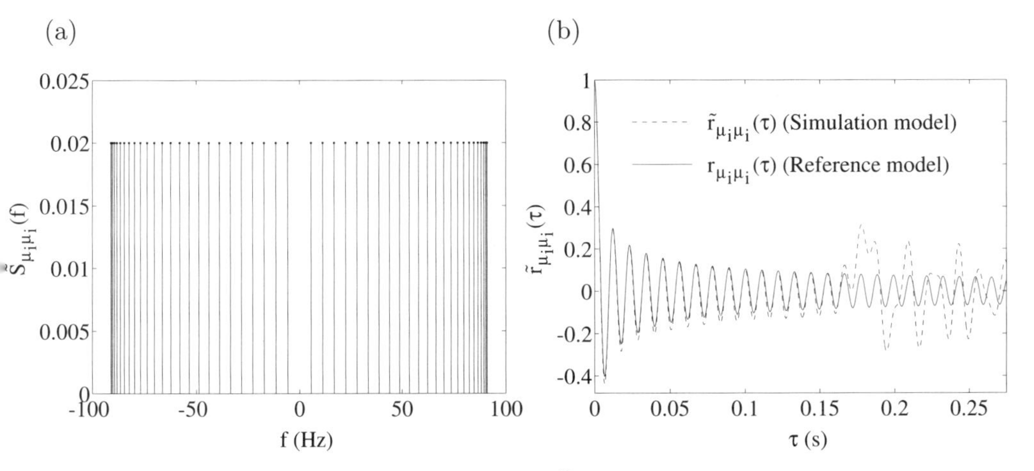

Figure 5.13: (a) Power spectral density $\tilde{S}_{\mu_i\mu_i}(f)$ and (b) autocorrelation function $\tilde{r}_{\mu_i\mu_i}(\tau)$ for $N_i = 25$ (MEA, Jakes PSD, $f_{max} = 91\,\mathrm{Hz}$, $\sigma_0^2 = 1$).

Without any difficulty, it can be proved that $\tilde{r}_{\mu_i\mu_i}(\tau) \to r_{\mu_i\mu_i}(\tau)$ holds as $N_i \to \infty$. To prove this property, we substitute (5.31) and (5.33) into (4.11), so that we may write

$$\begin{aligned}
\lim_{N_i\to\infty} \tilde{r}_{\mu_i\mu_i}(\tau) &= \lim_{N_i\to\infty} \sum_{n=1}^{N_i} \frac{c_{i,n}^2}{2} \cos(2\pi f_{i,n}\tau) \\
&= \lim_{N_i\to\infty} \sigma_0^2 \frac{1}{N_i} \sum_{n=1}^{N_i} \cos\left[2\pi f_{max}\tau \sin\left(\frac{\pi n}{2N_i}\right)\right] \\
&= \sigma_0^2 \frac{2}{\pi} \int_0^{\pi/2} \cos(2\pi f_{max}\tau \sin\alpha)\, d\alpha \\
&= \sigma_0^2 J_0(2\pi f_{max}\tau) \\
&= r_{\mu_i\mu_i}(\tau)\,.
\end{aligned} \tag{5.35}$$

In Subsection 4.3.1, we have furthermore proved that for $c_{i,n} = \sigma_0\sqrt{2/N_i}$ it holds: $\tilde{p}_{\mu_i}(x) \to p_{\mu_i}(x)$ as $N_i \to \infty$. Consequently, for an infinite number of harmonic functions, the deterministic Gaussian process $\tilde{\mu}_i(t)$ represents a sample function of the stochastic Gaussian random process $\mu_i(t)$. Note that the same relation also exists between the deterministic Rice process $\tilde{\xi}(t)$ and the stochastic Rice process $\xi(t)$.

A deeper insight into the performance of the MEA can again be gained by evaluating the performance criteria (4.83) and (4.84). Both of the resulting mean-square errors $E_{r_{\mu_i\mu_i}}$ and $E_{p_{\mu_i}}$ are shown in Figures 5.14(a) and 5.14(b), respectively.

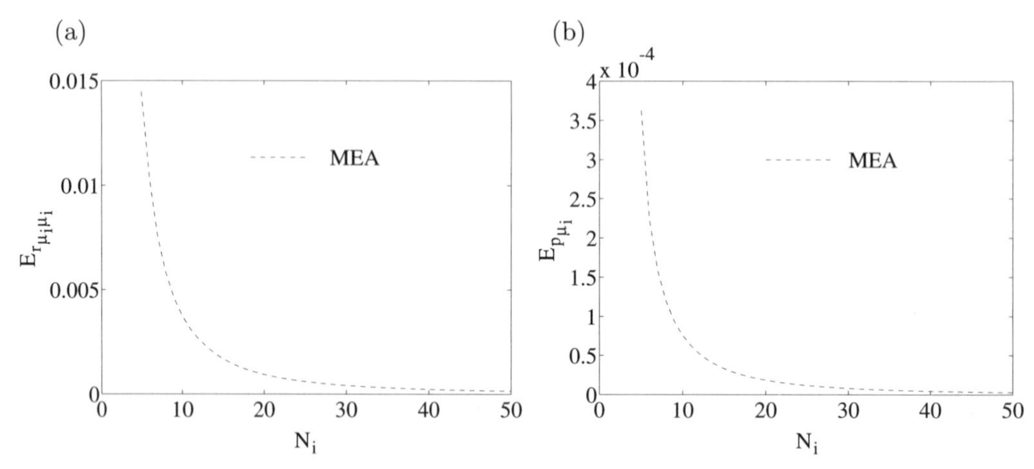

Figure 5.14: Mean-square errors: (a) $E_{r_{\mu_i\mu_i}}$ and (b) $E_{p_{\mu_i}}$ (MEA, Jakes PSD, $f_{max} =$ 91 Hz, $\sigma_0^2 = 1$, $\tau_{max} = N_i/(2f_{max})$).

Now, let us analyse the model error $\Delta\beta_i$ as well. With (5.31), (5.33), and (3.29), we at first find the following expression for $\tilde{\beta}_i$ [see (4.22)]

$$\tilde{\beta}_i = \beta \frac{2}{N_i} \sum_{n=1}^{N_i} \sin^2\left(\frac{\pi n}{2N_i}\right)$$

$$= \beta \left(1 + \frac{1}{N_i}\right) . \tag{5.36}$$

Since $\tilde{\beta}_i$ was introduced as $\tilde{\beta}_i = \beta + \Delta\beta_i$, we thus obtain a simple closed-form formula for the model error

$$\Delta\beta_i = \beta / N_i \, . \tag{5.37}$$

One may note that $\Delta\beta_i \to 0$ as $N_i \to \infty$. The convergence characteristic of the relative model error $\Delta\beta_i/\beta$ can be seen in Figure 5.15.

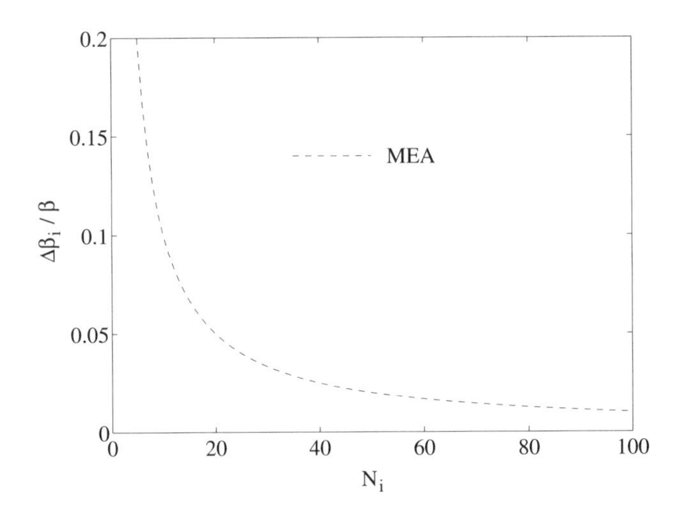

Figure 5.15: Relative model error $\Delta\beta_i/\beta$ (MEA, Jakes PSD).

We furthermore look at the relative error of the level-crossing rate $\tilde{N}_\xi(r)$, to which we want to refer as ϵ_{N_ξ} in the following, i.e.,

$$\epsilon_{N_\xi} = \frac{N_\xi(r) - \tilde{N}_\xi(r)}{N_\xi(r)} \, . \tag{5.38}$$

By using (4.66) and (5.37), we can approximate the relative error ϵ_{N_ξ} in the present case as follows

$$\epsilon_{N_\xi} \approx -\frac{\Delta\beta}{2\beta} \approx -\frac{\Delta\beta_i}{2\beta} = -\frac{1}{2N_i} \, . \tag{5.39}$$

This result makes clear that for a finite number of N_i, the level-crossing rate of the simulation model, which was designed with the MEA, is always greater than the level-crossing rate of the reference model. Obviously, $\epsilon_{N_\xi} \to 0$ holds as $N_i \to \infty$.

Analogously, for the relative error of the average duration of fades $\tilde{T}_{\xi_-}(r)$ one finds the approximate solution

$$\epsilon_{T_{\xi_-}} \approx \frac{\Delta\beta}{2\beta} \approx \frac{\Delta\beta_i}{2\beta} = \frac{1}{2N_i}. \tag{5.40}$$

The quasi-nonperiodic property of $\tilde{\mu}_i(t)$ now allows us to determine both the level-crossing rate and the average duration of fades of deterministic Rice processes by means of simulation. For this purpose, the parameters of the simulation model $\{c_{i,n}\}$ and $\{f_{i,n}\}$ were determined by applying the method of equal areas with $(N_1, N_2) = (10, 11)$. For the computation of the Doppler phases $\{\theta_{i,n}\}$, everything that we said at the beginning of this chapter also holds here. Just as in the previous examples, the Jakes power spectral density (3.8) was again characterized by $f_{max} = 91\,\text{Hz}$ and $\sigma_0^2 = 1$. For the sampling interval T_s of the deterministic Rice process $\tilde{\xi}(kT_s)$, the value $T_s = 10^{-4}$ s was chosen. The simulation time T_{sim} was determined for each individual signal level r in such a way that always 10^6 fading intervals or downwards (upwards) level crossings could be evaluated. The results found under these conditions are presented in Figures 5.16(a) and 5.16(b).

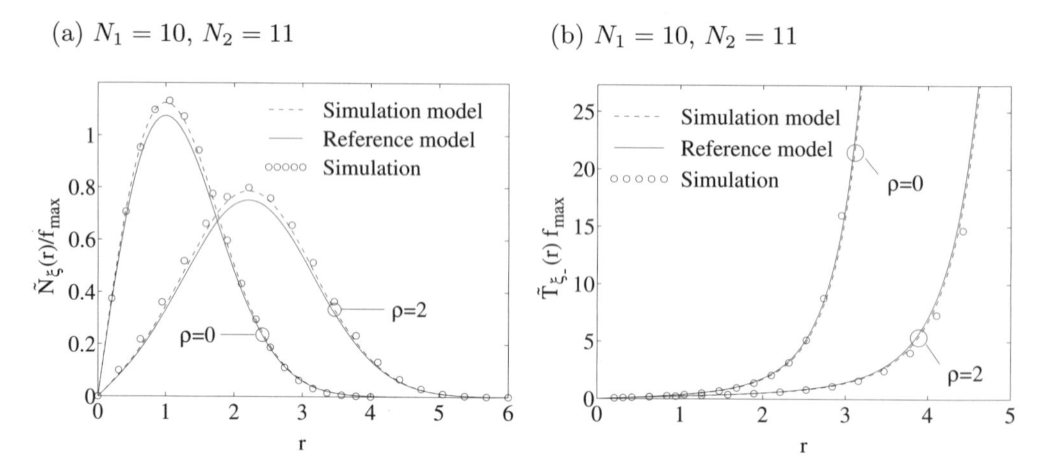

Figure 5.16: (a) Normalized level-crossing rate $\tilde{N}_\xi(r)/f_{max}$ and (b) normalized average duration of fades $\tilde{T}_{\xi_-}(r) \cdot f_{max}$ (MEA, Jakes PSD, $f_{max} = 91\,\text{Hz}$, $\sigma_0^2 = 1$).

These figures also show the analytical solutions previously found for the reference model and the simulation model. The quantities $\tilde{N}_\xi(r)$ and $\tilde{T}_{\xi_-}(r)$ were computed by using $\tilde{\beta} = \tilde{\beta}_1 = \beta(1 + 1/N_1)$ and by means of (4.64) and (4.69), respectively. These figures also demonstrate the excellent correspondence between the analytical expressions derived for the simulation model and the corresponding quantities determined from the measurement results of the simulated amplitude behaviour. Unfortunately, the statistical deviations between the reference model and the simulation model are comparatively high, which gives us a reason to search for a better parameter computation technique. For example, under the conditions given here, where $N_1 = 10$ and $N_2 = 11$, the percentage of the relative error of the level-crossing rate $\tilde{N}_\xi(r)$ and the average duration of fades $\tilde{T}_{\xi_-}(r)$ is about $\epsilon_{N_\xi} \approx -5$ per cent and $\epsilon_{T_{\xi_-}} \approx +5$ per cent, respectively.

Gaussian power spectral density: With the Gaussian power spectral density (3.11), we obtain the following expression for (5.27)

$$G_{\mu_i}(f_{i,n}) = \frac{\sigma_0^2}{2}\left[1 + \text{erf}\left(\frac{f_{i,n}}{f_c}\sqrt{\ln 2}\right)\right] \tag{5.41}$$

for all $n = 1, 2, \ldots, N_i$ and $i = 1, 2$. Since the inverse function $\text{erf}^{-1}(\cdot)$ of the Gaussian error function does not exist, the discrete Doppler frequencies $f_{i,n}$ cannot be in this case computed explicitly. Nevertheless, from the difference of both relations, (5.28) and (5.41), we obtain the equation

$$\frac{n}{N_i} - \text{erf}\left(\frac{f_{i,n}}{f_c}\sqrt{\ln 2}\right) = 0\,, \quad \forall\, n = 1, 2, \ldots, N_i \quad (i = 1, 2)\,, \tag{5.42}$$

from which the discrete Doppler frequencies $f_{i,n}$ can be determined by means of a proper numerical root-finding technique.

Since the difference between two neighbouring discrete Doppler frequencies $\Delta f_{i,n} = f_{i,n} - f_{i,n-1}$ depends on the index n over a strongly nonlinear relation, it can be assumed that the greatest common divisor $F_i = \gcd\{f_{i,n}\}_{n=1}^{N_i}$ is quite small, so that the period $T_i = 1/F_i$ of $\tilde{\mu}_i(t)$ is quite high. We can therefore assume that $\tilde{\mu}_i(t)$ is quasi-nonperiodic.

For the corresponding Doppler coefficients $c_{i,n}$, (5.31) moreover holds. Thus, the deterministic processes $\tilde{\mu}_i(t)$ designed in this way have the mean power $\tilde{\sigma}^2_{\mu_i} = \sigma_0^2$. In the same way as with the Jakes power spectral density, here also the uncorrelatedness of the deterministic processes $\tilde{\mu}_1(t)$ and $\tilde{\mu}_2(t)$ can be guaranteed sufficiently by defining N_2 according to $N_2 := N_1 + 1$.

For example, for $N_i = 25$, both the power spectral density $\tilde{S}_{\mu_i\mu_i}(f)$ [cf. (4.14)] and the corresponding autocorrelation function $\tilde{r}_{\mu_i\mu_i}(\tau)$ [cf. (4.11)] will be computed again. For these two functions, we obtain the results shown in Figures 5.17(a) and 5.17(b).

For the performance assessment, the criteria introduced by (4.84) and (4.83) will again be evaluated for $N_i = 5, 6, \ldots, 50$ at this point. The results obtained for $E_{r_{\mu_i\mu_i}}$ and $E_{p_{\mu_i}}$ are depicted in Figures 5.18(a) and 5.18(b), respectively. Figure 5.18(a) also shows the graph of $E_{r_{\mu_i\mu_i}}$ obtained by applying the modified method of equal areas (MMEA), which will be described below.

We now come to the analysis of the model error $\Delta\beta_i$. Since the discrete Doppler frequencies $f_{i,n}$ are not given in an explicit form, no closed-form solution can be derived for the model error $\Delta\beta_i$ either. Therefore, we proceed as follows. At first, the parameter sets $\{c_{i,n}\}$ and $\{f_{i,n}\}$ will be computed by using (5.31) and (5.42). Then, the quantity $\tilde{\beta}_i$ can be determined by means of (4.22). In connection with

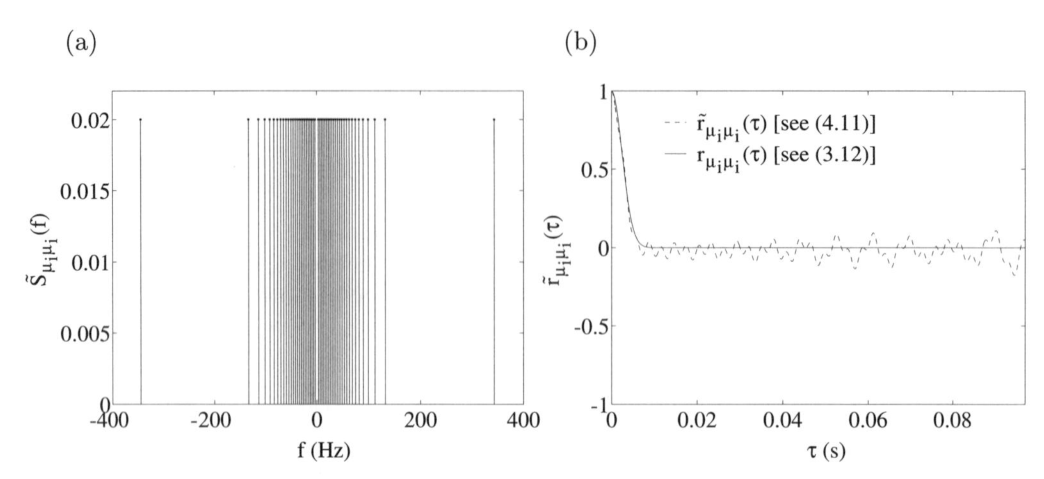

Figure 5.17: (a) Power spectral density $\tilde{S}_{\mu_i\mu_i}(f)$ and (b) autocorrelation function $\tilde{r}_{\mu_i\mu_i}(\tau)$ for $N_i = 25$ (MEA, Gaussian PSD, $f_c = \sqrt{\ln 2} f_{max}$, $f_{max} = 91\,\text{Hz}$, $\sigma_0^2 = 1$).

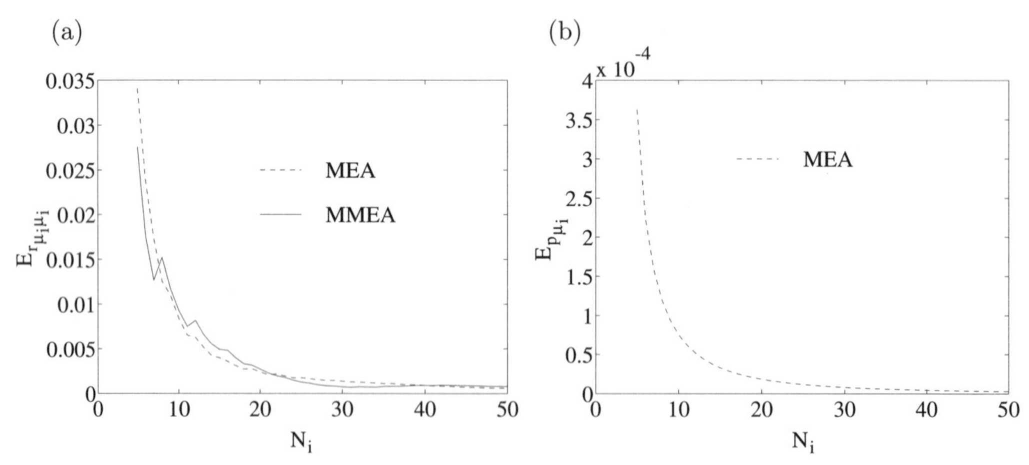

Figure 5.18: Mean-square errors: (a) $E_{r_{\mu_i\mu_i}}$ and (b) $E_{p_{\mu_i}}$ (MEA, Gaussian PSD, $f_c = \sqrt{\ln 2} f_{max}$, $f_{max} = 91\,\text{Hz}$, $\sigma_0^2 = 1$, $\tau_{max} = N_i/(2\kappa_c f_c)$, $\kappa_c = 2\sqrt{2/\ln 2}$).

$\beta = 2(\pi f_c \sigma_0)^2/\ln 2$, we are now able to evaluate the model error $\Delta\beta_i = \tilde{\beta}_i - \beta$. The results obtained for the relative model error $\Delta\beta_i/\beta$ are plotted in Figure 5.19 as a function of the number of sinusoids N_i.

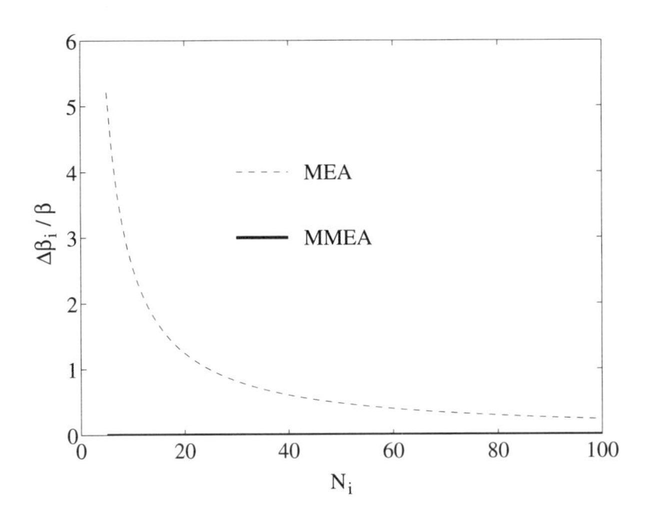

Figure 5.19: Relative model error $\Delta\beta_i/\beta$ (MEA, Gaussian PSD, $f_c = \sqrt{\ln 2} f_{max}$, $f_{max} = 91\,\text{Hz}$, $\sigma_0^2 = 1$).

Taking Figure 5.19 into account, it turns out that the percentage of the relative model error $\Delta\beta_i/\beta$ is only lower than 50 per cent if $N_i \geq 49$. Hence, the MEA is totally unsuitable for the Gaussian power spectral density. Since the main reason for this is the bad positioning of the discrete Doppler frequency $f_{i,n}$ for the value $n = N_i$ [see also Figure 5.17(a)], this imperfect adaptation can be avoided by a simple modification of the procedure. Instead of computing the complete set $\{f_{i,n}\}_{n=1}^{N_i}$ of the discrete Doppler frequencies according to (5.42), as done before, we will now only use the root-finding algorithm for the computation of $\{f_{i,n}\}_{n=1}^{N_i-1}$ and determine the remaining discrete Doppler frequency f_{i,N_i} in such a way that $\tilde{\beta}_i = \beta$ holds.

For this so-called modified method of equal areas (MMEA), one obtains the following set of equations:

$$\frac{n}{N_i} - \text{erf}\left(\frac{f_{i,n}}{f_c}\sqrt{\ln 2}\right) = 0\,, \quad \forall n = 1, 2, \ldots, N_i - 1\,, \tag{5.43a}$$

$$f_{i,N_i} = \sqrt{\frac{\beta N_i}{(2\pi\sigma_0)^2} - \sum_{n=1}^{N_i-1} f_{i,n}^2}\,. \tag{5.43b}$$

The corresponding Doppler coefficients $c_{i,n}$ are of course still given by (5.31). The advantage of the modified method of equal areas is that the relative model error $\Delta\beta_i/\beta$ is always equal to zero for all given values of $N_i = 1, 2, \ldots$ $(i = 1, 2)$. This is graphically demonstrated in Figure 5.19. However, the effects on the mean-square error $E_{r_{\mu_i\mu_i}}$ are small, as can be seen from Figure 5.18(a).

For the determination of the level-crossing rate $\tilde{N}_\xi(r)$ and the average duration of fades $\tilde{T}_{\xi_-}(r)$, we proceed in the same way as described in connection with the Jakes power spectral density before. The results obtained for $\tilde{N}_\xi(r)$ and $\tilde{T}_{\xi_-}(r)$ by choosing $(N_1, N_2) = (10, 11)$ are presented in Figures 5.20(a) and 5.20(b), respectively. Here, the modified method of equal areas was used for the computation of the model parameters.

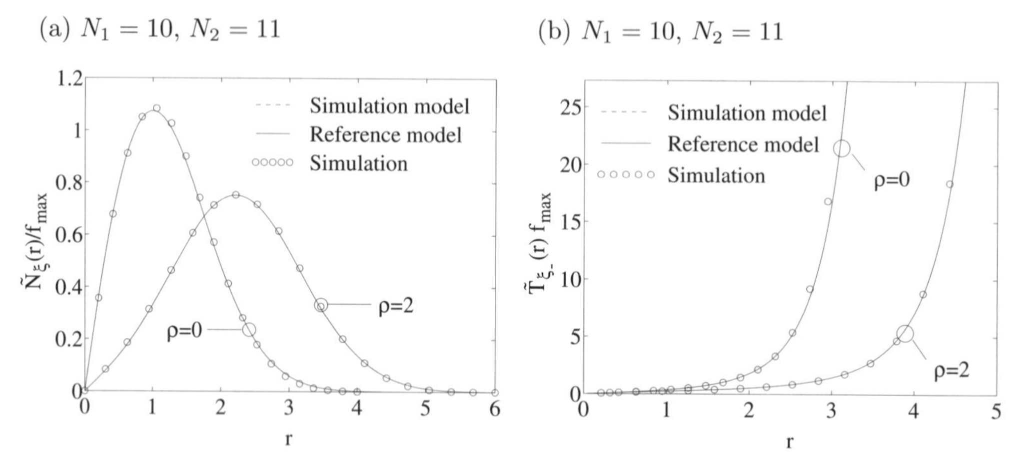

Figure 5.20: (a) Normalized level-crossing rate $\tilde{N}_\xi(r)/f_{max}$ and (b) normalized average duration of fades $\tilde{T}_{\xi_-}(r) \cdot f_{max}$ (MMEA, Gaussian PSD, $f_c = \sqrt{\ln 2} f_{max}$, $f_{max} = 91\,\text{Hz}$, $\sigma_0^2 = 1$).

Closed-form analytical expressions for the relative error of both the level-crossing rate $\tilde{N}_\xi(r)$ and the average duration of fades $\tilde{T}_{\xi_-}(r)$ cannot be derived for the MEA in the case of the Gaussian power spectral density. The reason for this is to be sought in the implicit equation (5.42) for the determination of the discrete Doppler frequencies $f_{i,n}$. For the MMEA, however, both relative errors ϵ_{N_ξ} and $\epsilon_{T_{\xi_-}}$ are equal to zero.

5.1.4 Monte Carlo Method (MCM)

The Monte Carlo method was first proposed in [Schu89] for the stochastic modelling and the digital simulation of mobile radio channels. Based on this paper, a model for the equivalent discrete-time channel [For72] in the complex baseband was introduced in [Hoe90, Hoe92]. In the following, we will use this method for the design of deterministic processes and will afterwards analyse their statistical properties.

The principle of the Monte Carlo method is based on the realization of the discrete Doppler frequencies $f_{i,n}$ according to a given probability density function $p_{\mu_i}(f)$, which is related to the power spectral density $S_{\mu_i\mu_i}(f)$ of the coloured Gaussian random process $\mu_i(t)$ by

$$p_{\mu_i}(f) = \frac{1}{\sigma_0^2} S_{\mu_i \mu_i}(f) . \tag{5.44}$$

Again, σ_0^2 here denotes the mean power (variance) of the Gaussian random process $\mu_i(t)$.

For the computation of the discrete Doppler frequencies $f_{i,n}$, we will be guided by the procedure presented in [Hoe90, Hoe92]. Let u_n be a random variable uniformly distributed within the interval (0,1]. Furthermore, let $g_{\mu_i}(u_n)$ be a mapping that is chosen in such a way that the distribution of the discrete Doppler frequencies $f_{i,n} = g_{\mu_i}(u_n)$ is equal to the desired cumulative distribution function

$$F_{\mu_i}(f_{i,n}) = \int_{-\infty}^{f_{i,n}} p_{\mu_i}(f) \, df . \tag{5.45}$$

According to [Pap91], $g_{\mu_i}(u_n)$ can then be identified with the inverse function of $F_{\mu_i}(f_{i,n}) = u_n$. Consequently, for the discrete Doppler frequencies $f_{i,n}$, the relation

$$f_{i,n} = g_{\mu_i}(u_n) = F_{\mu_i}^{-1}(u_n) \tag{5.46}$$

holds for all $n = 1, 2, \ldots, N_i$ $(i = 1, 2)$. Generally, we obtain positive as well as negative values for $f_{i,n}$. In cases where the probability density function $p_{\mu_i}(f)$ is an even function, i.e., $p_{\mu_i}(f) = p_{\mu_i}(-f)$, we can confine ourselves to positive values for $f_{i,n}$ without restriction of generality. This will be made possible by substituting the uniformly distributed random variable $u_n \in (0, 1]$ in (5.45) by $(1 + u_n)/2 \in (\frac{1}{2}, 1]$.

Since it follows from $u_n > 0$ that $f_{i,n} > 0$ holds, the time average of $\tilde{\mu}_i(t)$ is equal to zero, i.e., $\tilde{m}_{\mu_i} = m_{\mu_i} = 0$.

The Doppler coefficients $c_{i,n}$ are chosen so that the mean power of $\tilde{\mu}_i(t)$ is identical to the variance of $\mu_i(t)$, i.e., $\tilde{\sigma}_0^2 = \tilde{r}_{\mu_i \mu_i}(0) = \sigma_0^2$, which is guaranteed by choosing $c_{i,n}$ according to (5.31). Hence, it then follows

$$c_{i,n} = \sigma_0 \sqrt{\frac{2}{N_i}} \tag{5.47}$$

for all $n = 1, 2, \ldots, N_i$ $(i - 1, 2)$.

One may consider that with the Monte Carlo method, not only the Doppler phases $\theta_{i,n}$, but also the discrete Doppler frequencies $f_{i,n}$ are random variables. In principle, there is no difference whether statistical or deterministic methods are applied for the determination of the model parameters ($c_{i,n}$, $f_{i,n}$, $\theta_{i,n}$), because the process $\tilde{\mu}_i(t)$, which matters here, is a deterministic function per definition. (We refer to Section 4.1, where deterministic processes $\tilde{\mu}_i(t)$ have been introduced as sample functions or as realizations of stochastic processes $\hat{\mu}_i(t)$.) However, especially for a small number of harmonic functions, the ergodic properties of the stochastic process $\hat{\mu}_i(t)$ are poor, if the Monte Carlo method is applied for the computation of the discrete Doppler frequencies $f_{i,n}$ [Pae96e]. The consequence is that many important characteristic quantities of the deterministic process $\tilde{\mu}_i(t)$, like the Doppler spread, the level-crossing rate, and the average duration of fades become random values, which in particular cases can considerably deviate from the desired characteristic quantities prescribed by the reference model. In the following, we want to put this into concrete terms with the example of the Jakes and Gaussian power spectral density.

Jakes power spectral density: The application of the Monte Carlo method in connection with the Jakes power spectral density (3.8) results in the following expression for the discrete Doppler frequencies $f_{i,n}$

$$f_{i,n} = f_{max} \sin\left(\frac{\pi}{2} u_n\right), \tag{5.48}$$

where $u_n \in (0, 1]$ for all $n = 1, 2, \ldots, N_i$ $(i = 1, 2)$. For the Doppler coefficients $c_{i,n}$, furthermore (5.47) holds. It should be observed that the substitution of u_n in (5.48) by the deterministic quantity n/N_i exactly leads to the relation (5.33) that we found for the method of equal areas.

Since the discrete Doppler frequencies $f_{i,n}$ are random variables, the greatest common divisor $F_i = \gcd\{f_{i,n}\}_{n=1}^{N_i}$ is a random variable as well. However, for a given realization of the set $\{f_{i,n}\}$ with N_i elements, the greatest common divisor F_i is a constant that can be determined by applying the Euclidian algorithm on $\{f_{i,n}\}$, where we have to take into account that the discrete Doppler frequencies $f_{i,n}$ are real numbers. Generally, one can assume that the greatest common divisor F_i is very small, and, thus, the period $T_i = 1/F_i$ is very large, so that $\tilde{\mu}_i(t)$ can be considered as a quasi-nonperiodical function. Obviously, this holds even more, the greater the number of harmonic functions N_i is chosen.

The demand for uncorrelatedness of the real and the imaginary part generally also does not cause any difficulty when designing complex deterministic processes $\tilde{\mu}(t) = \tilde{\mu}_1(t) + j\tilde{\mu}_2(t)$. The reason for this is that even for $N_1 = N_2$, the realized sets $\{f_{1,n}\}$ and $\{f_{2,n}\}$ are in general mutually exclusive events, leading to the result that $\tilde{\mu}_1(t)$ and $\tilde{\mu}_2(t)$ are uncorrelated with respect to time averaging.

An example of the power spectral density $\tilde{S}_{\mu_i\mu_i}(f)$, obtained with $N_i = 25$ harmonic functions, is shown in Figure 5.21(a). The autocorrelation function $\tilde{r}_{\mu_i\mu_i}(\tau)$, which was computed according to (4.11), is plotted in Figure 5.21(b) for two different realizations of the sets $\{f_{i,n}\}$.

Regarding Figure 5.21(b), one can see that even in the range $0 \leq \tau \leq \tau_{max}$, the autocorrelation function $\tilde{r}_{\mu_i\mu_i}(\tau)$ of the deterministic process $\tilde{\mu}_i(t)$ can deviate considerably from the ideal autocorrelation function $r_{\mu_i\mu_i}(\tau)$ of the stochastic process $\mu_i(t)$.[2]

On the other hand, if we analyse the autocorrelation function $\hat{r}_{\mu_i\mu_i}(\tau)$ of the stochastic process $\hat{\mu}_i(t)$, then, by using (4.82), we obtain

---

[2] Using the Jakes power spectral density, furthermore, let $\tau_{max} = N_i/(2f_{max})$.

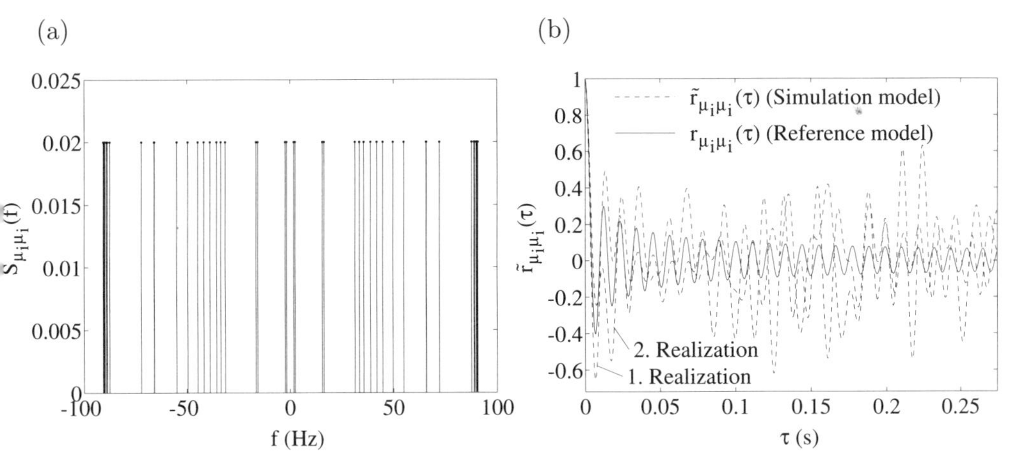

Figure 5.21: (a) Power spectral density $\tilde{S}_{\mu_i\mu_i}(f)$ and (b) autocorrelation function $\tilde{r}_{\mu_i\mu_i}(\tau)$ for $N_i = 25$ (MCM, Jakes PSD, $f_{max} = 91\,\text{Hz}$, $\sigma_0^2 = 1$).

$$\begin{aligned}
\hat{r}_{\mu_i\mu_i}(\tau) &:= E\{\hat{\mu}_i(t)\hat{\mu}_i(t+\tau)\} \\
&= \sum_{n=1}^{N_i} \frac{c_{i,n}^2}{2} E\{\cos(2\pi f_{i,n}\tau)\} \\
&= \sum_{n=1}^{N_i} \frac{c_{i,n}^2}{2} J_0(2\pi f_{max}\tau) \\
&= \sigma_0^2 J_0(2\pi f_{max}\tau) \\
&= r_{\mu_i\mu_i}(\tau)\,. \qquad (5.49)
\end{aligned}$$

Summarizing, we can say that the autocorrelation function $\hat{r}_{\mu_i\mu_i}(\tau)$ of the stochastic simulation model is equal to the ideal autocorrelation function $r_{\mu_i\mu_i}(\tau)$ of the reference model, whereas the autocorrelation function $\tilde{r}_{\mu_i\mu_i}(\tau)$ of the deterministic simulation model is different from both of the first-mentioned autocorrelation functions, i.e., $r_{\mu_i\mu_i}(\tau) = \hat{r}_{\mu_i\mu_i}(\tau) \neq \tilde{r}_{\mu_i\mu_i}(\tau)$ [Pae96e]. Due to $\hat{r}_{\mu_i\mu_i}(\tau) \neq \tilde{r}_{\mu_i\mu_i}(\tau)$, the stochastic process $\hat{\mu}_i(t)$ is therefore not ergodic with respect to the autocorrelation function [cf. Subsection 4.3.4].

The performance of the Monte Carlo method can again be assessed more precisely with the help of the mean-square error $E_{r_{\mu_i\mu_i}}$ [see (4.84)]. Figure 5.22 illustrates the evaluation of $E_{r_{\mu_i\mu_i}}$ as a function of N_i for a single realization of the autocorrelation function $\tilde{r}_{\mu_i\mu_i}(\tau)$ as well as for the expected value obtained by averaging $E_{r_{\mu_i\mu_i}}$ over a thousand realizations of $\tilde{r}_{\mu_i\mu_i}(\tau)$.

Figure 5.22 also shows the results found for the method of equal areas, which are obviously better compared to the results obtained for the Monte Carlo method. The relation (5.47) for the computation of the Doppler coefficients $c_{i,n}$ matches (5.31) exactly. Hence, for the mean-square error $E_{p_{\mu_i}}$ [cf. (4.83)], we obtain exactly the same results as presented in Figure 5.14(b).

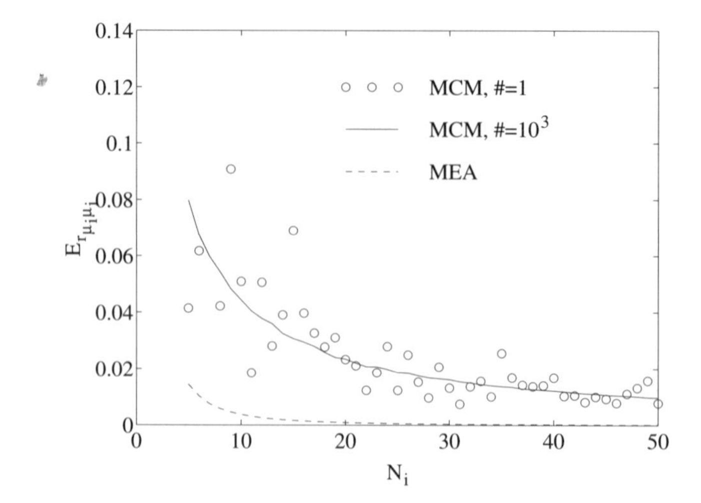

Figure 5.22: Mean-square error $E_{r_{\mu_i\mu_i}}$ (MCM, Jakes PSD, $f_{max} = 91\,\text{Hz}$, $\sigma_0^2 = 1$, $\tau_{max} = N_i/(2f_{max})$).

The discussion on the model error $\Delta\beta_i$ follows. Let us start with (4.22). Then, by using (3.29) and (5.47), the quantity $\tilde{\beta}_i = \beta + \Delta\beta_i$ can be expressed as a function of the discrete Doppler frequencies $f_{i,n}$ as follows

$$\tilde{\beta}_i = \frac{2\beta}{f_{max}^2 N_i} \sum_{n=1}^{N_i} f_{i,n}^2 \,. \tag{5.50}$$

For the Monte Carlo method, the discrete Doppler frequencies $f_{i,n}$ are random variables, so that $\tilde{\beta}_i$ is also a random variable. In what follows, we will determine the probability density function of $\tilde{\beta}_i$.

Starting from the uniform distribution of $u_n \in (0, 1]$ and noting that the mapping from u_n to $f_{i,n}$ is defined by (5.48), it follows that the probability density function of the discrete Doppler frequencies $f_{i,n}$ can be written as

$$p_{f_{i,n}}(f_{i,n}) = \begin{cases} \dfrac{2}{\pi f_{max}\sqrt{1-\left(\frac{f_{i,n}}{f_{max}}\right)^2}}\,, & 0 < f \leq f_{max}\,, \\ 0\,, & \text{else}\,. \end{cases} \tag{5.51}$$

Now, with the probability density function of $f_{i,n}$, the density of $f_{i,n}^2$ can easily be computed, and in order to compute the density of the sum of these squares, we preferably apply the concept of the characteristic function. After some straightforward computations, we obtain the result for the probability density function of $\tilde{\beta}_i$ in the following form [Pae96e]

$$p_{\tilde{\beta}_i}(\tilde{\beta}_i) = \begin{cases} 2\int\limits_0^\infty \left[J_0\left(\frac{2\pi\beta\nu}{N_i}\right)\right]^{N_i} \cos[2\pi(\tilde{\beta}_i - \beta)\nu]\, d\nu\,, & \text{if } \tilde{\beta}_i \in (0, 2\beta]\,, \\ 0\,, & \text{if } \tilde{\beta}_i \notin (0, 2\beta]\,. \end{cases} \tag{5.52}$$

By way of illustration, the probability density function $p_{\tilde{\beta}_i}(\tilde{\beta}_i)$ of $\tilde{\beta}_i$ is plotted in Figure 5.23 with N_i as a parameter.

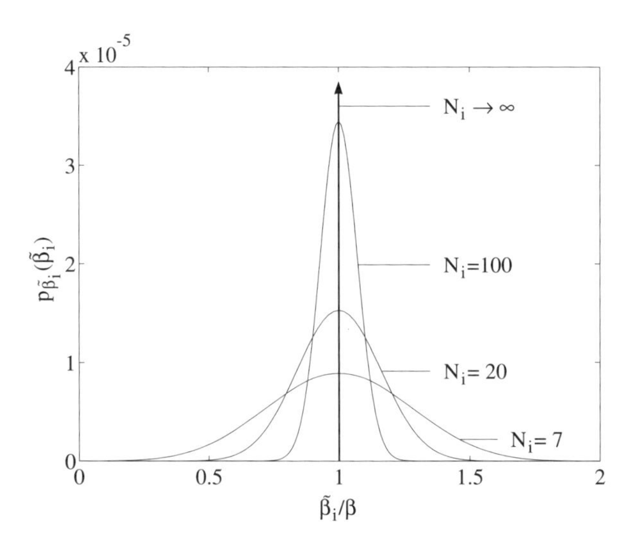

Figure 5.23: Probability density function $p_{\tilde{\beta}_i}(\tilde{\beta}_i)$ of $\tilde{\beta}_i$ by using the Monte Carlo method (Jakes PSD, $f_{max} = 91\,\text{Hz}$, $\sigma_0^2 = 1$).

The expected value $E\{\tilde{\beta}_i\}$ and the variance $\text{Var}\,\{\tilde{\beta}_i\}$ of $\tilde{\beta}_i$ are as follows:

$$E\{\tilde{\beta}_i\} = \beta\,, \tag{5.53a}$$

$$\text{Var}\,\{\tilde{\beta}_i\} = \frac{\beta^2}{2N_i}\,. \tag{5.53b}$$

It will also be shown that for large values of N_i, the random variable $\tilde{\beta}_i$ is approximately normally distributed with a mean value and a variance according to (5.53a) and (5.53b), respectively. Using the approximation for the Bessel function of 0th order [Abr72, eq. (9.1.12)]

$$J_0(x) \approx 1 - \frac{x^2}{4} \tag{5.54}$$

and taking into account that the relation [Abr72, eq. (4.2.21)]

$$e^{-x} = \lim_{N_i \to \infty} \left(1 - \frac{x}{N_i}\right)^{N_i} \tag{5.55}$$

can be approximated by $e^{-x} \approx (1 - x/N_i)$ if N_i is sufficiently large, we may express $p_{\tilde{\beta}_i}(\tilde{\beta}_i)$ [see (5.52)] approximately by

$$p_{\tilde{\beta}_i}(\tilde{\beta}_i) \approx \int_{-\infty}^{\infty} e^{-\frac{(\pi\beta\nu)^2}{N_i}} e^{-j2\pi(\tilde{\beta}_i - \beta)\nu} \, d\nu \,, \quad \tilde{\beta}_i \in (0, 2\beta] \,. \tag{5.56}$$

Finally, using the integral [Gra81, vol. I, eq. (3.323.2)]

$$\int_{-\infty}^{\infty} e^{-(ax)^2 \pm bx} \, dx = \frac{\sqrt{\pi}}{a} e^{(\frac{b}{2a})^2} \,, \quad a > 0 \,, \tag{5.57}$$

the desired approximation directly follows

$$p_{\tilde{\beta}_i}(\tilde{\beta}_i) \approx \frac{1}{\sqrt{2\pi}\beta/\sqrt{2N_i}} e^{-\frac{(\tilde{\beta}_i - \beta)^2}{2\beta^2/(2N_i)}} \,, \quad \tilde{\beta}_i \in (0, 2\beta] \,. \tag{5.58}$$

Hence, for large values of N_i, the quantity $\tilde{\beta}_i$ is approximately normally distributed, and we may write $\tilde{\beta}_i \sim N(\beta, \beta^2/(2N_i))$ without making too large an error. It should be observed that in the limit $N_i \to \infty$, it obviously follows $p_{\tilde{\beta}_i}(\tilde{\beta}_i) \to \delta(\tilde{\beta}_i - \beta)$. Evidently, the model error $\Delta\beta_i = \tilde{\beta}_i - \beta$ is likewise approximately normally distributed, i.e., $\Delta\beta_i \sim N(0, \beta^2/(2N_i))$, so that the random variable $\Delta\beta_i$ is in fact zero-mean, but unfortunately its variance merely behaves proportionally to the reciprocal value of the number of harmonic functions N_i. Finally, we also investigate the relative model error $\Delta\beta_i/\beta$, for which it approximately holds: $\Delta\beta_i/\beta \sim N(0, 1/(2N_i))$. Hence, the standard deviation of $\Delta\beta_i/\beta$ is equal to $1/\sqrt{2N_i}$ and for $N_i > 2$ it is thus always greater than the relative model error $\Delta\beta_i/\beta = 1/N_i$ obtained by using the method of equal areas [cf. (5.37)]. Figure 5.24 demonstrates the random behaviour of the relative model error $\Delta\beta_i/\beta$ in terms of the number of harmonic functions N_i. The evaluation of $\Delta\beta_i/\beta$ was performed here by means of (5.50), where five events of the set $\{f_{i,n}\}_{n=1}^{N_i}$ were processed for every value of $N_i \in \{5, 6, \ldots, 100\}$.

Due to (4.66), (4.70), and (4.78), it now becomes clear that the level-crossing rate $\tilde{N}_\xi(r)$, the average duration of fades $\tilde{T}_{\xi_-}(r)$, and the time intervals $\tilde{\tau}_q(r)$ likewise deviate in a random manner from the corresponding quantities of the reference model. For example, if we choose the pair (10, 11) for the couple (N_1, N_2), then for two different realizations of each of the sets $\{f_{1,n}\}_{n=1}^{N_1}$ and $\{f_{2,n}\}_{n=1}^{N_2}$, the behaviours shown in Figures 5.25(a) and 5.25(b) could occur for $\tilde{N}_\xi(r)$ and $\tilde{T}_{\xi_-}(r)$, respectively. Here, the simulations were carried out in the same way as previously described in Subsection 5.1.3.

With the Chebyshev inequality (2.15), one can show [see Appendix C] that even if $N_i = 2500$ harmonic functions are used, the probability that the absolute value of the relative model error $|\Delta\beta_i/\beta|$ exceeds a value of more than 2 per cent is merely less or equal to 50 per cent.

Gaussian power spectral density: If we apply the Monte Carlo method in connection with the Gaussian power spectral density (3.11), then it is not possible

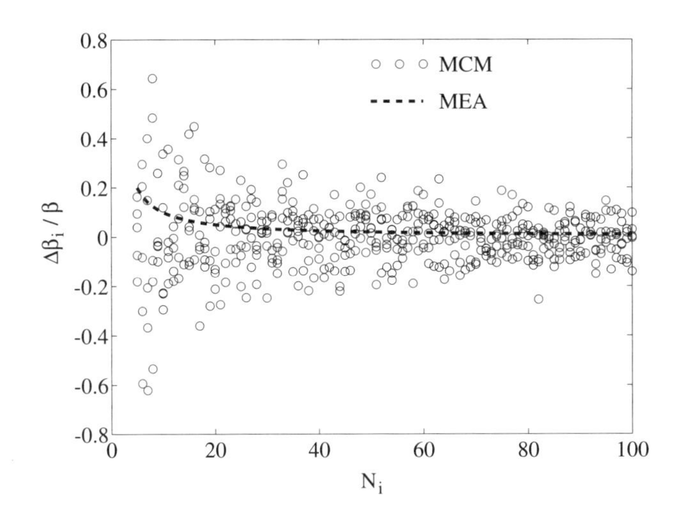

Figure 5.24: Relative model error $\Delta\tilde{\beta}_i/\beta$ (MCM, Jakes PSD).

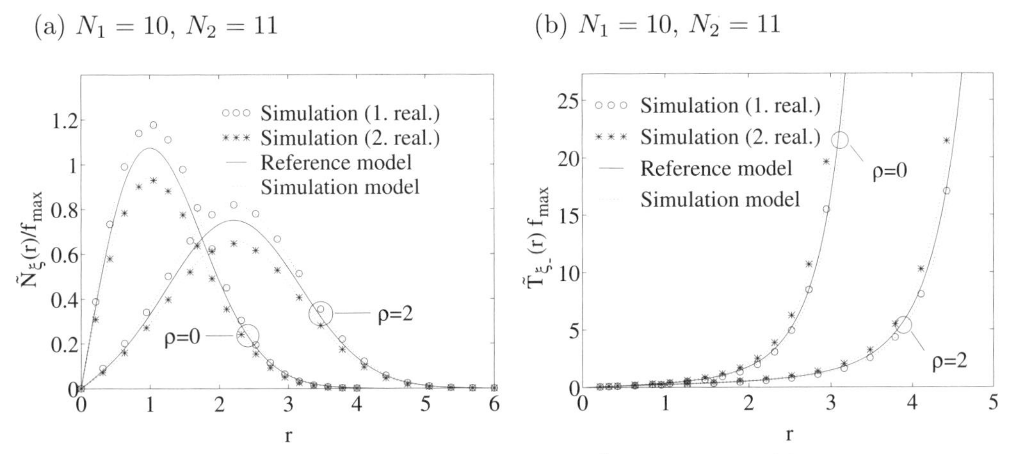

Figure 5.25: (a) Normalized level-crossing rate $\tilde{N}_\xi(r)/f_{max}$ and (b) normalized average duration of fades $\tilde{T}_{\xi_-}(r) \cdot f_{max}$ (MCM, Jakes PSD, $f_{max} = 91\,\text{Hz}$, $\sigma_0^2 = 1$).

to find a closed-form expression for the discrete Doppler frequencies $f_{i,n}$. In this case, however, the discrete Doppler frequencies $f_{i,n}$, are determined by the roots (zeros) of the following equations

$$u_n - \operatorname{erf}\left(\frac{f_{i,n}}{f_c}\sqrt{\ln 2}\right) = 0\,, \quad \forall n = 1, 2, \ldots, N_i \quad (i = 1, 2)\,. \tag{5.59}$$

Here, the corresponding Doppler coefficients $c_{i,n}$ are also available in the form (5.47). A comparison between the equation above and (5.42) again makes the close relation between the method of equal areas and the Monte Carlo method clear. If the uniformly distributed random variable u_n is substituted by the deterministic quantity n/N_i, then the latter statistical procedure turns into the former deterministic one. For any (arbitrary) event $\{f_{i,n}\}$, it turns out that the mean value $\tilde{m}_{\mu_i}$, the mean power $\tilde{\sigma}^2_{\mu_i}$, and the period T_i of the deterministic process $\tilde{\mu}_i(t)$ have the same properties as described before in connection with the Jakes power spectral density. The same also holds for the cross-correlation properties of the deterministic processes $\tilde{\mu}_1(t)$ and $\tilde{\mu}_2(t)$.

For an event $\{f_{i,n}\}$ with $N_i = 25$ outcomes, the power spectral density $\tilde{S}_{\mu_i\mu_i}(f)$ is depicted in Figure 5.26(a). Likewise for $N_i = 25$, Figure 5.26(b) shows two possible realizations of the autocorrelation function $\tilde{r}_{\mu_i\mu_i}(\tau)$.

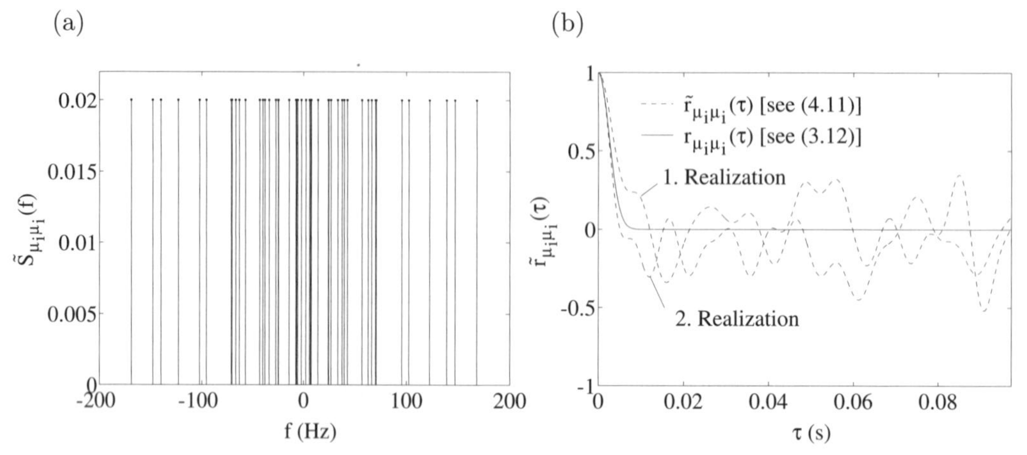

Figure 5.26: (a) Power spectral density $\tilde{S}_{\mu_i\mu_i}(f)$ and (b) autocorrelation function $\tilde{r}_{\mu_i\mu_i}(\tau)$ for $N_i = 25$ (MCM, Gaussian PSD, $f_c = \sqrt{\ln 2} f_{max}$, $f_{max} = 91\,\text{Hz}$, $\sigma_0^2 = 1$).

The large deviations between $\tilde{r}_{\mu_i\mu_i}(\tau)$ and $r_{\mu_i\mu_i}(\tau)$ within the range $0 \leq \tau \leq \tau_{max}$ ($\tau_{max} = N_i/(2\kappa_c f_c)$) are typical of the Monte Carlo method. This can be confirmed by evaluating the mean-square error $E_{r_{\mu_i\mu_i}}$ [see (4.84)]. Figure 5.27 shows the obtained results. In this figure, the mean-square error $E_{r_{\mu_i\mu_i}}$ is presented as a function of N_i for both a single realization of the autocorrelation function $\tilde{r}_{\mu_i\mu_i}(\tau)$ and for the average value of $E_{r_{\mu_i\mu_i}}$ obtained by averaging $E_{r_{\mu_i\mu_i}}$ over a thousand realizations of $\tilde{r}_{\mu_i\mu_i}(\tau)$.

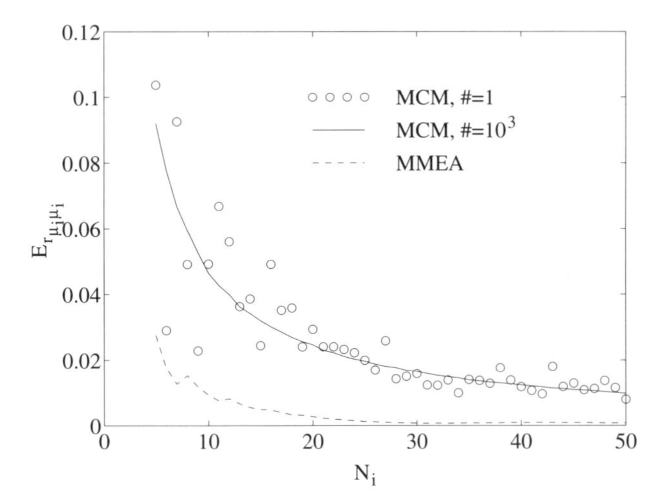

Figure 5.27: Mean-square error $E_{r_{\mu_i\mu_i}}$ (MCM, Gaussian PSD, $f_c = \sqrt{\ln 2} f_{max}$, $f_{max} =$ 91 Hz, $\sigma_0^2 = 1$, $\tau_{max} = N_i/(2\kappa_c f_c)$, $\kappa_c = 2\sqrt{2/\ln 2}$).

In this case, the analysis of the model error $\Delta\beta_i$ cannot be carried out analytically. Therefore, we proceed in such a way that for a given realization of $\{f_{i,n}\}$, at first the corresponding elementary event of the random variable $\tilde{\beta}_i$ is determined by means of (4.22). Afterwards, with $\beta = 2(\pi f_c \sigma_0)^2/\ln 2$, the computation of the model error $\Delta\beta_i = \tilde{\beta}_i - \beta$ will be performed. Figure 5.28 presents the evaluation of the relative model error $\Delta\beta_i/\beta$, where the obtained results are shown for each value of $N_i \in \{5, 6, \ldots, 100\}$ on the basis of five realizations of the set $\{f_{i,n}\}$.

The determination as well as the investigation of the properties of the level-crossing rate $\tilde{N}_\xi(r)$ and the average duration of fades $\tilde{T}_{\xi_-}(r)$ are also performed on the basis of several realizations of the set $\{f_{i,n}\}$. To illustrate the obtained results, we take a look at Figures 5.29(a) and 5.29(b), where two different realizations of $\tilde{N}_\xi(r)$ and $\tilde{T}_{\xi_-}(r)$ are shown, respectively. All the presented results have been obtained by using $(N_1, N_2) = (10, 11)$.

5.1.5 L_p-Norm Method (LPNM)

The L_p-norm method (LPNM) is based on the idea that the sets $\{c_{i,n}\}$ and $\{f_{i,n}\}$ are to be determined in such a way that the following requirements will be fulfilled for a given number of harmonic functions N_i [Pae98b, Pae96c]:

(i) With respect to the following L_p-norm

$$E_{p_{\mu_i}}^{(p)} := \left\{ \int_{-\infty}^{\infty} |p_{\mu_i}(x) - \tilde{p}_{\mu_i}(x)|^p \, dx \right\}^{1/p}, \quad p = 1, 2, \ldots, \tag{5.60}$$

the probability density function $\tilde{p}_{\mu_i}(x)$ of the deterministic process $\tilde{\mu}_i(t)$ will be an

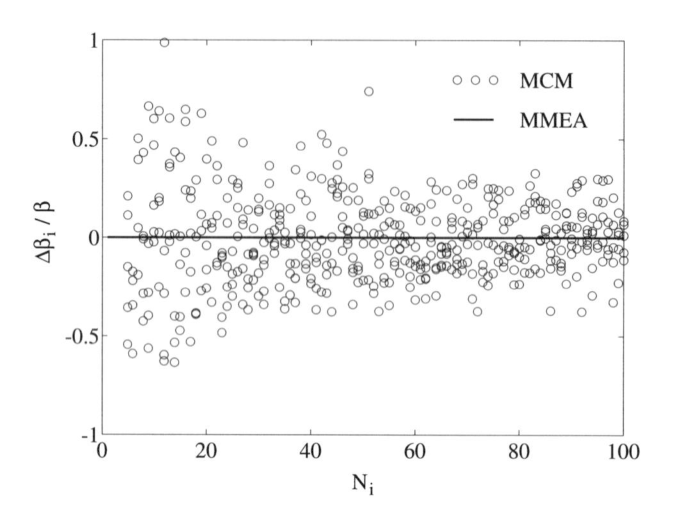

Figure 5.28: Relative model error $\tilde{\beta}_i/\beta$ (MCM, Gaussian PSD, $f_c = \sqrt{\ln 2} f_{max}$, $f_{max} = 91\,\text{Hz}$, $\sigma_0^2 = 1$).

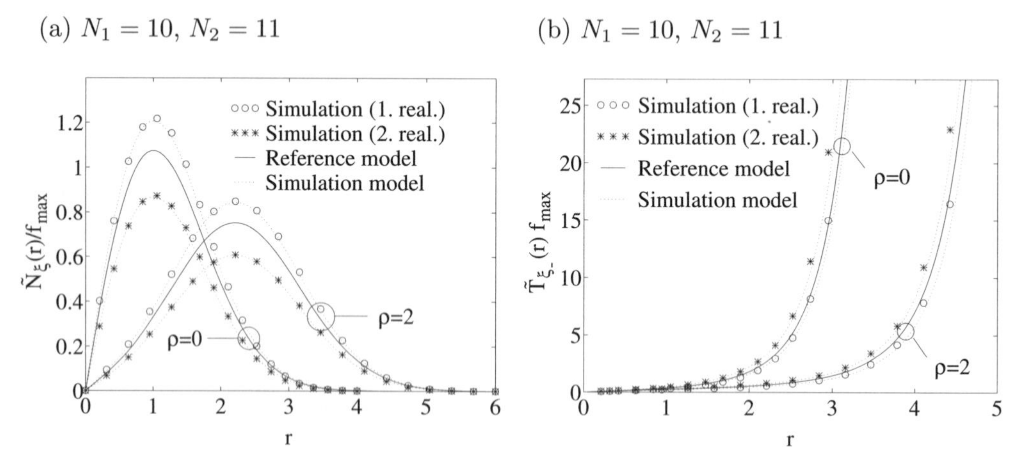

Figure 5.29: (a) Normalized level-crossing rate $\tilde{N}_\xi(r)/f_{max}$ and (b) normalized average duration of fades $\tilde{T}_{\xi_-}(r) \cdot f_{max}$ (MCM, Gaussian PSD, $f_c = \sqrt{\ln 2} f_{max}$, $f_{max} = 91\,\text{Hz}$, $\sigma_0^2 = 1$).

optimal approximation of the Gaussian distribution $p_{\mu_i}(x)$ of the stochastic process $\mu_i(t)$.

(ii) With respect to the following L_p-norm

$$E_{r_{\mu_i\mu_i}}^{(p)} := \left\{ \frac{1}{\tau_{max}} \int_0^{\tau_{max}} |r_{\mu_i\mu_i}(\tau) - \tilde{r}_{\mu_i\mu_i}(\tau)|^p \, d\tau \right\}^{1/p} , \quad p = 1, 2, \ldots , \qquad (5.61)$$

the autocorrelation function $\tilde{r}_{\mu_i\mu_i}(\tau)$ of the deterministic process $\tilde{\mu}_i(t)$ will be fitted as close as possible to a given (desired) autocorrelation function $r_{\mu_i\mu_i}(\tau)$ of the stochastic process $\mu_i(t)$, where τ_{max} again defines an appropriate time interval $[0, \tau_{max}]$ over which the approximation of $r_{\mu_i\mu_i}(\tau)$ is of interest.

We first pay attention to the requirement (i). Since, according to (4.34), $\tilde{p}_{\mu_i}(x)$ merely depends on the Doppler coefficients $c_{i,n}$, we ask ourselves: does an optimal solution for the set of Doppler coefficients $\{c_{i,n}\}$ exist, for which the L_p-norm $E_{p_{\mu_i}}^{(p)}$ becomes minimal? In order to answer this question, we first substitute (4.34) and (4.36) into (5.60), and afterwards perform a numerical optimization of the Doppler coefficients $c_{i,n}$, so that $E_{p_{\mu_i}}^{(p)}$ becomes minimal. As a numerical optimization technique, for example, the Fletcher-Powell algorithm [Fle63] is particularly well suited for this kind of problem. After the minimization of (5.60), the optimized Doppler coefficients $c_{i,n} = c_{i,n}^{(opt)}$ are available for the realization of deterministic simulation models. Figure 5.30(a) shows the resulting probability density function $\tilde{p}_{\mu_i}(x)$ by using the optimized quantities $c_{i,n}^{(opt)}$. For the choice of suitable starting values for the Doppler coefficients, we appropriately fall back to the quantities $c_{i,n} = \sigma_0\sqrt{2/N_i}$. For a better assessment of the obtained results, the probability density function $\tilde{p}_{\mu_i}(x)$, which is obtained by using the starting values $c_{i,n} = \sigma_0\sqrt{2/N_i}$ [cf. also Figure 4.4(a)], is again presented in Figure 5.30(b).

More meaningful than the comparison of Figures 5.30(a) and 5.30(b) are the results of Figure 5.31, where the mean-square error $E_{p_{\mu_i}}$ [see (4.39)] is presented for $c_{i,n} = c_{i,n}^{(opt)}$ as well as for $c_{i,n} = \sigma_0\sqrt{2/N_i}$. One can clearly realize that the optimization gain decreases strictly monotonously if the number of sinusoids N_i increases.

It is also worth mentioning that after the minimization of (5.60), all optimized Doppler coefficients $c_{i,n}^{(opt)}$ are in fact identical (due to the central limit theorem). But for a finite number of harmonic functions N_i, they are always smaller than the pre-set starting values, i.e., $c_{i,n}^{(opt)} < \sigma_0\sqrt{2/N_i}$, $\forall\, N_i = 1, 2, \ldots$ Since the optimized Doppler coefficients $c_{i,n}^{(opt)}$ are identical, which is even the case when arbitrary starting values are chosen, it is probable that the L_p-norm (5.60) has a global minimum at $c_{i,n} = c_{i,n}^{(opt)}$, and, thus, the Doppler coefficients $c_{i,n} = c_{i,n}^{(opt)}$ are optimal. For finite values of N_i, one may also take into account that due to $c_{i,n} = c_{i,n}^{(opt)} < \sigma_0\sqrt{2/N_i}$, the mean power of the deterministic process $\tilde{\mu}_i(t)$ is always smaller than the variance of the stochastic process $\mu_i(t)$, i.e., it holds $\tilde{\sigma}_0^2 < \sigma_0^2$. This can be realized by considering Figure 5.32.

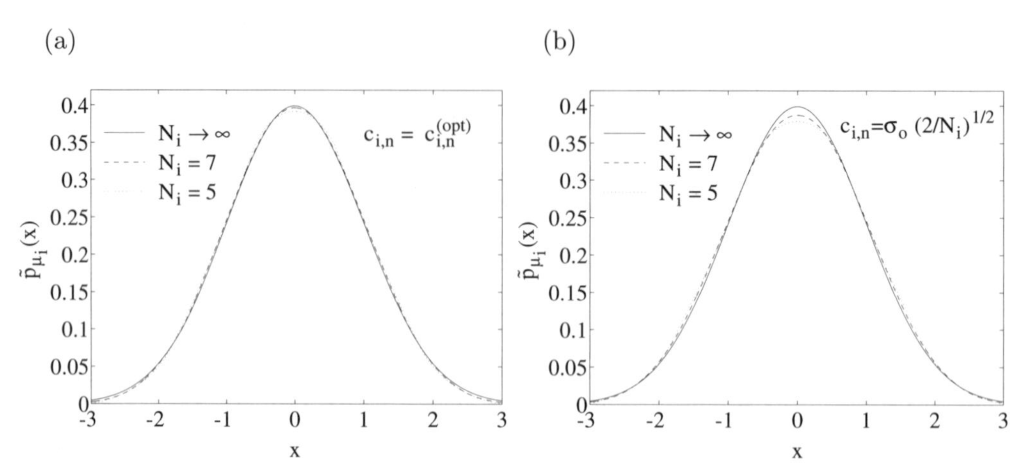

Figure 5.30: Probability density function $\tilde{p}_{\mu_i}(x)$ for $N_i \in \{5, 7, \infty\}$ by using: (a) $c_{i,n} = c_{i,n}^{(opt)}$ and (b) $c_{i,n} = \sigma_0\sqrt{2/N_i}$ ($\sigma_0^2 = 1$).

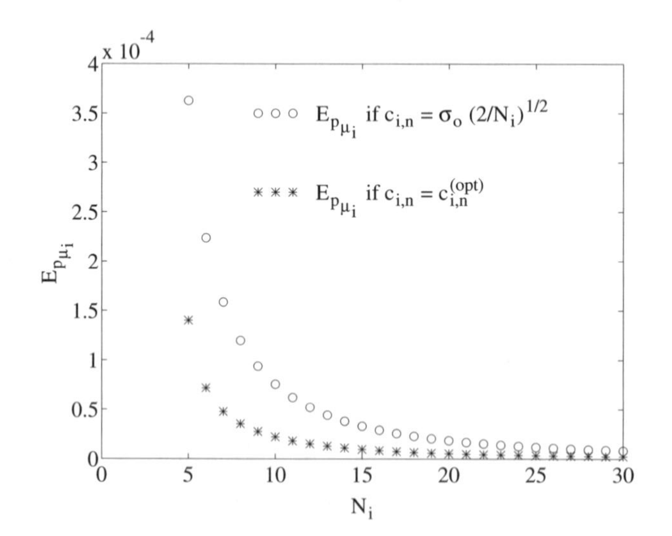

Figure 5.31: Mean-square error $E_{p_{\mu_i}}$, if $c_{i,n} = c_{i,n}^{(opt)}$ $(***)$ and $c_{i,n} = \sigma_0\sqrt{2/N_i}$ $(\circ\circ\circ)$ with $\sigma_0^2 = 1$.

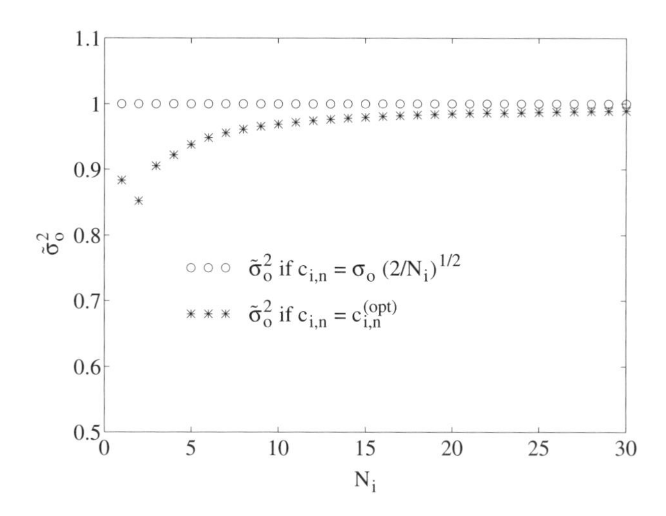

Figure 5.32: Mean power $\tilde{\sigma}_0^2$ of the deterministic process $\tilde{\mu}_i(t)$, if $c_{i,n} = c_{i,n}^{(opt)}$ $(* * *)$ and $c_{i,n} = \sigma_0\sqrt{2/N_i}$ $(\circ \circ \circ)$ with $\sigma_0^2 = 1$.

Obviously, a compromise between the attainable approximation precision of $\tilde{p}_{\mu_i}(x) \approx p_{\mu_i}(x)$ and $\tilde{\sigma}_0^2 \approx \sigma_0^2$ has to be made here. Let us try to avoid this compromise by imposing the so-called power constraint defined by $\tilde{\sigma}_0^2 = \sigma_0^2$ on the simulation model. Then, we have to optimize, e.g., the first $N_i - 1$ Doppler coefficients $c_{i,1}, c_{i,2}, \ldots, c_{i,N_{i-1}}$ and the remaining parameter c_{i,N_i} is determined in such a way that the imposed power constraint $\tilde{\sigma}_0^2 = \sigma_0^2$ is always fulfilled. In this case, the optimization results in $c_{i,n}^{(opt)} = \sigma_0\sqrt{2/N_i}$ for all $n = 1, 2, \ldots, N_i$. Thus, by including the power constraint $\tilde{\sigma}_0^2 = \sigma_0^2$ in the parameter design, an optimal approximation of the Gaussian distribution $p_{\mu_i}(x)$ for any number of harmonic functions N_i can only become possible if the Doppler coefficients $c_{i,n}$ are given by $c_{i,n} = c_{i,n}^{(opt)} = \sigma_0\sqrt{2/N_i}$. Therefore, when modelling Gaussian random processes and other processes derivable from these, such as Rayleigh processes, Rice processes, and lognormal processes, we will usually make use of the relation $c_{i,n} = \sigma_0\sqrt{2/N_i}$ in the following.

The suggested method is still quite useful and advantageous for the approximation of probability density functions which are not derivable from Gaussian distributions, e.g., like the Nakagami distribution (2.33). The Nakagami distribution [Nak60] is more flexible than the frequently used Rayleigh or Rice distribution and often enables a better adaptation to probability density functions which follow from experimental measurement results [Suz77].

In order to be able to determine the set of Doppler coefficients $\{c_{i,n}\}$ in such a way that the probability density function of the deterministic simulation model approximates the Nakagami distribution, we perform the optimization of the Doppler coefficients in a similar manner as described before in connection with the normal distribution. The only difference is that in (5.60) we have to substitute the Gaussian distribution

$p_{\mu_i}(x)$ by the Nakagami distribution $p_\omega(z)$ [see (2.33)] and $\tilde{p}_{\mu_i}(x)$ has to be replaced by $\tilde{p}_\xi(z)$ given by (4.50). Some optimization results obtained for various values of the parameter m are shown in Figure 5.33, where $N_1 = N_2 = 10$ harmonic functions have been used in all cases.

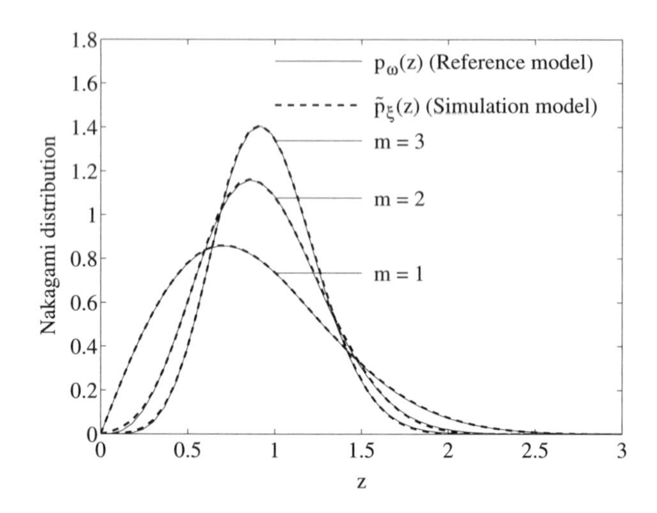

Figure 5.33: Approximation of the Nakagami distribution by using deterministic processes with $N_1 = N_2 = 10$ $(\Omega = 1)$.

Further details on the derivation and simulation of Nakagami fading channels can be found in [Bra91, Der93]. Results on the analysis of both the level-crossing rate and the average duration of fades of Nakagami processes were first published in [You96].

In [She77], the Weibull distribution was suggested for the approximation of the probability density function of real-world mobile radio channels in the 900 MHz frequency range. As is well known, the Weibull distribution can be derived by means of a nonlinear transformation of a uniformly distributed random variable [Joh94]. Since the uniform distribution can be determined from a further nonlinear transformation of two Gaussian distributed random variables [Joh94], the problem of modelling the Weibull distribution can thus be reduced to the problem of modelling Gaussian random processes, which we have already discussed. Therefore, we do not expect any essential new discoveries from further analysis of this matter.

Let us now consider the requirement (ii) [see (5.61)]. According to (4.11), the autocorrelation function $\tilde{r}_{\mu_i\mu_i}(\tau)$ depends on both the Doppler coefficients $c_{i,n}$ and the discrete Doppler frequencies $f_{i,n}$. Since the Doppler coefficients $c_{i,n}$ were already determined so that the probability density function $\tilde{p}_{\mu_i}(x)$ of the deterministic process $\tilde{\mu}_i(t)$ approximates the Gaussian distribution $p_{\mu_i}(x)$ of the stochastic process $\tilde{\mu}_i(t)$ as well as possible, only the discrete Doppler frequencies $f_{i,n}$ can be used for the minimization of the L_p-norm $E^{(p)}_{r_{\mu_i\mu_i}}$ defined by (5.61). The discrete Doppler frequencies

$f_{i,n}$ are now optimized, e.g., by applying the Fletcher-Powell algorithm again, so that $E^{(p)}_{r_{\mu_i\mu_i}}$ becomes as small as possible, and, hence, the autocorrelation function $\tilde{r}_{\mu_i\mu_i}(\tau)$ of the deterministic process $\tilde{\mu}_i(t)$ approximates the given autocorrelation function $r_{\mu_i\mu_i}(\tau)$ of the stochastic process $\mu_i(t)$ within the interval $[0, \tau_{max}]$. In general, we cannot guarantee that the Fletcher-Powell algorithm — like any other optimization algorithm suitable for this problem — finds the global minimum of $E^{(p)}_{r_{\mu_i\mu_i}}$, so that in most cases, we have to be satisfied with a local minimum. This property, which at first seems to be disadvantageous, could easily turn out to be an advantage, if we take into account that various local minima also lead to various sets of discrete Doppler frequencies $\{f_{i,n}\}$. For the generation of uncorrelated deterministic processes $\tilde{\mu}_1(t)$ and $\tilde{\mu}_2(t)$, we are therefore no longer restricted to the previous convention $N_2 := N_1 + 1$, but can now guarantee that the processes $\tilde{\mu}_1(t)$ and $\tilde{\mu}_2(t)$ are also uncorrelated for $N_1 = N_2$. However, the latter property can also be obtained by carrying out the optimizations with different values for the parameter p or by using different starting values for the discrete Doppler frequencies $f_{i,n}$.

In the following, we will apply the L_p-norm method to the Jakes and the Gaussian power spectral density, where it has to be taken into account that, in connection with the power constraint $\tilde{\sigma}_0^2 = \sigma_0^2$, the requirement (i) is already fulfilled by $c_{i,n} = c^{(opt)}_{i,n} = \sigma_0\sqrt{2/N_i}$ and, therefore, only the requirement (ii) has to be investigated in more detail.

Jakes power spectral density: By substituting (3.10) and (4.11) into (5.61), we obtain an optimized set $\{f^{(opt)}_{i,n}\}$ for the discrete Doppler frequencies after the numerical minimization of the L_p-norm $E^{(p)}_{r_{\mu_i\mu_i}}$. As starting values for the discrete Doppler frequencies $f_{i,n}$, for example, the quantities $f_{i,n} = f_{max}\sin[n\pi/(2N_i)]$, $\forall\, n = 1, 2, \ldots, N_i$ $(i = 1, 2)$, derived by using the method of equal areas, are suitable. For the Jakes power spectral density, the upper limit of the integral in (5.61) is given by the relation $\tau_{max} = N_i/(2f_{max})$, which we already know from Subsection 5.1.1. All of the following optimization results are based on the L_p-norm $E^{(p)}_{r_{\mu_i\mu_i}}$ with $p = 2$.

Generally valid statements on the greatest common divisor $F_i = \gcd\{f^{(opt)}_{i,n}\}_{n=1}^{N_i}$ cannot be made here. Numerical investigations, however, have shown that F_i is usually zero or at least extremely small. Therefore, the deterministic processes $\tilde{\mu}_i(t)$ designed with the L_p-norm method are nonperiodical or quasi-nonperiodical. For the time average $\tilde{m}_{\mu_i}$ and the mean power $\tilde{\sigma}^2_{\mu_i}$, again the relations $\tilde{m}_{\mu_i} = m_{\mu_i} = 0$ and $\tilde{\sigma}^2_{\mu_i} = \sigma_0^2$ follow, respectively.

As for the preceding methods, the power spectral density $\tilde{S}_{\mu_i\mu_i}(f)$ and the autocorrelation function $\tilde{r}_{\mu_i\mu_i}(\tau)$ have been evaluated exemplary for $N_i = 25$ here. One may study the results shown in Figures 5.34(a) and 5.34(b).

Due to $c_{i,n} = \sigma_0\sqrt{2/N_i}$, we again obtain the graph presented in Figure 5.14(b) for the mean-square error $E_{p_{\mu_i}}$ [see (4.83)]. The results of the evaluation of $E_{r_{\mu_i\mu_i}}$ [see (4.84)] are shown in Figure 5.35, where also the corresponding graph obtained by applying the method of equal areas is presented.

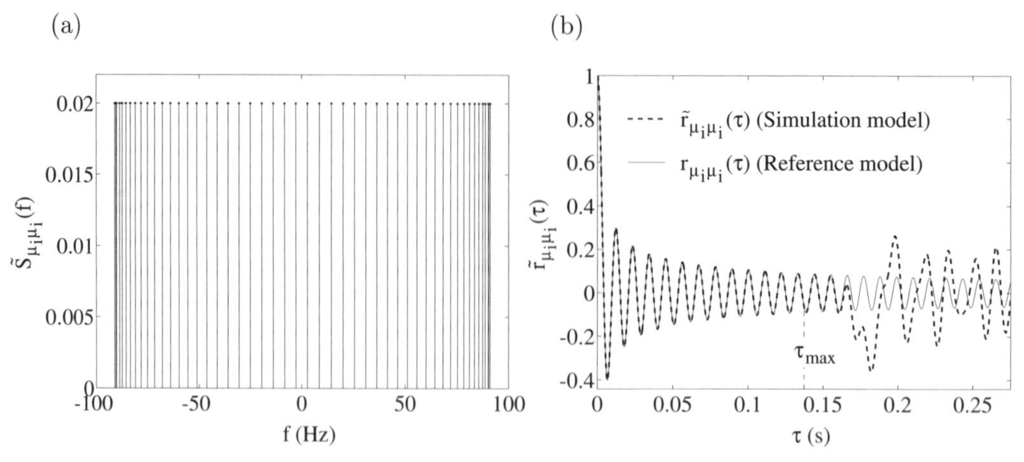

Figure 5.34: (a) Power spectral density $\tilde{S}_{\mu_i\mu_i}(f)$ and (b) autocorrelation function $\tilde{r}_{\mu_i\mu_i}(\tau)$ for $N_i = 25$ (LPNM, Jakes PSD, $f_{max} = 91\,\text{Hz}$, $\sigma_0^2 = 1$).

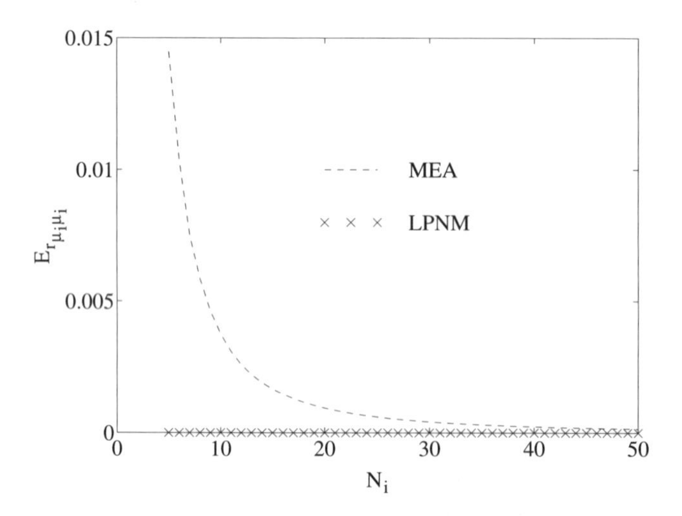

Figure 5.35: Mean-square error $E_{r_{\mu_i\mu_i}}$ (LPNM, Jakes PSD, $f_{max} = 91\,\text{Hz}$, $\sigma_0^2 = 1$, $\tau_{max} = N_i/(2f_{max})$).

In order to compute the model error $\Delta\beta_i = \tilde{\beta}_i - \beta$, the expression (4.22) has to be evaluated for $c_{i,n} = \sigma_0\sqrt{2/N_i}$ and $f_{i,n} = f_{i,n}^{(opt)}$. In comparison with the method of equal areas, we then obtain the graphs illustrated in Figure 5.36 for the relative model error $\Delta\beta_i/\beta$.

The simulation of the level-crossing rate and the average duration of fades is carried out in the same way as already described in Subsection 5.1.3. For reasons of unity, we here again choose the pair $(10, 11)$ for the couple (N_1, N_2). The simulation results for

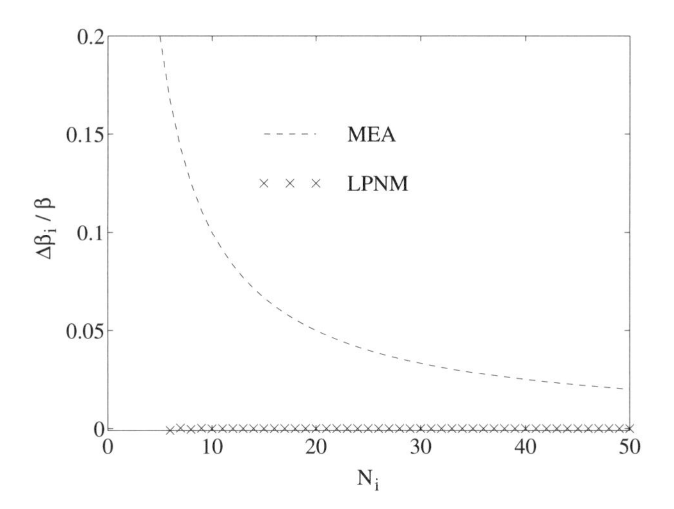

Figure 5.36: Relative model error $\Delta\beta_i/\beta$ (LPNM, Jakes PSD).

the normalized level-crossing rate $\tilde{N}_\xi(r)/f_{max}$ and the normalized average duration of fades $\tilde{T}_{\xi_-}(r) \cdot f_{max}$ are illustrated in Figures 5.37(a) and 5.37(b), respectively. The analytical results one finds for the reference model and the simulation model are also illustrated in these figures. Since the relative model errors $\Delta\beta_1$ and $\Delta\beta_2$ are extremely small for both cases $N_1 = 10$ and $N_2 = 11$, the individual curves can no longer be distinguished from each other in the presented graphs.

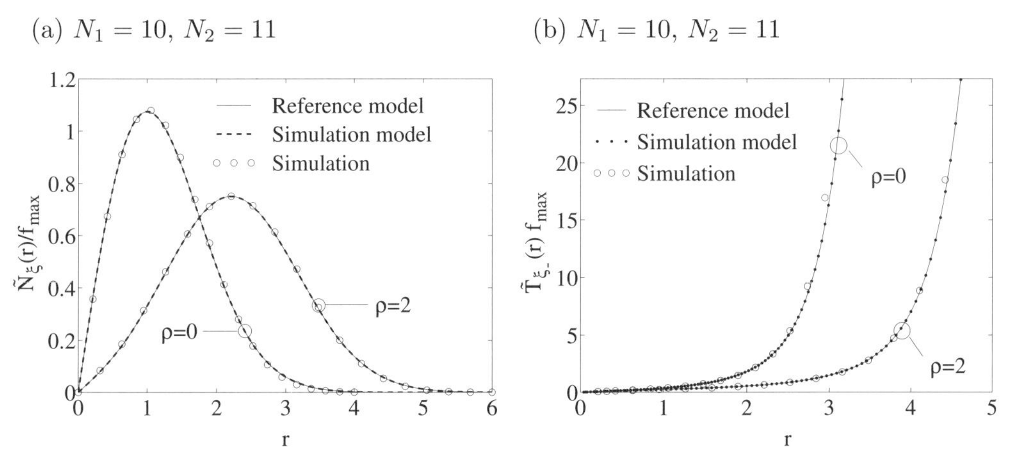

Figure 5.37: (a) Normalized level-crossing rate $\tilde{N}_\xi(r)/f_{max}$ and (b) normalized average duration of fades $\tilde{T}_\xi(r) \cdot f_{max}$ (LPNM, Jakes PSD, $f_{max} = 91\,\text{Hz}$, $\sigma_0^2 = 1$).

Gaussian power spectral density: The previously analysed methods for the determination of the model parameters of the deterministic processes have made it quite clear that the Gaussian power spectral density causes much greater problems than the Jakes power spectral density. In this subsection, we will find out how to get these problems under control by using the L_p-norm method. Therefore, we fully exploit the degrees of freedom which this method has to offer. All in all, this leads us to three fundamental variants [Pae97d] of the L_p-norm method. In the following, these variants will be briefly described and afterwards analysed with respect to their performances.

First variant of the L_p-norm method (LPNM I): In the first variant, the Doppler coefficients $c_{i,n}$ are again computed according to the equation $c_{i,n} = \sigma_0\sqrt{2/N_i}$ for all $n = 1, 2, \ldots, N_i$, whereas the discrete Doppler frequencies $f_{i,n}$ are optimized for $n = 1, 2, \ldots, N_i - 1$ in such a way that the L_p-norm $E^{(p)}_{r_{\mu_i\mu_i}}$ [see (5.61)] results in a (local) minimum, i.e.,

$$E^{(p)}_{r_{\mu_i\mu_i}}(\boldsymbol{f}_i) = \text{Min!}\,, \tag{5.62}$$

where $\boldsymbol{f}_i$ stands for the parameter vector $\boldsymbol{f}_i = (f_{i,1}, f_{i,2}, \ldots, f_{i,N_i-1})^T \in \mathbb{R}^{N_i-1}$. Boundary conditions, like the restriction that the components of the parameter vector $\boldsymbol{f}_i$ shall be positive, do not need to be imposed on the procedure, since the Gaussian power spectral density is symmetrical. The remaining discrete Doppler frequency f_{i,N_i} is defined by

$$f_{i,N_i} := \sqrt{\frac{\beta N_i}{(2\pi\sigma_0)^2} - \sum_{n=1}^{N_i-1} f^2_{i,n}}\,, \tag{5.63}$$

so that we have guaranteed in a simple manner that the model error $\Delta\beta_i$ is always zero for all chosen values of $N_i = 1, 2, \ldots$ $(i = 1, 2)$. With the corresponding quantity β, we can of course make use of this possibility when dealing with the Jakes power spectral density (or any other given power spectral density) as well. Quite suitable starting values for the optimization of the involved discrete Doppler frequencies are the quantities found with the method of equal areas [cf. Subsection 5.1.3]. For the evaluation of the L_p-norm $E^{(p)}_{r_{\mu_i\mu_i}}$, it is sufficient for our objectives to restrict ourselves to the case $p = 2$. In this connection, for the parameter τ_{max}, we return to the relation $\tau_{max} = N_i/(2\kappa_c f_c)$ with $\kappa_c = 2\sqrt{2/\ln 2}$ and $f_c = \sqrt{\ln 2} f_{max}$, which has already been employed several times in preceding investigations. For the quantities F_i, $\tilde{m}_{\mu_i}$, and $\tilde{\sigma}^2_{\mu_i}$, the statements made for the Jakes power spectral density are still valid in the present case. Uncorrelated deterministic processes $\tilde{\mu}_1(t)$ and $\tilde{\mu}_2(t)$ can also be obtained for $N_1 = N_2$ by optimizing the parameter vectors $\boldsymbol{f}_1$ and $\boldsymbol{f}_2$ under different conditions. Therefore, it is sufficient to change, for example, τ_{max} or p slightly and then to repeat the optimization once again.

We choose $N_i = 25$ and with the first variant of the L_p-norm method, we obtain the power spectral density $\tilde{S}_{\mu_i\mu_i}(f)$ presented in Figure 5.38(a). Figure 5.38(b) shows the corresponding autocorrelation function $\tilde{r}_{\mu_i\mu_i}(\tau)$.

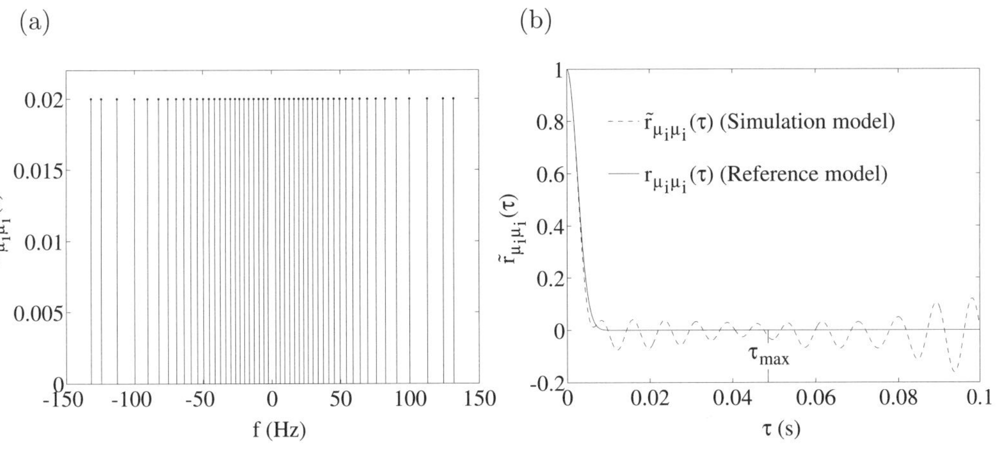

Figure 5.38: (a) Power spectral density $\tilde{S}_{\mu_i\mu_i}(f)$ and (b) autocorrelation function $\tilde{r}_{\mu_i\mu_i}(\tau)$ for $N_i = 25$ (LPNM I, Gaussian PSD, $f_c = \sqrt{\ln 2} f_{max}$, $f_{max} = 91\,\text{Hz}$, $\sigma_0^2 = 1$).

Second variant of the L_p-norm method (LPNM II): With the second variant of the L_p-norm method, our aim will be to fit the autocorrelation function $\tilde{r}_{\mu_i\mu_i}(\tau)$ within the interval $[0, \tau_{max}]$ far closer to $r_{\mu_i\mu_i}(\tau)$ than it is possible with the LPNM I. Therefore, we combine all parameters determining the behaviour of $\tilde{r}_{\mu_i\mu_i}(\tau)$ into the parameter vectors $\boldsymbol{c}_i = (c_{i,1}, c_{i,2}, \ldots, c_{i,N_i})^T \in \mathbb{R}^{N_i}$ and $\boldsymbol{f}_i = (f_{i,1}, f_{i,2}, \ldots, f_{i,N_i})^T \in \mathbb{R}^{N_i}$. Now, the task is actually to optimize the parameter vectors $\boldsymbol{c}_i$ and $\boldsymbol{f}_i$ in such a way that the L_p-norm $E^{(p)}_{r_{\mu_i\mu_i}}$ becomes minimal, i.e.,

$$E^{(p)}_{r_{\mu_i\mu_i}}(\boldsymbol{c}_i, \boldsymbol{f}_i) = \text{Min!}\,. \tag{5.64}$$

In this case as well, we again do not need to impose any boundary conditions on the components of the parameter vectors $\boldsymbol{c}_i$ and $\boldsymbol{f}_i$.

An example of the resulting power spectral density $\tilde{S}_{\mu_i\mu_i}(f)$ is depicted in Figure 5.39(a), where again $N_i = 25$ is chosen. In addition to that, Figure 5.39(b) shows the graph of the corresponding autocorrelation function $\tilde{r}_{\mu_i\mu_i}(\tau)$.

It cannot be missed that the approximation $r_{\mu_i\mu_i}(\tau) \approx \tilde{r}_{\mu_i\mu_i}(\tau)$ for $\tau \in [0, \tau_{max}]$ is extraordinary good. However, in order to obtain this advantage, we have to accept some disadvantages. Thus, for example, the power constraint $\tilde{\sigma}^2_{\mu_i} = \sigma_0^2$ is only fulfilled approximately; besides, the model error $\Delta\beta_i$ is unequal to zero. In general, the obtained approximations $\tilde{\sigma}^2_{\mu_i} \approx \sigma_0^2$ and $\tilde{\beta}_i \approx \beta$ or $\Delta\beta_i \approx 0$ are still very good and absolutely sufficient for most practical applications. A problem which should be considered as serious, however, occurs for the LPNM II when optimizing the Doppler coefficients $c_{i,n}$. The degradation of the probability density function $\tilde{p}_{\mu_i}(x)$, to which this problem leads, will be discussed further below. At this point, it is sufficient to mention that all of these disadvantages can be avoided with the third variant.

Third variant of the L_p-norm method (LPNM III): The third variant has the aim of optimizing both the autocorrelation function $\tilde{r}_{\mu_i\mu_i}(\tau)$ and the probability density

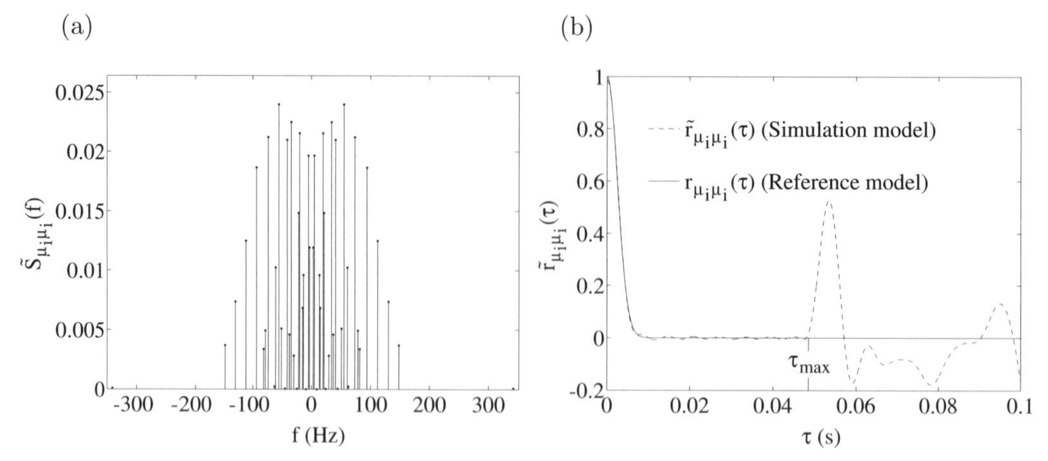

Figure 5.39: (a) Power spectral density $\tilde{S}_{\mu_i\mu_i}(f)$ and (b) autocorrelation function $\tilde{r}_{\mu_i\mu_i}(\tau)$ for $N_i = 25$ (LPNM II, Gaussian PSD, $f_c = \sqrt{\ln 2} f_{max}$, $f_{max} =$ 91 Hz, $\sigma_0^2 = 1$).

function $\tilde{p}_{\mu_i}(x)$. An error function suitable for this purpose has the form

$$E(\boldsymbol{c}_i, \boldsymbol{f}_i) = W_1 \cdot E^{(p)}_{r_{\mu_i\mu_i}}(\boldsymbol{c}_i, \boldsymbol{f}_i) + W_2 \cdot E^{(p)}_{p_{\mu_i}}(\boldsymbol{c}_i) , \tag{5.65}$$

where $E^{(p)}_{r_{\mu_i\mu_i}}(\cdot)$ and $E^{(p)}_{p_{\mu_i}}(\cdot)$ denote the L_p-norms introduced by (5.61) and (5.60), respectively. The quantities W_1 and W_2 are appropriate weighting factors, which will be defined by $W_1 = 1/4$ and $W_2 = 3/4$ in the sequel. To have both constraints $\tilde{\sigma}^2_{\mu_i} = \sigma_0^2$ and $\tilde{\beta}_i = \beta$ exactly fulfilled now, we will define the parameter vectors $\boldsymbol{c}_i$ and $\boldsymbol{f}_i$ by

$$\boldsymbol{c}_i = (c_{i,1}, c_{i,2}, \ldots, c_{i,N_i-1})^T \in \mathbb{R}^{N_i-1} \tag{5.66a}$$

and

$$\boldsymbol{f}_i = (f_{i,1}, f_{i,2}, \ldots, f_{i,N_i-1})^T \in \mathbb{R}^{N_i-1} , \tag{5.66b}$$

respectively, and calculate the remaining model parameters c_{i,N_i} and f_{i,N_i} as follows:

$$c_{i,N_i} = \sqrt{2\sigma_0^2 - \sum_{n=1}^{N_i-1} c_{i,n}^2} , \tag{5.67a}$$

$$f_{i,N_i} = \frac{1}{c_{i,N_i}} \sqrt{\frac{\beta}{2\pi^2} - \sum_{n=1}^{N_i-1} (c_{i,n} f_{i,n})^2} , \tag{5.67b}$$

where $\beta = -\ddot{r}_{\mu_i\mu_i}(0) = 2(\pi f_c \sigma_0)^2 / \ln 2$ $(i = 1, 2)$. Correlations between the deterministic processes $\tilde{\mu}_1(t)$ and $\tilde{\mu}_2(t)$ can now be avoided for $N_1 = N_2$ by performing the minimization of the error function (5.65) for $i = 1$ and $i = 2$ with different weighting factors of the respective L_p-norms $E^{(p)}_{r_{\mu_i\mu_i}}$ and $E^{(p)}_{p_{\mu_i}}$.

As in the preceding examples, we choose $N_i = 25$ and observe the resulting power spectral density $\tilde{S}_{\mu_i\mu_i}(f)$ in Figure 5.40(a). The corresponding autocorrelation function $\tilde{r}_{\mu_i\mu_i}(\tau)$ is shown in Figure 5.40(b).

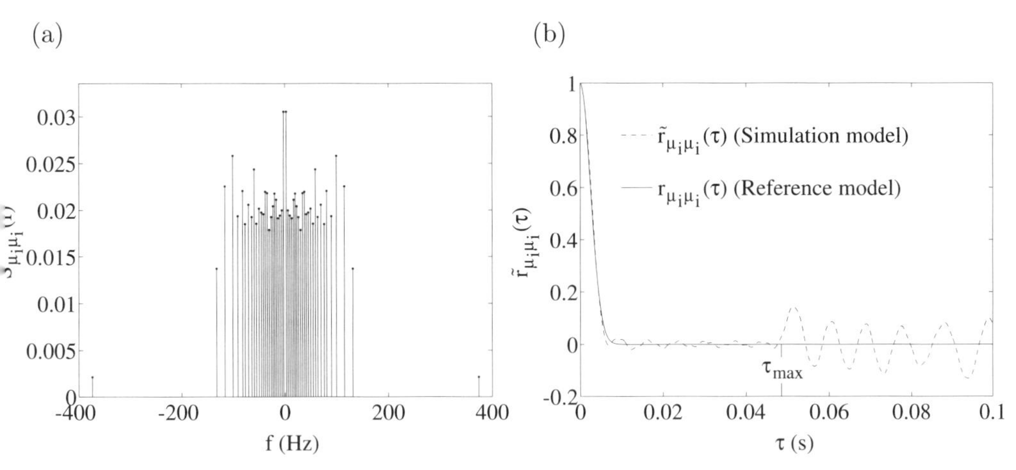

Figure 5.40: (a) Power spectral density $\tilde{S}_{\mu_i\mu_i}(f)$ and (b) autocorrelation function $\tilde{r}_{\mu_i\mu_i}(\tau)$ for $N_i = 25$ (LPNM III, Gaussian PSD, $f_c = \sqrt{\ln 2} f_{max}$, $f_{max} = 91\,\text{Hz}$, $\sigma_0^2 = 1$).

Finally, we will also analyse the performance of these three variants of the L_p-norm method. Here, we are especially interested in the mean-square errors $E_{r_{\mu_i\mu_i}}$ and $E_{p_{\mu_i}}$ [see (4.84) and (4.83), respectively], both of which are shown as a function of N_i in Figures 5.41(a) and 5.41(b), respectively. For the starting values, the parameters to be optimized were in all cases computed with the method of equal areas.

Studying Figure 5.41(a) it becomes obvious that the quality of the approximation $r_{\mu_i\mu_i}(\tau) \approx \tilde{r}_{\mu_i\mu_i}(\tau)$ can be improved enormously, if, besides the discrete Doppler frequencies $f_{i,n}$, also the Doppler coefficients $c_{i,n}$ are included in the optimization, as it is intended for the LPNM II and III. It should be noted that among the three variants of the L_p-norm method introduced here, the LPNM I has in fact the largest mean-square error $E_{r_{\mu_i\mu_i}}$ [see Figure 5.41(a)], but on the other hand the mean-square error $E_{p_{\mu_i}}$ [see Figure 5.41(b)] is the smallest. Exactly the opposite statement applies to the LPNM II. Only the LPNM III is a guarantee of a successful compromise between the minimization of both $E_{r_{\mu_i\mu_i}}$ and $E_{p_{\mu_i}}$. With a suitable choice of the weighting factors in (5.65), the minimization of $E^{(p)}_{r_{\mu_i\mu_i}}$ always turns out well with this method, i.e., we do not have to come to terms with the fact that considerable degradations concerning $E^{(p)}_{p_{\mu_i}}$ occur. Not only due to this property, but also because the boundary conditions $\tilde{\sigma}^2_{\mu_i} = \sigma_0^2$ and $\tilde{\beta}_i = \beta$ can be fulfilled exactly with the LPNM III, this variant of the L_p-norm method is without doubt the most efficient one.

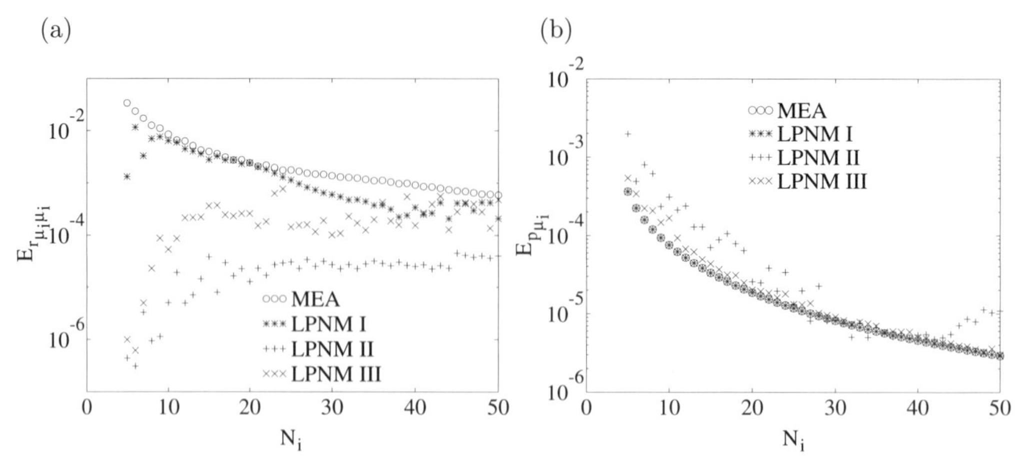

Figure 5.41: Mean-square errors: (a) $E_{r_{\mu_i\mu_i}}$ and (b) $E_{p_{\mu_i}}$ (LPNM I–III, Gaussian PSD, $f_c = \sqrt{\ln 2} f_{max}$, $f_{max} = 91\,\text{Hz}$, $\sigma_0^2 = 1$, $\tau_{max} = N_i/(2\kappa_c f_c)$, $\kappa_c = 2\sqrt{2/\ln 2}$).

Concerning the evaluation of the model error $\Delta\beta_i = \tilde{\beta}_i - \beta$ for the three variants of the L_p-norm method, we recall that during the introduction of the LPNM I and III we set great store on the fact that the model error $\Delta\beta_i$ is always equal to zero. This is guaranteed by (5.63) for the LPNM I and by (5.67b) for the LPNM III. In order to get the model error for the LPNM II, we substitute the optimized Doppler coefficients $c_{i,n} = c_{i,n}^{(opt)}$ and the optimized discrete Doppler frequencies $f_{i,n} = f_{i,n}^{(opt)}$ into (4.22). All the results obtained for the relative model error $\Delta\beta_i/\beta$ are presented in Figure 5.42. It can be seen that the model error $\Delta\beta_i$ corresponding to the LPNM II is different from zero. In the present case, the autocorrelation function $\tilde{r}_{\mu_i\mu_i}(\tau)$ was optimized over the interval $[0, \tau_{max}]$ with a constant weighting factor. However, if the approximation error of $\tilde{r}_{\mu_i\mu_i}(\tau)$ is weighted higher within an infinitesimal ϵ-interval around $\tau = 0$, then the model error $\Delta\beta_i$ can once more be reduced remarkably. Since it can still be seen clearly in Figure 5.42 that the relative model error $\Delta\beta_i/\beta$ is sufficiently small, we will accept the results found for this subject so far and continue with the analysis of the level-crossing rate and the average duration of fades.

For the analysis of the level-crossing rate $\tilde{N}_\xi(r)$ and the average duration of fades $\tilde{T}_{\xi_-}(r)$, we will confine ourselves to the LPNM III. Again, the simulation of the quantities $\tilde{N}_\xi(r)$ and $\tilde{T}_{\xi_-}(r)$ will be performed on the conditions described in Subsection 5.1.3. For the normalized level-crossing rate $\tilde{N}_\xi(r)/f_{max}$, the simulation results as well as the analytical results are depicted in Figure 5.43(a), where the pair $(10, 10)$ was chosen for the couple (N_1, N_2), exactly as in the preceding examples. Figure 5.43(b) next to it shows the corresponding normalized average duration of fades $\tilde{T}_{\xi_-}(r) f_{max}$.

We want to close this subsection with some general remarks about the L_p-norm method. The decisive advantage of this method lies in the possibility to design

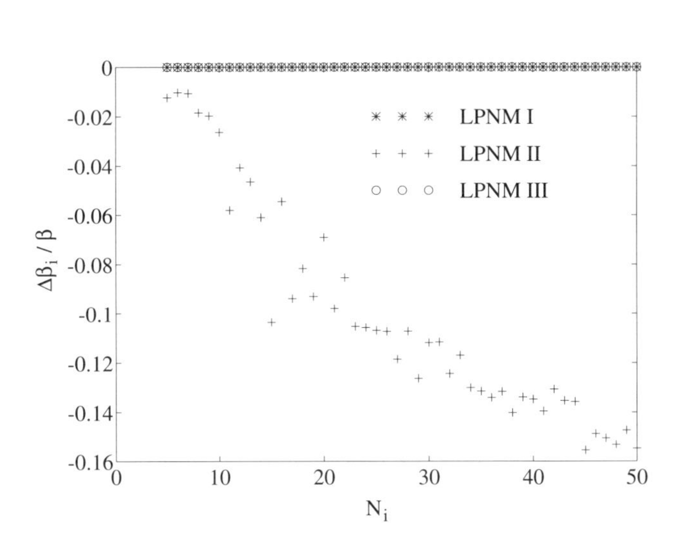

Figure 5.42: Relative model error $\Delta\beta_i/\beta$ (LPNM I–III, Gaussian PSD, $f_c = \sqrt{\ln 2} f_{max}$, $f_{max} = 91\,\text{Hz}$, $\sigma_0^2 = 1$).

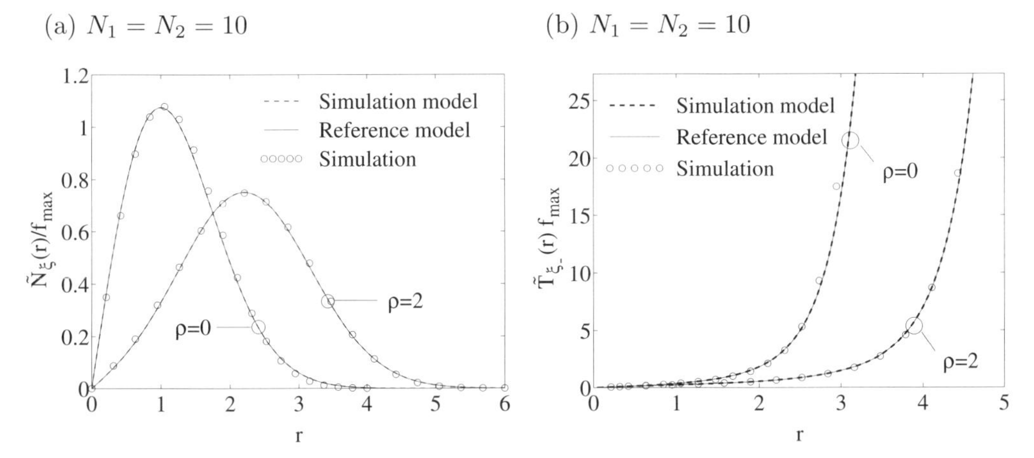

Figure 5.43: (a) Normalized level-crossing rate $\tilde{N}_\xi(r)/f_{max}$ and (b) normalized average duration of fades $\tilde{T}_{\xi_-}(r) \cdot f_{max}$ (LPNM III, Gaussian PSD, $f_c = \sqrt{\ln 2} f_{max}$, $f_{max} = 91\,\text{Hz}$, $\sigma_0^2 = 1$).

deterministic processes $\tilde{\mu}_i(t)$ or $\tilde{\xi}(t)$ so that they are able to reproduce the statistical properties of snapshot measurements taken from real-world mobile radio channels. Therefore, we only have to replace the probability density function $p_{\mu_i}(x)$ in (5.60) and the autocorrelation function $r_{\mu_i\mu_i}(\tau)$ in (5.61) with the corresponding measured quantities. The optimization can then be performed as described. Compared with other methods, the only disadvantage of the L_p-norm method is the relatively high numerical complexity. In fact, with modern computers, this is no longer a serious problem. But nevertheless, the expenditure is not worth it, at least not in connection with the Jakes power spectral density, since there exists a quite simple, elegant, and quasi-optimal solution, which we will discuss in the next subsection.

5.1.6 Method of Exact Doppler Spread (MEDS)

The method of exact Doppler spread (MEDS) was first introduced in [Pae96c] and has been developed especially for the often used Jakes power spectral density. Despite its simplicity, the method is distinguished for its high performance and enables a quasi-optimal approximation of the autocorrelation function corresponding to the Jakes power spectral density. In the following, we will first derive the method of exact Doppler spread in connection with the Jakes power spectral density and afterwards, we will investigate to what extent the method also offers advantages for the application on the Gaussian power spectral density.

Jakes power spectral density: Let us start with the integral presentation of the Bessel function of 0th order [Abr72, eq. (9.1.18)]

$$J_0(z) = \frac{2}{\pi}\int_0^{\pi/2} \cos(z \sin\alpha)\, d\alpha\,, \tag{5.68}$$

which can be expressed in form of an infinite series as

$$J_0(z) = \lim_{N_i\to\infty} \frac{2}{\pi}\sum_{n=1}^{N_i} \cos(z \sin\alpha_n)\Delta\alpha\,, \tag{5.69}$$

where $\alpha_n = \pi(2n-1)/(4N_i)$ and $\Delta\alpha = \pi/(2N_i)$. Hence, for (3.10) we can alternatively write

$$r_{\mu_i\mu_i}(\tau) = \lim_{N_i\to\infty} \frac{\sigma_0^2}{N_i}\sum_{n=1}^{N_i} \cos\left\{2\pi f_{max} \sin\left[\frac{\pi}{2N_i}\left(n-\frac{1}{2}\right)\right]\cdot\tau\right\}. \tag{5.70}$$

This relation describes the autocorrelation function of the stochastic reference model for a Gaussian random process $\mu_i(t)$, whose power spectral density is given by the Jakes power spectral density. Now, if we do not take the limit $N_i \to \infty$, then the stochastic reference model turns into the stochastic simulation model, as described in Section 4.1. Hence, the autocorrelation function of the stochastic simulation model for the process $\hat{\mu}_i(t)$ is

$$\hat{r}_{\mu_i\mu_i}(\tau) = \frac{\sigma_0^2}{N_i}\sum_{n=1}^{N_i} \cos\left\{2\pi f_{max} \sin\left[\frac{\pi}{2N_i}\left(n-\frac{1}{2}\right)\right]\cdot\tau\right\}. \tag{5.71}$$

The stochastic process $\hat{\mu}_i(t)$ will be ergodic with respect to the autocorrelation function. Then, regarding Subsection 4.3.4, it follows that $\hat{r}_{\mu_i\mu_i}(\tau) = \tilde{r}_{\mu_i\mu_i}(\tau)$ holds. Consequently, for the autocorrelation function of the deterministic process $\tilde{\mu}_i(t)$, we obtain the equation

$$\tilde{r}_{\mu_i\mu_i}(\tau) = \frac{\sigma_0^2}{N_i}\sum_{n=1}^{N_i}\cos\left\{2\pi f_{max}\sin\left[\frac{\pi}{2N_i}\left(n-\frac{1}{2}\right)\right]\cdot\tau\right\}. \tag{5.72}$$

If we now compare the above relation with the general expression (4.11), then the Doppler coefficients $c_{i,n}$ and the discrete Doppler frequencies $f_{i,n}$ can be identified with the equations

$$c_{i,n} = \sigma_0\sqrt{\frac{2}{N_i}} \tag{5.73}$$

and

$$f_{i,n} = f_{max}\sin\left[\frac{\pi}{2N_i}\left(n-\frac{1}{2}\right)\right], \tag{5.74}$$

respectively, for all $n = 1, 2, \ldots, N_i$ $(i = 1, 2)$. A deterministic process $\tilde{\mu}_i(t)$ designed with these parameters, has the time average $\tilde{m}_{\mu_i} = m_{\mu_i} = 0$ and the mean power $\tilde{\sigma}_{\mu_i}^2 = \sigma_0^2$. For all relevant values of N_i, the greatest common divisor $F_i = \gcd\{f_{i,n}\}_{n=1}^{N_i}$ is equal to zero (or very small), so that the period $T_i = 1/F_i$ becomes infinite (or very large). The uncorrelatedness of two deterministic processes $\tilde{\mu}_1(t)$ and $\tilde{\mu}_2(t)$ is again guaranteed by the convention $N_2 := N_1 + 1$.

The autocorrelation function $\tilde{r}_{\mu_i\mu_i}(\tau)$ computed according to (5.72) is presented in Figure 5.44(a) for $N_i = 7$ and in Figure 5.44(b) for $N_i = 21$.

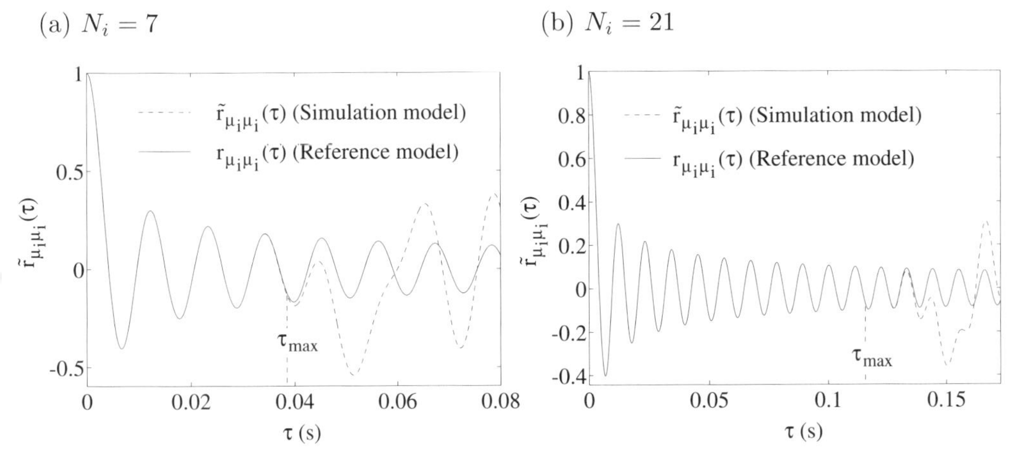

Figure 5.44: Autocorrelation function $\tilde{r}_{\mu_i\mu_i}(\tau)$ for (a) $N_i = 7$ and (b) $N_i = 21$ (MEDS, Jakes PSD, $f_{max} = 91\,\text{Hz}$, $\sigma_0^2 = 1$).

In connection with the Jakes power spectral density, the following rule of thumb applies: let there be N_i harmonic functions are given, then the approximation $r_{\mu_i\mu_i}(\tau) \approx \tilde{r}_{\mu_i\mu_i}(\tau)$ is quite good up to the N_ith zero-crossing of $r_{\mu_i\mu_i}(\tau)$.

Since the relation $c_{i,n} = \sigma_0\sqrt{2/N_i}$, which has already been obtained several times for the Doppler coefficients, also applies here, the mean-square error $E_{p_{\mu_i}}$ [see (4.83)] is again identical to the results shown in Figure 5.14(b). The evaluation of the mean-square error $E_{r_{\mu_i\mu_i}}$ [see (4.84)] in terms of N_i results in the graph depicted in Figure 5.45. As shown in this figure, the comparison with the L_p-norm method clearly demonstrates that even by applying numerical optimization techniques, only minor improvements can be achieved.

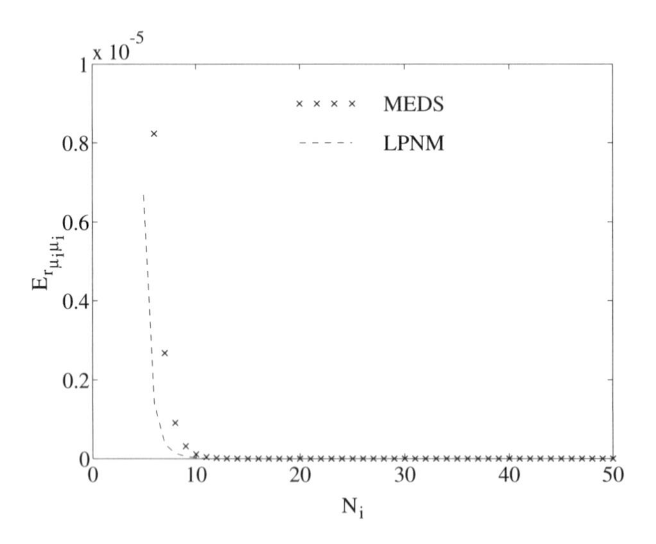

Figure 5.45: Mean-square error $E_{r_{\mu_i\mu_i}}$ (MEDS, Jakes PSD, $f_{max} = 91\,\text{Hz}$, $\sigma_0^2 = 1$, $\tau_{max} = N_i/(2f_{max})$).

Putting the equations (5.73) and (5.74) into (4.22) and making use of the relation (3.29), we can easily show that $\tilde{\beta}_i = \beta$ holds. In other words, the model error $\Delta\beta_i$ is equal to zero for all $N_i \in \mathbb{N} \setminus \{0\}$. Since we have $\tilde{\sigma}^2_{\mu_i} = \sigma_0^2$ and $\tilde{\beta}_i = \tilde{\beta} = \beta$ in the present case, it follows from (3.15b) and (4.25) that

$$\tilde{B}^{(2)}_{\mu\mu} = \tilde{B}^{(2)}_{\mu_i\mu_i} = B^{(2)}_{\mu_i\mu_i} = B^{(2)}_{\mu\mu} \tag{5.75}$$

holds. Hence, the Doppler spread of the simulation model is identical to that of the reference model. This is exactly the reason why this procedure is called the 'method of exact Doppler spread'.

The time-domain simulation will be restricted here to the emulation of the level-crossing rate, where we now choose the couple $(N_1, N_2) = (5, 6)$ and proceed, apart from that, exactly as in Subsection 5.1.3. Even for such small numbers of harmonic functions, the simulation results match the analytical results very well, as can be seen when considering the results shown in Figure 5.46.

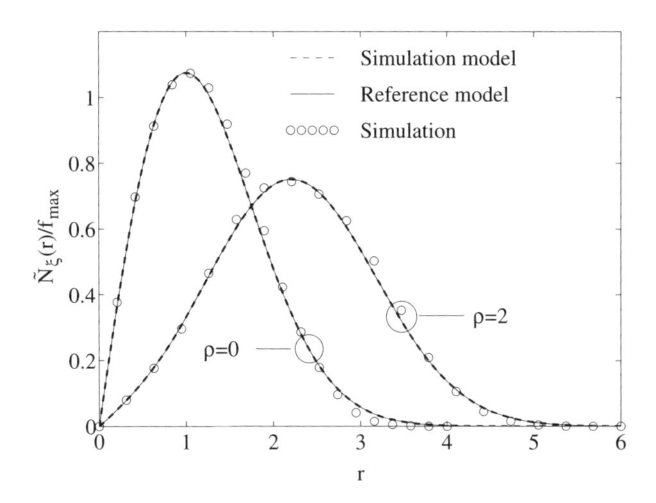

Figure 5.46: Normalized level-crossing rate $\tilde{N}_\xi(r)/f_{max}$ for $N_1 = 5$ and $N_2 = 6$ (MEDS, Jakes PSD, $f_{max} = 91\,\text{Hz}$, $\sigma_0^2 = 1$).

Gaussian power spectral density: From (5.33) and (5.74), we can see that the latter equation is obtained if n is substituted by $n - 1/2$ in the one mentioned first. This points out that a close relation exists between the method of equal areas and the method of exact Doppler spread. We will briefly return to this relation at the end of this subsection. At first, it seems obvious to make an attempt to apply the mapping $n \to n - 1/2$ to (5.43a) and (5.43b) as well, so that the discrete Doppler frequencies $f_{i,n}$ are now computed by means of the relations

$$\frac{2n-1}{2N_i} - \operatorname{erf}\left(\frac{f_{i,n}}{f_c}\sqrt{\ln 2}\right) = 0\,, \quad \forall n = 1, 2, \ldots, N_i - 1\,, \tag{5.76a}$$

and

$$f_{i,N_i} = \sqrt{\frac{\beta N_i}{(2\pi\sigma_0)^2} - \sum_{n=1}^{N_i-1} f_{i,n}^2}\,, \tag{5.76b}$$

where the latter equation again guarantees that the model error $\Delta\beta_i$ is equal to zero for all $N_i = 1, 2, \ldots$ ($i = 1, 2$). For the Doppler coefficients $c_{i,n}$, the expression (5.73) still remains valid.

The autocorrelation function $\tilde{r}_{\mu_i\mu_i}(\tau)$ can be computed according to (4.11) with the model parameters obtained in this way. Figures 5.47(a) and 5.47(b) give us an impression of the behaviour of $\tilde{r}_{\mu_i\mu_i}(\tau)$ for $N_i = 7$ and $N_i = 21$, respectively.

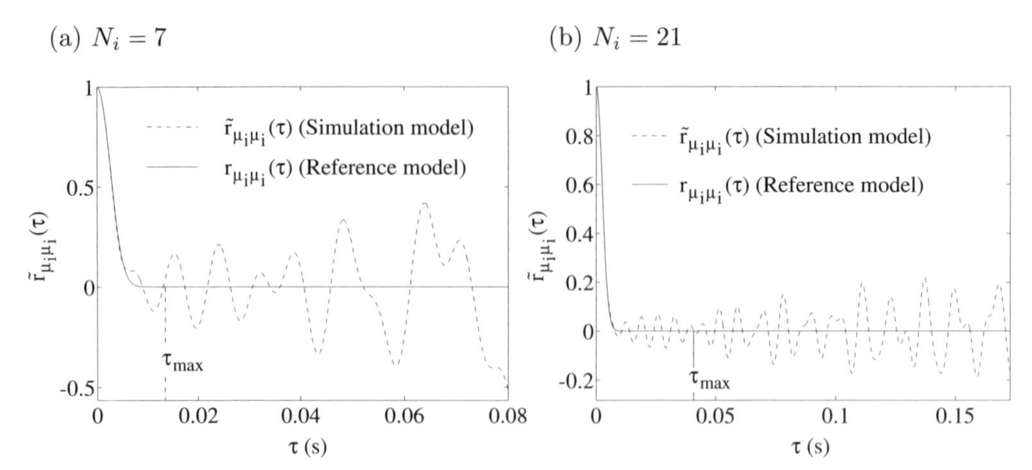

Figure 5.47: Autocorrelation function $\tilde{r}_{\mu_i\mu_i}(\tau)$ for (a) $N_i = 7$ and (b) $N_i = 21$ (MEDS, Gaussian PSD, $f_c = \sqrt{\ln 2} f_{max}$, $f_{max} = 91\,\text{Hz}$, $\sigma_0^2 = 1$).

The mean-square error $E_{r_{\mu_i\mu_i}}$ [see (4.84)], which results from the application of the present method and the standard L_p-norm method, is presented in the succeeding Figure 5.48 as a function of N_i. Unlike the case of the Jakes power spectral density, for a small number of harmonic functions N_i, the method of exact Doppler spread delivers clearly higher values for $E_{r_{\mu_i\mu_i}}$ than the L_p-norm method does. However, if $N_i \geq 25$, no considerable improvements are achievable by means of numerical optimization techniques.

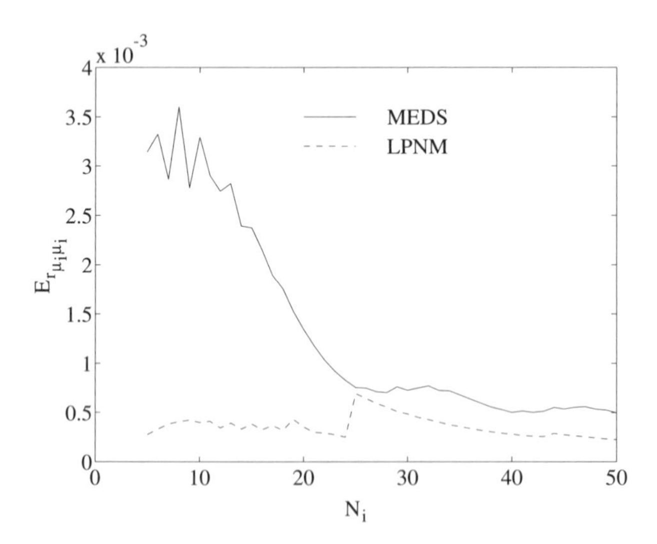

Figure 5.48: Mean-square error $E_{r_{\mu_i\mu_i}}$ (MEDS, Gaussian PSD, $f_c = \sqrt{\ln 2} f_{max}$, $f_{max} = 91\,\text{Hz}$, $\sigma_0^2 = 1$, $\tau_{max} = N_i/(2\kappa_c f_c)$, $\kappa_c = 2\sqrt{2/\ln 2}$).

Due to $\tilde{\sigma}^2_{\mu_i} = \sigma_0^2$ and $\Delta\beta_i = 0$, i.e., $\tilde{\beta}_i = \beta$, (5.75) holds here again.

It remains worth mentioning that the analytical results of the level-crossing rate can be confirmed very precisely by simulation, even if N_1 and N_2 are chosen very low, e.g., $(N_1, N_2) = (5, 6)$. One may therefore study the results shown in the following Figure 5.49.

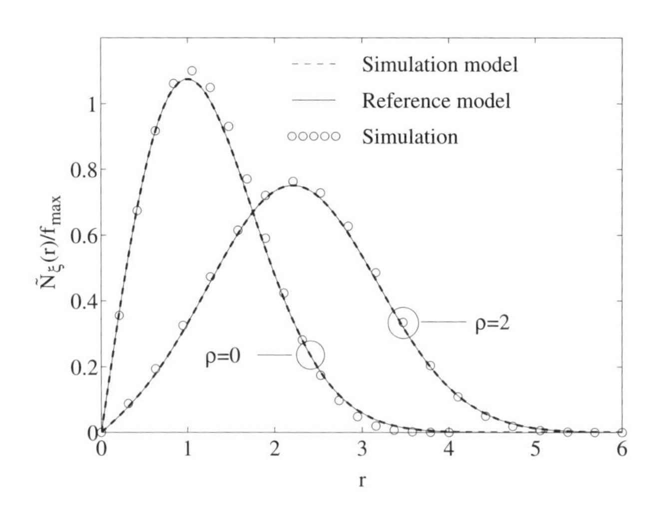

Figure 5.49: Normalized level-crossing rate $\tilde{N}_\xi(r)/f_{max}$ for $N_1 = 5$ and $N_2 = 6$ (MEDS, Gaussian PSD, $f_c = \sqrt{\ln 2} f_{max}$, $f_{max} = 91\,\text{Hz}$, $\sigma_0^2 = 1$).

As mentioned before, the method of equal areas is closely related to the method of exact Doppler spread. In fact, the former method can be transformed into the latter one and vice versa. For example, if we replace the right-hand side of (5.26) by $\sigma_0^2/(4N_i)$ and $f_{i,n}$ by $f_{i,2n-1}$ in (5.27), then we obtain (5.29), if n is replaced by $n - 1/2$ there. Consequently, for (5.33) and (5.43), we exactly obtain the equations (5.74) and (5.76), respectively. A similar relationship exists between the Monte Carlo method and the method of exact Doppler spread. For example, if we substitute the random variable $u_n \in (0, 1]$ in (5.48) by the deterministic quantity $(n - 1/2)/N_i$ for all $n = 1, 2, \ldots, N_i$ $(i = 1, 2)$, then we again obtain (5.74).

5.1.7 Jakes Method (JM)

The Jakes method (JM) [Jak93] has been developed exclusively for the Jakes power spectral density. Not only for completeness, but also due to its great popularity, this so-called classical method will be described here as well. A detailed description will not be given — this can be found in [Jak93, p. 67ff.]. Instead of this, we will primarily restrict ourselves to the analysis of the performance investigated in [Pae98e].

Jakes power spectral density: After rewriting the expressions for the parameters of the simulation model given in [Jak93, p. 70] by taking the notation chosen here into account, the following relations hold for the Doppler coefficients $c_{i,n}$, the discrete Doppler frequencies $f_{i,n}$, and the Doppler phases $\theta_{i,n}$:

$$c_{i,n} = \begin{cases} \dfrac{2\sigma_0}{\sqrt{N_i - \frac{1}{2}}} \sin\left(\dfrac{\pi n}{N_i - 1}\right), & n = 1, 2, \ldots, N_i - 1, \quad i = 1, \\ \dfrac{2\sigma_0}{\sqrt{N_i - \frac{1}{2}}} \cos\left(\dfrac{\pi n}{N_i - 1}\right), & n = 1, 2, \ldots, N_i - 1, \quad i = 2, \\ \dfrac{\sigma_0}{\sqrt{N_i - \frac{1}{2}}}, & n = N_i, \quad i = 1, 2, \end{cases} \tag{5.77}$$

$$f_{i,n} = \begin{cases} f_{max} \cos\left(\dfrac{n\pi}{2N_i - 1}\right), & n = 1, 2, \ldots, N_i - 1, \quad i = 1, 2, \\ f_{max}, & n = N_i, \quad i = 1, 2 \end{cases} \tag{5.78}$$

$$\theta_{i,n} = 0, \quad n = 1, 2, \ldots, N_i, \quad i = 1, 2, \tag{5.79}$$

where $N_1 = N_2$. The Doppler coefficients $c_{i,n}$ were scaled here in such a way that the mean power $\tilde{\sigma}^2_{\mu_i}$ of $\tilde{\mu}_i(t)$ meets the relation $\tilde{\sigma}^2_{\mu_i} = \sigma_0^2$ for $i = 1, 2$. Due to $f_{i,n} \neq 0$, the following relation holds for the time average: $\tilde{m}_{\mu_i} = m_{\mu_i} = 0$ $(i = 1, 2)$.

The resulting power spectral densities $\tilde{S}_{\mu_1\mu_1}(f)$ and $\tilde{S}_{\mu_2\mu_2}(f)$ as well as the corresponding autocorrelation functions $\tilde{r}_{\mu_1\mu_1}(\tau)$ and $\tilde{r}_{\mu_2\mu_2}(\tau)$ are shown in Figures 5.50(a)–5.50(d) for $N_1 = N_2 = 9$. Even for small values of τ, it can be seen from Figures 5.50(b) and 5.50(d) that the autocorrelation functions of the deterministic processes $\tilde{\mu}_1(t)$ and $\tilde{\mu}_2(t)$ strongly deviate from the ideal autocorrelation function $r_{\mu_i\mu_i}(\tau) = \sigma_0^2 J_0(2\pi f_{max}\tau)$. On the other hand, as Figure 5.50(f) reveals, the autocorrelation function $\tilde{r}_{\mu\mu}(\tau)$ of the complex deterministic process $\tilde{\mu}(t)$ over the interval $\tau \in [0, \tau_{max}]$ matches $r_{\mu\mu}(\tau) = 2\sigma_0^2 J_0(2\pi f_{max}\tau)$ very well. Figure 5.50(e) shows the power spectral density $\tilde{S}_{\mu\mu}(f)$ that corresponds to the autocorrelation function $\tilde{r}_{\mu\mu}(\tau)$.

It is interesting that even for $N_i \to \infty$, the autocorrelation function $\tilde{r}_{\mu_i\mu_i}(\tau)$ does not tend to $r_{\mu_i\mu_i}(\tau)$. Instead of this, after substituting (5.77) and (5.78) into (4.11) and taking the limit $N_i \to \infty$ afterwards, we rather obtain the functions

$$\lim_{N_1 \to \infty} \tilde{r}_{\mu_1\mu_1}(\tau) = \frac{2\sigma_0^2}{\pi} \int_0^{\pi/2} [1 - \cos(4z)] \cos(2\pi f_{max}\tau \cos z) \, dz \tag{5.80a}$$

and

$$\lim_{N_2 \to \infty} \tilde{r}_{\mu_2\mu_2}(\tau) = \frac{2\sigma_0^2}{\pi} \int_0^{\pi/2} [1 + \cos(4z)] \cos(2\pi f_{max}\tau \cos z) \, dz, \tag{5.80b}$$

which, by making use of [Gra81, eq. (3.715.19)]

$$\int_0^{\pi/2} \cos(z \cos x) \cos(2nx) \, dx = (-1)^n \cdot \frac{\pi}{2} J_{2n}(z), \tag{5.81}$$

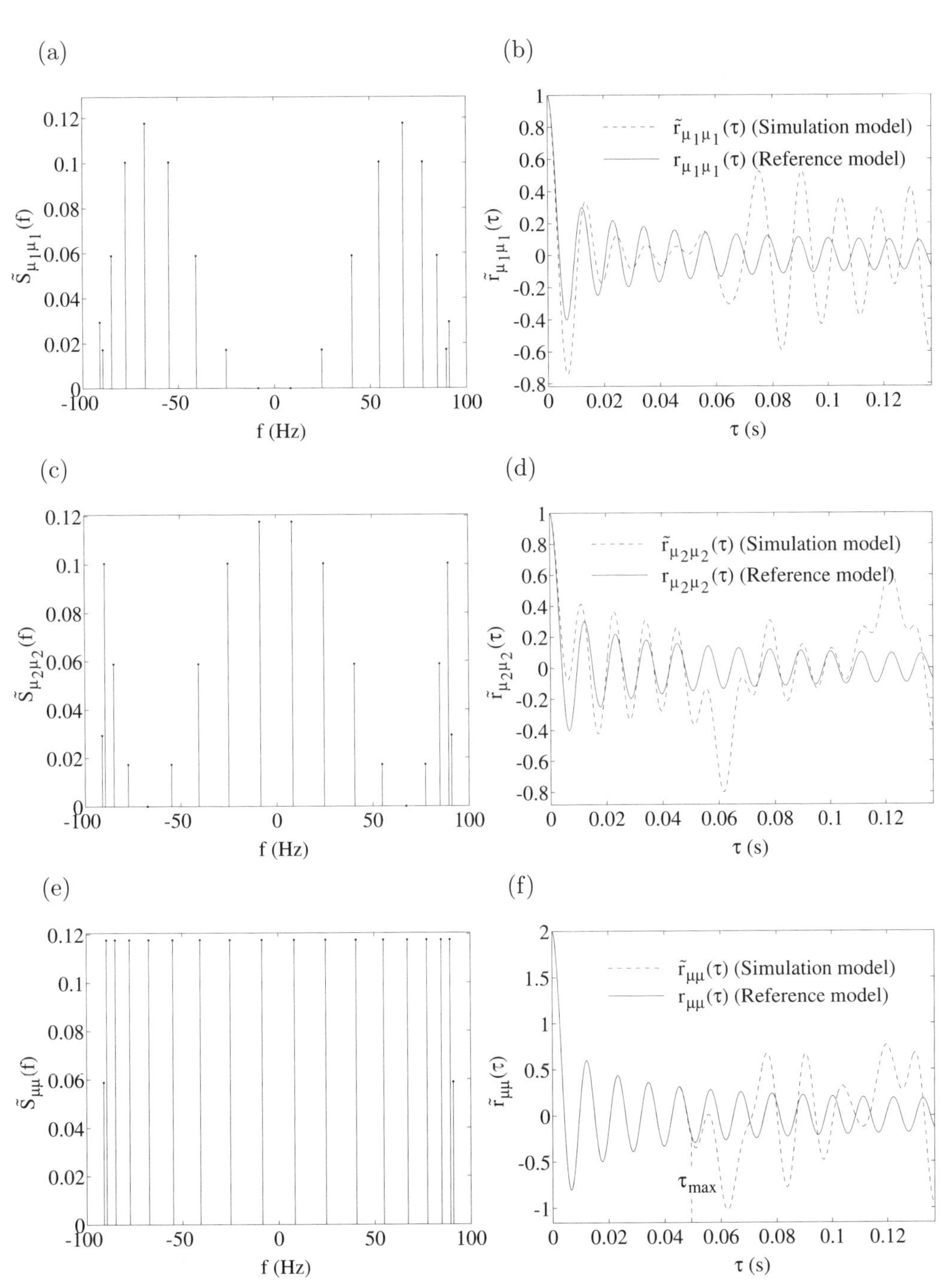

Figure 5.50: Power spectral densities and corresponding autocorrelation functions for $N_1 = N_2 = 9$: (a) $\tilde{S}_{\mu_1\mu_1}(f)$, (b) $\tilde{r}_{\mu_1\mu_1}(\tau)$, (c) $\tilde{S}_{\mu_2\mu_2}(f)$, (d) $\tilde{r}_{\mu_2\mu_2}(\tau)$, (e) $\tilde{S}_{\mu\mu}(f)$, and (f) $\tilde{r}_{\mu\mu}(\tau)$ (JM, Jakes PSD, $f_{max} = 91\,\text{Hz}$, $\sigma_0^2 = 1$).

can also be brought into the form

$$\lim_{N_1 \to \infty} \tilde{r}_{\mu_1\mu_1}(\tau) = \sigma_0^2 \big[J_0(2\pi f_{max}\tau) - J_4(2\pi f_{max}\tau) \big] \tag{5.82a}$$

and

$$\lim_{N_2 \to \infty} \tilde{r}_{\mu_2\mu_2}(\tau) = \sigma_0^2 \big[J_0(2\pi f_{max}\tau) + J_4(2\pi f_{max}\tau) \big] , \tag{5.82b}$$

as shown in [Pae98e]. Thus, even after taking the limit $N_i \to \infty$, the inequality $\tilde{r}_{\mu_i\mu_i}(\tau) \neq r_{\mu_i\mu_i}(\tau)$ $(i = 1, 2)$ holds. On the contrary, the autocorrelation function $\tilde{r}_{\mu\mu}(\tau)$ of the complex deterministic process $\tilde{\mu}(t) = \tilde{\mu}_1(t) + j\tilde{\mu}_2(t)$ does tend very well to the autocorrelation function $r_{\mu\mu}(\tau)$ of the reference model, as $N_i \to \infty$. This fact becomes immediately evident after substituting (5.82a) and (5.82b) into the general expression

$$\tilde{r}_{\mu\mu}(\tau) = \tilde{r}_{\mu_1\mu_1}(\tau) + \tilde{r}_{\mu_2\mu_2}(\tau) + j\big(\tilde{r}_{\mu_1\mu_2}(\tau) - \tilde{r}_{\mu_2\mu_1}(\tau)\big) \tag{5.83}$$

following from (2.71), and then making use of the relation $\tilde{r}_{\mu_1\mu_2}(\tau) = \tilde{r}_{\mu_2\mu_1}(\tau)$, which holds here, as we will see subsequently. Hence, one directly realizes that

$$\lim_{N_i \to \infty} \tilde{r}_{\mu\mu}(\tau) = r_{\mu\mu}(\tau) = 2\sigma_0^2 \, J_0(2\pi f_{max}\tau) \tag{5.84}$$

holds.

We furthermore want to analyse to which functions the power spectral densities $\tilde{S}_{\mu_1\mu_1}(f)$ and $\tilde{S}_{\mu_2\mu_2}(f)$ tend to in the limits $N_1 \to \infty$ and $N_2 \to \infty$, respectively. Therefore, we transform (5.82a) and (5.82b) into the spectral domain by means of the Fourier transform and obtain

$$\lim_{N_1 \to \infty} \tilde{S}_{\mu_1\mu_1}(f) = \begin{cases} \sigma_0^2 \cdot \dfrac{1 - \cos\left[4 \arcsin(f/f_{max})\right]}{\pi f_{max}\sqrt{1 - (f/f_{max})^2}} , & |f| \leq f_{max} , \\ 0 , & |f| > f_{max} , \end{cases} \tag{5.85a}$$

$$\lim_{N_2 \to \infty} \tilde{S}_{\mu_2\mu_2}(f) = \begin{cases} \sigma_0^2 \cdot \dfrac{1 + \cos\left[4 \arcsin(f/f_{max})\right]}{\pi f_{max}\sqrt{1 - (f/f_{max})^2}} , & |f| \leq f_{max} , \\ 0 , & |f| > f_{max} . \end{cases} \tag{5.85b}$$

Of course, we might as well have substituted (5.77) and (5.78) in $\tilde{S}_{\mu_i\mu_i}(f)$ given by (4.14). Then, after taking the limit $N_i \to \infty$, we would have obtained the results (5.85a) and (5.85b) stated above in an alternative way. If we now put these results into the Fourier transform of (5.83), then we obtain the Jakes power spectral density as expected, i.e.,

$$\lim_{N_i \to \infty} \tilde{S}_{\mu\mu}(f) = S_{\mu\mu}(f) = \begin{cases} \dfrac{2\sigma_0^2}{\pi f_{max}\sqrt{1 - (f/f_{max})^2}} , & |f| \leq f_{max} , \\ 0 , & |f| > f_{max} . \end{cases} \tag{5.86}$$

Consequently, as $N_i \to \infty$, it follows $\tilde{S}_{\mu\mu}(f) \to S_{\mu\mu}(f)$ but not $\tilde{S}_{\mu_i\mu_i}(f) \to S_{\mu_i\mu_i}(f)$ $(i = 1, 2)$.

To illustrate the results given above, we study Figure 5.51, where the power spectral densities $\tilde{S}_{\mu_1\mu_1}(f)$, $\tilde{S}_{\mu_2\mu_2}(f)$, and $\tilde{S}_{\mu\mu}(f)$ are presented together with the corresponding autocorrelation functions for the limit $N_i \to \infty$.

Using the Jakes method it has to be taken into account that the deterministic processes $\tilde{\mu}_1(t)$ and $\tilde{\mu}_2(t)$ are correlated, because $f_{1,n} = f_{2,n}$ holds according to (5.78) for all $n = 1, 2, \ldots, N_1$ $(N_1 = N_2)$. After substituting (5.77)–(5.79) into (4.13), we find the following expression for the cross-correlation function $\tilde{r}_{\mu_1\mu_2}(\tau)$

$$\tilde{r}_{\mu_1\mu_2}(\tau) = \frac{\sigma_0^2}{N_i - \frac{1}{2}} \left\{ \sum_{n=1}^{N_i - 1} \sin\left(\frac{2\pi n}{N_i - 1}\right) \cos\left[2\pi f_{max} \cos\left(\frac{n\pi}{2N_i - 1}\right)\tau\right] + \frac{1}{2}\cos(2\pi f_{max}\tau) \right\}. \tag{5.87}$$

Since $\tilde{r}_{\mu_1\mu_2}(\tau)$ is a real-valued and even function, it can be shown, by using (2.49), that $\tilde{r}_{\mu_2\mu_1}(\tau) = \tilde{r}^*_{\mu_1\mu_2}(-\tau) = \tilde{r}_{\mu_1\mu_2}(\tau)$ holds. Figure 5.52(b) conveys an impression of the behaviour of the cross-correlation function $\tilde{r}_{\mu_1\mu_2}(\tau)$ computed according to (5.87). Figure 5.52(a) next to it shows the corresponding cross-power spectral density $\tilde{S}_{\mu_1\mu_2}(f)$ computed by using (4.16). The results in these figures have been obtained by choosing $N_1 = N_2 = 9$.

One can see that a strong correlation between $\tilde{\mu}_1(t)$ and $\tilde{\mu}_2(t)$ exists. This problem was taken up in [Den93], where a modification for the Jakes method was suggested, which is essentially based on a modification of the relation (5.77). However, this variant merely guarantees that $\tilde{r}_{\mu_1\mu_2}(\tau)$ is 0 at the origin $\tau = 0$. In order to guarantee that $\tilde{r}_{\mu_1\mu_2}(\tau) = 0$ holds for all values of τ, the deterministic processes $\tilde{\mu}_1(t)$ and $\tilde{\mu}_2(t)$ have to be realized with disjoint sets $\{f_{1,n}\}$ and $\{f_{2,n}\}$.

The question, whether the cross-correlation function $\tilde{r}_{\mu_1\mu_2}(\tau)$ vanishes for $N_i \to \infty$, will be answered in the following. Therefore, we let N_i tend to infinity in (5.87), so that we find the integral

$$\lim_{N_i \to \infty} \tilde{r}_{\mu_1\mu_2}(\tau) = \frac{2\sigma_0^2}{\pi} \int_0^{\pi/2} \sin(4z) \cos(2\pi f_{max}\tau \cos z)\, dz\,, \tag{5.88}$$

which has to be solved numerically. The result of the numerical integration is shown in Figure 5.53(b). Obviously, even when the number of harmonic functions N_i is infinite, the correlation between $\tilde{\mu}_1(t)$ and $\tilde{\mu}_2(t)$ does not vanish. Consequently, $\tilde{r}_{\mu_1\mu_2}(\tau) \to r_{\mu_1\mu_2}(\tau)$ does not hold for $N_i \to \infty$.

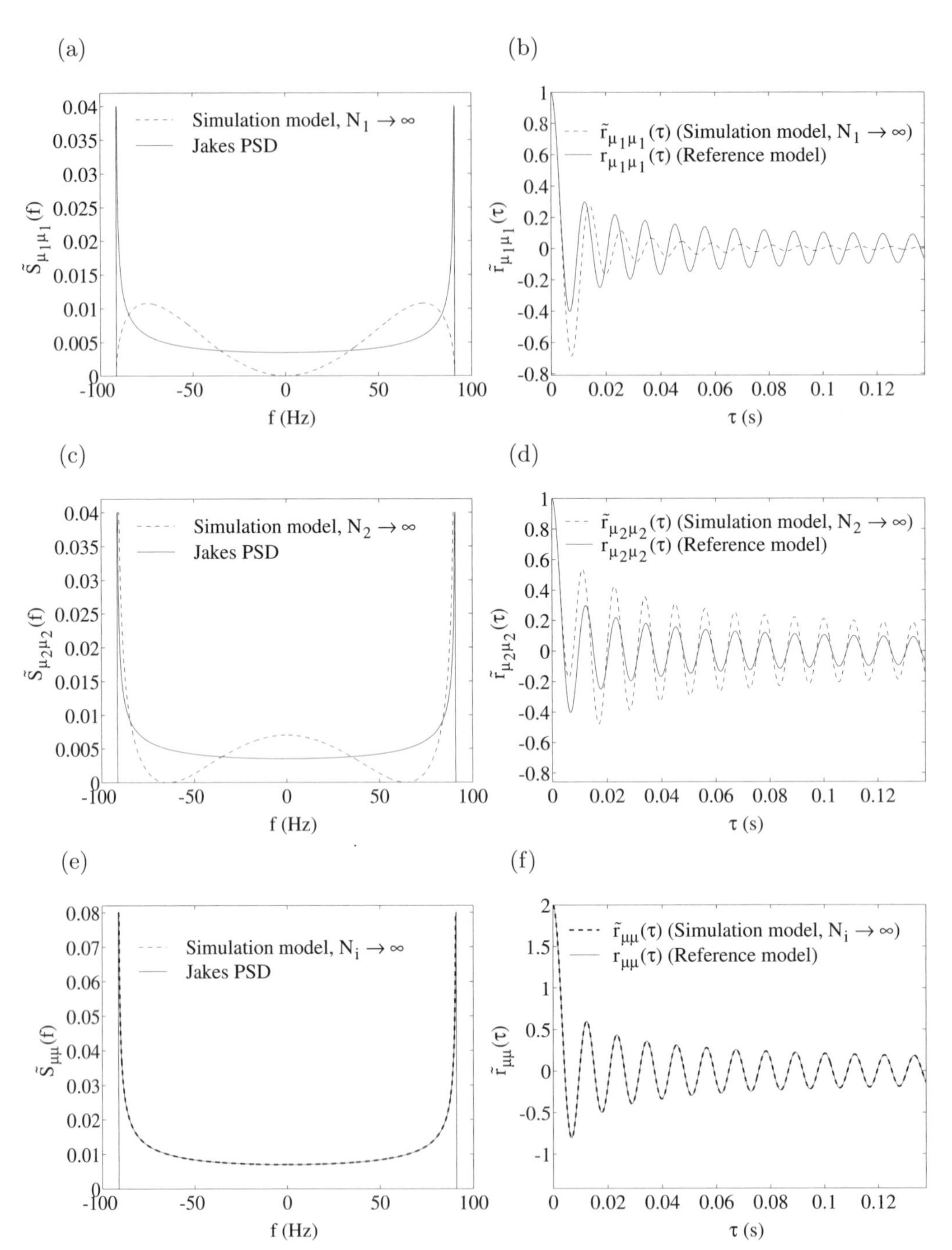

Figure 5.51: Power spectral densities and corresponding autocorrelation functions for $N_1 \to \infty$ and $N_2 \to \infty$: (a) $\tilde{S}_{\mu_1\mu_1}(f)$, (b) $\tilde{r}_{\mu_1\mu_1}(\tau)$, (c) $\tilde{S}_{\mu_2\mu_2}(f)$, (d) $\tilde{r}_{\mu_2\mu_2}(\tau)$, (e) $\tilde{S}_{\mu\mu}(f)$, and (f) $\tilde{r}_{\mu\mu}(\tau)$ (JM, Jakes PSD, $f_{max} = 91\,\text{Hz}$, $\sigma_0^2 = 1$).

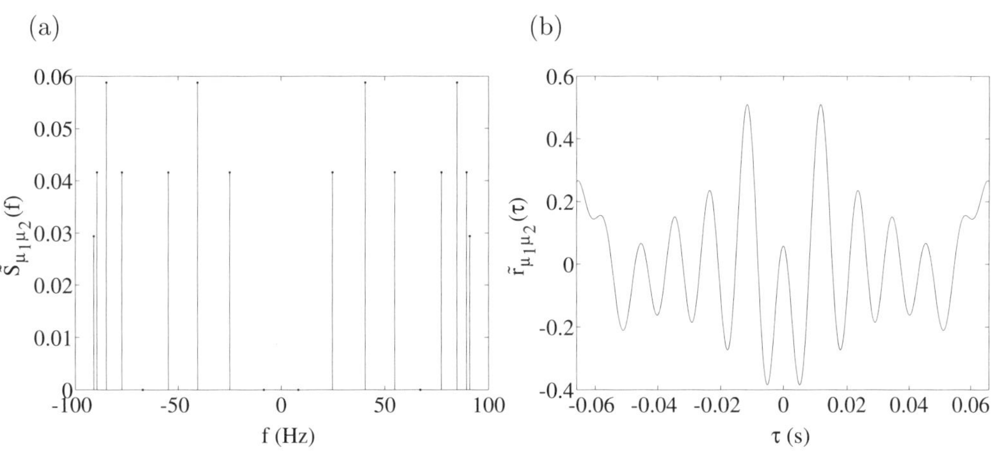

Figure 5.52: (a) Cross-power spectral density $\tilde{S}_{\mu_1\mu_2}(f)$ and (b) cross-correlation function $\tilde{r}_{\mu_1\mu_2}(\tau)$ for $N_1 = N_2 = 9$ (JM, Jakes PSD, $f_{max} = 91\,\text{Hz}$, $\sigma_0^2 = 1$).

In case of the limit $N_i \to \infty$, we obtain the following closed-form expression for the cross-power spectral density $\tilde{S}_{\mu_1\mu_2}(f)$ after performing the Fourier transform of (5.88)

$$\lim_{N_i\to\infty} \tilde{S}_{\mu_1\mu_2}(f) = \begin{cases} \sigma_0^2 \cdot \dfrac{\sin\left[4\arccos(|f|/f_{max})\right]}{\pi f_{max}\sqrt{1-(f/f_{max})^2}}\,, & |f| \leq f_{max}\,, \\ 0\,, & |f| > f_{max}\,. \end{cases} \tag{5.89}$$

The evaluation result of this equation is illustrated in Figure 5.53(a). In contrast to the Jakes power spectral density (3.8), which becomes singular at $f = \pm f_{max}$, the cross-power spectral density (5.89) takes on the finite value $4/(\pi f_{max})$ at these points, i.e., $\tilde{S}_{\mu_1\mu_2}(\pm f_{max}) = 4/(\pi f_{max})$ holds, which can easily be proven by using the rule of de l'Hospital [Bro91].

Expediently, the mean-square error (4.84) of the autocorrelation function is in this case evaluated with respect to $\tilde{r}_{\mu_1\mu_1}(\tau)$, $\tilde{r}_{\mu_2\mu_2}(\tau)$, and $\tilde{r}_{\mu\mu}(\tau)$. The obtained results are depicted in Figure 5.54(a) as a function of N_i. Using the Jakes method, the Doppler coefficients $c_{i,n}$ partly deviate considerably from the (quasi-)optimal quantities $c_{i,n} = \sigma_0\sqrt{2/N_i}$. This inevitably leads to an increase in the mean-square error $E_{p_{\mu_i}}$ [see (4.83)], as can clearly be seen in Figure 5.54(b).

For the Jakes method, we have: $N_1 = N_2$ and $f_{1,n} = f_{2,n}\,\forall\, n = 1, 2, \ldots, N_1\,(N_2)$. But, on the other hand, $c_{1,n} \neq c_{2,n}$ still holds for almost all $n = 1, 2, \ldots, N_1\,(N_2)$. Thus, it is to be expected that the model errors $\Delta\beta_1$ and $\Delta\beta_2$ are different for a given number of harmonic functions N_i. One should therefore study Figure 5.55, where the relative model errors $\Delta\beta_1/\beta$ and $\Delta\beta_2/\beta$ are presented.

Due to $\Delta\beta_1 \neq \Delta\beta_2$, we must use the expression (B.13) for the computation of the level-crossing rate $\tilde{N}_\xi(r)$, where β_1 and β_2 have to be substituted by $\tilde{\beta}_1 = \beta + \Delta\beta_1$ and

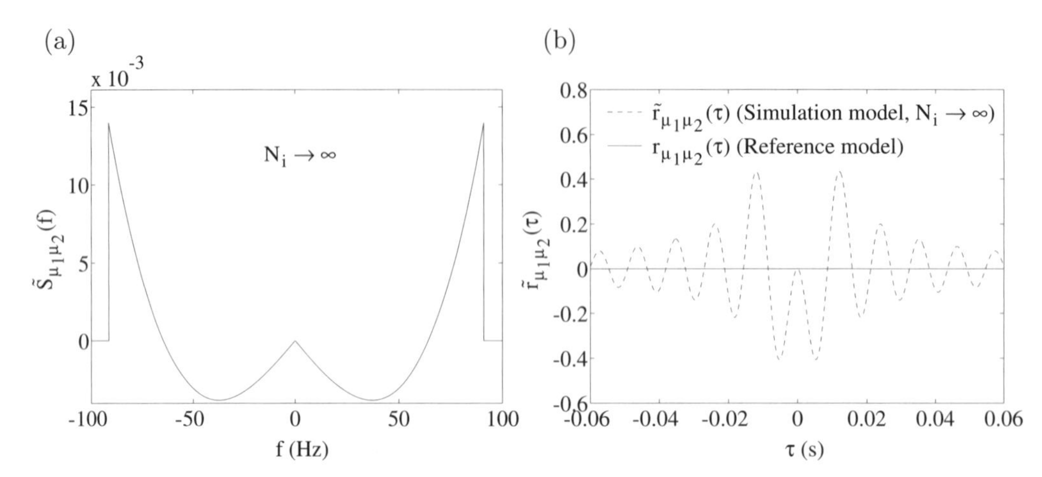

Figure 5.53: (a) Cross-power spectral density $\tilde{S}_{\mu_1\mu_2}(f)$ and (b) cross-correlation function $\tilde{r}_{\mu_1\mu_2}(\tau)$ for $N_i \rightarrow \infty$ (JM, Jakes PSD, $f_{max} = 91\,\text{Hz}$, $\sigma_0^2 = 1$).

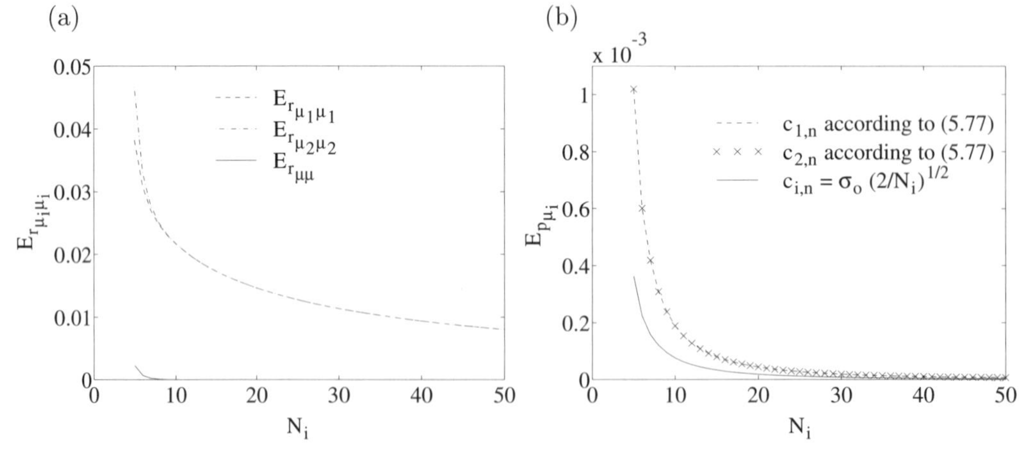

Figure 5.54: (a) Mean-square errors $E_{r_{\mu_1\mu_1}}$, $E_{r_{\mu_2\mu_2}}$, and $E_{r_{\mu\mu}}$ with $\tau_{max} = N_i/(2f_{max})$ and (b) mean-square error $E_{p_{\mu_i}}$ (JM, Jakes PSD, $f_{max} = 91\,\text{Hz}$, $\sigma_0^2 = 1$).

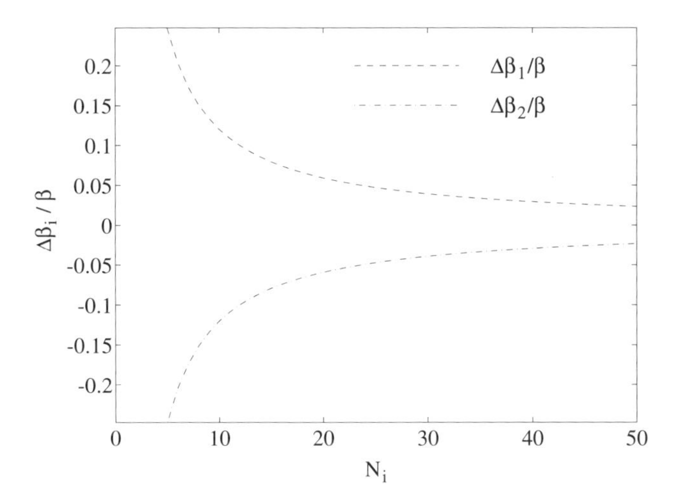

Figure 5.55: Relative model errors $\Delta\beta_1/\beta$ and $\Delta\beta_2/\beta$ (JM, Jakes PSD, $f_{max} = 91\,\text{Hz}$, $\sigma_0^2 = 1$).

$\tilde{\beta}_2 = \beta + \Delta\beta_2$, respectively. Generally speaking, in case of a correlation between $\tilde{\mu}_1(t)$ and $\tilde{\mu}_2(t)$, the level-crossing rate $\tilde{N}_\xi(r)$ also depends on the quantities $\dot{\tilde{r}}_{\mu_1\mu_2}(0)$ and $\ddot{\tilde{r}}_{\mu_1\mu_2}(0)$. Since it follows immediately from (5.87) that $\dot{\tilde{r}}_{\mu_1\mu_2}(0)$ is equal to zero, and furthermore the influence of $\ddot{\tilde{r}}_{\mu_1\mu_2}(0)$ on $\tilde{N}_\xi(r)$ is quite small, this dependency will be ignored here. Figure 5.56(a) presents the analytical results for $N_\xi(r)/f_{max}$ as well as for $\tilde{N}_\xi(r)/f_{max}$ with $N_1 = N_2 = 9$. This figure also shows the corresponding simulation results, which match the analytical solutions for $N_\xi(r)/f_{max}$ and $\tilde{N}_\xi(r)/f_{max}$ very well.

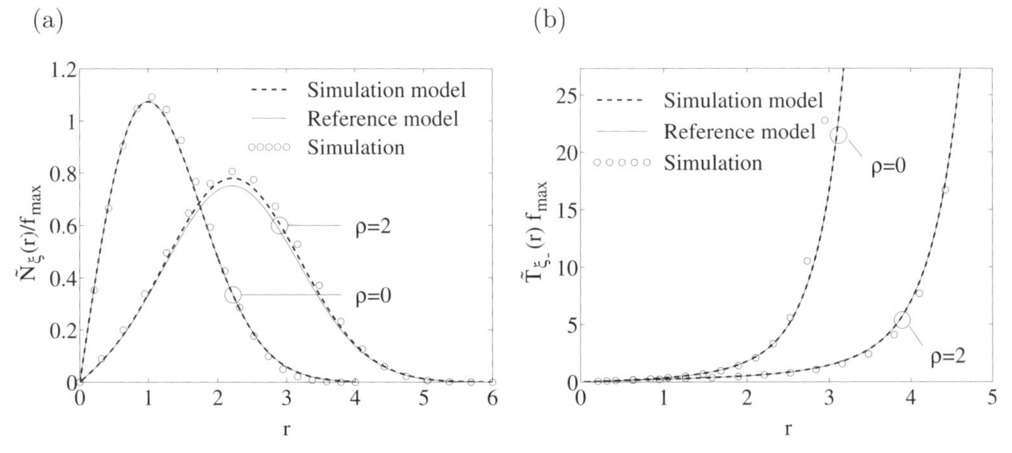

Figure 5.56: (a) Normalized level-crossing rate $\tilde{N}_\xi(r)/f_{max}$ and (b) normalized average duration of fades $\tilde{T}_{\xi_-}(r) \cdot f_{max}$ for $N_1 = N_2 = 9$ (JM, Jakes PSD, $f_{max} = 91\,\text{Hz}$, $\sigma_0^2 = 1$).

When using the Jakes method, it seems that the relatively large deviations between the autocorrelation functions $\tilde{r}_{\mu_i\mu_i}(\tau)$ and $r_{\mu_i\mu_i}(\tau)$ as well as the cross-correlation function $\tilde{r}_{\mu_1\mu_2}(\tau)$ different from zero, do not have an excessively negative influence on $\tilde{N}_\xi(r)$. By the way, this also holds for the average duration of fades $\tilde{T}_{\xi_-}(r)$ [see Figure 5.56(b)] as well as for the probability density function $\tilde{p}_{0_-}(\tau_-; r)$ of the fading intervals τ_- at low levels r [see Figure 5.57]. From this, however, we may not conclude that especially the influence of the cross-correlation function $\tilde{r}_{\mu_1\mu_2}(\tau)$ on $\tilde{N}_\xi(r)$ and therefore also on $\tilde{T}_{\xi_-}(r)$ can be ignored in all cases. This rather depends on the specific type of the cross-correlation function $\tilde{r}_{\mu_1\mu_2}(\tau)$. In the following Chapter 6, we will see that certain classes of cross-correlation functions really exist, which not only affect the statistical properties of higher orders, but also have an influence on the probability density function of the signal amplitude. In this way, it is possible to increase the flexibility of the statistical properties of mobile fading channel models considerably.

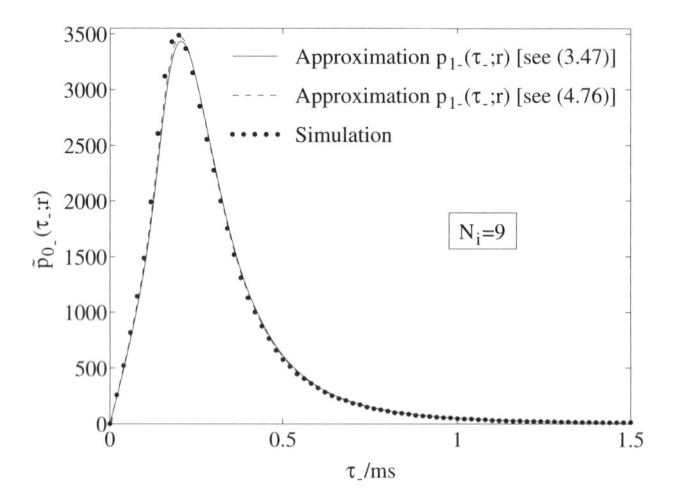

Figure 5.57: Probability density function $\tilde{p}_{0_-}(\tau_-; r)$ of the fading intervals τ_- at the level $r = 0.1$ (JM, Jakes PSD, $f_{max} = 91\,\text{Hz}$, $\sigma_0^2 = 1$).

Summarizing, we can record the fact that the essential disadvantage of the Jakes method is not to be seen in the cross-correlation function different from zero, but in the fact that the deterministic processes $\tilde{\mu}_1(t)$ and $\tilde{\mu}_2(t)$ are not optimally Gaussian distributed [cf. Figure 5.54(b)] for a given number of harmonic functions N_i. Since the loss of performance is not too high and can easily be compensated by a slight increase of the number of harmonic functions N_i, we can all in all say that in case N_i is chosen higher or equal to nine, the Jakes method is quite suitable for the modelling of Rayleigh and Rice processes with the classical Doppler power spectral density given by (3.8) [Pae98e]. Finally, it should also be noted that an implementation technique of the Jakes method on a signal processor has been described in [Cas88, Cas90].

5.2 METHODS FOR THE COMPUTATION OF THE DOPPLER PHASES

In this section, we will briefly deal with the significance of the Doppler phases $\theta_{i,n}$. Moreover, we will also make a suggestion for the deterministic computation of these quantities.

Except for the Jakes method, where the Doppler phases $\theta_{i,n}$ are equal to zero per definition, we have assumed that for all other parameter computation methods treated in Section 5.1 the Doppler phases $\theta_{i,n}$ are realizations of a random variable uniformly distributed within the interval $(0, 2\pi]$. Let us in the following assume that the set of Doppler coefficients $\{c_{i,n}\}$ and the set of discrete Doppler frequencies $\{f_{i,n}\}$ have been computed with the method of exact Doppler spread. Then, for two determined events $\{\theta_{1,n}\}_{n=1}^{N_1}$ and $\{\theta_{2,n}\}_{n=1}^{N_2}$ of the size $N_1 = 7$ and $N_2 = 8$, respectively, the behaviour of the resulting deterministic Rayleigh process $\tilde{\zeta}(t)$ is as shown in Figure 5.58(a). Here, it has to be taken into account that different events $\{\theta_{i,n}\}_{n=1}^{N_i}$ always result in different realizations for $\tilde{\zeta}(t)$. However, all of these different realizations have the same statistical properties, since the underlying stochastic processes $\hat{\mu}_1(t)$ and $\hat{\mu}_2(t)$ are ergodic with respect to the autocorrelation function. Moreover, with the method of exact Doppler spread it is guaranteed that due to the definition $N_2 := N_1 + 1$, the relation $f_{1,n} \neq \pm f_{2,m}$ holds for all $n = 1, 2, \ldots, N_1$ and $m = 1, 2, \ldots, N_2$, so that the cross-correlation function $\tilde{r}_{\mu_1\mu_2}(\tau)$ [cf. (4.13)], which in general depends on $\theta_{i,n}$, is equal to zero. Since the Doppler phases $\theta_{i,n}$ have no influence on the statistical properties of $\tilde{\zeta}(t)$ if the underlying deterministic Gaussian processes $\tilde{\mu}_1(t)$ and $\tilde{\mu}_2(t)$ are uncorrelated, we are inclined to set the Doppler phases $\theta_{i,n}$ equal to zero. In this case, however, we obtain $\tilde{\mu}_i(0) = \sigma_0\sqrt{2N_i}$ $(i = 1, 2)$, so that the deterministic Rayleigh process $\tilde{\zeta}(t)$ takes its maximum value $2\sigma_0\sqrt{N_1 + 1/2}$ at the time-instant $t = 0$, i.e., $\tilde{\zeta}(0) = 2\sigma_0\sqrt{N_1 + 1/2}$. This leads to the typical transient behaviour depicted in Figure 5.58(b). As we can see in Figure 5.58(c), a similar effect is also obtained, if the Doppler phases $\theta_{i,n}$ are computed deterministically, according to $\theta_{i,n} = 2\pi n/N_i$ $(n = 1, 2, \ldots, N_i$ and $i = 1, 2)$. A simple possibility to avoid the transient behaviour around the origin is to substitute the time variable t by $t + T_0$, where T_0 is a positive real-valued quantity, which has to be chosen sufficiently high. It should therefore be noted that the substitution $t \to t + T_0$ is equivalent to the substitution $\theta_{i,n} \to \theta_{i,n} + 2\pi f_{i,n} T_0$, which leads to the fact that for different values of n, the transformed Doppler phases now no longer have any rational relation to each other.

A further possibility would be [Pae98b] to introduce a standard phase vector $\vec{\Theta}_i$ with N_i deterministic components according to

$$\vec{\Theta}_i = \left(2\pi\frac{1}{N_i + 1}, 2\pi\frac{2}{N_i + 1}, \ldots, 2\pi\frac{N_i}{N_i + 1}\right) \tag{5.90}$$

and to regard the Doppler phases $\theta_{i,n}$ as components of the so-called Doppler phase vector

$$\vec{\theta}_i = (\theta_{i,1}, \theta_{i,2}, \ldots, \theta_{i,N_i}) \,. \tag{5.91}$$

By identifying the components of the Doppler phase vector $\vec{\theta}_i$ with the permutated components of the standard phase vector $\vec{\Theta}_i$, the transient behaviour located around the origin of the time axis can be avoided from the start. In this connection, one may observe the simulation results of $\tilde{\zeta}(t)$ shown in Figure 5.58(d).

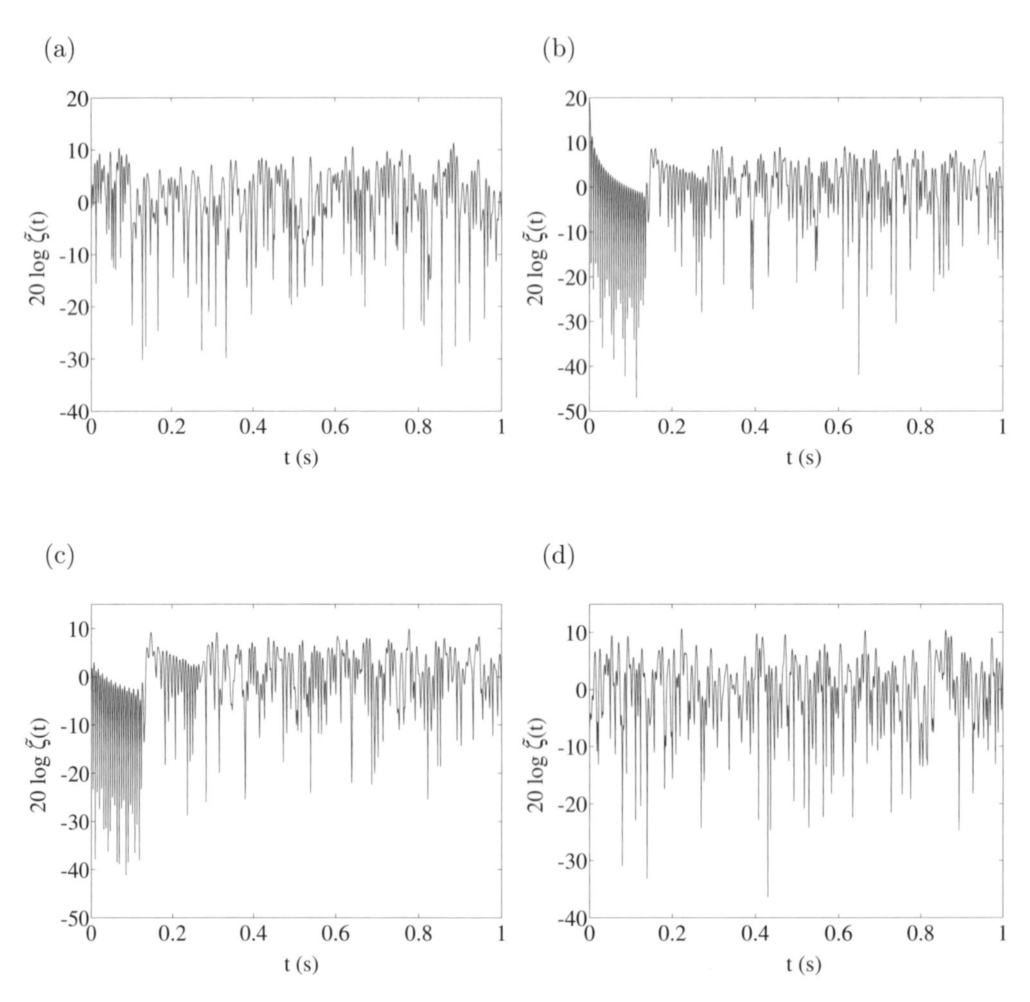

Figure 5.58: Influence of the phases $\theta_{i,n}$ on the transient behaviour of $\tilde{\zeta}(t)$ around the origin: (a) random phases $\theta_{i,n} \in (0, 2\pi]$, (b) $\theta_{i,n} = 0$, (c) $\theta_{i,n} = 2\pi n/N_i$ ($n = 1, 2, \ldots, N_i$), and (d) permutated phases (MEDS, Jakes PSD, $f_{max} = 91\,\text{Hz}$, $\sigma_0^2 = 1$, $N_1 = 21$, $N_2 = 22$).

By permuting the components of (5.90), it is possible to construct $N_i!$ different sets $\{\theta_{i,n}\}$ of Doppler phases. Thus, for any given sets of $\{c_{i,n}\}$ and $\{f_{i,n}\}$, altogether $N_1! \cdot N_2!$ deterministic Rayleigh processes $\tilde{\zeta}(t)$ with different time behaviour but identical statistical properties can be realized.

5.3 FADING INTERVALS OF DETERMINISTIC RAYLEIGH PROCESSES

The statistical properties of deterministic Rayleigh and Rice processes analysed so far, such as the probability density function of the amplitude and phase, the level-crossing rate, and the average duration of fades, are independent of the behaviour of the autocorrelation function $\tilde{r}_{\mu_i\mu_i}(\tau)$ $(i = 1, 2)$ for $\tau > 0$. In the following, we will again follow the question, which statistical properties depend on $\tilde{r}_{\mu_i\mu_i}(\tau)$ $(i = 1, 2)$ for $\tau > 0$ at all. Related to this question is the open problem of determining the size of the interval $[0, \tau_{max}]$ over which the approximation of $r_{\mu_i\mu_i}(\tau)$ by $\tilde{r}_{\mu_i\mu_i}(\tau)$ is relevant. Hence, we have to find a value for τ_{max} in such a way that further statistical properties of the simulation system can also hardly be distinguished from those of the reference system. In case of the Jakes power spectral density, where τ_{max} is related with N_i over $\tau_{max} = N_i/(2f_{max})$, we will see in the following that the number of harmonic functions N_i necessary for the simulation system can — at least for this kind of power spectral density — easily be determined.

We will therefore once again return to the probability density function $\tilde{p}_{0_-}(\tau_-; r)$ of the fading intervals of deterministic Rayleigh processes. Since an approximate solution with sufficient precision neither exists for $\tilde{p}_{0_-}(\tau_-; r)$ nor for $p_{0_-}(\tau_-; r)$ at medium and especially at high levels of r, this problem can only solved by means of simulation.

We will first carry out the simulation for the Jakes power spectral density with $f_{max} = 91\,\text{Hz}$ and $\sigma_0^2 = 1$ and determine the parameters of the simulation model by making use of the method of exact Doppler spread. Due to the advantages of this method (very good approximation of the autocorrelation function $r_{\mu_i\mu_i}(\tau) = \sigma_0^2 J_0(2\pi f_{max}\tau)$ from $\tau = 0$ to $\tau = \tau_{max} = N_i/(2f_{max})$, no model error, no correlation between $\tilde{\mu}_1(t)$ and $\tilde{\mu}_2(t)$, and, last but not least, the very good periodicity properties), the resulting deterministic simulation model will fulfil all essential demands. In this specific case, we may regard the simulation model designed with the couple $(N_1, N_2) = (100, 101)$ as the reference model. The simulation of the resulting discrete deterministic process $\tilde{\zeta}(kT)$ has been performed with the sampling interval $T_s = 0.5 \cdot 10^{-4}\,\text{s}$. The simulated samples of $\tilde{\zeta}(kT_s)$ have then been used to measure the probability density function $\tilde{p}_{0_-}(\tau_-; r)$ at a low level $(r = 0.1)$, a medium level $(r = 1)$, and a high level $(r = 2.5)$. All obtained results are shown in Figures 5.59(a)–(c) for various couples (N_1, N_2), where 10^7 fading intervals τ_- have been used for the determination of each probability density function $\tilde{p}_{0_-}(\tau_-; r)$. As can be seen from Figure 5.59(a), there is an excellent accordance between the results obtained for $\tilde{p}_{0_-}(\tau_-; r)$ and the theoretical approximation $p_{1_-}(\tau_-; r)$ [cf. (3.47)] at the low level $r = 0.1$. That was to be expected, since at deep fades, the probability that a fading interval is long, is very low, consequently, the probability that further level-crossings occur between t_1 and $t_2 = t_1 + \tau_-$ is negligible. Exactly for this case, the approximation $p_{0_-}(\tau_-; r) \approx p_{1_-}(\tau_-; r)$ turns out to be very useful. Figures 5.59(a) and (b) clearly depict that with $N_1 = 7$ and $N_2 = 8$, the number of harmonic functions chosen is sufficiently large, so that at least at low and medium levels, the obtained probability density functions $\tilde{p}_{0_-}(\tau_-; r)$ are hardly to be distinguished from the ones of the reference model $(N_1 = 100,\ N_2 = 101)$. As shown in Figure 5.59(c), clear differences in comparison with the reference model first occur if the level r is high $(r = 2.5)$ and if the simulation model is designed with $N_1 = 7$ and $N_2 = 8$ harmonic functions.

However, if the differences from the reference model will be negligible for this level too, then at least $N_1 = 21$ and $N_2 = 22$ harmonic functions will be required. A further increase of N_i makes no sense!

At this point, it should be noted that $N_1 = 7$ and $N_2 = 8$ harmonic functions are in general sufficient for the modelling of mobile radio channels, where channel models are often required to determine the bit error probability of digital transmission systems consisting of a transmitter, a channel model, and a receiver. Of course, $N_1 = 7$ and $N_2 = 8$ are only sufficient, if the parameter design of the harmonic functions has been carried out correctly. This can be attributed to the fact that the bit error probability is essentially determined by the statistical properties (i.e., the probability density function of the amplitude, the level-crossing rate, the average duration of fades, and the probability density function of the fading intervals) of $\tilde{\zeta}(t)$ at low levels r. The behaviour of $\tilde{\zeta}(t)$ at high levels is in this case not of particular importance.

A comparison of Figures 3.7(a)–(c) and Figures 5.59(a)–(c) shows that the theoretical approximation $p_{1_-}(\tau_-; r)$ at all levels r only fits the measured probability densities $\tilde{p}_0(\tau_-; r)$ at small fading intervals τ_- very well. One should also note that for $\tau_- \to \infty$, it always follows $\tilde{p}_{0_-}(\tau_-; r) \to 0 \, \forall \, r \in \{0.1, 1, 2.5\}$. However, this convergence property is not fulfilled by $p_{1_-}(\tau_-; r)$ for the levels $r = 1$ and $r = 2.5$ [see Figures 3.7(b) and 3.7(c)]. From the convergence behaviour of $\tilde{p}_{0_-}(\tau_-; r)$, we can now approximately conclude the interval $[0, \tau_{max}]$, over which the approximation $r_{\mu_i\mu_i}(\tau) \approx \tilde{r}_{\mu_i\mu_i}(\tau)$ has to be as good as possible. We will therefore make use of the quantity $\tau_q = \tau_q(r)$, introduced in Subsection 3.3.3, where we substitute $p_{0_-}(\tau_-; r)$ into (3.49) by the probability density function $\tilde{p}_{0_-}(\tau_-; r)$ of the fading intervals of the reference model ($N_1 = 100$, $N_2 = 101$) and choose q so large that for all fading intervals $\tau_- \geq \tau_q$, the probability density function $\tilde{p}_{0_-}(\tau_-; r)$ becomes sufficiently small. Furthermore, we demand that τ_{max} must fulfil the inequality $\tau_{max} \geq \tau_q$. We remember that by using the method of exact Doppler spread, $\tilde{r}_{\mu_i\mu_i}(\tau)$ represents a very good approximation for $r_{\mu_i\mu_i}(\tau)$ within the range $0 \leq \tau \leq \tau_{max}$, where τ_{max} is related with N_i over the equation $\tau_{max} = N_i/(2f_{max})$. Therefore, for the estimation of the required number of harmonic functions N_i, we can obtain the following simple formula by using $\tau_{max} = N_i/(2f_{max}) \geq \tau_q(r)$

$$N_i \geq \lceil 2 f_{max} \tau_q(r) \rceil \, . \tag{5.92}$$

For example, if we choose $q = 90$, then we find the value 135.7 ms for $\tau_{90} = \tau_{90}(r)$ at the high level $r = 2.5$ [see Figure 5.59(c)]. With respect to (5.92) it then follows $N_i \geq 25$. This result matches the one obtained before by experimental means very well. Now, in reverse order, let us assume that N_i is given (for example, by $N_i \geq 7$), then the resulting probability density function $\tilde{p}_0(\tau_-; r)$ matches the corresponding probability density function of the reference model within the range $0 \leq \tau_- \leq 38.5$ ms very well. This is also confirmed by considering Figure 5.59(c).

For low and medium levels, where usually $\tau_{90} < 1/f_{max}$ holds, (5.92) does not provide any admissible values for N_i, since in these cases the obtained values fall below the lower limit $N_i = 7$, which is considered to be the necessary number of harmonic functions for a sufficient approximation of the Gaussian probability density function

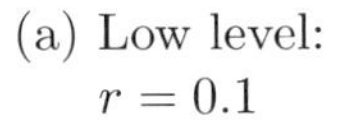

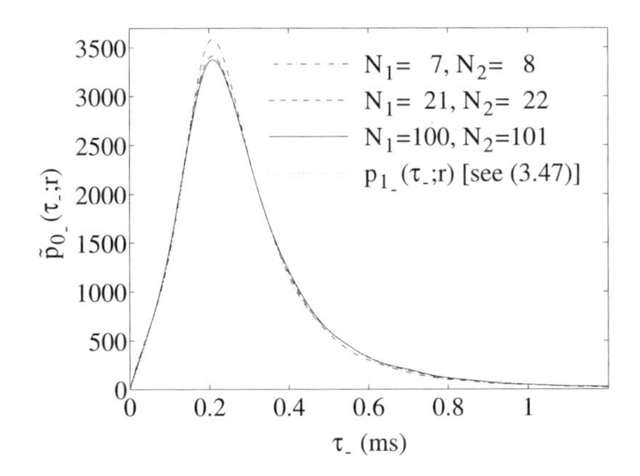

(b) Medium level:
$r = 1$

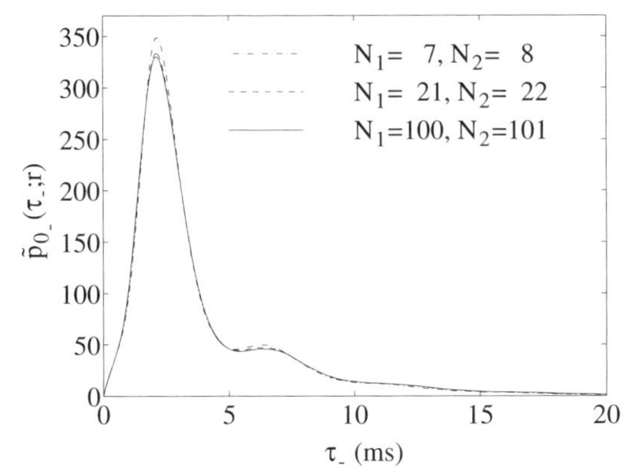

(c) High level:
$r = 2.5$

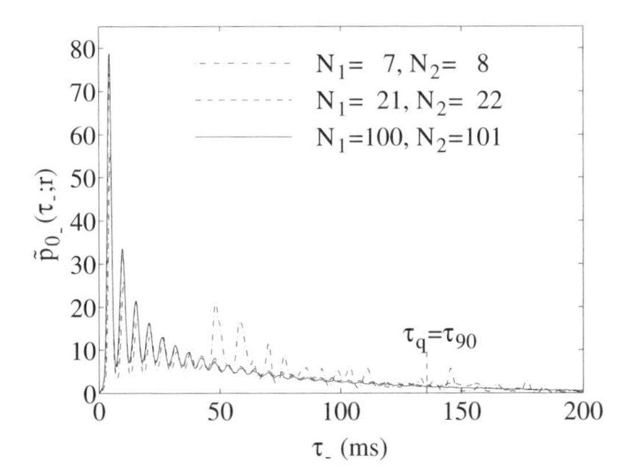

Figure 5.59: Probability density function $\tilde{p}_{0_-}(\tau_-; r)$ of the fading intervals of deterministic Rayleigh processes $\tilde{\zeta}(t)$: (a) $r = 0.1$, (b) $r = 1$, and (c) $r = 2.5$ (MEDS, Jakes PSD, $f_{max} = 91\,\text{Hz}$, $\sigma_0^2 = 1$).

$p_{\mu_i}(x)$. Therefore, as a useful estimation, which is valid for all levels $r \geq 0$, the inequality

$$N_i \geq \max\{7, \lceil 2 f_{max} \tau_q(r) \rceil\} \tag{5.93}$$

is suggested.

Next, we will study the statistics of the fading intervals of deterministic Rayleigh processes $\tilde{\zeta}(t)$ with underlying Gaussian shaped coloured processes $\tilde{\mu}_1(t)$ and $\tilde{\mu}_2(t)$. The Gaussian power spectral density (3.11) is characterized by the parameters $f_c = \sqrt{\ln 2} f_{max}$, $f_{max} = 91\,\text{Hz}$, and $\sigma_0^2 = 1$. The parameters of the simulation model are again determined with respect to the method of exact Doppler spread. Exactly as in the preceding case, we consider the simulation model for the couple $(N_1, N_2) = (100, 101)$ as reference model. The repetition of the measurement of the probability density function $\tilde{p}_{0_-}(\tau_-; r)$, which was carried out on the simulated fading behaviour of $\tilde{\zeta}(kT_s)$ for low, medium, and high levels r, now leads to the results shown in Figures 5.60(a)–(c). For all levels, the sampling interval T_s was thereby again given by the constant quantity $T_s = 0.5 \cdot 10^{-4}\,\text{s}$. Here again 10^7 fading intervals τ_- were evaluated for the determination of each probability density function $\tilde{p}_{0_-}(\tau_-; r)$.

From the comparison of Figures 5.60(a) and 5.59(a), it follows that the respective probability density functions $\tilde{p}_{0_-}(\tau_-; r)$ are identical. That was to be expected, since the exact shape of the power spectral density of the processes $\tilde{\mu}_1(t)$ and $\tilde{\mu}_2(t)$ has no influence on the density $\tilde{p}_0(\tau_-; r)$ at low levels r. Only the values of the quantities $\tilde{\sigma}_0^2 = \tilde{r}_{\mu_i\mu_i}(0)$ and $\tilde{\beta}_i = -\ddot{\tilde{r}}_{\mu_i\mu_i}(0)$ are of importance here. In the present case, they are identical for the Jakes and the Gaussian power spectral density. Only with an increasing level r, the behaviour of $\tilde{r}_{\mu_i\mu_i}(\tau)$ gains more and more influence on the density $\tilde{p}_{0_-}(\tau_-; r)$ for values $\tau > 0$. Therefore, one should compare Figures 5.60(b) and 5.60(c) with Figures 5.59(b) and 5.59(c), respectively. Obviously, the following fundamental relation exists between $\tilde{p}_{0_-}(\tau_-; r)$ and $\tilde{r}_{\mu\mu}(\tau)$: the probability density function $\tilde{p}_{0_-}(\tau_-; r)$ only has several maxima if this also holds for the autocorrelation function $\tilde{r}_{\mu\mu}(\tau)$ of the underlying complex deterministic process $\tilde{\mu}(t) = \tilde{\mu}_1(t) + j\tilde{\mu}_2(t)$.

At the end of this section, we will study the two-dimensional joint probability density function of the fading and connecting intervals, which will be denoted here by $\tilde{p}_{0_{-+}}(\tau_-, \tau_+; r)$. The function $\tilde{p}_{0_{-+}}(\tau_-, \tau_+; r)$ describes the density of the joint probability that the fading interval τ_- and the connecting interval τ_+ occur in pairs. This is the probability density for the case that a deterministic Rayleigh process $\tilde{\zeta}(t)$ crosses a constant level r upwards after the time duration τ_- for the first time within the interval $(t + \tau_-, t + \tau_- + d\tau_-)$ and afterwards falls below that level again after the duration τ_+ within the interval $(t + \tau_- + \tau_+, t + \tau_- + \tau_+ + d\tau_+)$ for the first time, provided that a level-crossing through r has already appeared from up to down at the time instant t.

Some simulation results for the two-dimensional joint probability density function $\tilde{p}_{0_{-+}}(\tau_-, \tau_+; r)$ are shown in Figures 5.61(a)–(c) and 5.62(a)–(c) for the case of Jakes and Gaussian shaped coloured deterministic processes $\tilde{\mu}_i(t)$, respectively. The numerical integration of these probability densities over the connecting interval τ_+,

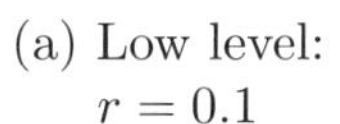
(a) Low level: $r = 0.1$

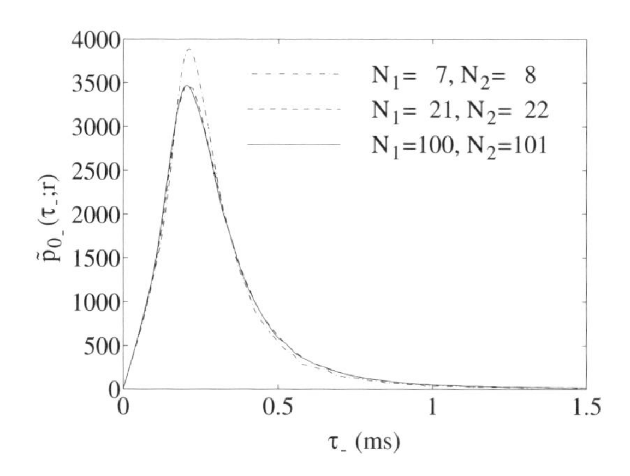

(b) Medium level: $r = 1$

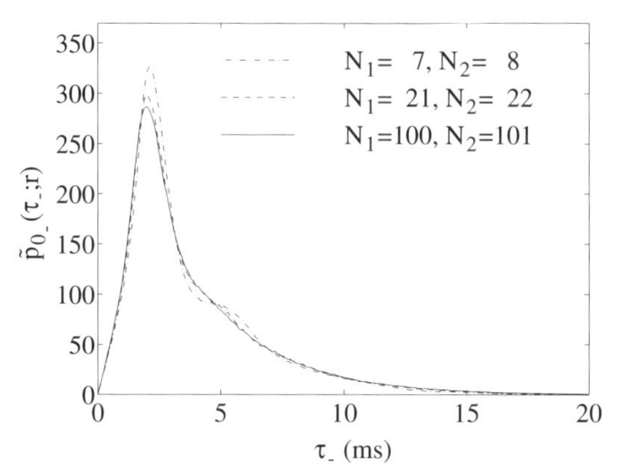

(c) High level: $r = 2.5$

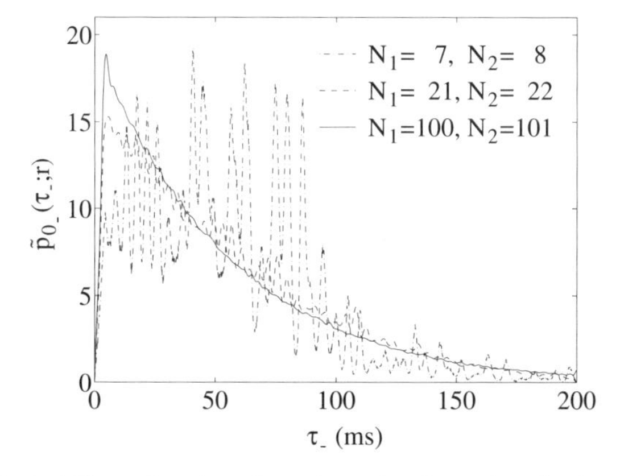

Figure 5.60: Probability density function $\tilde{p}_{0_-}(\tau_-; r)$ of the fading intervals of deterministic Rayleigh processes $\tilde{\zeta}(t)$: (a) $r = 0.1$, (b) $r = 1$, and (c) $r = 2.5$ (MEDS, Gaussian PSD, $f_c = \sqrt{\ln 2} f_{max}$, $f_{max} = 91\,\text{Hz}$, $\sigma_0^2 = 1$).

i.e., $\tilde{p}_{0_-}(\tau_-; r) = \int_0^\infty \tilde{p}_{0_{-+}}(\tau_-, \tau_+; r)\, d\tau_+$, again results in the graphs depicted in Figures 5.59(a)–(c) and 5.60(a)–(c) before.

At the end of this chapter, we want to return to the Monte Carlo method and the Jakes method again. We will therefore repeat the previously described simulations for the determination of the probability density function $\tilde{p}_{0_-}(\tau_-; r)$, where we now determine the parameters of the simulation model first with respect to the Monte Carlo method and afterwards by applying the Jakes method. In order to shorten matters, we will here only apply both methods on the Jakes power spectral density (3.8) with the parameters $f_{max} = 91\,\text{Hz}$ and $\sigma_0^2 = 1$. The probability densities $\tilde{p}_{0_-}(\tau_-; r)$ which were found based on the Monte Carlo method are shown in Figures 5.63(a)–(c) for various levels r with two different realizations of the respective sets of discrete Doppler frequencies $\{f_{i,n}\}$. Although in this case, the quantities N_1 and N_2 with $N_1 = 21$ and $N_2 = 22$ are chosen relatively high, one can clearly recognize the random behaviour of the probability density $\tilde{p}_{0_-}(\tau_-; r)$, which partly deviates from the desired density of the reference model (MEDS with $N_1 = 100$ and $N_2 = 101$) considerably.

Finally, Figures 5.64(a)–(c) depict the results obtained by applying the Jakes method.

(a) Low level:
$r = 0.1$

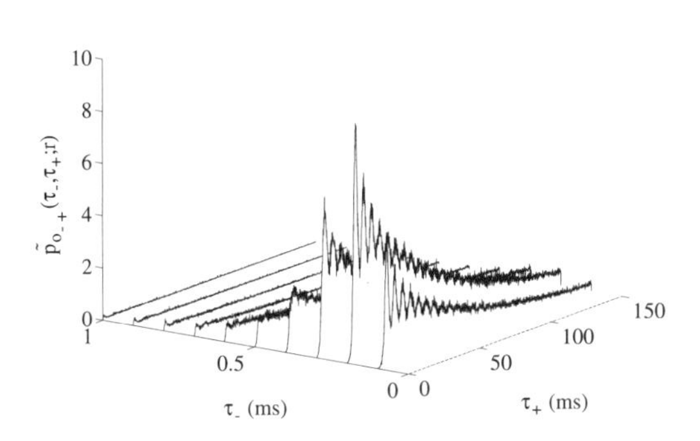

(b) Medium level:
$r = 1$

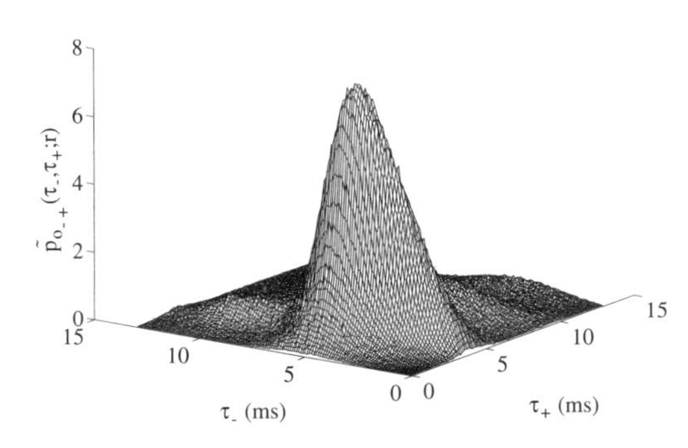

(c) High level:
$r = 2.5$

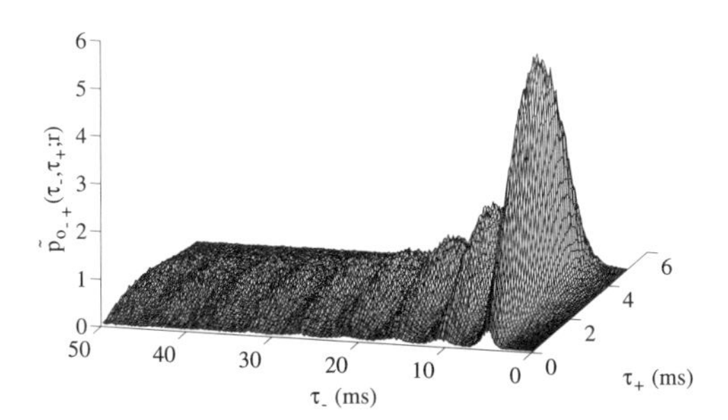

Figure 5.61: Joint probability density function $\tilde{p}_{0_-+}(\tau_-, \tau_+; r)$ of the fading and connecting intervals of deterministic Rayleigh processes $\tilde{\zeta}(t)$: (a) $r = 0.1$, (b) $r = 1$, and (c) $r = 2.5$ (MEDS, Jakes PSD, $f_{max} = 91\,\text{Hz}$, $\sigma_0^2 = 1$).

(a) Low level:
$r = 0.1$

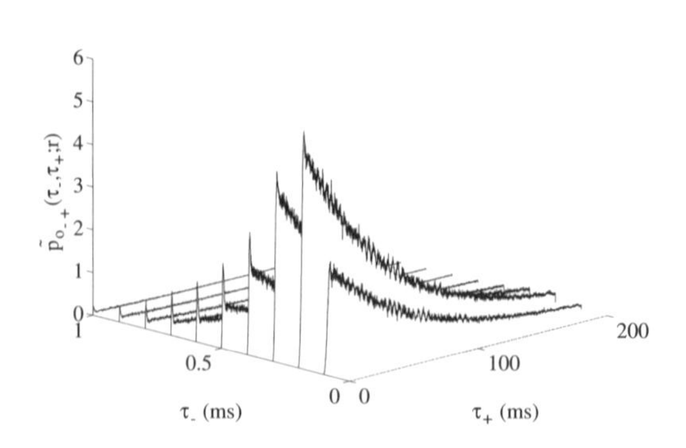

(b) Medium level:
$r = 1$

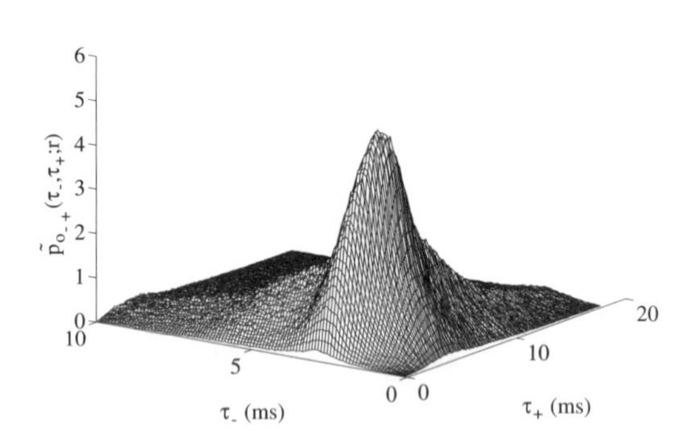

(c) High level:
$r = 2.5$

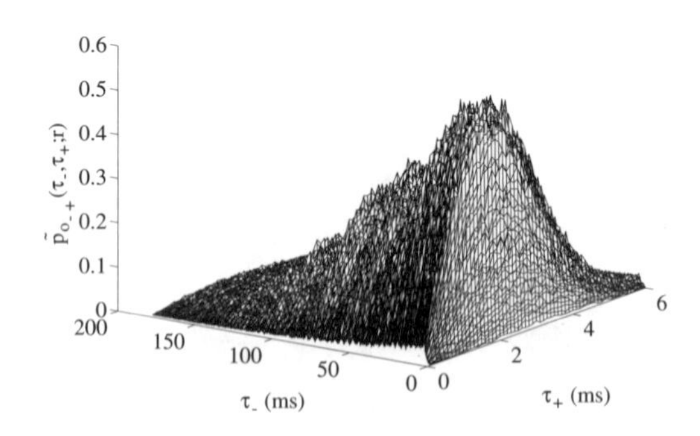

Figure 5.62: Joint probability density function $\tilde{p}_{0_-+}(\tau_-, \tau_+; r)$ of the fading and connecting intervals of deterministic Rayleigh processes $\tilde{\zeta}(t)$: (a) $r = 0.1$, (b) $r = 1$, and (c) $r = 2.5$ (MEDS, Gaussian PSD, $f_c = \sqrt{\ln 2} f_{max}$, $f_{max} = 91\,\text{Hz}$, $\sigma_0^2 = 1$).

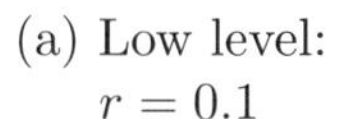

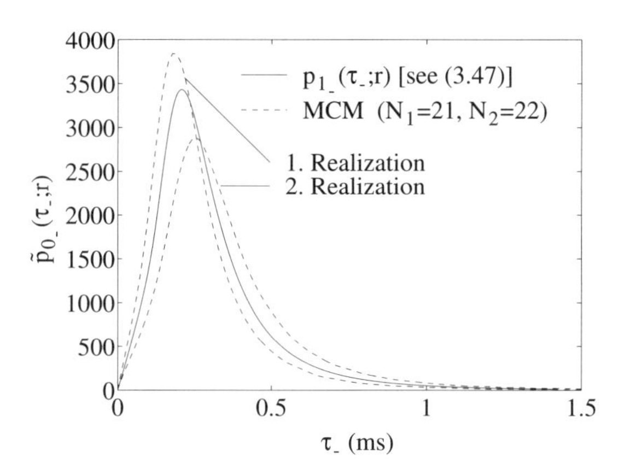

(b) Medium level:
$r = 1$

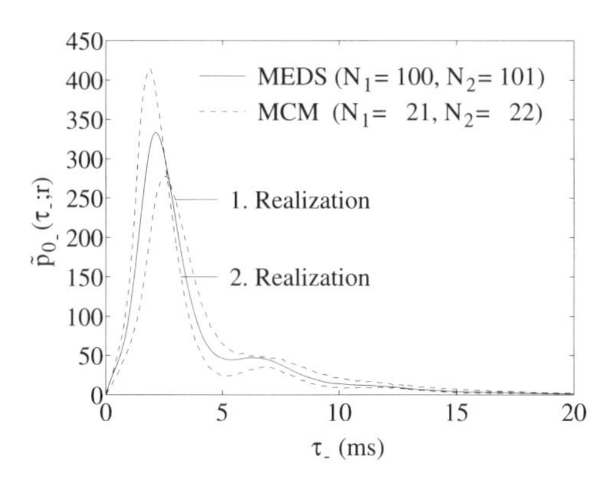

(c) High level:
$r = 2.5$

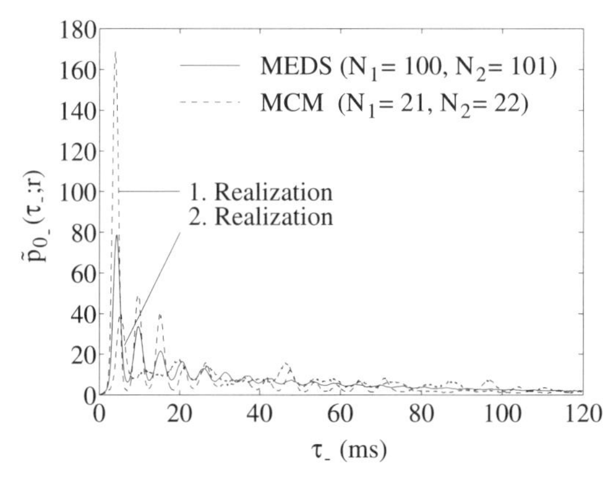

Figure 5.63: Probability density function $\tilde{p}_{0_-}(\tau_-; r)$ of the fading intervals of deterministic Rayleigh processes $\tilde{\zeta}(t)$: (a) $r = 0.1$, (b) $r = 1$, and (c) $r = 2.5$ (MCM, Jakes PSD, $f_{max} = 91\,\text{Hz}$, $\sigma_0^2 = 1$).

(a) Low level:
$r = 0.1$

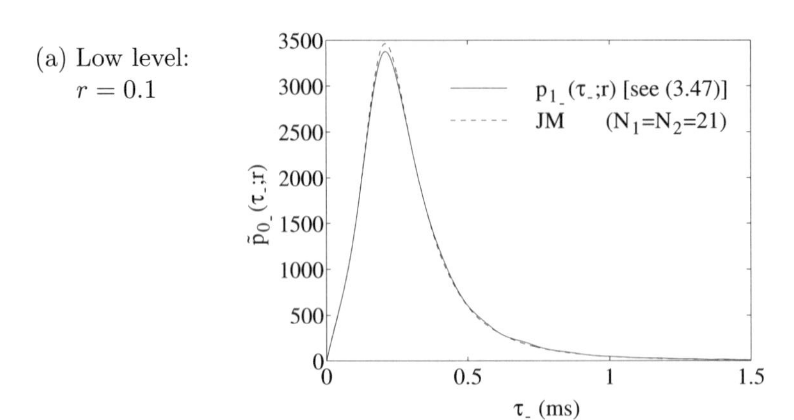

(b) Medium level:
$r = 1$

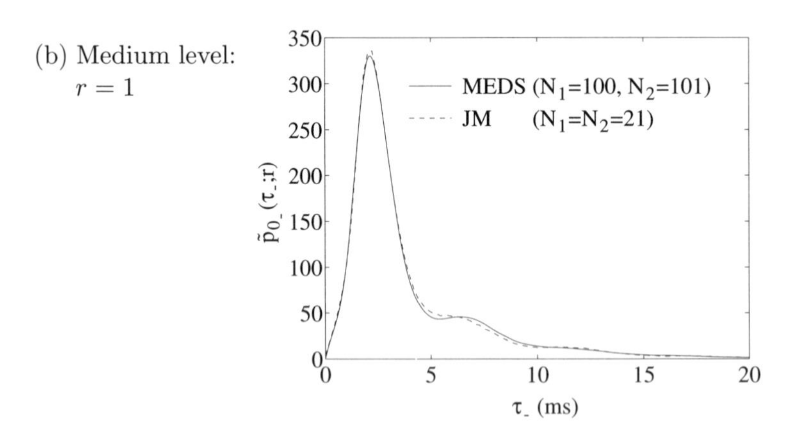

(c) High level:
$r = 2.5$

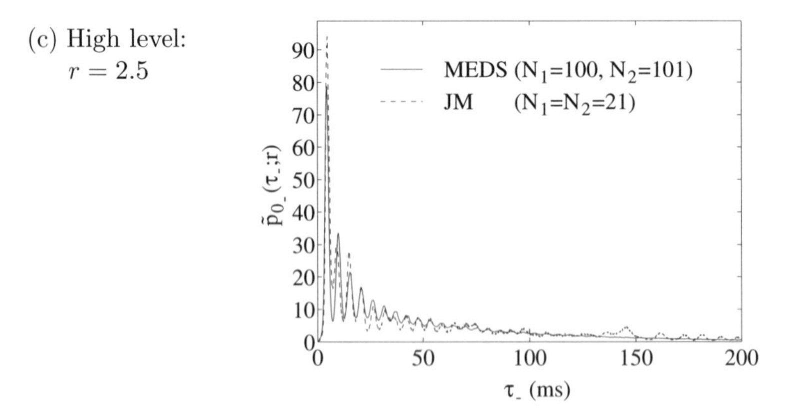

Figure 5.64: Probability density function $\tilde{p}_{0_-}(\tau_-; r)$ of the fading intervals of deterministic Rayleigh processes $\tilde{\zeta}(t)$: (a) $r = 0.1$, (b) $r = 1$, and (c) $r = 2.5$ (JM, Jakes PSD, $f_{max} = 91\,\text{Hz}$, $\sigma_0^2 = 1$).

6

FREQUENCY-NONSELECTIVE STOCHASTIC AND DETERMINISTIC CHANNEL MODELS

For frequency-nonselective, terrestrial, cellular land mobile radio channels and frequency-nonselective satellite mobile radio channels, meaning channels, in which the propagation delay differences of the reflected and scattered signal components at the receiver antenna are negligible in comparison with the symbol interval, the random fluctuations of the received signal can be modelled by a multiplication of the transmitted signal with a suitable stochastic model process. The discovery and the description of suitable stochastic model processes and their adaptation to real-world channels have been the subject of research [Suz77, Loo85, Loo91, Lut91, Cor94] for a considerable time.

The simplest stochastic model processes to be applied to this are Rayleigh and Rice processes described in the third chapter. The flexibility of these models is, however, too limited and often not large enough for a sufficient adaptation to the statistics of real-world channels. For the frequency-nonselective land mobile radio channel, it has turned out that the *Suzuki process* [Suz77, Han77] is a more suitable stochastic model in many cases. The Suzuki process is a product process of a Rayleigh process and a lognormal process. The slow signal fading, stated for real-world channels, is here modelled by the lognormal process taking the slow time variation of the average local received power into account. The Rayleigh process here models the fast fading as always. By modelling the channel on the basis of the Suzuki process, it is assumed that no line-of-sight component exists due to shadowing. Usually, it is also assumed that the two narrow-band real-valued Gaussian random processes, which produce the Rayleigh process, are uncorrelated. If we drop the last assumption, then this leads to the so-called *modified Suzuki process* analysed in [Kra90a, Kra90b].

Although the Suzuki process and its modified version were originally suggested as a model for the terrestrial, cellular land mobile radio channel, these stochastic processes are also quite suitable for modelling satellite mobile radio channels in urban regions, where the assumption that the line-of-sight signal component is shadowed, is justified for most of the time. Suburban and rural regions or even open areas with partial or no shadowing of the line-of-sight component, however, make further model extensions necessary. A contribution to this was made in the publication [Cor94]. The stochastic model introduced there is based on a product process of a Rice process and a lognormal process. Such a product process is suitable for modelling a large class of environments (urban, suburban, rural, open). Here, the two real-valued Gaussian random processes producing the Rice process are again assumed to be uncorrelated. If this assumption is dropped, then the flexibility of this model can be improved considerably with respect to the statistics of higher order. According to the specification of the cross-correlation, we distinguish between *extended Suzuki processes of Type I* [Pae98d] and such of *Type II* [Pae97a].

Moreover, in [Pae97c] a so-called *generalized Suzuki process* was suggested, which contains the classical Suzuki process [Suz77, Han77], the modified Suzuki process [Kra90a, Kra90b], as well as the two extended Suzuki processes [Pae98d, Pae97a] of Type I and Type II as special cases. As a rule, the first and second order statistical properties of generalized Suzuki processes are very flexible and can therefore be adapted to given measurement results of real-world channels very well.

A further stochastic model was introduced by Loo [Loo85, Loo87, Loo90, Loo91]. Loo's model is designated for a satellite mobile radio channel in rural environments, where a line-of-sight component between the satellite and the vehicle exists for most of the time of the transmission. The model is based on a Rayleigh process with constant mean power for the absolute value of the sum of all scattered and reflected multipath components. For the amplitude of the line-of-sight component, it is assumed that the statistics of this amplitude behaves like a lognormal process. In this way, the slow amplitude variations of the line-of-sight component caused by foliage (shadowing) are taken into account.

All the stochastic channel models described up to now have one property in common: They are stationary, i.e., they are based on stationary stochastic processes with constant parameters. A non-stationary model, which is valid for very large areas, was introduced by Lutz et al. [Lut91]. This model has especially been developed for frequency-nonselective, land mobile satellite channels. One distinguishes between regions, in which the line-of-sight component is shadowed (bad channel state), and regions without shadowing (good channel state). The important thing is that the proposed channel model is a 2-state model, for which the amplitude of the fading signal is modelled by the classical Suzuki process in the bad channel state and by a Rice process in the good channel state. This procedure can easily be generalized and leads to an M-state model, where each state is represented by a specific stationary stochastic model process. In this sense, the fading behaviour of non-stationary channels can be approximated by M stationary channel models [Vuc92, Mil95]. Experimental measurements have shown that a 4-state model is sufficient for most channels [Vuc90]. After all, as it has been shown in [Pae99a], even one and the same stationary channel model can be applied for each state, provided that the flexibility of this model is sufficiently high. Then, a specific set of coefficients will be assigned to each channel state. A change of a channel state is thus equivalent to a new configuration of a universal stationary channel model.

In this chapter, we will in detail deal with the description of the extended Suzuki process of Type I (Section 6.1) and of Type II (Section 6.2) as well as with the generalized Suzuki process (Section 6.3). Also, we will get to know a modified version of the Loo model in Section 6.4. The modified Loo model contains the classical Loo model as a special case. Moreover, in Section 6.5, some methods for the modelling of nonstationary mobile radio channels will be introduced. In each section, we will always proceed in such a way that a description of the respective reference model takes place at first. Afterwards, the corresponding deterministic simulation model derivable from the reference model will be presented. For the purpose of demonstrating the usefulness of the suggested reference models, the statistical properties like the probability density function of the amplitude, the level-crossing rate, and the average duration of fades are always fitted to measurement results available from the literature. All conformities achieved between the reference model, the simulation model, and the underlying measurements are usually astonishingly good, as will be clearly demonstrated by various examples.

6.1 THE EXTENDED SUZUKI PROCESS OF TYPE I

As mentioned at the beginning, the product process of a Rayleigh process and a lognormal process is said to be a Suzuki process. For these kind of processes, an extension is suggested in the text that follows. The Rayleigh process is in this case substituted by a Rice process taking the influence of a line-of-sight component into account. In the proposed model, the line-of-sight component can definitely be Doppler-shifted. Also, a cross-correlation between the two real-valued Gaussian random processes determining the Rice process can be admitted. In this way, the number of the grades of freedom increases, which in fact increases the mathematical complexity to be pursued, but on the other hand clearly improves the flexibility of the stochastic model in the end. The resulting product process from a Rice process with cross-correlated underlying Gaussian random processes and a lognormal process was introduced as *extended Suzuki process (of Type I)* [Pae95a, Pae98d]. This process is suitable as a stochastic model for a large class of satellite and land mobile radio channels in environments, where a direct line-of-sight connection between the transmitter and the receiver cannot be ignored.

The description of the reference model and the derivation of the statistical properties are carried out here by using the (complex) baseband notation as usual. At first, we will deal with the Rice process, which is used for the modelling of the short-term fading.

6.1.1 Modelling and Analysis of the Short-Term Fading

For the modelling of the short-term fading, thus, the fast fading, we will consider the Rice process (3.6), i.e.,

$$\xi(t) = |\mu_\rho(t)| = |\mu(t) + m(t)|\,, \tag{6.1}$$

where the line-of-sight component $m(t)$ will again be described according to (3.2), and $\mu(t)$ is the narrow-band complex-valued Gaussian random process introduced by (3.1), whose real and imaginary parts have zero-mean and identical variances $\sigma_{\mu_1}^2 = \sigma_{\mu_2}^2 = \sigma_0^2$.

We have assumed until now that the angles of arrival of the electromagnetic waves arriving the antenna of the receiver are uniformly distributed within the interval $[0, 2\pi)$, and that the antenna has a circular-symmetrical radiation pattern. The Doppler power spectral density $S_{\mu\mu}(f)$ of the complex-valued process $\mu(t)$ then has a symmetrical form (see (3.8)), which has the consequence that the two real-valued Gaussian random processes $\mu_1(t)$ and $\mu_2(t)$ are uncorrelated. In the following, we will drop this assumption. Instead of this, we assume that by spatially limited obstacles or by using directional antennas or sector antennas, i.e., antennas with noncircular-symmetrical radiation patterns, no electromagnetic waves with angles of arrival within the interval from α_0 to $2\pi - \alpha_0$ can arrive at the receiver, where α_0 shall be restricted to the interval $[\pi/2, 3\pi/2]$. The resulting unsymmetrical Doppler power spectral density $S_{\mu\mu}(f)$ is then described as follows

$$S_{\mu\mu}(f) = \begin{cases} \dfrac{2\sigma_0^2}{\pi f_{max}\sqrt{1-(f/f_{max})^2}}\,, & -f_{min} \leq f \leq f_{max}\,, \\ 0\,, & \text{else}\,, \end{cases} \tag{6.2}$$

where f_{max} again denotes the maximum Doppler frequency, and $f_{min} = -f_{max}\cos\alpha_0$ lies within the range $0 \leq f_{min} \leq f_{max}$. Only for the special case $\alpha_0 = \pi$, i.e., $f_{min} = f_{max}$, do we obtain the symmetrical Doppler power spectral density according to Jakes again. In general, however, the shape of (6.2) is unsymmetrical, which results in a cross-correlation of the real-valued Gaussian random processes $\mu_1(t)$ and $\mu_2(t)$. In the following, we denote the Doppler power spectral density according to (6.2) as *left-sided restricted Jakes power spectral density*. With a given value for f_{max} and a suitable choice of f_{min}, this often allows a better fitting to the Doppler spread of measured fading signals than the conventional Jakes power spectral density whose Doppler spread is often too large in comparison with reality (see Subsection 6.1.5).

Figure 6.1 depicts the reference model for the Rice process $\xi(t)$, whose underlying complex-valued Gaussian random process is characterized by the left-sided restricted Jakes power spectral density (6.2).

From this figure, we conclude the relations

$$\mu_1(t) = \nu_1(t) + \nu_2(t) \tag{6.3}$$

and

$$\mu_2(t) = \check{\nu}_1(t) - \check{\nu}_2(t)\,, \tag{6.4}$$

where $\nu_i(t)$ represents a coloured Gaussian random process, and its Hilbert transform is denoted by $\check{\nu}_i(t)$ (i=1,2). Here, the spectral shaping of $\nu_i(t)$ is based on filtering of white Gaussian noise $n_i(t) \sim N(0,1)$ by using an ideal filter whose transfer function

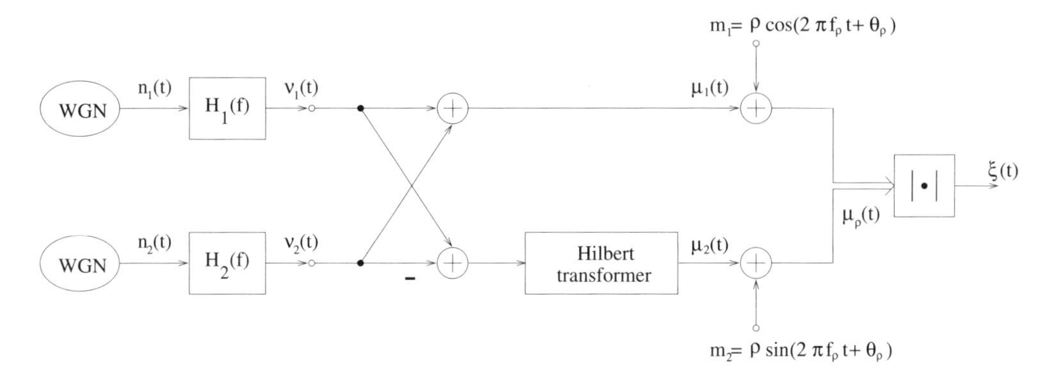

Figure 6.1: Reference model for Rice processes $\xi(t)$ with cross-correlated underlying Gaussian random processes $\mu_1(t)$ and $\mu_2(t)$.

is given by $H_i(f) = \sqrt{S_{\nu_i\nu_i}(f)}$. In the following, we will assume that the white Gaussian random processes $n_1(t)$ and $n_2(t)$ are uncorrelated.

The autocorrelation function of $\mu(t) = \mu_1(t) + j\mu_2(t)$, which is generally defined by (2.48), can be expressed in terms of the autocorrelation and cross-correlation functions of $\mu_1(t)$ and $\mu_2(t)$ as follows [Kam96]

$$r_{\mu\mu}(\tau) = r_{\mu_1\mu_1}(\tau) + r_{\mu_2\mu_2}(\tau) + j(r_{\mu_1\mu_2}(\tau) - r_{\mu_2\mu_1}(\tau)) . \tag{6.5}$$

Using the relations $r_{\nu_i\nu_i}(\tau) = r_{\check{\nu}_i\check{\nu}_i}(\tau)$ and $r_{\nu_i\check{\nu}_i}(\tau) = r_{\check{\nu}_i\nu_i}(-\tau) = -r_{\check{\nu}_i\nu_i}(\tau)$ (cf. also (2.56e) and (2.56c), respectively), we may write:

$$r_{\mu_1\mu_1}(\tau) = r_{\nu_1\nu_1}(\tau) + r_{\nu_2\nu_2}(\tau) = r_{\mu_2\mu_2}(\tau) , \tag{6.6a}$$

$$r_{\mu_1\mu_2}(\tau) = r_{\nu_1\check{\nu}_1}(\tau) - r_{\nu_2\check{\nu}_2}(\tau) = -r_{\mu_2\mu_1}(\tau) , \tag{6.6b}$$

so that (6.5) can be expressed by

$$r_{\mu\mu}(\tau) = 2[r_{\nu_1\nu_1}(\tau) + r_{\nu_2\nu_2}(\tau) + j(r_{\nu_1\check{\nu}_1}(\tau) - r_{\nu_2\check{\nu}_2}(\tau))] . \tag{6.7}$$

After the Fourier transform of (6.5) and (6.7), we obtain the following expressions for the Doppler power spectral density

$$S_{\mu\mu}(f) = S_{\mu_1\mu_1}(f) + S_{\mu_2\mu_2}(f) + j(S_{\mu_1\mu_2}(f) - S_{\mu_2\mu_1}(f)) , \tag{6.8a}$$

$$S_{\mu\mu}(f) = 2[S_{\nu_1\nu_1}(f) + S_{\nu_2\nu_2}(f) + j(S_{\nu_1\check{\nu}_1}(f) - S_{\nu_2\check{\nu}_2}(f))] . \tag{6.8b}$$

For the Doppler power spectral densities $S_{\nu_i\nu_i}(f)$ and $S_{\nu_i\check{\nu}_i}(f)$ as well as for the corresponding autocorrelation functions $r_{\nu_i\nu_i}(\tau)$ and $r_{\nu_i\check{\nu}_i}(\tau)$, the following relations hold:

$$S_{\nu_1\nu_1}(f) \quad = \quad \frac{\sigma_0^2}{2\pi f_{max}\sqrt{1-(f/f_{max})^2}} , \tag{6.9a}$$

$$\begin{array}{c} \bullet \\ | \\ \circ \end{array}$$

$$r_{\nu_1\nu_1}(\tau) = \frac{\sigma_0^2}{2} J_0(2\pi f_{max}\tau), \tag{6.9b}$$

$$S_{\nu_2\nu_2}(f) = \text{rect}\,(f/f_{min}) \cdot S_{\nu_1\nu_1}(f), \tag{6.9c}$$

$$\begin{array}{c} \bullet \\ | \\ \circ \end{array}$$

$$r_{\nu_2\nu_2}(\tau) = f_{min}\sigma_0^2 J_0(2\pi f_{max}\tau) * \text{sinc}\,(2\pi f_{min}\tau), \tag{6.9d}$$

$$S_{\nu_1\check{\nu}_1}(f) = -j\,\text{sgn}\,(f) \cdot S_{\nu_1\nu_1}(f), \tag{6.9e}$$

$$\begin{array}{c} \bullet \\ | \\ \circ \end{array}$$

$$r_{\nu_1\check{\nu}_1}(\tau) = \frac{\sigma_0^2}{2} H_0(2\pi f_{max}\tau), \tag{6.9f}$$

$$S_{\nu_2\check{\nu}_2}(f) = -j\,\text{sgn}\,(f) \cdot S_{\nu_2\nu_2}(f), \tag{6.9g}$$

$$\begin{array}{c} \bullet \\ | \\ \circ \end{array}$$

$$r_{\nu_2\check{\nu}_2}(\tau) = f_{min}\sigma_0^2 H_0(2\pi f_{max}\tau) * \text{sinc}\,(2\pi f_{min}\tau), \tag{6.9h}$$

where $J_0(\cdot)$ and $H_0(\cdot)$ denote the 0th order Bessel function of the first kind and the Struve's function of 0th order, respectively.[1] If we now substitute (6.9e) and (6.9g) into (6.8b), then we can express $S_{\mu\mu}(f)$ in terms of $S_{\nu_i\nu_i}(f)$ as follows

$$S_{\mu\mu}(f) = 2[(1 + \text{sgn}\,(f)) \cdot S_{\nu_1\nu_1}(f) + (1 - \text{sgn}\,(f)) \cdot S_{\nu_2\nu_2}(f)]\,. \tag{6.10}$$

Figure 6.2 illustrates the shapes of $S_{\nu_1\nu_1}(f)$ and $S_{\nu_2\nu_2}(f)$ as well as the corresponding left-sided restricted Jakes power spectral density $S_{\mu\mu}(f)$.

In the following derivation of the statistical properties of $\xi(t) = |\mu_\rho(t)|$ and $\vartheta(t) = \arg\{\mu_\rho(t)\}$, we often make use of the abbreviations

$$\psi_0^{(n)} := \left.\frac{d^n}{d\tau^n} r_{\mu_1\mu_1}(\tau)\right|_{\tau=0} = \left.\frac{d^n}{d\tau^n} r_{\mu_2\mu_2}(\tau)\right|_{\tau=0} \tag{6.11a}$$

and

$$\phi_0^{(n)} := \left.\frac{d^n}{d\tau^n} r_{\mu_1\mu_2}(\tau)\right|_{\tau=0} \tag{6.11b}$$

for $n = 0, 1, 2$. Using (6.6) and (6.9), these characteristic quantities can be expressed as follows:

---

[1] The rectangular function used in (6.9c) is defined by

$$\text{rect}\,(x) = \begin{cases} 1 & \text{for} \quad |x| < 1 \\ 1/2 & \text{for} \quad x = \pm 1 \\ 0 & \text{for} \quad |x| > 1 \end{cases}$$

and $\text{sinc}\,(x) = \sin(x)/x$ in (6.9d) denotes the sinc function.

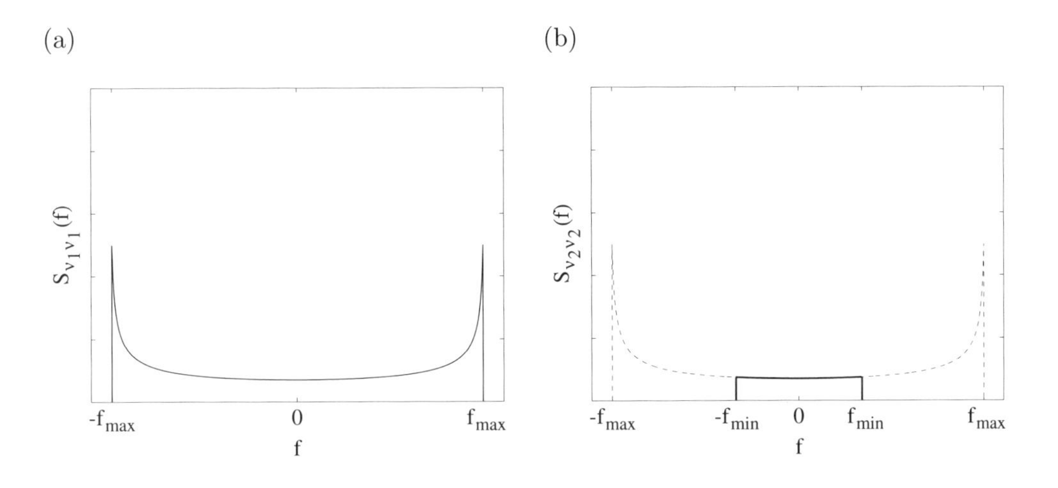

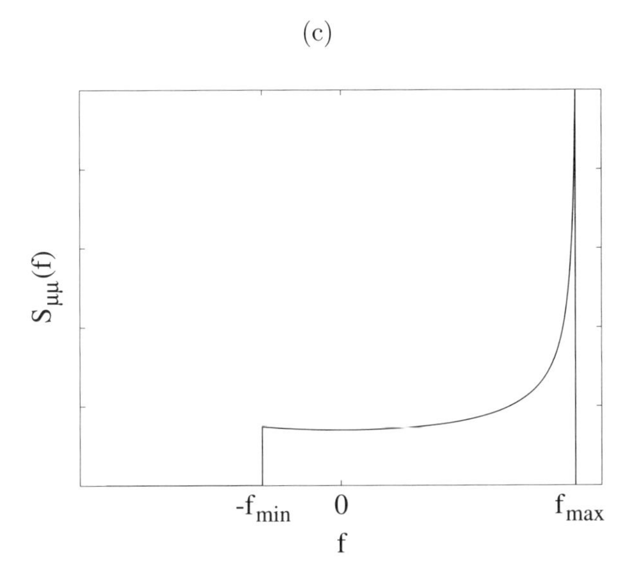

Figure 6.2: Doppler power spectral densities: (a) $S_{\nu_1\nu_1}(f)$, (b) $S_{\nu_2\nu_2}(f)$, and (c) the resulting left-sided restricted Jakes power spectral density.

$$\psi_0^{(0)} = \psi_0 = \frac{\sigma_0^2}{2}\left[1 + \frac{2}{\pi}\arcsin(\kappa_0)\right], \tag{6.12a}$$

$$\psi_0^{(1)} = \dot{\psi}_0 = 0, \tag{6.12b}$$

$$\psi_0^{(2)} = \ddot{\psi}_0 = -(\pi\sigma_0 f_{max})^2 \left\{1 + \frac{2}{\pi}\left[\arcsin(\kappa_0) - \frac{1}{2}\sin(2\arcsin(\kappa_0))\right]\right\}, \tag{6.12c}$$

$$\phi_0^{(0)} = \phi_0 = 0, \tag{6.12d}$$

$$\phi_0^{(1)} = \dot{\phi}_0 = 2\sigma_0^2 f_{max}\sqrt{1-\kappa_0^2}, \tag{6.12e}$$

$$\phi_0^{(2)} = \ddot{\phi}_0 = 0, \tag{6.12f}$$

where the overdot indicates the time derivative, and the parameter κ_0 denotes the frequency ratio

$$\kappa_0 = f_{min}/f_{max}, \quad 0 \leq \kappa_0 \leq 1. \tag{6.13}$$

One should note that the shape of $S_{\mu\mu}(f)$ is only symmetrical for the special case $\kappa_0 = 1$. In this case, the processes $\mu_1(t)$ and $\mu_2(t)$ are uncorrelated, and from (6.12a)–(6.12f), the relations $\psi_0 = \sigma_0^2$, $\ddot{\psi}_0 = -2(\pi\sigma_0 f_{max})^2$, and $\dot{\phi}_0 = 0$, which we already know from Subsection 3.3.2, follow.

A starting point for the derivation of the statistical properties of Rice processes $\xi(t)$ with unsymmetrical Doppler power spectral densities is given by the joint probability density function of the processes $\mu_{\rho_1}(t)$, $\mu_{\rho_2}(t)$, $\dot{\mu}_{\rho_1}(t)$, and $\dot{\mu}_{\rho_2}(t)$ [see (3.4)] at the same point within the time t. This joint probability density function will be denoted by $p_{\mu_{\rho_1}\mu_{\rho_2}\dot{\mu}_{\rho_1}\dot{\mu}_{\rho_2}}(x_1, x_2, \dot{x}_1, \dot{x}_2)$ here. It should be noted that $\mu_{\rho_i}(t)$ is a real-valued Gaussian random process with the time variant mean value $E\{\mu_{\rho_i}(t)\} = m_i(t)$ and the variance $\text{Var}\{\mu_{\rho_i}(t)\} = \text{Var}\{\mu_i(t)\} = r_{\mu_i\mu_i}(0) = \psi_0$. Consequently, its time derivative $\dot{\mu}_{\rho_i}(t)$ is a real-valued Gaussian random process too. However, this process is characterized by the mean value $E\{\dot{\mu}_{\rho_i}(t)\} = \dot{m}_i(t)$ and the variance $\text{Var}\{\dot{\mu}_{\rho_i}(t)\} = \text{Var}\{\dot{\mu}_i(t)\} = r_{\dot{\mu}_i\dot{\mu}_i}(0) = -\ddot{r}_{\mu_i\mu_i}(0) = -\ddot{\psi}_0$. It is also worth mentioning that the processes $\mu_{\rho_i}(t)$ and $\dot{\mu}_{\rho_i}(t)$ are in pairs correlated at the same time instant t. The joint probability density function $p_{\mu_{\rho_1}\mu_{\rho_2}\dot{\mu}_{\rho_1}\dot{\mu}_{\rho_2}}(x_1, x_2, \dot{x}_1, \dot{x}_2)$ can therefore be expressed by the multivariate Gaussian distribution (2.20), i.e.,

$$p_{\mu_{\rho_1}\mu_{\rho_2}\dot{\mu}_{\rho_1}\dot{\mu}_{\rho_2}}(x_1, x_2, \dot{x}_1, \dot{x}_2) = \frac{e^{-\frac{1}{2}(\boldsymbol{x}-\boldsymbol{m})^T \boldsymbol{C}_{\mu_\rho}^{-1}(\boldsymbol{x}-\boldsymbol{m})}}{(2\pi)^2\sqrt{\det \boldsymbol{C}_{\mu_\rho}}}, \tag{6.14}$$

where $\boldsymbol{x}$ and $\boldsymbol{m}$ are the column vectors defined by

$$\boldsymbol{x} = \begin{pmatrix} x_1 \\ x_2 \\ \dot{x}_1 \\ \dot{x}_2 \end{pmatrix} \tag{6.15}$$

and

$$\boldsymbol{m} = \begin{pmatrix} E\{\mu_{\rho_1}(t)\} \\ E\{\mu_{\rho_2}(t)\} \\ E\{\dot{\mu}_{\rho_1}(t)\} \\ E\{\dot{\mu}_{\rho_2}(t)\} \end{pmatrix} = \begin{pmatrix} m_1(t) \\ m_2(t) \\ \dot{m}_1(t) \\ \dot{m}_2(t) \end{pmatrix} = \begin{pmatrix} \rho\cos(2\pi f_\rho t + \theta_\rho) \\ \rho\sin(2\pi f_\rho t + \theta_\rho) \\ -2\pi f_\rho \rho \sin(2\pi f_\rho t + \theta_\rho) \\ 2\pi f_\rho \rho \cos(2\pi f_\rho t + \theta_\rho) \end{pmatrix}, \qquad (6.16)$$

respectively, and det $\mathbf{C}_{\mu_\rho}$ ($\boldsymbol{C}_{\mu_\rho}^{-1}$) denotes the determinant (inverse) of the covariance matrix

$$\boldsymbol{C}_{\mu_\rho} = \begin{pmatrix} C_{\mu_{\rho_1}\mu_{\rho_1}} & C_{\mu_{\rho_1}\mu_{\rho_2}} & C_{\mu_{\rho_1}\dot{\mu}_{\rho_1}} & C_{\mu_{\rho_1}\dot{\mu}_{\rho_2}} \\ C_{\mu_{\rho_2}\mu_{\rho_1}} & C_{\mu_{\rho_2}\mu_{\rho_2}} & C_{\mu_{\rho_2}\dot{\mu}_{\rho_1}} & C_{\mu_{\rho_2}\dot{\mu}_{\rho_2}} \\ C_{\dot{\mu}_{\rho_1}\mu_{\rho_1}} & C_{\dot{\mu}_{\rho_1}\mu_{\rho_2}} & C_{\dot{\mu}_{\rho_1}\dot{\mu}_{\rho_1}} & C_{\dot{\mu}_{\rho_1}\dot{\mu}_{\rho_2}} \\ C_{\dot{\mu}_{\rho_2}\mu_{\rho_1}} & C_{\dot{\mu}_{\rho_2}\mu_{\rho_2}} & C_{\dot{\mu}_{\rho_2}\dot{\mu}_{\rho_1}} & C_{\dot{\mu}_{\rho_2}\dot{\mu}_{\rho_2}} \end{pmatrix}. \qquad (6.17)$$

The entries of the covariance matrix $\boldsymbol{C}_{\mu_\rho}$ can be calculated as follows

$$\begin{aligned} \boldsymbol{C}_{\mu_{\rho_i}^{(k)}\mu_{\rho_j}^{(\ell)}} &= \boldsymbol{C}_{\mu_{\rho_i}^{(k)}\mu_{\rho_j}^{(\ell)}}(t_i, t_j) && (6.18a) \\ &= E\{\left(\mu_{\rho_i}^{(k)}(t_i) - m_i^{(k)}(t_i)\right)\left(\mu_{\rho_j}^{(\ell)}(t_j) - m_j^{(\ell)}(t_j)\right)\} && (6.18b) \\ &= E\{\mu_i^{(k)}(t_i)\mu_j^{(\ell)}(t_j)\} && (6.18c) \\ &= r_{\mu_i^{(k)}\mu_j^{(\ell)}}(t_i, t_j) && (6.18d) \\ &= r_{\mu_i^{(k)}\mu_j^{(\ell)}}(\tau), && (6.18e) \end{aligned}$$

for all $i, j = 1, 2$ and $k, \ell = 0, 1$. The transition from (6.18d) to (6.18e) is possible if we take into account that $\mu_i(t)$ and $\dot{\mu}_i(t)$ are Gaussian random processes, which are strict-sense stationary per definition. As a consequence, for the autocorrelation and cross-correlation functions, it follows that these correlation functions only depend on the time difference $\tau = t_j - t_i$, i.e., $r_{\mu_i^{(k)}\mu_j^{(\ell)}}(t_i, t_j) = r_{\mu_i^{(k)}\mu_j^{(\ell)}}(t_i, t_i + \tau) = r_{\mu_i^{(k)}\mu_j^{(\ell)}}(\tau)$. Studying the equations (6.17) and (6.18e), it now becomes clear that the covariance matrix $\boldsymbol{C}_{\mu_\rho}$ of the processes $\mu_{\rho_1}(t)$, $\mu_{\rho_2}(t)$, $\dot{\mu}_{\rho_1}(t)$, and $\dot{\mu}_{\rho_2}(t)$ is identical to the correlation matrix $\boldsymbol{R}_\mu$ of the processes $\mu_1(t)$, $\mu_2(t)$, $\dot{\mu}_1(t)$, and $\dot{\mu}_2(t)$, i.e., we may write

$$\boldsymbol{C}_{\mu_\rho}(\tau) = \boldsymbol{R}_\mu(\tau) = \begin{pmatrix} r_{\mu_1\mu_1}(\tau) & r_{\mu_1\mu_2}(\tau) & r_{\mu_1\dot{\mu}_1}(\tau) & r_{\mu_1\dot{\mu}_2}(\tau) \\ r_{\mu_2\mu_1}(\tau) & r_{\mu_2\mu_2}(\tau) & r_{\mu_2\dot{\mu}_1}(\tau) & r_{\mu_2\dot{\mu}_2}(\tau) \\ r_{\dot{\mu}_1\mu_1}(\tau) & r_{\dot{\mu}_1\mu_2}(\tau) & r_{\dot{\mu}_1\dot{\mu}_1}(\tau) & r_{\dot{\mu}_1\dot{\mu}_2}(\tau) \\ r_{\dot{\mu}_2\mu_1}(\tau) & r_{\dot{\mu}_2\mu_2}(\tau) & r_{\dot{\mu}_2\dot{\mu}_1}(\tau) & r_{\dot{\mu}_2\dot{\mu}_2}(\tau) \end{pmatrix}. \qquad (6.19)$$

For the entries of the correlation matrix $\boldsymbol{R}_\mu(\tau)$, the following relations hold [Pap91]:

$$r_{\mu_j\mu_i}(\tau) = r_{\mu_i\mu_j}(-\tau), \quad r_{\mu_i\dot{\mu}_j}(\tau) = \dot{r}_{\mu_i\mu_j}(\tau), \qquad (6.20a, b)$$

$$r_{\dot{\mu}_i\mu_j}(\tau) = -\dot{r}_{\mu_i\mu_j}(\tau), \quad r_{\dot{\mu}_i\dot{\mu}_j}(\tau) = -\ddot{r}_{\mu_i\mu_j}(\tau), \qquad (6.20c, d)$$

for all $i, j = 1, 2$.

For the derivation of the level-crossing rate and the average duration of fades, we have to consider the correlation properties of the processes $\mu_{\rho_i}^{(k)}(t_i)$ and $\mu_{\rho_j}^{(\ell)}(t_j)$ at the same time instant, i.e., $t_i = t_j$, and, thus, the time-difference variable $\tau = t_j - t_i$ is equal to zero. Therefore, in connection with (6.12a)–(6.12f), we can profit from the notation (6.11) enabling us to present the covariance matrix and the correlation matrix (6.19) as follows

$$\boldsymbol{C}_{\mu_\rho}(0) = \boldsymbol{R}_\mu(0) = \begin{pmatrix} \psi_0 & 0 & 0 & \dot{\phi}_0 \\ 0 & \psi_0 & -\dot{\phi}_0 & 0 \\ 0 & -\dot{\phi}_0 & -\ddot{\psi}_0 & 0 \\ \dot{\phi}_0 & 0 & 0 & -\ddot{\psi}_0 \end{pmatrix} . \tag{6.21}$$

After substituting (6.21) into the relation (6.14), we can now express the joint probability density function $p_{\mu_{\rho_1}\mu_{\rho_2}\dot{\mu}_{\rho_1}\dot{\mu}_{\rho_2}}(x_1, x_2, \dot{x}_1, \dot{x}_2)$ in terms of the quantities (6.12a)–(6.12f). For our intention, however, it is advisable to perform a transformation of the Cartesian coordinates (x_1, x_2) to polar coordinates (z, θ) first. For that purpose, we consider the following system of equations:

$$z = \sqrt{x_1^2 + x_2^2}\,, \qquad \dot{z} = \frac{x_1\dot{x}_1 + x_2\dot{x}_2}{\sqrt{x_1^2 + x_2^2}}\,, \tag{6.22a}$$

$$\theta = \arctan\left(\frac{x_2}{x_1}\right), \qquad \dot{\theta} = \frac{x_1\dot{x}_2 - x_2\dot{x}_1}{x_1^2 + x_2^2}\,. \tag{6.22b}$$

For $z > 0$, $|\dot{z}| < \infty$, $|\theta| \leq \pi$, and $|\dot{\theta}| < \infty$, this system of equations has the real-valued solutions

$$x_1 = z\cos\theta\,, \qquad \dot{x}_1 = \dot{z}\cos\theta - \dot{\theta}z\sin\theta\,, \tag{6.23a}$$

$$x_2 = z\sin\theta\,, \qquad \dot{x}_2 = \dot{z}\sin\theta + \dot{\theta}z\cos\theta\,. \tag{6.23b}$$

Applying the transformation rule (2.38) leads to the joint probability density function

$$\begin{aligned} p_{\xi\dot{\xi}\vartheta\dot{\vartheta}}(z, \dot{z}, \theta, \dot{\theta}) \quad &= \quad |J|^{-1} p_{\mu_{\rho_1}\mu_{\rho_2}\dot{\mu}_{\rho_1}\dot{\mu}_{\rho_2}}(z\cos\theta,\; z\sin\theta, \\ &\qquad \dot{z}\cos\theta - \dot{\theta}z\sin\theta,\; \dot{z}\sin\theta + \dot{\theta}z\cos\theta)\,, \end{aligned} \tag{6.24}$$

where J denotes the Jacobian determinant

$$J = \begin{vmatrix} \frac{\partial z}{\partial x_1} & \frac{\partial z}{\partial x_2} & \frac{\partial z}{\partial \dot{x}_1} & \frac{\partial z}{\partial \dot{x}_2} \\ \frac{\partial \dot{z}}{\partial x_1} & \frac{\partial \dot{z}}{\partial x_2} & \frac{\partial \dot{z}}{\partial \dot{x}_1} & \frac{\partial \dot{z}}{\partial \dot{x}_2} \\ \frac{\partial \theta}{\partial x_1} & \frac{\partial \theta}{\partial x_2} & \frac{\partial \theta}{\partial \dot{x}_1} & \frac{\partial \theta}{\partial \dot{x}_2} \\ \frac{\partial \dot{\theta}}{\partial x_1} & \frac{\partial \dot{\theta}}{\partial x_2} & \frac{\partial \dot{\theta}}{\partial \dot{x}_1} & \frac{\partial \dot{\theta}}{\partial \dot{x}_2} \end{vmatrix} = \begin{vmatrix} \frac{\partial x_1}{\partial z} & \frac{\partial x_1}{\partial \dot{z}} & \frac{\partial x_1}{\partial \theta} & \frac{\partial x_1}{\partial \dot{\theta}} \\ \frac{\partial x_2}{\partial z} & \frac{\partial x_2}{\partial \dot{z}} & \frac{\partial x_2}{\partial \theta} & \frac{\partial x_2}{\partial \dot{\theta}} \\ \frac{\partial \dot{x}_1}{\partial z} & \frac{\partial \dot{x}_1}{\partial \dot{z}} & \frac{\partial \dot{x}_1}{\partial \theta} & \frac{\partial \dot{x}_1}{\partial \dot{\theta}} \\ \frac{\partial \dot{x}_2}{\partial z} & \frac{\partial \dot{x}_2}{\partial \dot{z}} & \frac{\partial \dot{x}_2}{\partial \theta} & \frac{\partial \dot{x}_2}{\partial \dot{\theta}} \end{vmatrix}^{-1} = -\frac{1}{z^2}\,. \tag{6.25}$$

After some further algebraic manipulations, we are now in the position to bring the desired joint probability density function $p_{\xi\dot{\xi}\vartheta\dot{\vartheta}}(z, \dot{z}, \theta, \dot{\theta})$ into the following form [Pae98d]

$$
\begin{aligned}
p_{\xi\dot{\xi}\vartheta\dot{\vartheta}}(z, \dot{z}, \theta, \dot{\theta}) \quad = \quad & \frac{z^2}{(2\pi)^2 \psi_0 \beta} e^{-\frac{z^2+\rho^2}{2\psi_0}} \cdot e^{\frac{z\rho}{\psi_0}\cos(\theta - 2\pi f_\rho t - \theta_\rho)} \\
& \cdot e^{-\frac{1}{2\beta}\left[\dot{z} - \sqrt{2\beta}\alpha\rho\sin(\theta - 2\pi f_\rho t - \theta_\rho)\right]^2} \\
& \cdot e^{-\frac{z^2}{2\beta}\left\{\dot{\theta} - \frac{\dot{\phi}_0}{\psi_0} - \sqrt{2\beta}\frac{\alpha\rho}{z}\cos(\theta - 2\pi f_\rho t - \theta_\rho)\right\}^2}
\end{aligned}
\qquad (6.26)
$$

for $z \geq 0$, $|\dot{z}| < \infty$, $|\theta| \leq \pi$, and $|\dot{\theta}| < \infty$, where

$$\alpha = \left(2\pi f_\rho - \frac{\dot{\phi}_0}{\psi_0}\right) \Big/ \sqrt{2\beta}\,, \qquad (6.27)$$

$$\beta = -\ddot{\psi}_0 - \dot{\phi}_0^2/\psi_0\,. \qquad (6.28)$$

The joint probability density function (6.26) represents a fundamental equation. With this, we will at first determine the probability density function of the amplitude and the phase of the process $\mu_\rho(t)$ in the following subsection, and will then proceed with the derivation of the level-crossing rate and the average duration of fades of the process $\xi(t) = |\mu_\rho(t)|$ by again making use of (6.26).

6.1.1.1 Probability Density Function of the Amplitude and the Phase

Employing the rule (2.40) now allows us to calculate the probability density $p_\xi(z)$ of the process $\xi(t)$ from the joint probability density function $p_{\xi\dot{\xi}\vartheta\dot{\vartheta}}(z, \dot{z}, \theta, \dot{\theta})$. We therefore consider the threefold integral

$$p_\xi(z) = \int_{-\infty}^{\infty} \int_{-\pi}^{\pi} \int_{-\infty}^{\infty} p_{\xi\dot{\xi}\vartheta\dot{\vartheta}}(z, \dot{z}, \theta, \dot{\theta})\, d\dot{\theta}\, d\theta\, d\dot{z}\,, \quad z \geq 0\,. \qquad (6.29)$$

Putting (6.26) in the above expression results in the well-known Rice distribution

$$p_\xi(z) = \begin{cases} \dfrac{z}{\psi_0}\, e^{-\frac{z^2+\rho^2}{2\psi_0}}\, I_0\left(\dfrac{z\rho}{\psi_0}\right), & z \geq 0\,, \\ 0\,, & z < 0\,. \end{cases} \qquad (6.30)$$

Due to the correlation of the processes $\mu_1(t)$ and $\mu_2(t)$, this result cannot be regarded as a matter of course, as we will see later in Subsection 6.2. Since the probability density (6.30) is independent of the quantity $\dot{\phi}_0$ in the present case, it follows that the correlation between the processes $\mu_1(t)$ and $\mu_2(t)$ has no influence on the probability density function of the amplitude $\xi(t)$. However, one should note that the parameter κ_0 determining the Doppler bandwidth exerts an influence on the variance ψ_0 of the processes $\mu_1(t)$ and $\mu_2(t)$ [cf. (6.12a)] and consequently determines the behaviour of (6.30) decisively.

The probability density function of the phase $\vartheta(t)$, denoted by $p_\vartheta(\theta)$, can be calculated in a similar way. Substituting (6.26) into

$$p_\vartheta(\theta) = \int_0^\infty \int_{-\infty}^\infty \int_{-\infty}^\infty p_{\xi\dot{\xi}\vartheta\dot{\vartheta}}(z,\dot{z},\theta,\dot{\theta})\, d\dot{\theta}\, d\dot{z}\, dz\,, \quad -\pi \leq \theta \leq \pi\,, \tag{6.31}$$

results in

$$p_\vartheta(\theta) = p_\vartheta(\theta;t) = \frac{e^{-\frac{\rho^2}{2\psi_0}}}{2\pi}\left\{1 + \sqrt{\frac{\pi}{2\psi_0}}\,\rho\cos(\theta - 2\pi f_\rho t - \theta_\rho)\cdot e^{\frac{\rho^2\cos^2(\theta-2\pi f_\rho t-\theta_\rho)}{2\psi_0}}\left[1 + \operatorname{erf}\left(\frac{\rho\cos(\theta - 2\pi f_\rho t - \theta_\rho)}{\sqrt{2\psi_0}}\right)\right]\right\}, \quad -\pi \leq \theta \leq \pi\,. \tag{6.32}$$

One observes that also in this case, the cross-correlation function $r_{\mu_1\mu_2}(\tau)$ has no influence on the probability density function $p_\vartheta(\theta)$, since $p_\vartheta(\theta)$ is independent of $\dot{\phi}_0$. For the special case $\kappa_0 = 1$, we have $\psi_0 = \sigma_0^2$, and, thus, from (6.32) it follows (3.21). The investigation of further special cases, for example: (i) $f_\rho = 0$, (ii) $\rho \to 0$, and (iii) $\rho \to \infty$ leads to the statements made below the equation (3.21), which will not be revised again at this point.

6.1.1.2 *Level-Crossing Rate and Average Duration of Fades*

The derivation of the level-crossing rate using

$$N_\xi(r) = \int_0^\infty \dot{z}\, p_{\xi\dot{\xi}}(r,\dot{z})\, d\dot{z} \tag{6.33}$$

requires the knowledge of the joint probability density function $p_{\xi\dot{\xi}}(z,\dot{z})$ of the stationary processes $\xi(t)$ and $\dot{\xi}(t)$ at the same time instant t at the level $z = r$. For the joint probability density function $p_{\xi\dot{\xi}}(z,\dot{z})$, one finds, after substituting (6.26) into

$$p_{\xi\dot{\xi}}(z,\dot{z}) = \int_{-\pi}^{\pi} \int_{-\infty}^\infty p_{\xi\dot{\xi}\vartheta\dot{\vartheta}}(z,\dot{z},\theta,\dot{\theta})\, d\dot{\theta}\, d\theta\,, \quad z \geq 0\,, \quad |\dot{z}| < \infty\,, \tag{6.34}$$

the result

$$p_{\xi\dot{\xi}}(z,\dot{z}) = \frac{z}{\psi_0\sqrt{\beta}(2\pi)^{3/2}}\cdot e^{-\frac{z^2+\rho^2}{2\psi_0}} \int_{-\pi}^{\pi} e^{\frac{z\rho}{\psi_0}\cos\theta} \cdot e^{-\frac{1}{2\beta}\left[\dot{z} - \sqrt{2\beta}\alpha\rho\sin\theta\right]^2} d\theta\,, \quad z \geq 0\,, \quad |\dot{z}| < \infty\,, \tag{6.35}$$

where α, β, and ψ_0 are the quantities introduced by (6.27), (6.28), and (6.12a), respectively. Obviously, the processes $\xi(t)$ and $\dot{\xi}(t)$ are in general statistically dependent, because $p_{\xi\dot{\xi}}(z,\dot{z}) \neq p_\xi(z)\cdot p_{\dot{\xi}}(\dot{z})$ holds. Only for the special case $\alpha = 0$, i.e., (i) if the two real-valued Gaussian random processes $\mu_1(t)$ and $\mu_2(t)$ are uncorrelated and f_ρ is equal to zero, or, (ii) if f_ρ and $\dot{\phi}_0$ are related by $f_\rho = \dot{\phi}_0/(2\pi\psi_0)$, then we obtain statistically independent processes $\xi(t)$ and $\dot{\xi}(t)$, since from (6.35) it now follows

$$\begin{aligned} p_{\xi\dot{\xi}}(z,\dot{z}) &= p_\xi(z)\cdot p_{\dot{\xi}}(\dot{z}) \\ &= \frac{z}{\psi_0} e^{-\frac{z^2+\rho^2}{2\psi_0}} I_0\left(\frac{z\rho}{\psi_0}\right)\cdot \frac{e^{-\frac{\dot{z}^2}{2\beta}}}{\sqrt{2\pi\beta}}, \end{aligned} \tag{6.36}$$

where β in this case again represents $\beta = -\ddot{\psi}_0 - \dot{\phi}_0^2/\psi_0 \geq 0$. Hence, for $\alpha = 0$, the joint probability density function $p_{\xi\dot{\xi}}(z,\dot{z})$ is equal to the product of the probability density functions of the stochastic processes $\xi(t)$ and $\dot{\xi}(t)$, which are Rice and Gaussian distributed, respectively.

With the joint probability density function (6.35), we are now able to calculate the level-crossing rate of Rice processes whose underlying complex-valued Gaussian process has cross-correlated in-phase and quadrature components. In this way, after substituting (6.35) in the definition (6.33) and performing some tedious algebraic manipulations, we finally obtain the result [Pae98d]

$$\begin{aligned} N_\xi(r) &= \frac{r\sqrt{2\beta}}{\pi^{3/2}\psi_0} e^{-\frac{r^2+\rho^2}{2\psi_0}} \int_0^{\pi/2} \cosh\left(\frac{r\rho}{\psi_0}\cos\theta\right) \\ &\quad \cdot \left\{ e^{-(\alpha\rho\sin\theta)^2} + \sqrt{\pi}\alpha\rho\sin(\theta)\cdot \operatorname{erf}(\alpha\rho\sin\theta) \right\} d\theta, \quad r \geq 0, \end{aligned} \tag{6.37}$$

where the characteristic quantities α, β, and ψ_0 are given in the form (6.27), (6.28), and (6.12a), respectively. Further simplifications are not possible; the remaining integral has to be solved numerically. Let us again consider the special case $\kappa_0 = 1$. Then, we obtain: $\alpha = 2\pi f_\rho/\sqrt{2\beta}$, $\beta = -\ddot{\psi}_0 = -\ddot{r}_{\mu_i\mu_i}(0)$, and $\psi_0 = \sigma_0^2$, so that the level-crossing rate $N_\xi(r)$ given above results in the expression introduced by (3.24), as was to be expected.

Let us assume that the line-of-sight component tends to zero, i.e., $\rho \to 0$, which leads to $\xi(t) \to \zeta(t)$. Then, it follows that (6.37) tends to

$$N_\xi(r) = \sqrt{\frac{\beta}{2\pi}}\cdot\frac{r}{\psi_0} e^{-\frac{r^2}{2\psi_0}}, \quad r \geq 0, \tag{6.38}$$

where the quantity β is given by (6.28). The above result shows us that the level-crossing rate is proportional to the Rayleigh distribution. This property has also been mentioned in [Kra90b]. Due to (6.28), the proportionality factor $\sqrt{\beta/(2\pi)}$ is not only determined by the curvature of the autocorrelation function at the origin $\tau = 0$ ($\ddot{\psi}_0 = \ddot{r}_{\mu_i\mu_i}(0)$), but also decisively by the gradient of the cross-correlation function at $\tau = 0$ ($\dot{\phi}_0 = \dot{r}_{\mu_1\mu_2}(0)$).

Now, let $\rho \neq 0$ and $f_\rho = \dot{\phi}_0/(2\pi\psi_0)$. Then, it follows $\alpha = 0$ [see (6.27)], and from (6.37) we obtain the level-crossing rate $N_\xi(r)$ according to (3.27), if σ_0^2 is substituted by ψ_0 in that equation, i.e.,

$$N_\xi(r) = \sqrt{\frac{\beta}{2\pi}} \cdot \frac{r}{\psi_0} e^{-\frac{r^2+\rho^2}{2\psi_0}} I_0\left(\frac{r\rho}{\psi_0}\right), \quad r \geq 0, \tag{6.39}$$

where β is again given by (6.28).

In connection with the Jakes power spectral density, the level-crossing rate $N_\xi(r)$ described by (6.37) is always proportional to the maximum Doppler frequency f_{max}. The normalization of $N_\xi(r)$ to f_{max} therefore eliminates the influence of both the velocity of the vehicle and the carrier frequency. The influence of the parameters κ_0 and σ_0^2 on the normalized level-crossing rate $N_\xi(r)/f_{max}$ is illustrated in Figure 6.3(a) and in Figure 6.3(b), respectively.

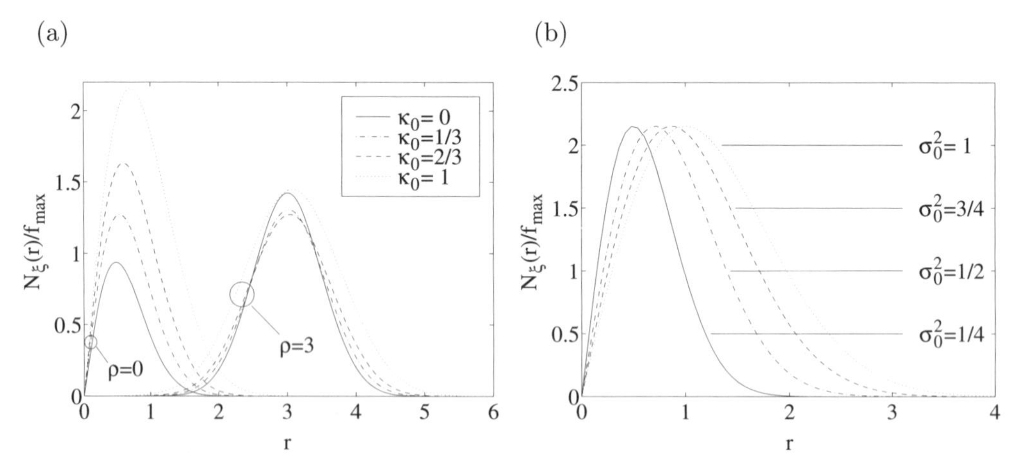

Figure 6.3: Normalized level-crossing rate $N_\xi(r)/f_{max}$ of Rice processes (with cross-correlated underlying Gaussian random processes): (a) $\kappa_0 = f_{min}/f_{max}$ ($\sigma_0^2 = 1$) and (b) σ_0^2 ($\rho = 0$, $\kappa_0 = 1$).

For the calculation of the average duration of fades $T_{\xi_-}(r)$, we will be guided by the basic relation (2.63), i.e.,

$$T_{\xi_-}(r) = \frac{F_{\xi_-}(r)}{N_\xi(r)}, \tag{6.40}$$

where $F_{\xi_-}(r)$ denotes the cumulative distribution function of the Rice process $\xi(t)$ and therefore states the probability that $\xi(t)$ takes a value which is lower or equal to the signal level r. Using (6.30), the following integral expression can be derived for $F_{\xi_-}(r)$

$$F_{\xi_-}(r) = P(\xi(t) \leq r)$$

$$= \int_0^r p_\xi(z)dz$$

$$= \frac{e^{-\frac{\rho^2}{2\psi_0}}}{\psi_0} \int_0^r z e^{-\frac{z^2}{2\psi_0}} I_0\left(\frac{z\rho}{\psi_0}\right) dz \,. \tag{6.41}$$

The average duration of fades of Rice processes $\xi(t)$ with cross-correlated in-phase and quadrature components $\mu_1(t)$ and $\mu_2(t)$ is, thus, the quotient (6.40) of the integral expressions (6.41) and (6.37), which have to be solved numerically.

The influence of the parameters κ_0 and σ_0^2 on the normalized average duration of fades $T_{\xi_-}(r) \cdot f_{max}$ is depicted in Figures 6.4(a) and 6.4(b), respectively.

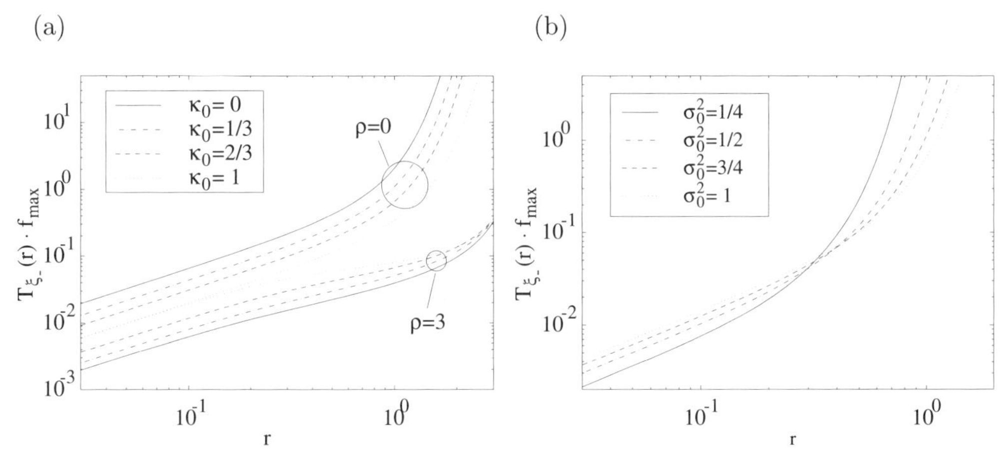

Figure 6.4: Normalized average duration of fades $T_\xi(r) \cdot f_{max}$ of Rice processes (with cross-correlated underlying Gaussian random processes): (a) $\kappa_0 = f_{min}/f_{max}$ ($\sigma_0^2 = 1$) and (b) σ_0^2 ($\rho = 0$, $\kappa_0 = 1$).

6.1.2 Modelling and Analysis of the Long-Term Fading

Measurements have shown that the slow fading behaves in its statistical properties quite similar to a lognormal process [Reu72, Bla72, Oku68]. With such a process, the slow fluctuations of the local mean value of the received signal, which are determined by shadowing effects, can be reproduced. In the following, we will denote lognormal processes by $\lambda(t)$. Lognormal processes can be derived by means of the nonlinear transform

$$\lambda(t) = e^{\sigma_3 \nu_3(t) + m_3} \tag{6.42}$$

from a third real-valued Gaussian random process $\nu_3(t)$ with the expected value $E\{\nu_3(t)\} = 0$ and the variance $\text{Var}\{\nu_3(t)\} = 1$. Fitting the model behaviour to the statistics of real-world channels, the model parameters m_3 and σ_3 can be used in connection with the parameters of the Rice process ($\sigma_0^2, f_{max}, f_{min}, \rho, f_\rho$). We assume henceforth that the stochastic process $\nu_3(t)$ is statistically independent of the processes $\nu_1(t)$ and $\nu_2(t)$. Figure 6.5 illustrates the reference model for the lognormal process $\lambda(t)$ introduced this way.

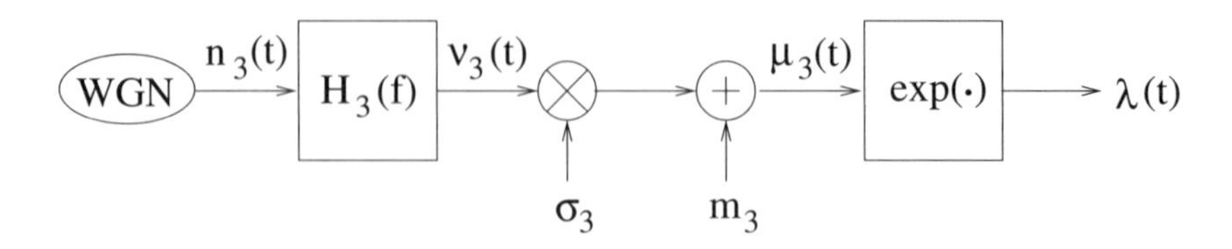

Figure 6.5: Reference model for lognormal processes $\lambda(t)$.

Here, the process $\nu_3(t)$ is obtained by filtering white Gaussian noise $n_3(t) \sim N(0,1)$ with a real-valued low-pass filter, whose transfer function $H_3(f)$ is related to the power spectral density $S_{\nu_3\nu_3}(f)$ of the process $\nu_3(t)$ according to (2.52f), i.e., $H_3(f) = \sqrt{S_{\nu_3\nu_3}(f)}$. For $S_{\nu_3\nu_3}(f)$, the Gaussian power spectral density is assumed in the form [cf. also (3.11)]

$$S_{\nu_3\nu_3}(f) = \frac{1}{\sqrt{2\pi}\sigma_c} e^{-\frac{f^2}{2\sigma_c^2}} , \tag{6.43}$$

where the 3-dB-cut-off frequency $f_c = \sigma_c\sqrt{2\ln 2}$ is in general much smaller than the maximum Doppler frequency f_{max}. In order to simplify the notation, we introduce the symbol κ_c for the frequency ratio f_{max}/f_c, i.e., $\kappa_c = f_{max}/f_c$. A study on modified Suzuki processes has shown [Kra90b] that both the parameter κ_c as well as the exact shape of the power spectral density of $\nu_3(t)$ have no considerable influence on the relevant statistical properties of modified Suzuki processes, if $\kappa_c > 10$. Other types of power spectral densities $S_{\nu_3\nu_3}(f)$ than the form (6.43) studied here have been proposed, for example, in [Kra90a, Kra90b] and [Loo91], where RC-low-pass filters and Butterworth filters of third order have been applied, respectively.

The autocorrelation function $r_{\nu_3\nu_3}(\tau)$ of the process $\nu_3(t)$ can be described after calculating the inverse Fourier transform of (6.43) by

$$r_{\nu_3\nu_3}(\tau) = e^{-2(\pi\sigma_c\tau)^2} . \tag{6.44}$$

Next, let us consider the lognormal process $\lambda(t)$ [see (6.42)]. The autocorrelation function $r_{\lambda\lambda}(\tau)$ of this process can be expressed in terms of $r_{\nu_3\nu_3}(\tau)$ as follows

$$\begin{aligned} r_{\lambda\lambda}(\tau) &= E\{\lambda(t)\cdot\lambda(t+\tau)\} \\ &= E\{e^{2m_3+\sigma_3[\nu_3(t)+\nu_3(t+\tau)]}\} \\ &= \int_{-\infty}^{\infty}\int_{-\infty}^{\infty} e^{2m_3+\sigma_3(x_1+x_2)} \cdot p_{\nu_3\nu_3'}(x_1,x_2)\, dx_1\, dx_2 , \end{aligned} \tag{6.45}$$

where

$$p_{\nu_3 \nu_3'}(x_1, x_2) = \frac{1}{2\pi\sqrt{1 - r^2_{\nu_2\nu_3}(\tau)}} e^{-\frac{x_1^2 - 2r_{\nu_3\nu_3}(\tau)\, x_1 x_2 + x_2^2}{2[1 - r^2_{\nu_3\nu_3}(\tau)]}} \tag{6.46}$$

describes the joint probability density function of the Gaussian random process $\nu_3(t)$ at two different time instants $t_1 = t$ and $t_2 = t + \tau$. After substituting (6.46) in (6.45) and solving the double integral, the autocorrelation function $r_{\lambda\lambda}(\tau)$ can be expressed in a closed form by

$$r_{\lambda\lambda}(\tau) = e^{2m_3 + \sigma_3^2[1 + r_{\nu_3\nu_3}(\tau)]} \, . \tag{6.47}$$

With this relation, the mean power of the lognormal process $\lambda(t)$ can easily be determined. We obtain $r_{\lambda\lambda}(0) = e^{2(m_3 + \sigma_3^2)}$.

The power spectral density $S_{\lambda\lambda}(f)$ of the lognormal process $\lambda(t)$ can now be expressed in terms of the power spectral density $S_{\nu_3\nu_3}(f)$ of $\nu_3(t)$ as follows [Pae98c]

$$\begin{aligned} S_{\lambda\lambda}(f) &= \int_{-\infty}^{\infty} r_{\lambda\lambda}(\tau) e^{-j2\pi f\tau} d\tau \\ &= e^{2m_3+\sigma_3^2} \cdot \{\delta(f) + \int_{-\infty}^{\infty} \left(e^{\sigma_3^2 r_{\nu_3\nu_3}(\tau)} - 1 \right) e^{-j2\pi f\tau} \} d\tau \\ &= e^{2m_3+\sigma_3^2} \cdot \left[\delta(f) + \sum_{n=1}^{\infty} \frac{\sigma_3^{2n}}{n!} \cdot \frac{S_{\nu_3\nu_3}\left(\frac{f}{\sqrt{n}}\right)}{\sqrt{n}} \right] . \end{aligned} \tag{6.48}$$

This result shows us that the power spectral density $S_{\lambda\lambda}(f)$ of the lognormal process $\lambda(t)$ consists of a weighted delta function at the origin $f = 0$ and of an infinite sum of strictly monotonously decreasing power spectral densities $S_{\nu_3\nu_3}(f/\sqrt{n})/\sqrt{n}$. One should note that $S_{\nu_3\nu_3}(f/\sqrt{n})/\sqrt{n}$ follows directly from (6.43), if the quantity σ_c is replaced by $\sqrt{n}\sigma_3$ there.

The probability density function $p_\lambda(y)$ of the lognormal process $\lambda(t)$ is described by the lognormal distribution (2.28), i.e.,

$$p_\lambda(y) = \begin{cases} \dfrac{1}{\sqrt{2\pi}\sigma_3 y} e^{-\frac{(\ln y - m_3)^2}{2\sigma_3^2}} , & y \geq 0 \, , \\ 0 \, , & y < 0 \, , \end{cases} \tag{6.49}$$

with the expected value and the variance according to (2.29a) and (2.29b), respectively.

For the calculation of the level-crossing rate and the average duration of fades of (extended) Suzuki processes, we require the knowledge of the joint probability density function of the lognormal process $\lambda(t)$ and its corresponding time derivative $\dot{\lambda}(t)$ at the same time instant t. This joint probability density function will be denoted by $p_{\lambda\dot{\lambda}}(y, \dot{y})$ here and will briefly be derived in the following. We start from the underlying Gaussian random process $\nu_3(t)$ and its time derivative $\dot{\nu}_3(t)$. For the cross-correlation function of these two processes it follows that $r_{\nu_3\dot{\nu}_3}(0) = 0$ holds, i.e., $\nu_3(t_1)$ and $\dot{\nu}_3(t_2)$ are uncorrelated at the same time instant $t = t_1 = t_2$. Since $\nu_3(t)$ and, hence, also $\dot{\nu}_3(t)$ are Gaussian random processes, it follows from the uncorrelatedness that these processes are statistically independent. For the joint probability density function $p_{\nu_3\dot{\nu}_3}(x, \dot{x})$ of the processes $\nu_3(t)$ and $\dot{\nu}_3(t)$, we can therefore write

$$p_{\nu_3\dot{\nu}_3}(x, \dot{x}) = p_{\nu_3}(x) \cdot p_{\dot{\nu}_3}(\dot{x}) = \frac{e^{-\frac{x^2}{2}}}{\sqrt{2\pi}} \cdot \frac{e^{-\frac{\dot{x}^2}{2\gamma}}}{\sqrt{2\pi\gamma}}, \tag{6.50}$$

where

$$\gamma = r_{\dot{\nu}_3\dot{\nu}_3}(0) = -\ddot{r}_{\nu_3\nu_3}(0) = (2\pi\sigma_c)^2 \tag{6.51}$$

denotes the variance of the process $\dot{\nu}_3(t)$.

Similar to the scheme described in detail in Subsection 6.1.1, we can take $p_{\nu_3\dot{\nu}_3}(x, \dot{x})$ as our starting point to determine the desired joint probability density function $p_{\lambda\dot{\lambda}}(y, \dot{y})$. The nonlinear mapping (6.42) in connection with the following substitution of variables

$$x = \frac{\ln y - m_3}{\sigma_3}, \quad \dot{x} = \frac{\dot{y}}{\sigma_3 y} \tag{6.52a, b}$$

yields the expression $J = (\sigma_3 y)^2$ for the Jacobian determinant (6.25). With the transformation rule (2.38), we then obtain the following result for the joint probability density function $p_{\lambda\dot{\lambda}}(y, \dot{y})$

$$p_{\lambda\dot{\lambda}}(y, \dot{y}) = \frac{e^{-\frac{(\ln y - m_3)^2}{2\sigma_3^2}}}{\sqrt{2\pi}\,\sigma_3 y} \cdot \frac{e^{-\frac{\dot{y}^2}{2\gamma(\sigma_3 y)^2}}}{\sqrt{2\pi\gamma}\,\sigma_3 y}. \tag{6.53}$$

This result shows us that the processes $\lambda(t)$ and $\dot{\lambda}(t)$ are statistically dependent, although the underlying Gaussian processes $\nu_3(t)$ and $\dot{\nu}_3(t)$ are statistically independent.

6.1.3 The Stochastic Extended Suzuki Process of Type I

The extended Suzuki process (Type I), denoted by $\eta(t)$, was introduced in [Pae98d] as a product process of a Rice process $\xi(t)$ [see (6.1)] with cross-correlated underlying Gaussian random processes $\mu_1(t)$ and $\mu_2(t)$ and a lognormal process $\lambda(t)$ [see (6.42)], i.e.,

$$\eta(t) = \xi(t) \cdot \lambda(t). \tag{6.54}$$

Figure 6.6 shows the structure of the reference model belonging to $\eta(t)$ for a frequency-nonselective mobile radio channel.

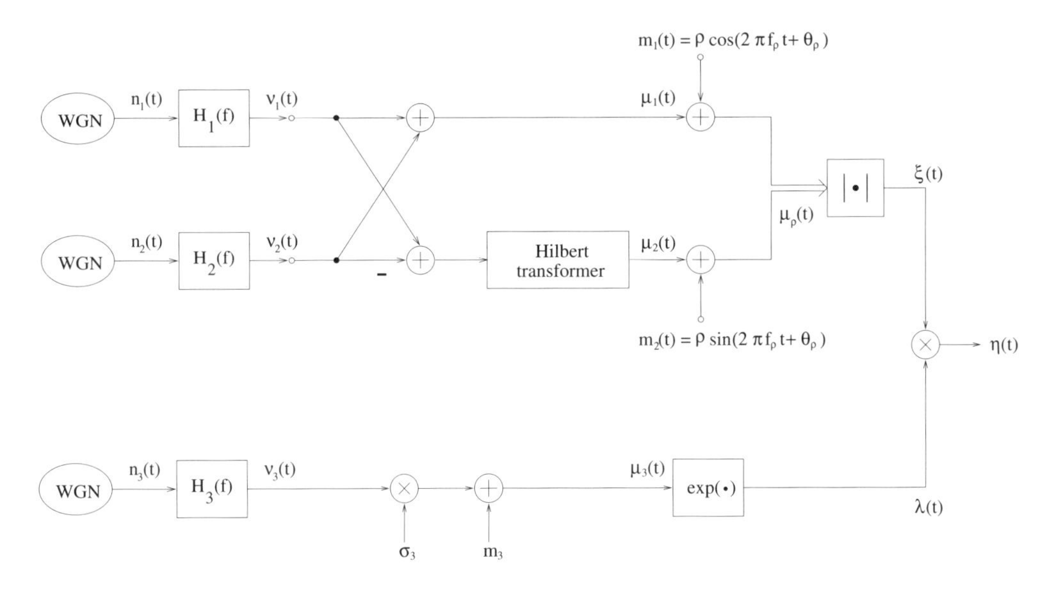

Figure 6.6: Reference model for extended Suzuki processes (Type I).

The probability density function $p_\eta(z)$ of the extended Suzuki process $\eta(t)$ can be calculated by means of the relation [Pap91]

$$p_\eta(z) = \int_{-\infty}^{\infty} \frac{1}{|y|} \, p_{\xi\lambda}\left(\frac{z}{y}, y\right) \, dy \,, \tag{6.55}$$

where $p_{\xi\lambda}(x, y)$ is the joint probability density function of the processes $\xi(t)$ and $\lambda(t)$ at the same time instant t. According to our assumption, the coloured Gaussian random processes $\nu_1(t)$, $\nu_2(t)$, and $\nu_3(t)$ are in pairs statistically independent. Consequently, the Rice process $\xi(t)$ and the lognormal process $\lambda(t)$ are also statistically independent, so that for the joint probability density function $p_{\xi\lambda}(x, y)$ it follows: $p_{\xi\lambda}(x, y) = p_\xi(x) \cdot p_\lambda(y)$. Hence, the multiplicative relation between the processes $\xi(t)$ and $\lambda(t)$ leads to the following integral equation for the probability density function of extended Suzuki processes

$$p_\eta(z) = \frac{z}{\sqrt{2\pi}\psi_0\sigma_3} \int_0^{\infty} \frac{1}{y^3} \, e^{-\frac{(z/y)^2+\rho^2}{2\psi_0}} \, I_0\left(\frac{z\rho}{y\psi_0}\right) \, e^{-\frac{(\ln y - m_3)^2}{2\sigma_3^2}} \, dy \,, \quad z \geq 0 \,. \tag{6.56}$$

For $\rho = 0$, it should be noted that the probability density function (6.56) can be reduced to the (classical) Suzuki distribution (2.30) introduced in [Suz77]. The influence of the parameters ρ and σ_3 on the behaviour of $p_\xi(z)$ can be concluded from Figure 6.7.

Studying (6.56), one can clearly see that $p_\eta(z)$ merely depends on the quantities ψ_0, ρ, σ_3, and m_3. Accordingly, the exact shape of the power spectral density of the complex-valued Gaussian random process $\mu(t)$ and especially the cross-correlation of the processes $\mu_1(t)$ and $\mu_2(t)$ have no influence on the probability density function

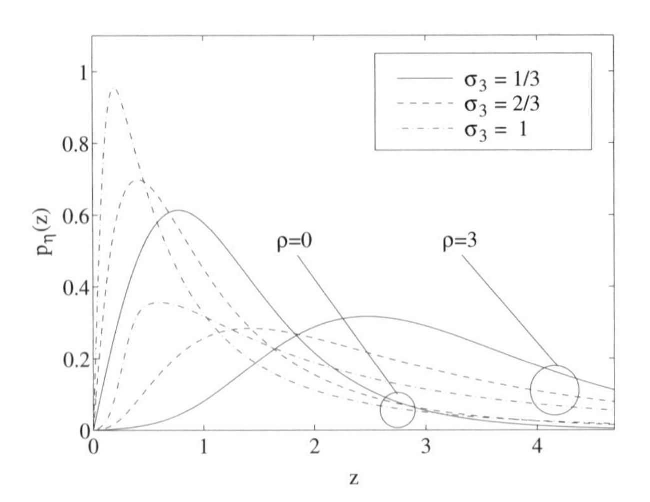

Figure 6.7: Probability density function $p_\eta(z)$ for various values of the parameters ρ and σ_3 ($\psi_0 = 1$, $m_3 = -\sigma_3^2/2$).

of the extended Suzuki process. For the adaptation of (6.56) to a given measured probability density function by merely optimizing these model parameters, there is a risk that the statistics of real-world channels are reproduced insufficiently by the channel model.

In the following, we therefore study the level-crossing rate $N_\eta(r)$ of the process $\eta(t)$, i.e.,

$$N_\eta(r) = \int_0^\infty \dot{z}\, p_{\eta\dot{\eta}}(r, \dot{z})\, d\dot{z}\,, \tag{6.57}$$

which requires the knowledge of the joint probability density function $p_{\eta\dot{\eta}}(z, \dot{z})$ of the process $\eta(t)$ and its time derivative $\dot{\eta}(t)$ at the same time instant t. This joint probability density can be derived by substituting the equations (6.35) and (6.53) obtained for $p_{\xi\dot{\xi}}(x, \dot{x})$ and $p_{\lambda\dot{\lambda}}(y, \dot{y})$, respectively, into the relation [Kra90a]

$$p_{\eta\dot{\eta}}(z, \dot{z}) = \int\limits_0^\infty \int\limits_{-\infty}^\infty \frac{1}{y^2} p_{\xi\dot{\xi}}\left(\frac{z}{y}, \frac{\dot{z}}{y} - \frac{z}{y^2}\dot{y}\right) p_{\lambda\dot{\lambda}}(y, \dot{y})\, d\dot{y}\, dy\,, \quad z \geq 0,\ |\dot{z}| < \infty\,. \tag{6.58}$$

Hence, after some tedious algebraic manipulations, we find the expression

$$\begin{aligned} p_{\eta\dot{\eta}}(z, \dot{z}) \;&=\; \frac{z}{(2\pi)^{\frac{3}{2}}\psi_0\sqrt{\beta}} \int_0^\infty \frac{e^{-\frac{(z/y)^2+\rho^2}{2\psi_0}}}{y^3 K(z, y)} \cdot \frac{e^{-\frac{(\ln y - m_3)^2}{2\sigma_3^2}}}{\sqrt{2\pi}\,\sigma_3 y} \cdot \\ &\int_0^{2\pi} e^{\frac{z\rho}{y\psi_0}\cos\theta} \cdot e^{-\frac{\left(\dot{z} - \sqrt{2\beta}\alpha y \rho \sin\theta\right)^2}{2\beta y^2 K^2(z,y)}}\, d\theta\, dy\,, \quad z \geq 0\,, \quad |\dot{z}| < \infty\,, \end{aligned} \tag{6.59}$$

where

$$K(z,y) = \sqrt{1 + \frac{\gamma}{\beta}\left(\frac{z\sigma_3}{y}\right)^2} \,. \tag{6.60}$$

After substituting (6.59) into (6.57), we then obtain the following final result for the level-crossing rate $N_\eta(r)$ of the extended Suzuki process of Type I

$$\begin{aligned} N_\eta(r) &= \frac{r\sqrt{2\beta}}{\pi^{3/2}\psi_0} \cdot \int_0^\infty \frac{K(r,y)}{y} \cdot \frac{e^{-\frac{(\ln y - m_3)^2}{2\sigma_3^2}}}{\sqrt{2\pi}\,\sigma_3 y} \, e^{-\frac{(r/y)^2+\rho^2}{2\psi_0}} \cdot \int_0^{\pi/2} \cosh\left(\frac{r\rho}{y\psi_0}\cos\theta\right) \\ & \left\{ e^{-\left(\alpha\rho\frac{\sin\theta}{K(r,y)}\right)^2} + \sqrt{\pi}\,\alpha\rho\frac{\sin\theta}{K(r,y)} \;\text{erf}\left[\alpha\rho\frac{\sin\theta}{K(r,y)}\right]\right\} d\theta\, dy, \end{aligned} \tag{6.61}$$

where α, β, and γ again are the quantities introduced by (6.27), (6.28), and (6.51), respectively, and ψ_0 is determined by (6.12a). Exactly due to α and β, the influence of the shape of the Doppler power spectral density is now taken into consideration, because α depends on $\dot\phi_0$ and β is a function of $\dot\phi_0$ and $\ddot\psi_0$. A detailed analysis of (6.61) here also shows that $N_\eta(r)$ is again proportional to the maximum Doppler frequency and, thus, to the speed of the vehicle as well.

Moreover, we are interested in some special cases. Assuming $\sigma_3 \to 0$, then the lognormal distribution (6.49) converges to the probability density function $p_\lambda(y) = \delta(y - e^{m_3})$. Consequently, especially in case $m_3 = 0$, the level-crossing rate $N_\eta(r)$ tends to $N_\xi(r)$ according to (6.37).

In case of a missing line-of-sight component, i.e., $\rho = 0$, the level-crossing rate of modified Suzuki processes follows from (6.61)

$$\begin{aligned} N_\eta(r)\big|_{\rho=0} &= \sqrt{\frac{\beta}{2\pi}}\frac{r}{\psi_0}\int_0^\infty \frac{K(r,y)}{y}\, p_\lambda(y)\, e^{-\frac{r^2}{2\psi_0 y^2}}\, dy \\ &= \sqrt{\frac{\beta}{2\pi}}\int_0^\infty K(r,y)\, p_\zeta(r/y)\, p_\lambda(y)\, dy \end{aligned} \tag{6.62}$$

as stated in [Kra90a, Kra90b]. It should also be mentioned that for $\rho \neq 0$, the two cases

$$\text{(i)} \quad f_\rho = \dot\phi_0/(2\pi\psi_0)\,, \tag{6.63a}$$

$$\text{(ii)} \quad f_\rho = 0 \text{ and } \dot\phi_0 = 0 \tag{6.63b}$$

are equivalent with respect to the level-crossing rate $N_\eta(r)$, because we then have $\alpha = 0$ due to (6.27), and from (6.61) the same expression

$$\begin{aligned} N_\eta(r)\big|_{\alpha=0} &= \sqrt{\frac{\beta}{2\pi}}\frac{r}{\psi_0}\int_0^\infty \frac{K(r,y)}{y}\, p_\lambda(y)\, e^{-\frac{(r/y)^2+\rho^2}{2\psi_0}}\, I_0\left(\frac{r\rho}{y\psi_0}\right) dy \\ &= \sqrt{\frac{\beta}{2\pi}}\int_0^\infty K(r,y)\, p_\xi(r/y)\, p_\lambda(y)\, dy \end{aligned} \tag{6.64}$$

always follows. One should note, however, that the cases (i) and (ii) result in different values for β. Under the condition (i), the general relation (6.28) is valid for β, whereas this equation can be simplified in case of (ii) to $\beta = -\ddot{\psi}_0$.

At the end of this subsection, we also derive the cumulative distribution function $F_{\eta_-}(r) = P(\eta(t) \leq r)$, which is required for the calculation of the average duration of fades of extended Suzuki processes of Type I

$$T_{\eta_-}(r) = \frac{F_{\eta_-}(r)}{N_\eta(r)} \,. \tag{6.65}$$

Thus, using (6.56), we obtain

$$\begin{aligned} F_{\eta_-}(r) &= \int_0^r p_\eta(z)\, dz \\ &= \frac{1}{\sqrt{2\pi}\,\psi_0\sigma_3} \int_0^\infty \int_0^r \frac{z}{y^3} e^{-\frac{(z/y)^2+\rho^2}{2\psi_0}} I_0\left(\frac{z\rho}{y\psi_0}\right) e^{-\frac{(\ln y - m_3)^2}{2\sigma_3^2}} \, dz\, dy \\ &= 1 - \int_0^\infty Q_1\left(\frac{\rho}{\sqrt{\psi_0}}, \frac{r}{y\sqrt{\psi_0}}\right) p_\lambda(y)\, dy \,, \end{aligned} \tag{6.66}$$

where $Q_1(.,.)$ (see [Pro95, p. 44]) is the *generalized Marcum's Q-function* defined by

$$Q_m(a,b) = \int_b^\infty z \left(\frac{z}{a}\right)^{m-1} e^{-\frac{z^2+a^2}{2}} I_{m-1}(az)\, dz \,, \quad m = 1, 2, \ldots \tag{6.67}$$

In order to illustrate the results found in this section, let us consider the parameter study shown in Figures 6.8(a)–6.8(d). Figures 6.8(a) and 6.8(b) depict the normalized level-crossing rate $N_\eta(r)/f_{max}$ calculated according to (6.61) for several values of the parameters m_3 and σ_3. The graphs of the corresponding normalized average duration of fades $T_{\eta_-}(r) \cdot f_{max}$ are presented in Figures 6.8(c) and 6.8(d).

6.1.4 The Deterministic Extended Suzuki Process of Type I

In the preceding subsection, we have seen that the reference model for the extended Suzuki process of Type I is based on the use of three real-valued coloured Gaussian random processes $\nu_i(t)$ or $\mu_i(t)$ $(i = 1, 2, 3)$ (see Figure 6.6). We now make use of the principle of deterministic channel modelling explained in Section 4.1, and approximate the ideal Gaussian random processes $\nu_i(t)$ by

$$\tilde{\nu}_i(t) = \sum_{n=1}^{N_i} c_{i,n} \cos(2\pi f_{i,n} t + \theta_{i,n}) \,, \quad i = 1, 2, 3 \,. \tag{6.68}$$

In the following, we therefore assume that the processes $\tilde{\nu}_1(t)$, $\tilde{\nu}_2(t)$, and $\tilde{\nu}_3(t)$ are uncorrelated in pairs. The uncorrelatedness property can be guaranteed without problems for nearly all parameter design methods discussed in Section 6.1. After

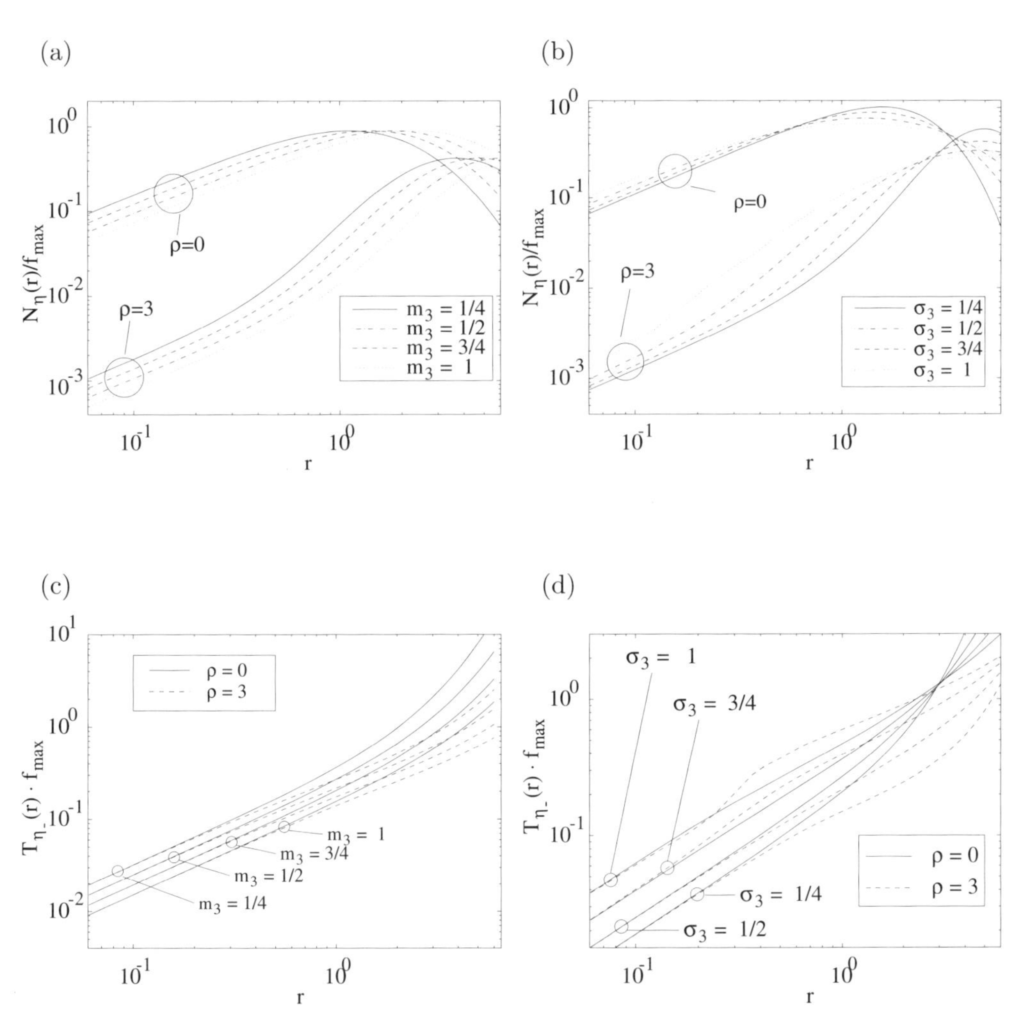

Figure 6.8: Normalized level-crossing rate $N_\eta(r)/f_{max}$ of extended Suzuki processes (Type I) for several values of (a) $m_3$gl6-68 ($\sigma_3 = 1/2$) and (b) σ_3 ($m_3 = 1/2$) as well as (c) and (d) the corresponding normalized average duration of fades $T_{\eta_-}(r) \cdot f_{max}$ ($\kappa_0 = 1,\ \psi_0 = 1$).

a few elementary network transformations, the continuous-time structure shown in Figure 6.9 enabling the simulation of *deterministic extended Suzuki processes of Type I* follows from the stochastic reference model (see Figure 6.6).

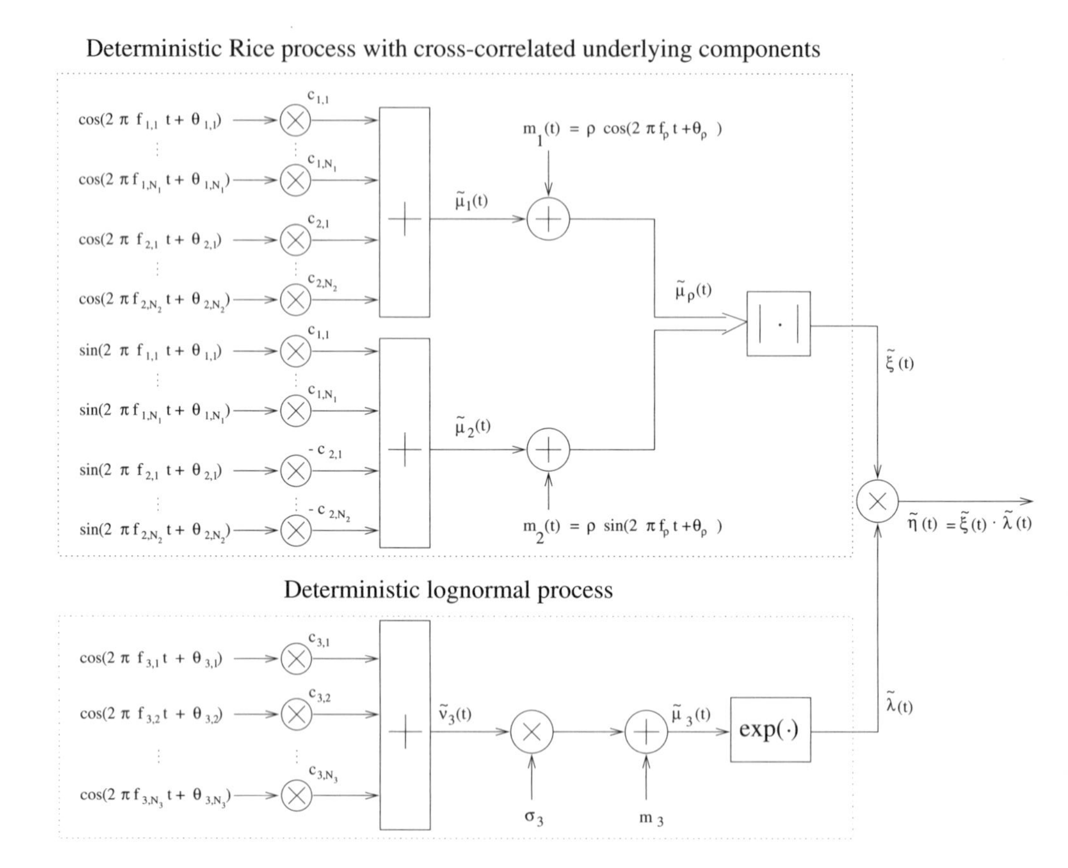

Figure 6.9: Deterministic simulation model for extended Suzuki processes (Type I).

Studying Figure 6.9, we notice that not only the design of the digital filters, which are usually employed for spectral shaping, but also the realization of the Hilbert transformer can be avoided. Moreover, deterministic simulation models offer the advantage that all relations derived for the reference model before such as, for instance, the expressions for the probability density function $p_\eta(z)$, the level-crossing rate $N_\eta(r)$, and the average duration of fades $T_{\eta_-}(r)$ can be used to approximately describe the behaviour of deterministic extended Suzuki process $\tilde{\eta}(t)$. In all those expressions, which are of interest for us, we therefore only have to replace the characteristic quantities of the reference model ψ_0, $\ddot{\psi}_0$, and $\dot{\phi}_0$ by the corresponding quantities of the simulation model, i.e.,

$$\tilde{\psi}_0 = \tilde{r}_{\mu_1\mu_1}(0) = \tilde{r}_{\nu_1\nu_1}(0) + \tilde{r}_{\nu_2\nu_2}(0) = \tilde{r}_{\mu_2\mu_2}(0)\,, \tag{6.69a}$$

$$\ddot{\tilde{\psi}}_0 = \ddot{\tilde{r}}_{\mu_1\mu_1}(0) = \ddot{\tilde{r}}_{\nu_1\nu_1}(0) + \ddot{\tilde{r}}_{\nu_2\nu_2}(0) = \ddot{\tilde{r}}_{\mu_2\mu_2}(0)\,, \tag{6.69b}$$

$$\dot{\tilde{\phi}}_0 = \dot{\tilde{r}}_{\mu_1\mu_2}(0) = \dot{\tilde{r}}_{\nu_1\check{\nu}_1}(0) - \dot{\tilde{r}}_{\nu_2\check{\nu}_2}(0) = -\dot{\tilde{r}}_{\mu_2\mu_1}(0)\,, \tag{6.69c}$$

where the tilde ($\sim$) refers to the fact that the underlying processes are deterministic processes. These quantities determine the statistical behaviour of $\tilde{\eta}(t)$ decisively and can explicitly be calculated in a simple way. With the autocorrelation function

$$\tilde{r}_{\nu_i\nu_i}(\tau) = \sum_{n=1}^{N_i} \frac{c_{i,n}^2}{2} \cos(2\pi f_{i,n}\tau)\,, \quad i = 1, 2, 3\,, \tag{6.70}$$

and the property (2.56a), it then follows from (6.69a)–(6.69c) [Pae95a]:

$$\tilde{\psi}_0 = \sum_{n=1}^{N_1} \frac{c_{1,n}^2}{2} + \sum_{n=1}^{N_2} \frac{c_{2,n}^2}{2}\,, \tag{6.71a}$$

$$\ddot{\tilde{\psi}}_0 = -2\pi^2 \left[\sum_{n=1}^{N_1} (c_{1,n} f_{1,n})^2 + \sum_{n=1}^{N_2} (c_{2,n} f_{2,n})^2 \right], \tag{6.71b}$$

$$\dot{\tilde{\phi}}_0 = \pi \left[\sum_{n=1}^{N_1} c_{1,n}^2 f_{1,n} - \sum_{n=1}^{N_2} c_{2,n}^2 f_{2,n} \right]. \tag{6.71c}$$

Throughout Chapter 6, we will exclusively employ the method of exact Doppler spread described in detail in Subsection 5.1.6 for the computation of the model parameters $c_{i,n}$ and $f_{i,n}$. The Doppler phases $\theta_{i,n} \in (0, 2\pi]$ are assumed to be realizations (outcomes) of a uniformly distributed random generator. For the method of exact Doppler spread, however, we have to take into account that this procedure was originally derived for the classical Jakes power spectral density ($\kappa_0 = 1$). Its application on the restricted Jakes power spectral density ($\kappa_0 \leq 1$) makes a slight modification necessary. For the discrete Doppler frequencies $f_{i,n}$, we now have [Pae98d]

$$f_{i,n} = \begin{cases} f_{max} \sin\left[\dfrac{\pi}{2N_1}\left(n - \dfrac{1}{2}\right)\right], & i = 1\,, \quad n = 1, 2, \ldots, N_1\,, \\ f_{max} \sin\left[\dfrac{\pi}{2N_2'}\left(n - \dfrac{1}{2}\right)\right], & i = 2\,, \quad n = 1, 2, \ldots, N_2\,, \end{cases} \tag{6.72}$$

where

$$N_2' = \left\lceil \frac{N_2}{\frac{2}{\pi}\arcsin(\kappa_0)} \right\rceil \tag{6.73}$$

is an auxiliary variable that depends on the frequency ratio $\kappa_0 = f_{min}/f_{max}$. In connection with (6.72), the quantity N_2' restricts the discrete Doppler frequencies

$f_{2,n}$ to the relevant interval $(0, f_{min}]$. One should take note of the fact that the actual required number of harmonic functions N_2 $(\leq N_2')$, which is necessary for the realization of $\tilde{\nu}_2(t)$, is still defined by the user. We therefore call the auxiliary variable N_2' the *virtual number of harmonic functions* of $\tilde{\nu}_2(t)$. Moreover, the Doppler coefficients $c_{i,n}$ are also affected by this modification, particularly since a power adaptation is necessary. Now, the Doppler coefficients read as follows

$$c_{i,n} = \begin{cases} \sigma_0\sqrt{1/N_1}\,, & i=1\,, \quad n=1,2,\ldots,N_1\,, \\ \sigma_0\sqrt{1/N_2'}\,, & i=2\,, \quad n=1,2,\ldots,N_2\,. \end{cases} \tag{6.74}$$

The computation of the discrete Doppler frequencies $f_{3,n}$ of the third deterministic Gaussian process $\tilde{\nu}_3(t)$, whose power spectral density is Gaussian shaped, can be accomplished by means of (5.76a) and (5.76b). After the adaptation of these equations to the notation used here, we obtain the following set of equations

$$\frac{2n-1}{2N_3} - \operatorname{erf}\left(\frac{f_{3,n}}{\sqrt{2}\sigma_c}\right) = 0\,, \quad \forall n = 1,2,\ldots,N_3-1\,, \tag{6.75a}$$

and

$$f_{3,N_3} = \sqrt{\frac{\gamma N_3}{(2\pi)^2} - \sum_{n=1}^{N_3-1} f_{3,n}^2}\,, \tag{6.75b}$$

where the meaning of $\sigma_c = f_{max}/(\kappa_c\sqrt{2\ln 2})$ follows from (6.43), and the parameter γ is defined by (6.51). Due to $\nu_3(t) \sim N(0,1)$, we compute $c_{3,n}$ according to the formula $c_{3,n} = \sqrt{2/N_3}$ for all $n = 1,2,\ldots,N_3$.

When using the method of exact Doppler spread, we obtain the results shown in Figure 6.10 as a function of $N_1 = N_2 = N_i$ for the convergence behaviour and for the approximation quality of the normalized characteristic quantities $\ddot{\tilde{\psi}}_0/f_{max}^2$ and $\dot{\tilde{\phi}}_0/f_{max}$. Figures 6.10(a) and 6.10(b) also show us that in all cases where $N_i \geq 10$ holds, the deviations between the presented characteristic quantities of the simulation model and the ones of the reference model can be ignored.

Let $N_i \geq 7$, then (6.61) can be considered an excellent approximation for the level-crossing rate of the simulation model $\tilde{N}_\eta(r)$, if the characteristic quantities of the reference model $(\psi_0, \ddot{\psi}_0, \dot{\phi}_0)$ and (α, β, γ) are replaced there by the corresponding quantities of the simulation model $(\tilde{\psi}_0, \ddot{\tilde{\psi}}_0, \dot{\tilde{\phi}}_0)$ and $(\tilde{\alpha}, \tilde{\beta}, \tilde{\gamma})$, respectively. The same of course also holds for the average duration of fades $\tilde{T}_{\eta_-}(r)$ of the simulation model. Hence, $\tilde{N}_\eta(r)$ and $\tilde{T}_{\eta_-}(r)$ must not necessarily be determined from lengthy and time-consuming simulation runs, but they can be determined directly by solving the integral equation (6.61) and making use of (6.65) in conjunction with (6.71a)–(6.71c). Nevertheless, if $\tilde{N}_\eta(r)$ $(\tilde{T}_{\eta_-}(r))$ is determined by means of simulation of the fading envelope $\tilde{\eta}(t)$ in the following, then this only serves to support the obtained theoretical results. We will see subsequently that the deviations between $\tilde{N}_\eta(r)$ and $N_\eta(r)$ are actually extremely small, so that a further analysis of the facts of the matter is unnecessary at this point.

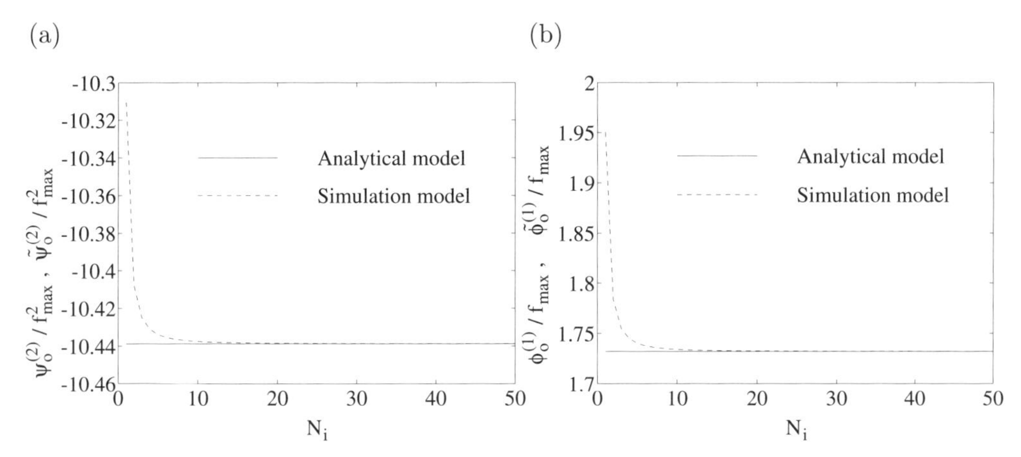

Figure 6.10: Illustration of (a) $\ddot{\psi}_0/f_{max}^2$ and $\ddot{\tilde{\psi}}_0/f_{max}^2$ as well as (b) $\dot{\phi}_0/f_{max}$ and $\dot{\tilde{\phi}}_0/f_{max}$ (MEDS, $\sigma_0^2 = 1$, $\kappa_0 = 1/2$).

6.1.5 Applications and Simulation Results

In this subsection, we will show how the statistics of the channel model can be adapted to the statistics of real-world channels by optimizing the relevant parameters of the reference model. Here we are not satisfied with an adaptation of the first order statistics, but we include the statistics of the second order in the design procedure as well. Starting from the fitted reference model, the parameters of the corresponding deterministic simulation model will be determined afterwards. At the end of this subsection, the verification of the proposed procedure will then follow by means of simulation as well.

The measurement results of the complementary cumulative distribution function[2] $F^{\star}_{\eta_+}(r)$ [Figure 6.11(a)] and the level-crossing rate $N^{\star}_{\eta}(r)$ [Figure 6.11(b)] considered here were taken from the literature [But83]. For the measurement experiments carried out therein, a helicopter equipped with an 870 MHz transmitter and a vehicle with a receiver were used to simulate a real-world satellite channel. Concerning the relative location of the helicopter and the mobile receiver, the elevation angle was held constant at 15°. One test route led through regions, in which the line-of-sight component was heavily shadowed, another one through regions with light shadowing. The measurement results of this so-called *equivalent satellite channel* have also been used in [Loo91]. Therefore, they offer a suitable basis for a fair comparison of the procedures. Further reports on measurement results of real-world satellite channels can be found, e.g., in [Huc83, Vog88, Vog90, Vog95].

Now let us combine all relevant model parameters, which decisively determine the statistical properties of the extended Suzuki process (Type I), into a parameter vector denoted and defined by $\mathbf{\Omega} := (\sigma_0,\, \kappa_0,\, \rho,\, f_\rho, \sigma_3,\, m_3)$. In practice, the frequency ratio

---

[2] We mention here that the complementary cumulative distribution function $F_{\eta_+}(r) = P(\eta(t) > r)$ and the cumulative distribution function $F_{\eta_-}(r) = P(\eta(t) \leq r)$ are generally related by $F_{\eta_+}(r) = 1 - F_{\eta_-}(r)$.

$\kappa_c = f_{max}/f_c$ is in general greater than 10. According to the statements made in Subsection 6.1.2, that parameter in this case exerts no influence on the statistics of the first and second order of $\eta(t)$. This is the reason, why κ_c has not been included in the parameter vector $\mathbf{\Omega}$. Without restriction of generality, we will therefore arbitrarily fix the value of the frequency ratio κ_c to 20. Moreover, without having to take any further restrictions into account, we also choose *a priori* $\theta_\rho = 0$.

As suitable measure of the deviations between the complementary distribution functions $F_{\eta_+}(r/\rho)$ and $F^\star_{\eta_+}(r/\rho)$ as well as between the normalized level-crossing rates $N_\eta(r/\rho)/f_{max}$ and $N^\star_\eta(r/\rho)/f_{max}$, we introduce the following error function

$$E_2(\mathbf{\Omega}) := \left\{\sum_{m=1}^{M}\left[W_1\left(\frac{r_m}{\rho}\right)\left(F^\star_{\eta_+}\left(\frac{r_m}{\rho}\right) - F_{\eta_+}\left(\frac{r_m}{\rho}\right)\right)\right]^2\right\}^{1/2} + \frac{1}{f_{max}}\left\{\sum_{m=1}^{M}\left[W_2\left(\frac{r_m}{\rho}\right)\left(N^\star_\eta\left(\frac{r_m}{\rho}\right) - N_\eta\left(\frac{r_m}{\rho}\right)\right)\right]^2\right\}^{1/2}, \quad (6.76)$$

where M is the number of different levels r_m at which the measurements were taken. In addition, $W_1(\cdot)$ and $W_2(\cdot)$ denote two weighting functions, which we here want to choose proportionally to the reciprocals of $F^\star_{\eta_+}(\cdot)$ and $N^\star_\eta(\cdot)$, respectively. The optimization of the components of the parameter vector $\mathbf{\Omega}$ is carried out numerically by applying the quasi-Newton procedure according to Fletcher-Powell [Fle63].

We first perform the optimization by using the classical Jakes power spectral density. Therefore, we keep the parameter $\kappa_0 = f_{min}/f_{max}$ constant at the value $\kappa_0 = 1$ during the minimization. Furthermore, we also fix f_ρ to the value $f_\rho = 0$, so that the extended Suzuki model simplifies to the conventional Rice-lognormal model. Now, we are confronted with the problem that there are no free parameters available for the optimization of the normalized level-crossing rate $N_\eta(r/\rho)/f_{max}$, because all the remaining model parameters (σ_0, ρ, σ_3, m_3) are completely used for the optimization of the complementary cumulative distribution function $F_{\eta_+}(r/\rho)$. In other words, a better approximation of the second order statistics is only possible at the expense of a worse approximation of the first order statistics. We will not yet make this compromise at this point. For the moment, we will be content with the approximation of $F_{\eta_+}(r/\rho)$, and temporarily put $W_2(r/\rho)$ equal to 0. The results of the parameter optimization for regions with light and heavy shadowing are listed in Table 6.1.

Table 6.1: The optimized parameters of the reference model for areas with heavy and light shadowing (without optimization of κ_0 and f_ρ).

Shadowing	σ_0	ρ	σ_3	m_3
heavy	0.1847	0.0554	0.1860	0.3515
light	0.3273	0.9383	0.0205	0.1882

Studying Figure 6.11(a), where the resulting complementary cumulative distribution function $F_{\eta_+}(r/\rho)$ is depicted, one can see that this function can be fitted very closely to the given measurement results. However, it becomes clear from Figure 6.11(b) that these satisfying results cannot be obtained for the normalized level-crossing rate $N_\eta(r/\rho)/f_{max}$. The deviations from the measurement results are partly more than 300 per cent in this case. The deeper reason of this mismatching is due to the far too high Doppler spread of the Jakes power spectral density.

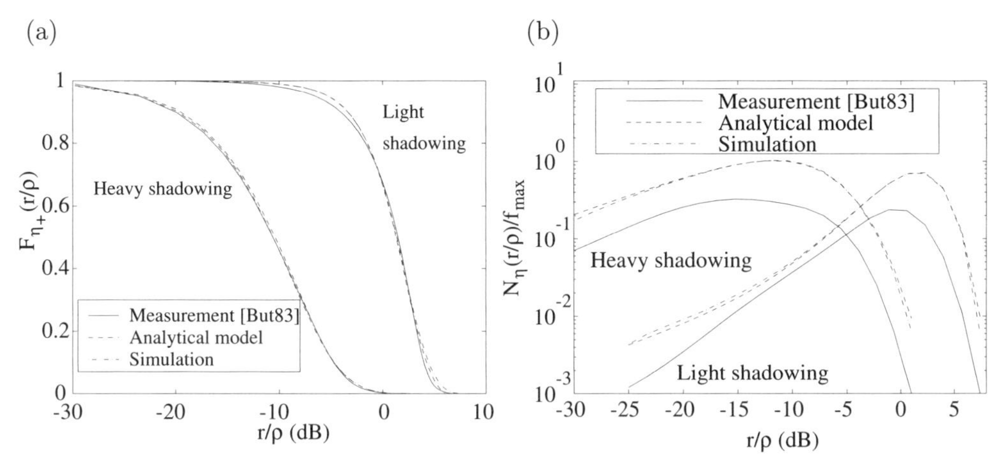

Figure 6.11: (a) Complementary cumulative distribution function $F_{\eta_+}(r/\rho)$ and (b) normalized level-crossing rate $N_\eta(r/\rho)/f_{max}$ for regions with heavy and light shadowing (without optimization of κ_0 and f_ρ).

For comparison and in order to confirm the results found, Figures 6.11(a) and 6.11(b) also show the results obtained from a discrete-time simulation of the extended Suzuki process. The deterministic processes $\tilde{\nu}_1(t)$, $\tilde{\nu}_2(t)$, and $\tilde{\nu}_3(t)$ were in this case designed by applying the techniques described in the preceding Subsection 6.1.4 (MEDS with $N_1 = 15$, $N_2 = 16$, and $N_3 = 15$).

The next step is to enable a reduction of the Doppler bandwidth and, thus, of the Doppler spread as well, by including the parameter κ_0 in the optimization procedure. In order to exploit the full flexibility of the channel model, the optimization of the parameter f_ρ will now also be permitted within the range $-f_{min} \leq f_\rho \leq f_{max}$. The numerical minimization of the error function (6.76) then yields for the components of the parameter vector $\boldsymbol{\Omega}$ to the results presented in Table 6.2. With these parameters, which have been optimized with respect to both $F_{\eta_+}(r/\rho)$ and $N_\eta(r/\rho)/f_{max}$, the behaviour of $F_{\eta_+}(r/\rho)$ remains almost unchanged (cf. Figures 6.11(a) and 6.12(a).) However, the actual advantages of the extended Suzuki model of Type I first become apparent by studying the statistics of second order. Observe that due to the model extension, the normalized level-crossing rate $N_\eta(r/\rho)/f_{max}$ of the reference model can now obviously be fitted to the measurement results better than in the case $\kappa_0 = 1$ and $f_\rho = 0$ (cf. Figures 6.11(b) and 6.12(b)).

Table 6.2: The optimized parameters of the reference channel model for areas with heavy and light shadowing (with optimization of κ_0 and f_ρ).

Shadowing	σ_0	κ_0	ρ	σ_3	m_3	f_ρ/f_{max}
heavy	0.2022	4.4E-11	0.1118	0.1175	0.4906	0.6366
light	0.4497	5.9E-08	0.9856	0.0101	0.0875	0.7326

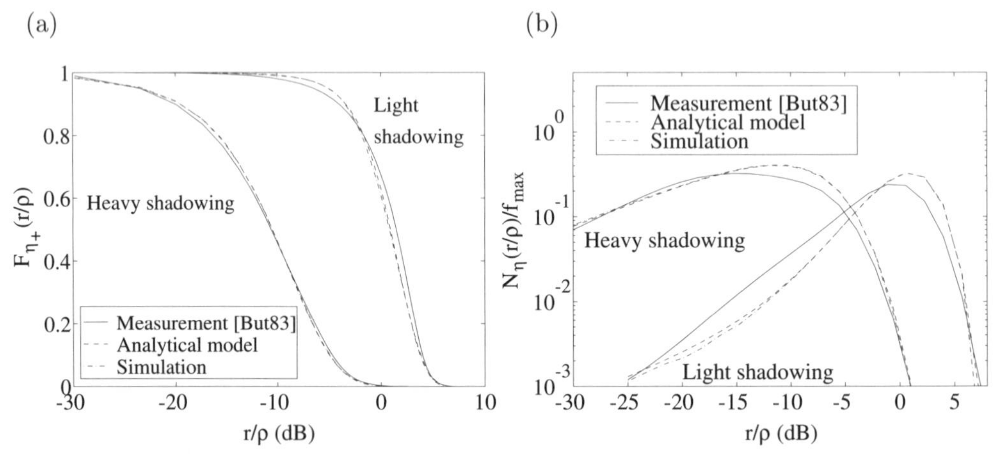

Figure 6.12: (a) Complementary cumulative distribution function $F_{\eta_+}(r/\rho)$ and (b) normalized level-crossing rate $N_\eta(r/\rho)/f_{max}$ for regions with heavy and light shadowing (with optimization of κ_0 and f_ρ).

It should also be mentioned that the Rice factor (3.18) of the extended Suzuki model (Type I) now reads as

$$c_R = \frac{\rho^2}{2\psi_0} = \frac{\rho^2}{\sigma_0^2[1 + \frac{2}{\pi}\arcsin(\kappa_0)]} \,. \tag{6.77}$$

Using the parameters listed in Table 6.2, we obtain the values $c_R = -5.15\,\text{dB}$ (heavy shadowing) and $c_R = 6.82\,\text{dB}$ (light shadowing) for the Rice factor.

The verification of the analytical results is now again established by means of simulation. Due to the fact that $\kappa_0 = f_{min}/f_{max}$ is very small in both cases determined by light and heavy shadowing (see Table 6.2), the influence of $\nu_2(t)$ or $\tilde{\nu}_2(t)$ can be neglected, and, consequently, N_2 can be set to zero, which is synonymous with an additional drastic reduction concerning the realization expenditure. The other processes $\tilde{\nu}_1(t)$ and $\tilde{\nu}_3(t)$ are again realized by employing the method of exact Doppler spread with $N_1 = 15$ and $N_3 = 15$ cosine functions, respectively. The simulation results are also depicted in Figures 6.12(a) and 6.12(b). From these figures, it can be realized that there is nearly an absolute correspondence between the reference model and the simulation model.

In order to illustrate the results, the Figures 6.13(a) and 6.13(b) both show us a part of the simulated sequence of the deterministic extended Suzuki process $\tilde{\eta}(t)$ for regions with heavy and light shadowing, respectively. One recognizes that for a heavily shadowed line-of-sight component (see Figure 6.13(a)), the average signal level is, all in all, obviously smaller than for an only lightly shadowed line-of-sight component (see Figure 6.13(b)). Also, the deep fades for heavy shadowing are much deeper than for light shadowing.

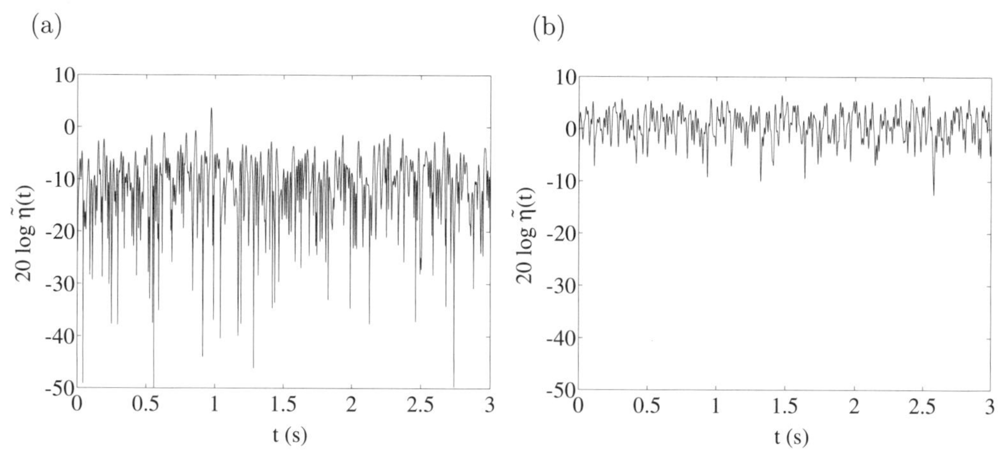

Figure 6.13: Simulation of deterministic extended Suzuki processes $\tilde{\eta}(t)$ of Type I for regions with (a) heavy shadowing and (b) light shadowing (MEDS, $N_1 = 15$, $N_2 = 0$, $N_3 = 15$, $f_{max} = 91\,\text{Hz}$, $\kappa_c = 20$).

6.2 THE EXTENDED SUZUKI PROCESS OF TYPE II

In the preceding Section 6.1, it has been shown how a higher model class can be created by introducing a correlation between the two Gaussian random processes determining the Rice process. In this way, the flexibility of the statistical properties of the second order could be increased. On the other hand, the statistical properties of the first order were not influenced. The model described in Section 6.1, however, is not the only possible one for which cross-correlated Gaussian random processes can be used. A further possibility, which was first introduced in [Pae97a], will be discussed in this section. We will see that a special type can be found for the cross-correlation function of the real part and the imaginary part of a complex-valued Gaussian random process, which not only increases the flexibility of the statistical properties of the second order of the stochastic model for modelling the short-term fading, but also the ones of the first order. In this model, the Rice, Rayleigh, and one-sided Gaussian random processes are included as special cases. The long-term fading is again modelled by means of a lognormal process as usual. The product of both processes, which is useful for modelling short-term and long-term fading, is called *extended Suzuki process of Type II.*

The description of the extended Suzuki process (of Type II) and the derivation of its statistical properties of first and second order will be the aim of this section. We will at first deal with the modelling and analysis of short-term fading.

6.2.1 Modelling and Analysis of the Short-Term Fading

The modelling of short-term fading will be performed by considering the reference model depicted in Figure 6.14. In the following, this model will be described.

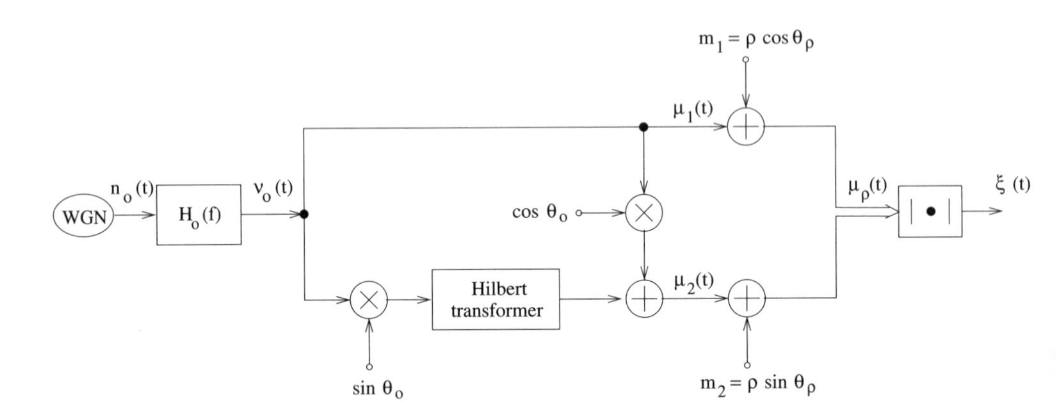

Figure 6.14: Reference model for stochastic processes $\xi(t)$ with cross-correlated Gaussian random processes $\mu_1(t)$ and $\mu_2(t)$.

Regarding this figure, one should notice that the complex-valued Gaussian random process

$$\mu(t) = \mu_1(t) + j\mu_2(t) \tag{6.78}$$

with the cross-correlated components $\mu_1(t)$ and $\mu_2(t)$ is derived from a single real-valued zero-mean Gaussian random process $\nu_0(t)$. In order to simplify the model, we will in the following assume that the Doppler frequency of the line-of-sight component is equal to zero, and, thus, the line-of-sight component is described by the time-invariant expression (3.3), i.e.,

$$m = m_1 + jm_2 = \rho e^{j\theta_\rho} . \tag{6.79}$$

As for the preceding models, we will also derive a further stochastic process for this one by taking the absolute value of the complex-valued Gaussian random process $\mu_\rho(t) = \mu(t) + m$, i.e.,

$$\xi(t) = |\mu_\rho(t)| = \sqrt{(\mu_1(t) + m_1)^2 + (\mu_2(t) + m_2)^2} . \tag{6.80}$$

We will see in Subsection 6.2.1.1 that Rice, Rayleigh, and one-sided Gaussian random processes are merely special cases of this process. To do justice to this property, the output process of the model shown in Figure 6.14 will be called *extended Rice process* in the following.

The Doppler power spectral density $S_{\nu_0\nu_0}(f)$ of the process $\nu_0(t)$ is described by the function (see Figure 6.15(a))

$$S_{\nu_0\nu_0}(f) = \begin{cases} \dfrac{\sigma_0^2}{\pi f_{max}\sqrt{1-(f/f_{max})^2}}, & |f| \leq \kappa_0 \cdot f_{max}, \\ 0, & |f| > \kappa_0 \cdot f_{max}, \end{cases} \tag{6.81}$$

where $0 < \kappa_0 \leq 1$. The symmetrical Doppler power spectral density $S_{\nu_0\nu_0}(f)$ as defined above is called the *restricted Jakes power spectral density.* We note that for the special case $\kappa_0 = 1$, the (classical) Jakes power spectral density (3.8) follows from the restricted Jakes power spectral density (6.81). The underlying physical model of the restricted Jakes power spectral density is based on the simplified assumption that in the presence of spatially limited obstacles or if sector antennas are used, the electromagnetic waves, whose angles of arrival lie within the intervals $(-\alpha_0, \alpha_0)$ and $(\pi - \alpha_0, \pi + \alpha_0)$, do not make a contribution to the received signal. Here, α_0 will be restricted to the range $(0, \pi/2]$. Furthermore, α_0 can be related to the parameter κ_0 via the equation $\kappa_0 = f_{min}/f_{max} = \cos\alpha_0$. All angles of arrival, which do not lie in any of the intervals just mentioned, are again assumed to be uniformly distributed. The actual reason for introducing the restricted Jakes power spectral density in our model is not to be found in the fitting of the theoretical Doppler power spectral density to power spectral densities rarely seen in practice. Instead of this, the variable κ_0 will give us a simple and an effective chance to reduce the Doppler spread of the Jakes power spectral density, which is often too large compared with practice.

From Figure 6.14, we can read the relations

$$\mu_1(t) = \nu_0(t) \tag{6.82}$$

and

$$\mu_2(t) = \cos\theta_0 \cdot \nu_0(t) + \sin\theta_0 \cdot \check{\nu}_0(t), \tag{6.83}$$

where the parameter θ_0 will be kept restricted to the interval $[-\pi, \pi)$, and $\check{\nu}_0(t)$ denotes the Hilbert transform of the coloured Gaussian random process $\mu_0(t)$. The spectral shaping of $\nu_0(t)$ in the reference model is obtained by filtering of white Gaussian noise $n_0(t) \sim N(0,1)$, where we again assume that the filter is real-valued and completely described by the transfer function $H_0(f) = \sqrt{S_{\nu_0\nu_0}(f)}$.

The autocorrelation functions $r_{\mu_1\mu_1}(\tau)$ and $r_{\mu_2\mu_2}(\tau)$, as well as the cross-correlation functions $r_{\mu_1\mu_2}(\tau)$ and $r_{\mu_2\mu_1}(\tau)$ can be expressed in terms of the autocorrelation function $r_{\nu_0\nu_0}(\tau)$ of the process $\nu_0(t)$ and the cross-correlation function $r_{\check{\nu}_0\nu_0}(\tau)$ of the processes $\check{\nu}_0(t)$ and $\nu_0(t)$ as follows:

$$r_{\mu_1\mu_1}(\tau) = r_{\mu_2\mu_2}(\tau) = r_{\nu_0\nu_0}(\tau), \tag{6.84a}$$

$$r_{\mu_1\mu_2}(\tau) = \cos\theta_0 \cdot r_{\nu_0\nu_0}(\tau) - \sin\theta_0 \cdot r_{\check{\nu}_0\nu_0}(\tau), \tag{6.84b}$$

$$r_{\mu_2\mu_1}(\tau) = \cos\theta_0 \cdot r_{\nu_0\nu_0}(\tau) + \sin\theta_0 \cdot r_{\check{\nu}_0\nu_0}(\tau). \tag{6.84c}$$

One should be aware of the influence of the parameter θ_0 here. Note that this parameter does not have any influence on the autocorrelation functions $r_{\mu_1\mu_1}(\tau)$ and $r_{\mu_2\mu_2}(\tau)$, but on the cross-correlation functions $r_{\mu_1\mu_2}(\tau)$ and $r_{\mu_2\mu_1}(\tau)$.

Substituting the relations (6.84a)–(6.84c) into (6.5), we obtain the following expression for the autocorrelation function $r_{\mu\mu}(\tau)$ of the complex-valued process $\mu(t) = \mu_1(t) + j\mu_2(t)$

$$r_{\mu\mu}(\tau) = 2r_{\nu_0\nu_0}(\tau) - j2\sin\theta_0 \cdot r_{\check{\nu}_0\nu_0}(\tau). \tag{6.85}$$

The Fourier transform of the above result gives us the power spectral density in the form

$$S_{\mu\mu}(f) = 2S_{\nu_0\nu_0}(f) - j2\sin\theta_0 \cdot S_{\check{\nu}_0\nu_0}(f). \tag{6.86}$$

From (2.56b) and (2.56d), we obtain the relation $S_{\check{\nu}_0\nu_0}(f) = j\,\mathrm{sgn}\,(f) \cdot S_{\nu_0\nu_0}(f)$, so that $S_{\mu\mu}(f)$ can now be expressed in terms of the restricted Jakes power spectral density $S_{\nu_0\nu_0}(f)$ as follows

$$S_{\mu\mu}(f) = 2[1 + \mathrm{sgn}\,(f)\sin\theta_0] \cdot S_{\nu_0\nu_0}(f). \tag{6.87}$$

Note that $S_{\mu\mu}(f)$ is an unsymmetrical function for all values of $\theta_0 \in (-\pi, \pi)\backslash\{0\}$. An example of the power spectral density $S_{\mu\mu}(f)$ is depicted in Figure 6.15(b), where the value 19.5° has been chosen for the parameter θ_0.

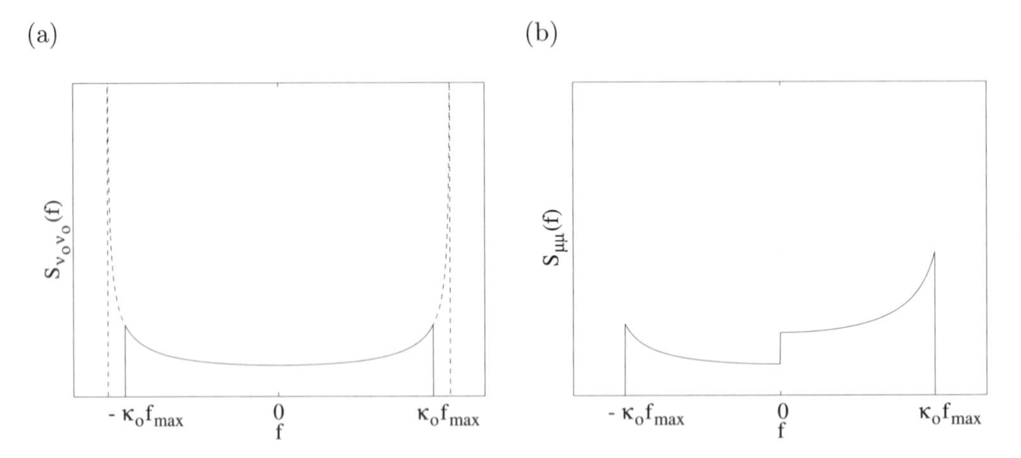

Figure 6.15: Doppler power spectral densities: (a) restricted Jakes PSD $S_{\nu_0\nu_0}(f)$ and (b) $S_{\mu\mu}(f)$ ($\theta_0 = 19.5°$).

Deriving the statistical properties of $\xi(t) = |\mu_\rho(t)|$ and $\vartheta(t) = \arg\{\mu_\rho(t)\}$, we will again make use of the abbreviations (6.11a) and (6.11b). Therefore, we substitute (6.84a) into (6.11a) and (6.84b) into (6.11b), so that after some lengthy but simple algebraic computations, the characteristic quantities of the extended Rice model can be written as follows:

$$\psi_0^{(0)} = \psi_0 = \frac{2}{\pi}\sigma_0^2 \arcsin(\kappa_0)\,, \tag{6.88a}$$

$$\psi_0^{(1)} = \dot{\psi}_0 = 0\,, \tag{6.88b}$$

$$\psi_0^{(2)} = \ddot{\psi}_0 = -\psi_0 \cdot 2(\pi f_{max})^2 \left\{1 - \frac{\sin[2\arcsin(\kappa_0)]}{2\arcsin(\kappa_0)}\right\}, \tag{6.88c}$$

$$\phi_0^{(0)} = \phi_0 = \psi_0 \cdot \cos\theta_0\,, \tag{6.88d}$$

$$\phi_0^{(1)} = \dot{\phi}_0 = 4\sigma_0^2 f_{max}(1 - \sqrt{1-\kappa_0^2}\,) \cdot \sin\theta_0\,, \tag{6.88e}$$

$$\phi_0^{(2)} = \ddot{\phi}_0 = \ddot{\psi}_0 \cdot \cos\theta_0\,, \tag{6.88f}$$

where $0 < \kappa_0 \leq 1$ and $-\pi \leq \theta_0 < \pi$. A comparison between the equations (6.88a)–(6.88f) and (6.12a)–(6.12f) shows us that for the present model even the quantities ϕ_0 and $\ddot{\phi}_0$ are in general different from zero. Only for the special case $\theta_0 = \pm\pi/2$, do we have $\phi_0 = \ddot{\phi}_0 = 0$. Hence, there are reasons for supposing that the statistical properties of the extended Rice process are different from those of the classical Rice process.

The starting point, which enables the analysis of the statistical properties of extended Rice processes, is again the multivariate Gaussian distribution of the processes $\mu_{\rho_1}(t)$, $\mu_{\rho_2}(t)$, $\dot{\mu}_{\rho_1}(t)$, and $\dot{\mu}_{\rho_2}(t)$ at the same time instant t [see (6.14)]. For the present model, where it was assumed for simplification that $f_\rho = 0$, the multivariate Gaussian distribution (6.14) is completely described by the column vectors

$$\boldsymbol{x} = \begin{pmatrix} x_1 \\ x_2 \\ \dot{x}_1 \\ \dot{x}_2 \end{pmatrix} \quad \text{and} \quad \boldsymbol{m} = \begin{pmatrix} m_1 \\ m_2 \\ \dot{m}_1 \\ \dot{m}_2 \end{pmatrix} = \begin{pmatrix} \rho\cos\theta_\rho \\ \rho\sin\theta_\rho \\ 0 \\ 0 \end{pmatrix} \tag{6.89a, b}$$

as well as by the covariance or correlation matrix

$$\boldsymbol{C}_{\mu_\rho}(0) = \boldsymbol{R}_\mu(0) = \begin{pmatrix} \psi_0 & \phi_0 & 0 & \dot{\phi}_0 \\ \phi_0 & \psi_0 & -\dot{\phi}_0 & 0 \\ 0 & -\dot{\phi}_0 & -\ddot{\psi}_0 & -\ddot{\phi}_0 \\ \dot{\phi}_0 & 0 & -\ddot{\phi}_0 & -\ddot{\psi}_0 \end{pmatrix}. \tag{6.90}$$

Employing the relations (6.88d) and (6.88f) results in

$$\boldsymbol{C}_{\mu_\rho}(0) = \boldsymbol{R}_\mu(0) = \begin{pmatrix} \psi_0 & \psi_0\cos\theta_0 & 0 & \dot{\phi}_0 \\ \psi_0\cos\theta_0 & \psi_0 & -\dot{\phi}_0 & 0 \\ 0 & -\dot{\phi}_0 & -\ddot{\psi}_0 & -\ddot{\psi}_0\cos\theta_0 \\ \dot{\phi}_0 & 0 & -\ddot{\psi}_0\cos\theta_0 & -\ddot{\psi}_0 \end{pmatrix}. \tag{6.91}$$

Now, after substituting (6.89a, b) and (6.91) into (6.14), the desired joint probability density function $p_{\mu_{\rho 1}\mu_{\rho 2}\dot{\mu}_{\rho 1}\dot{\mu}_{\rho 2}}(x_1, x_2, \dot{x}_1, \dot{x}_2)$ of our model can be calculated. We then transform the Cartesian coordinates (x_1, x_2) of this density to polar coordinates (z, θ) by means of (6.22a,b). After some further algebraic manipulations, we will succeed in converting the joint probability density function (6.24) to the following form [Pae97a]

$$
\begin{aligned}
p_{\xi\dot{\xi}\vartheta\dot{\vartheta}}(z, \dot{z}, \theta, \dot{\theta}) \quad = \quad & \frac{z^2}{(2\pi)^2 \beta \psi_0 \sin^2 \theta_0} \cdot e^{-\frac{1}{2\psi_0 \sin^2 \theta_0}[z^2+\rho^2-2z\rho\cos(\theta-\theta_\rho)]} \\
& \cdot e^{\frac{\cos\theta_0}{2\psi_0 \sin^2\theta_0}[z^2 \sin 2\theta + \rho^2 \sin 2\theta_\rho - 2z\rho \sin(\theta+\theta_\rho)]} \\
& \cdot e^{-\frac{1}{2\beta(1+\cos\theta_0 \cdot \sin 2\theta)} \cdot \left(\dot{z} + \frac{\dot{\phi}_0[\rho \sin(\theta-\theta_\rho) - \cos\theta_0 (z \cos 2\theta - \rho\cos(\theta+\theta_\rho))]}{\psi_0 \sin^2 \theta_0} \right)^2} \\
& \cdot e^{-\frac{z^2(1+\cos\theta_0 \cdot \sin 2\theta)}{2\beta \sin^2\theta_0} \cdot \left(\dot{\theta} + \frac{\dot{\phi}_0[\rho\cos(\theta-\theta_\rho) - z] - \psi_0 \dot{z} \cos\theta_0 \cdot \cos 2\theta}{\psi_0 z (1+\cos\theta_0 \cdot \sin 2\theta)} \right)^2}
\end{aligned} \tag{6.92}
$$

for $z \geq 0$, $|\dot{z}| < \infty$, $|\theta| \leq \pi$, and $|\dot{\theta}| < \infty$. Here, one should note that the quantity β in (6.92) is no longer given by (6.28), but is defined by the extended expression

$$
\beta = -\ddot{\psi}_0 - \frac{\dot{\phi}_0^2}{\psi_0 \sin^2 \theta_0} \,. \tag{6.93}
$$

In the following subsection, we will derive the probability density function of the amplitude $\xi(t)$ and the phase $\vartheta(t)$ from the joint probability density function (6.92). The analysis of the level-crossing rate and the average duration of fades of $\xi(t)$ will then follow subsequently.

6.2.1.1 Probability Density Function of the Amplitude and the Phase

For the probability density function of the extended Rice process $\xi(t)$, denoted by $p_\xi(z)$, we obtain the following result after substituting (6.92) into (6.29)

$$
\begin{aligned}
p_\xi(z) \quad = \quad & \frac{z}{2\pi\psi_0 |\sin\theta_0|} e^{-\frac{z^2+\rho^2}{2\psi_0 \sin^2\theta_0}} \\
& \cdot \int_{-\pi}^{\pi} e^{\frac{z\rho\cos(\theta-\theta_\rho)}{\psi_0 \sin^2\theta_0}} \cdot e^{\frac{\cos\theta_0}{2\psi_0\sin^2\theta_0}[z^2\sin 2\theta + \rho^2 \sin 2\theta_\rho - 2z\rho\sin(\theta+\theta_\rho)]} d\theta, \quad z \geq 0 \,.
\end{aligned} \tag{6.94}
$$

Exactly as for conventional Rice processes, the probability density function $p_\xi(z)$ in this case also depends on the mean power of the processes $\mu_1(t)$ and $\mu_2(t)$, i.e., ψ_0, as well as on the amplitude ρ of the line-of-sight component. Moreover, the density of the extended Rice process is also determined by the parameter θ_0 and — what is surprising at first — by the phase θ_ρ of the line-of-sight component. We will understand this property as soon as we have derived the corresponding simulation model (see Subsection 6.2.3). In order to illustrate the results, we will study Figures 6.16(a) and 6.16(b), where the probability density function (6.94) is shown for various values of the parameters θ_0 and θ_ρ, respectively. It should be pointed out that even for this model, the density $p_\xi(z)$ neither depends on the first and second time derivative of the autocorrelation function (6.11a), i.e., $\dot{\psi}_0$ and $\ddot{\psi}_0$, nor on the first and second time derivative of the cross-correlation function (6.11b), i.e., $\dot{\phi}_0$ and $\ddot{\phi}_0$.

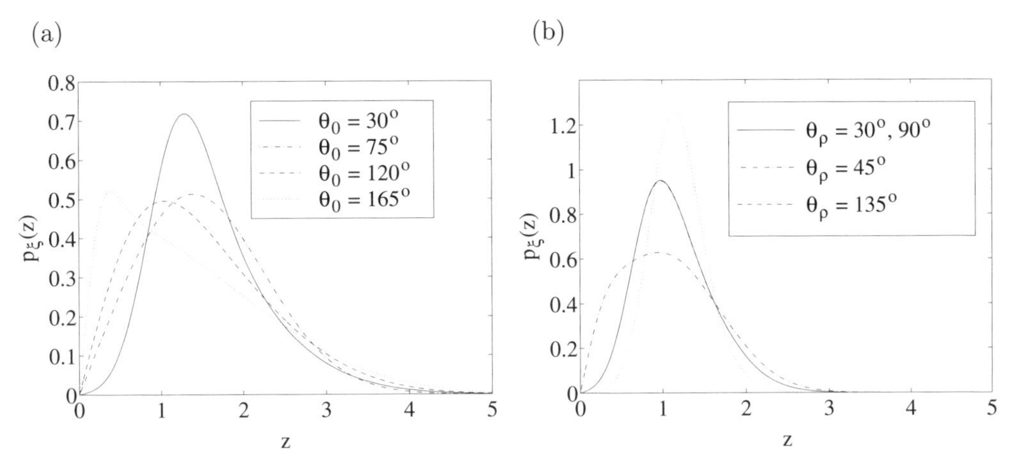

Figure 6.16: Probability density function $p_\xi(z)$ of extended Rice processes $\xi(t)$ for various values of the parameters (a) θ_0 ($\psi_0 = 1$, $\rho = 1$, $\theta_\rho = 127°$) and (b) θ_ρ ($\psi_0 = 1$, $\rho = 1$, $\theta_0 = 45°$).

In the text that follows, we will study some special cases. Especially, if we have $\theta_0 = \pm\pi/2$, then the integral in (6.94) can be solved explicitly, and it again follows the Rice distribution

$$p_\xi(z) = \frac{z}{\psi_0} e^{-\frac{z^2+\rho^2}{2\psi_0}} I_0\left(\frac{z\rho}{\psi_0}\right), \quad z \geq 0, \tag{6.95}$$

with ψ_0 according to (6.88a). For a shadowed line-of-sight component, i.e., $\rho = 0$, and, at first, for arbitrary values of $\theta_0 \in [-\pi, \pi)$, the following density follows from (6.94)

$$p_\xi(z) = \frac{z}{\psi_0 |\sin\theta_0|} e^{-\frac{z^2}{2\psi_0 \sin^2\theta_0}} I_0\left(\frac{z^2\cos\theta_0}{2\psi_0 \sin^2\theta_0}\right), \quad z \geq 0, \tag{6.96}$$

from which, especially for $\theta_0 = \pm\pi/2$, the Rayleigh distribution

$$p_\xi(z) = \frac{z}{\psi_0} e^{-\frac{z^2}{2\psi_0}}, \quad z \geq 0, \tag{6.97}$$

and for $\theta_0 \to 0$, the one-sided Gaussian distribution

$$p_\xi(z) = \frac{1}{\sqrt{\pi\psi_0}} e^{-\frac{z^2}{4\psi_0}}, \quad z \geq 0, \tag{6.98}$$

follows. Consequently, the Rice distribution, the Rayleigh distribution, and the one-sided Gaussian distribution are special cases of the extended Rice distribution

(6.94).

For the probability density function of the phase $\vartheta(t)$, denoted by $p_\vartheta(\theta)$, we obtain, after substituting (6.92) into (6.31), the expression

$$
\begin{aligned}
p_\vartheta(\theta) &= \frac{|\sin\theta_0|}{2\pi(1-\cos\theta_0\cdot\sin 2\theta)}\cdot e^{-\frac{\rho^2(1-\cos\theta_0\cdot\sin 2\theta_\rho)}{2\psi_0\sin^2\theta_0}} \\
&\quad \cdot\left\{1+\sqrt{\pi}f(\theta)e^{f^2(\theta)}[1+\operatorname{erf}(f(\theta))]\right\}, \quad -\pi\le\theta\le\pi, \qquad (6.99)
\end{aligned}
$$

where

$$
f(\theta)=\frac{\rho[\cos(\theta-\theta_\rho)-\cos\theta_0\cdot\sin(\theta+\theta_\rho)]}{|\sin\theta_0|\sqrt{2\psi_0(1-\cos\theta_0\cdot\sin 2\theta)}}\,. \qquad (6.100)
$$

Exactly like the probability density function of the amplitude [see (6.94)], the probability density function of the phase merely depends on the parameters ψ_0, ρ, θ_0, and θ_ρ, and not on the quantities $\dot{\psi}_0$, $\ddot{\psi}_0$, $\dot{\phi}_0$, and $\ddot{\phi}_0$.

The same probability density function of the phase, that we became acquainted with during the analysis of Rice processes with uncorrelated Gaussian random processes $\mu_1(t)$ and $\mu_2(t)$ in Subsection 3.3.1 [see there (3.22)], also follows from (6.99) for the special case $\theta_0 = \pm\pi/2$. If the parameters ρ and θ_0 are determined by $\rho = 0$ and $\theta_0 = \pm\pi/2$, then the phase $\vartheta(t)$ is uniformly distributed within the interval $[-\pi, \pi]$.

Finally, the influence of the parameters θ_0 and θ_ρ on the behaviour of the density $p_\vartheta(\theta)$ shall be made clear by the Figures 6.17(a) and 6.17(b), respectively.

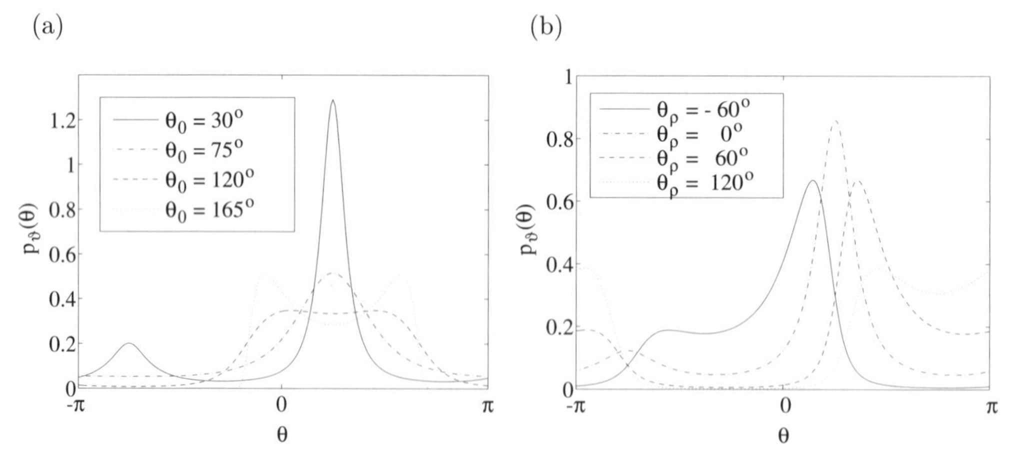

Figure 6.17: Probability density function $p_\vartheta(\theta)$ of the phase $\vartheta(t)$ for various values of the parameters (a) θ_0 ($\psi_0 = 1$, $\rho = 1$, $\theta_\rho = 45°$) and (b) θ_ρ ($\psi_0 = 1$, $\rho = 1$, $\theta_0 = 45°$).

6.2.1.2 *Level-Crossing Rate and Average Duration of Fades*

For the calculation of the level-crossing rate $N_\xi(r)$, the joint probability density function $p_{\xi\dot{\xi}}(z,\dot{z})$ of the stochastic processes $\xi(t)$ and $\dot{\xi}(t)$ have to be known at the same time instant t. For this density, we obtain the following integral expression after substituting (6.92) into (6.34)

$$
\begin{aligned}
p_{\xi\dot{\xi}}(z,\dot{z}) \;=\; & \frac{z}{(2\pi)^{3/2}\psi_0\sqrt{\beta}|\sin\theta_0|} e^{-\frac{z^2+\rho^2}{2\psi_0\sin^2\theta_0}} \int_{-\pi}^{\pi} \frac{1}{\sqrt{1+\cos\theta_0\cdot\sin 2\theta}} \\
& \cdot e^{\frac{z\rho\cos(\theta-\theta_\rho)}{\psi_0\sin^2\theta_0}} \cdot e^{\frac{\cos\theta_0}{2\psi_0\sin^2\theta_0}[z^2\sin 2\theta+\rho^2\sin 2\theta_\rho-2z\rho\sin(\theta+\theta_\rho)]} \\
& \cdot e^{-\frac{1}{2\beta(1+\cos\theta_0\cdot\sin 2\theta)}\left(\dot{z}+\frac{\dot{\phi}_0[\rho\sin(\theta-\theta_\rho)-\cos\theta_0(z\cos 2\theta-\rho\cos(\theta+\theta_\rho))]}{\psi_0\sin^2\theta_0}\right)^2} d\theta\,, \quad (6.101)
\end{aligned}
$$

for $z \geq 0$ and $|\dot{z}| < \infty$. Here, ψ_0, $\dot{\phi}_0$, and β again are the quantities defined by (6.88a), (6.88e), and (6.93), respectively. Within the interval $(-\pi,\pi)\backslash\{0\}$, no value can be found for the parameter θ_0 in such a way that the stochastic processes $\xi(t)$ and $\dot{\xi}(t)$ become statistically independent, because $p_{\xi\dot{\xi}}(z,\dot{z}) \neq p_\xi(z)\cdot p_{\dot{\xi}}(\dot{z})$ always holds. Even for the special case $\theta_0 = \pm\pi/2$, the equation (6.35) may follow from (6.101), but here it has to be taken into consideration that the relations (6.88a)–(6.88f) hold now, so that we have $\dot{\phi}_0 \neq 0$ ($\alpha \neq 0$), and, thus, (6.101) can never be brought into the form (6.36).

With the joint probability density function (6.101), all assumptions for the derivation of the level-crossing rate $N_\xi(r)$ of extended Rice processes $\xi(t)$ are made. We substitute (6.101) into the definition (6.33), and, after some algebraic manipulations, obtain the result

$$
\begin{aligned}
N_\xi(r) \;=\; & \frac{r\sqrt{\beta}}{(2\pi)^{3/2}\psi_0|\sin\theta_0|}\cdot e^{-\frac{r^2+\rho^2}{2\psi_0\sin^2\theta_0}} \int_{-\pi}^{\pi} \sqrt{1+\cos\theta_0\cdot\sin 2\theta} \\
& \cdot e^{\frac{r\rho\cos(\theta-\theta_\rho)}{\psi_0\sin^2\theta_0}} e^{\frac{\cos\theta_0}{2\psi_0\sin^2\theta_0}[r^2\sin 2\theta+\rho^2\sin 2\theta_\rho-2r\rho\sin(\theta+\theta_\rho)]} \\
& \cdot\left\{e^{-g^2(r,\theta)}+\sqrt{\pi}g(r,\theta)[1+\operatorname{erf}(g(r,\theta))]\right\} d\theta\,, \quad r \geq 0\,, \qquad (6.102)
\end{aligned}
$$

where the function $g(r,\theta)$ stands for

$$
g(r,\theta) = -\frac{\dot{\phi}_0\{\rho\sin(\theta-\theta_\rho)-\cos\theta_0[r\cos 2\theta-\rho\cos(\theta+\theta_\rho)]\}}{\psi_0\sin^2\theta_0\sqrt{2\beta(1+\cos\theta_0\cdot\sin 2\theta)}}\,. \qquad (6.103)
$$

The quantities ψ_0, $\dot{\phi}_0$, and β are again defined by (6.88a), (6.88e), and (6.93), respectively. It should be noted that we have made use of the integral [Gra81, vol. I, eq. (3.462.5)]

$$
\int_0^\infty x\,e^{-ax^2-2bx}\,dx = \frac{1}{2a}\left\{1-b\sqrt{\frac{\pi}{a}}\,e^{\frac{b^2}{a}}\left[1-\operatorname{erf}\left(\frac{b}{\sqrt{a}}\right)\right]\right\}, \quad a>0\,, \quad (6.104)
$$

for the derivation of (6.102). Using (6.88a)–(6.88f), we easily find out that (6.102) is proportional to the maximum Doppler frequency f_{max}, i.e., the normalised level-crossing rate $N_\xi(r)/f_{max}$ is independent of the speed of the vehicle and the carrier frequency, just as in the previous case. A brief parameter study, which makes the influence of the parameters κ_0, ρ, θ_0, and θ_ρ on the normalised level-crossing rate $N_\xi(r)/f_{max}$ clear, is depicted in Figures 6.18(a)–6.18(d). The variation of κ_0 (Figure 6.18(a)) and ρ (Figure 6.18 (b)) leads to curves, which are in principle similar to those shown in Figure 6.3(a) and Figure 3.5(b), respectively. A further, more powerful parameter exists, namely θ_0, which has a decisive influence on the behaviour of $N_\xi(r)/f_{max}$, as can be seen in Figure 6.18(c). For the parameters (ψ_0, κ_0, ρ, θ_0), which give rise to the results shown in Figure 6.18(d), the value of the quantity θ_ρ is only of secondary importance.

Now, attention is given to some special cases. On the assumption that $\theta_0 = \pm\pi/2$ holds, the level-crossing rate described by (6.37) follows from (6.102). If, in addition, $\rho = 0$ holds, then $N_\xi(r)$ becomes directly proportional to the Rayleigh distribution and can be brought into the form (6.38). Moreover, for the special case $\rho = 0$ and $\theta_0 \to 0°$, one can show that the level-crossing rate of one-sided Gaussian random processes follows from (6.102), i.e.,

$$N_\xi(r) = \frac{\sqrt{\beta}}{\pi\sqrt{\psi_0}} e^{-\frac{r^2}{4\psi_0}}, \quad r \geq 0, \tag{6.105}$$

where β is given by $\beta = -\ddot{\psi}_0 > 0$ in the present case. Further special cases such as, e.g., $\rho = 0$ in connection with arbitrary values for $\theta_0 \in [-\pi, \pi)$ can also be analysed easily with the help of (6.102).

When calculating the average duration of fades $T_{\xi_-}(r)$ [see (6.40)], we also need to know the cumulative distribution function $F_{\xi_-}(r)$ of the extended Rice process $\xi(t)$, besides the level-crossing rate $N_\xi(r)$. For the former, we obtain the following double integral by using the probability density function (6.94)

$$\begin{aligned} F_{\xi_-}(r) &= \int_0^r p_\xi(z)\, dz \\ &= \int_0^r \frac{z}{2\pi\psi_0 |\sin\theta_0|} e^{-\frac{z^2+\rho^2}{2\psi_0 \sin^2\theta_0}} \cdot \int_{-\pi}^{\pi} e^{\frac{z\rho\cos(\theta-\theta_\rho)}{\psi_0\sin^2\theta_0}} \\ &\quad \cdot e^{\frac{\cos\theta_0}{2\psi_0\sin^2\theta_0}\cdot[z^2\sin 2\theta + \rho^2 \sin 2\theta_\rho - 2z\rho\sin(\theta+\theta_\rho)]}\, d\theta\, dz, \quad r \geq 0. \end{aligned} \tag{6.106}$$

According to (6.40), the average duration of fades $T_{\xi_-}(r)$ of extended Rice processes $\xi(t)$ is then the quotient of (6.106) and (6.102).

Figures 6.19(a) to 6.19(d) clearly show the influence which the parameters κ_0, ρ, θ_0, and θ_ρ have on the normalised average duration of fades $T_{\xi_-}(r) \cdot f_{max}$. The model parameters, which lead to the results shown in Figures 6.18(a) to 6.18(d), have also been used for the calculation of $T_{\xi_-}(r) \cdot f_{max}$ here. Varying the parameter κ_0, we

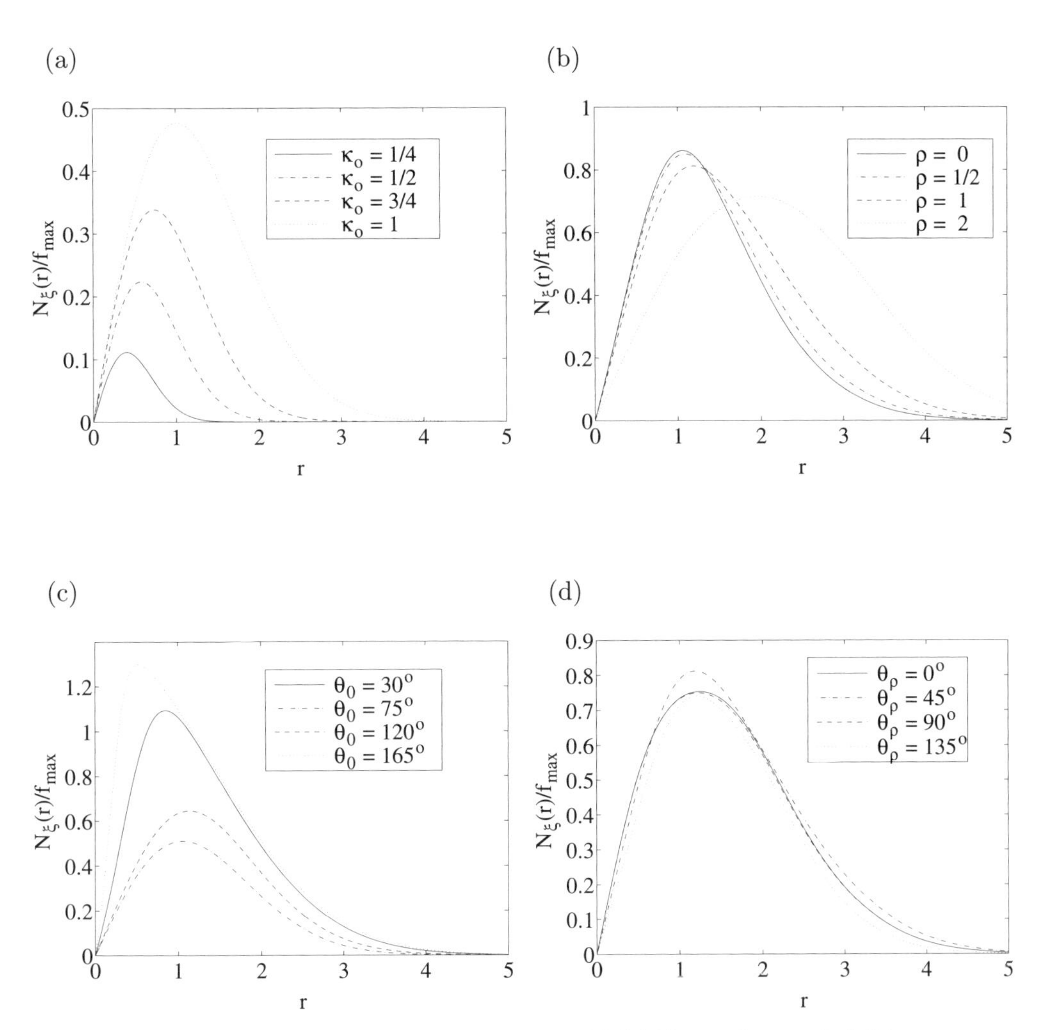

Figure 6.18: Normalized level-crossing rate $N_\xi(r)/f_{max}$ of extended Rice processes (Type II) depending on: (a) κ_0 ($\sigma_0^2 = 1$, $\rho = 0$, $\theta_0 = 45°$), (b) ρ ($\psi_0 = 1$, $\kappa_0 = 1$, $\theta_\rho = 45°$, $\theta_0 = 45°$), (c) θ_0 ($\psi_0 = 1$, $\kappa_0 = 1$, $\rho = 0$), and (d) θ_ρ ($\psi_0 = 1$, $\kappa_0 = 1$, $\rho = 1$, $\theta_0 = 45°$).

can recognize similar effects on $T_{\xi_-}(r) \cdot f_{max}$ in Figure 6.19(a) as in Figure 6.4(a). Figure 6.19(b) shows that even at low levels r, an increase of ρ results in a reduction in the normalised average duration of fades $T_{\xi_-}(r) \cdot f_{max}$. This is obviously in contrast to the results depicted in Figure 3.6(b), where no considerable effects on $T_{\xi_-}(r) \cdot f_{max}$ can be observed at low levels r by a variation of ρ. From Figure 6.19(c), it can be realized that the parameter θ_0 affects the behaviour of $T_{\xi_-}(r) \cdot f_{max}$ at medium and high levels r, whereas its influence at low levels r can be ignored (at least if the parameters are chosen as in the present example: $\psi_0 = 1$, $\kappa_0 = 1$, and $\rho = 0$). Similarly, the exact opposite relations hold for the variation of the parameter θ_ρ (see Figure 6.19(d)).

6.2.2 The Stochastic Extended Suzuki Process of Type II

In [Pae97a], the extended Suzuki process of Type II, denoted by $\eta(t)$, was introduced as product process of the extended Rice process $\xi(t)$ studied before and the lognormal process $\lambda(t)$ described in Subsection 6.1.2, i.e., $\eta(t) = \xi(t) \cdot \lambda(t)$. The structure of the reference model corresponding to this process is depicted in Figure 6.20.

In the following, we will analyse the probability density function of the amplitude, the level-crossing rate, and the average duration of fades of this model.

Let us assume that the coloured Gaussian random processes $\nu_0(t)$ and $\nu_3(t)$ are statistically independent, which leads to the fact that the extended Rice process $\xi(t)$ and the lognormal process $\lambda(t)$ are statistically independent as well. Due to the multiplicative relation between the two statistically independent processes $\xi(t)$ and $\lambda(t)$, the probability density function $p_\eta(z)$ of the extended Suzuki process of Type II can be derived by using (6.94) and (6.49) as follows:

$$p_\eta(z) = \int_{-\infty}^{\infty} \frac{1}{|y|} p_\xi\left(\frac{z}{y}\right) \cdot p_\lambda(y)\, dy\,, \tag{6.107a}$$

$$= \int_{-\infty}^{\infty} \frac{1}{|y|} p_\xi(y) \cdot p_\lambda\left(\frac{z}{y}\right)\, dy\,, \tag{6.107b}$$

$$= \frac{1}{2\pi\psi_0 |\sin\theta_0|} \int_0^{\infty} \frac{e^{-\frac{[\ln(z/y)-m_3]^2}{2\sigma_3^2}}}{\sqrt{2\pi}\sigma_3 (z/y)} \cdot e^{-\frac{y^2+\rho^2}{2\psi_0 \sin^2\theta_0}} \cdot \int_{-\pi}^{\pi} e^{\frac{y\rho\cos(\theta-\theta_\rho)}{\psi_0 \sin^2\theta_0}}$$
$$\cdot e^{\frac{\cos\theta_0}{2\psi_0\sin^2\theta_0}[y^2\sin 2\theta+\rho^2\sin 2\theta_\rho - 2y\rho\sin(\theta+\theta_\rho)]} d\theta\, dy, \quad z \geq 0\,. \tag{6.107c}$$

Here, we deliberately preferred the relation (6.107b) to (6.107a), because the solution of (6.107c) can then be performed more advantageously by means of numerical integration techniques. For $\sigma_3 \to 0$ and $m_3 \to 0$, it follows $p_\lambda(z/y) \to |y|\delta(z-y)$ and, thus, $p_\eta(z) \to p_\xi(z)$, where $p_\xi(z)$ is described by (6.94). In general, the

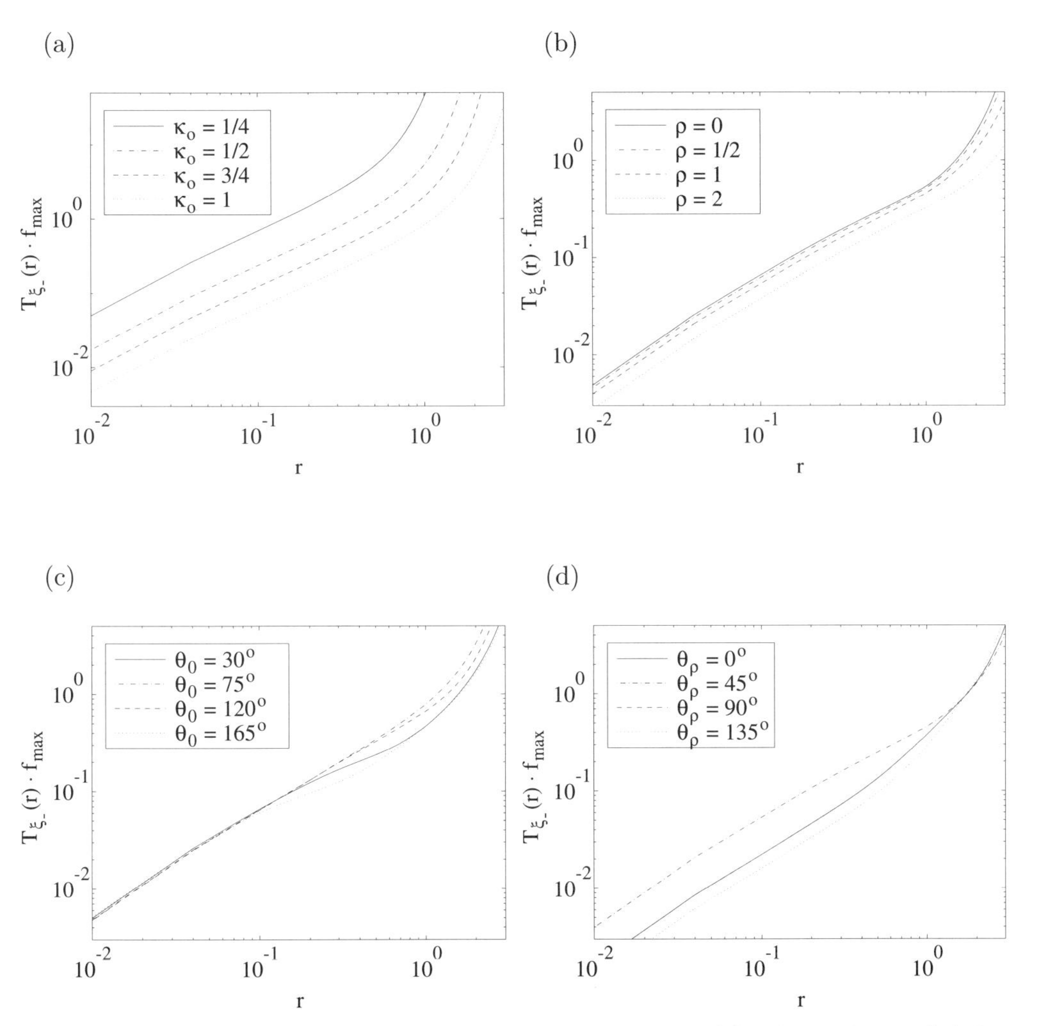

Figure 6.19: Normalized average duration of fades $T_{\xi_-}(r) \cdot f_{max}$ of extended Rice processes (Type II) depending on: (a) κ_0 ($\sigma_0^2 = 1,\ \rho = 0,\ \theta_0 = 45°$), (b) ρ ($\psi_0 = 1,\ \kappa_0 = 1,\ \theta_\rho = 45°,\ \theta_0 = 45°$), (c) θ_0 ($\psi_0 = 1,\ \kappa_0 = 1,\ \rho = 0$), and (d) θ_ρ ($\psi_0 = 1,\ \kappa_0 = 1,\ \rho = 1,\ \theta_0 = 45°$).

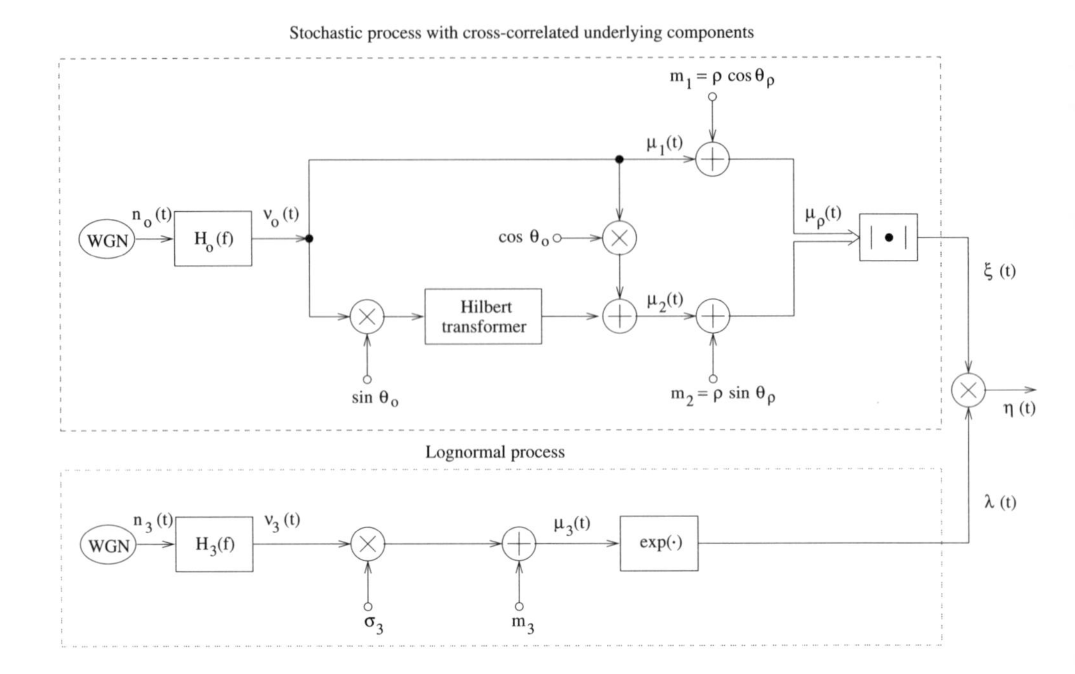

Figure 6.20: Reference model for extended Suzuki processes (Type II).

probability density function (6.107c) depends on the mean power ψ_0, the parameters σ_3, m_3, ρ, θ_ρ, and, last but not least, on θ_0. Figures 6.21(a) and 6.21(b) let us imagine what influence the parameters σ_3 and m_3, respectively, have on the behaviour of the probability density function $p_\eta(z)$.

Next, we will calculate the level-crossing rate $N_\eta(r)$ of extended Suzuki processes (Type II). Since the joint probability density function $p_{\eta\dot{\eta}}(z,\dot{z})$ of the processes $\eta(t)$ and $\dot{\eta}(t)$ at the same time t is required for our purpose, we at first substitute the relations (6.101) and (6.53) found for $p_{\xi\dot{\xi}}(z,\dot{z})$ and $p_{\lambda\dot{\lambda}}(y,\dot{y})$, respectively, into (6.58). Thus,

$$
\begin{aligned}
p_{\eta\dot{\eta}}(z,\dot{z}) \;=\; & \frac{1}{(2\pi)^{3/2}\psi_0\sqrt{\beta}|\sin\theta_0|} \cdot \int_0^\infty \frac{e^{-\frac{[\ln(z/y)-m_3]^2}{2\sigma_3^2}}}{\sqrt{2\pi}\sigma_3 (z/y)^2} \cdot e^{-\frac{y^2+\rho^2}{2\psi_0 \sin^2\theta_0}} \\
& \cdot \int_{-\pi}^{\pi} \frac{e^{\frac{y\rho\cos(\theta-\theta_\rho)}{\psi_0\sin^2\theta_0}}\, e^{\frac{\cos\theta_0}{2\psi_0\sin^2\theta_0}[y^2\sin 2\theta+\rho^2\sin 2\theta_\rho - 2y\rho\sin(\theta+\theta_\rho)]}}{h(y,\theta)\sqrt{1+\cos\theta_0\cdot\sin 2\theta}} \\
& \cdot e^{-\frac{\left\{\dot{z}+\frac{\dot{\phi}_0(z/y)[\rho\sin(\theta-\theta_\rho)-\cos\theta_0(y\cos 2\theta-\rho\cos(\theta+\theta_\rho))]}{\psi_0\sin^2\theta_0}\right\}}{2\beta(z/y)^2 h^2(y,\theta)(1+\cos\theta_0\cdot\sin 2\theta)}}\, d\theta\, dy, \quad z\geq 0,\; |\dot{z}|<\infty, \quad (6.108)
\end{aligned}
$$

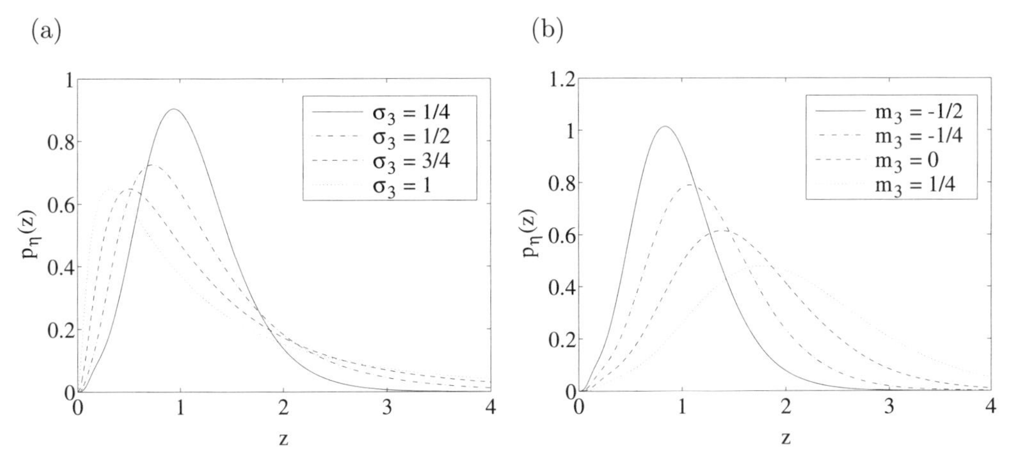

Figure 6.21: Probability density function $p_\eta(z)$ of extended Suzuki processes (Type II) for various values of the parameters (a) σ_3 ($m_3 = 1$, $\psi_0 = 0.0412$, $\rho = 0.918$, $\theta_\rho = 86°$, $\theta_0 = 97°$) and (b) m_3 ($\sigma_3 = 0.5$, $\psi_0 = 0.0412$, $\rho = 0.918$, $\theta_\rho = 86°$, $\theta_0 = 97°$).

where

$$h(y, \theta) = \sqrt{1 + \frac{\gamma(\sigma_3 y)^2}{\beta(1 + \cos\theta_0 \cdot \sin 2\theta)}} . \tag{6.109}$$

Here, ψ_0, $\dot{\phi}_0$, β, and γ again are the quantities introduced by (6.88a), (6.88e), (6.93), and (6.51), respectively. If we furthermore substitute (6.108) into (6.57), we obtain the result for the level-crossing rate $N_\eta(r)$ of extended Suzuki processes of Type II as follows

$$\begin{aligned} N_\eta(r) \;=\; & \frac{\sqrt{\beta}}{(2\pi)^2 \sigma_3 \psi_0 |\sin\theta_0|} \cdot \int_0^\infty e^{-\frac{[\ln(r/y) - m_3]^2}{2\sigma_3^2}} \cdot e^{-\frac{y^2+\rho^2}{2\psi_0 \sin^2\theta_0}} \\ & \cdot \int_{-\pi}^{\pi} h(y,\theta) \sqrt{1 + \cos\theta_0 \cdot \sin 2\theta} \\ & \cdot e^{\frac{y\rho\cos(\theta - \theta_\rho)}{\psi_0 \sin^2\theta_0}} \cdot e^{\frac{\cos\theta_0}{2\psi_0 \sin^2\theta_0}[y^2 \sin 2\theta + \rho^2 \sin 2\theta_\rho - 2y\rho\sin(\theta + \theta_\rho)]} \\ & \cdot \left\{ e^{-\left[\frac{g(y,\theta)}{h(y,\theta)}\right]^2} + \sqrt{\pi}\frac{g(y,\theta)}{h(y,\theta)} \left[1 + \operatorname{erf}\left(\frac{g(y,\theta)}{h(y,\theta)} \right) \right] \right\} d\theta \, dy , \quad r \geq 0 , \end{aligned} \tag{6.110}$$

where the functions $g(y, \theta)$ and $h(y, \theta)$ are given by (6.103) and (6.109), respectively.

In the case $\sigma_3 \to 0$ and $m_3 \to 0$, it follows $p_\lambda(r/y) \to |y|\delta(r - y)$ and $h(y, \theta) \to 1$, so that $N_\eta(r)$, according to (6.110), converges towards the expression (6.102), which describes the level-crossing rate of extended Rice processes. This result was to be expected. Furthermore, it should be taken into account that although (6.110) can be brought into the form (6.61) for the special case $\theta_0 = \pm\pi/2$, however, the definitions (6.88a)–(6.88f) still hold and not (6.12a)–(6.12f), so that, generally speaking, the level-crossing rate of extended Suzuki processes of Type II cannot be mapped exactly onto that of Type I. The maximum Doppler frequency f_{max} is again proportional to the level-crossing rate $N_\eta(r)$, as can easily be shown by substituting (6.88a)–(6.88f) into (6.110).

For the computation of the average duration of fades $T_{\eta_-}(r)$, we again make use of the definition (6.65). For the necessary cumulative distribution function $F_{\eta_-}(r) = P(\eta(t) \leq r)$, we obtain the following double integral by means of (6.107c)

$$\begin{aligned} F_{\eta_-}(r) &= \int_0^r p_\eta(z)\,dz \\ &= \frac{1}{2\pi\psi_0|\sin\theta_0|}\int_0^\infty \frac{y}{2}\left\{1+\operatorname{erf}\left[\frac{\ln(r/y)-m_3}{\sigma_3}\right]\right\}\cdot e^{-\frac{y^2+\rho^2}{2\psi_0\sin^2\theta_0}} \\ &\cdot \int_{-\pi}^{\pi} e^{\frac{y\rho\cos(\theta-\theta_\rho)}{\psi_0\sin^2\theta_0}}\cdot e^{\frac{\cos\theta_0}{2\psi_0\sin^2\theta_0}[y^2\sin 2\theta+\rho^2\sin 2\theta_\rho-2y\rho\sin(\theta+\theta_\rho)]}\,d\theta\,dy\,. \quad (6.111) \end{aligned}$$

According to (6.65), the quotient of (6.111) and (6.110) results in the average duration of fades $T_{\eta_-}(r)$ of extended Suzuki processes of Type II.

A few examples, which should help illustrate the results found for $N_\eta(r)$ and $T_{\eta_-}(r)$ are depicted in Figures 6.22(a) to 6.22(d). Figures 6.22(a) and 6.22(b) show the normalized level-crossing rate $N_\eta(r)/f_{max}$, calculated according to (6.110) for various values of the parameter m_3 and $\kappa_c = f_{max}/f_c$, respectively. In the logarithmic representation of Figure 6.22(a), one can see that a change of the parameter m_3 essentially causes a horizontal shift of the normalized level-crossing rate. Figure 6.22(b) makes it clear that the influence of the parameter κ_c is absolutely negligible, if κ_c takes on realistic values, i.e., $\kappa_c > 10$. The normalized average duration of fades $T_{\eta_-}(r)\cdot f_{max}$, which was calculated according to (6.65), is depicted in Figures 6.22(c) and 6.22(d) for different values of m_3 and κ_c, respectively.

6.2.3 The Deterministic Extended Suzuki Process of Type II

Referring to the stochastic model of the extended Suzuki process of Type II described in the subsection before, we will now derive the corresponding deterministic model. Therefore, we again make use of the principle of deterministic channel modelling (see Section 4.1), and approximate the coloured zero-mean Gaussian random process $\nu_0(t)$ by a finite sum of weighted harmonic functions

$$\tilde{\nu}_0(t) = \sum_{n=1}^{N_1} c_{1,n}\cos(2\pi f_{1,n}t + \theta_{1,n})\,. \quad (6.112)$$

With the Hilbert transform of the deterministic process above

$$\check{\tilde{\nu}}_0(t) = \sum_{n=1}^{N_1} c_{1,n}\sin(2\pi f_{1,n}t + \theta_{1,n})\,, \quad (6.113)$$

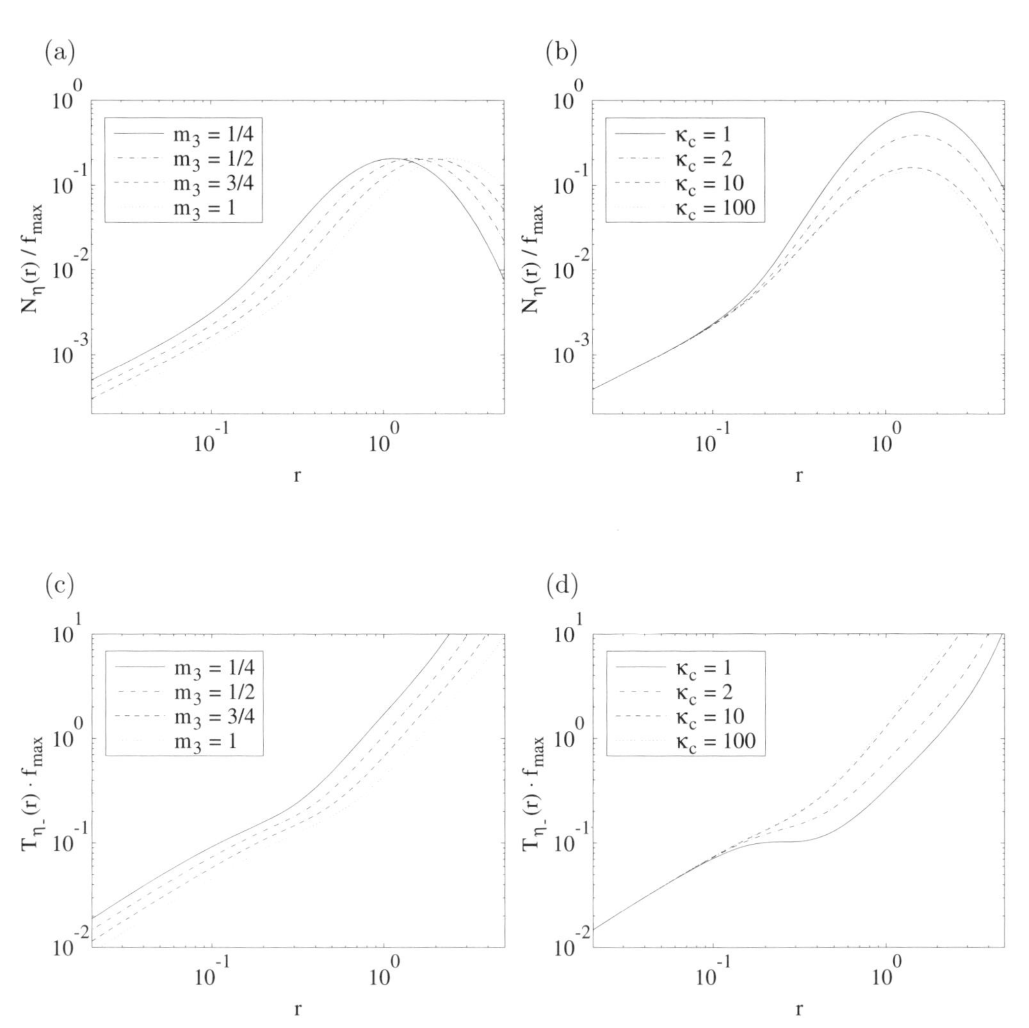

Figure 6.22: Normalized level-crossing rate $N_\eta(r)/f_{max}$ of extended Suzuki processes (Type II) for various values of the parameters: (a) m_3 ($\kappa_c = 5$) and (b) κ_c ($m_3 = 0.5$), as well as (c) and (d) the corresponding normalized average duration of fades $T_{\eta_-}(r) \cdot f_{max}$ ($\psi_0 = 0.0412$, $\kappa_0 = 0.4553$, $\rho = 0.918$, $\theta_\rho = 86°$, $\theta_0 = 97°$, $\sigma_3 = 0.5$).

we can transform the two relations (6.82) and (6.83) to the deterministic model. Accordingly, we obtain

$$\tilde{\mu}_1(t) = \tilde{\nu}_0(t) \tag{6.114}$$

and

$$\tilde{\mu}_2(t) = \cos\theta_0 \cdot \tilde{\nu}_0(t) + \sin\theta_0 \cdot \check{\tilde{\nu}}_0(t) \, . \tag{6.115}$$

If we now substitute the deterministic process $\tilde{\nu}_0(t)$ and its Hilbert transform $\check{\tilde{\nu}}_0(t)$ by the respective right-hand side of (6.112) and (6.113), then the generating deterministic components can be written as follows:

$$\tilde{\mu}_1(t) = \sum_{n=1}^{N_1} c_{1,n} \cos(2\pi f_{1,n} t + \theta_{1,n}) \, , \tag{6.116}$$

$$\tilde{\mu}_2(t) = \sum_{n=1}^{N_1} c_{1,n} \cos(2\pi f_{1,n} t + \theta_{1,n} - \theta_0) \, . \tag{6.117}$$

At this point, the role of θ_0 becomes clear: The parameter θ_0 describes the phase shift between the elementary harmonic functions $\tilde{\mu}_{1,n}(t)$ and $\tilde{\mu}_{2,n}(t)$ [see (4.27)]. Therefore, the Doppler phases $\theta_{2,n}$ of the second deterministic process $\tilde{\mu}_2(t)$ depend on the Doppler phases $\theta_{1,n}$ of the first deterministic process $\tilde{\mu}_1(t)$, because $\theta_{2,n} = \theta_{1,n} - \theta_0$ holds.

One may also take into account that for the Doppler coefficients $c_{i,n}$ and the Doppler frequencies $f_{i,n}$ the relations $c_{1,n} = c_{2,n}$ and $f_{1,n} = f_{2,n}$ hold. In particular, for the special case $\theta_0 = \pm 90°$, the complex-valued deterministic process $\tilde{\mu}(t) = \tilde{\mu}_1(t) + j\tilde{\mu}_2(t)$ can be represented as

$$\tilde{\mu}(t) = \sum_{n=1}^{N_1} c_{1,n} e^{\pm j(2\pi f_{1,n} t + \theta_{1,n})} \, . \tag{6.118}$$

The deterministic lognormal process $\tilde{\lambda}(t)$, which models the slow fading, is realized exactly as shown in the bottom part of Figure 6.9. Accordingly, a further deterministic process $\tilde{\nu}_3(t)$ is necessary, which has to be designed in such a way that it does not correlate with the process $\tilde{\nu}_0(t)$. Since these two processes are (approximately) Gaussian distributed, the statistical independence of $\tilde{\nu}_0(t)$ and $\tilde{\nu}_3(t)$ follows from the uncorrelatedness. As a result, the deterministic processes $\tilde{\xi}(t)$ and $\tilde{\lambda}(t)$ derived from these are also statistically independent.

By using (6.116) and (6.117), the stochastic reference model for the extended Suzuki process of Type II (see Figure 6.20) can now easily be transformed into the deterministic simulation model shown in Figure 6.23.

The statistical properties of deterministic extended Suzuki processes $\tilde{\eta}(t)$ of Type II can approximately be described by the relations $p_\eta(z)$, $N_\eta(r)$, and $T_{\eta_-}(r)$ derived for

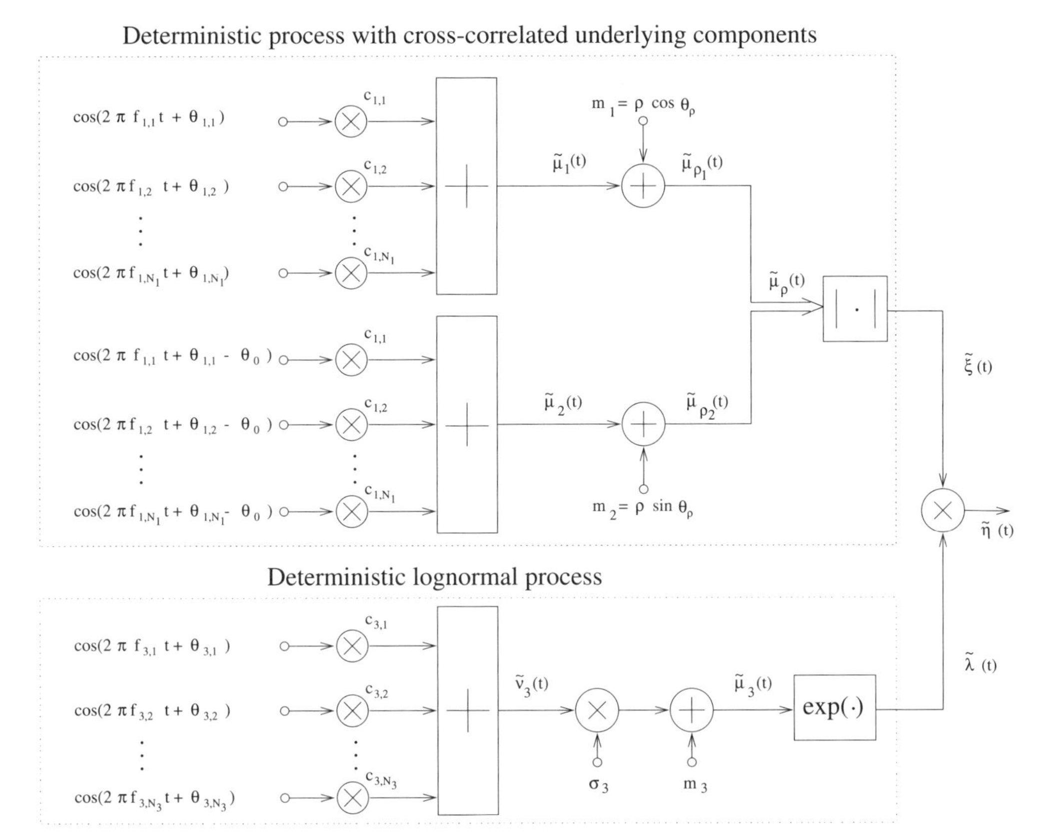

Figure 6.23: Deterministic simulation model for extended Suzuki processes (Type II).

the reference model before, if the characteristic quantities (6.88a)–(6.88f) are there replaced by those corresponding to the simulation model. In the following, we will derive the characteristic quantities of the simulation model. We therefore need the autocorrelation functions of the processes $\tilde{\mu}_1(t)$ and $\tilde{\mu}_2(t)$

$$\tilde{r}_{\mu_1\mu_1}(\tau) = \tilde{r}_{\mu_2\mu_2}(\tau) = \sum_{n=1}^{N_1} \frac{c_{1,n}^2}{2} \cos(2\pi f_{1,n}\tau) \tag{6.119}$$

as well as the cross-correlation function calculated according to (4.13)

$$\tilde{r}_{\mu_1\mu_2}(\tau) = \tilde{r}_{\mu_2\mu_1}(-\tau) = \sum_{n=1}^{N_1} \frac{c_{1,n}^2}{2} \cos(2\pi f_{1,n}\tau - \theta_0)\,. \tag{6.120}$$

With these two functions, the characteristic quantities of the simulation model $\tilde{\psi}_0^{(n)} = \tilde{r}_{\mu_1\mu_1}^{(n)}(0)$ and $\tilde{\phi}_0^{(n)} = \tilde{r}_{\mu_1\mu_2}^{(n)}(0)$ can easily be determined for $n = 0, 1, 2$. We immediately obtain the following closed-form expressions:

$$\tilde{\psi}_0^{(0)} = \tilde{\psi}_0 = \sum_{n=1}^{N_1} \frac{c_{1,n}^2}{2}, \tag{6.121a}$$

$$\tilde{\psi}_0^{(1)} = \dot{\tilde{\psi}}_0 = 0, \tag{6.121b}$$

$$\tilde{\psi}_0^{(2)} = \ddot{\tilde{\psi}}_0 = -2\pi^2 \sum_{n=1}^{N_1} (c_{1,n} f_{1,n})^2, \tag{6.121c}$$

$$\tilde{\phi}_0^{(0)} = \tilde{\phi}_0 = \tilde{\psi}_0 \cdot \cos\theta_0, \tag{6.121d}$$

$$\tilde{\phi}_0^{(1)} = \dot{\tilde{\phi}}_0 = \pi \sum_{n=1}^{N_1} (c_{1,n}^2 f_{1,n}) \cdot \sin\theta_0, \tag{6.121e}$$

$$\tilde{\phi}_0^{(2)} = \ddot{\tilde{\phi}}_0 = \ddot{\tilde{\psi}}_0 \cdot \cos\theta_0. \tag{6.121f}$$

Since this model uses the restricted Jakes power spectral density ($\kappa_0 \leq 1$), we appropriately return to the modified method of exact Doppler spread described in Subsection 6.4.1 in order to calculate the discrete Doppler frequencies $f_{1,n}$ and the Doppler coefficients $c_{1,n}$. After adjusting the equations (6.72)–(6.74) to the present model, we obtain

$$f_{1,n} = f_{max} \sin\left[\frac{\pi}{2N_1'}\left(n - \frac{1}{2}\right)\right] \quad \text{and} \quad c_{1,n} = \sigma_0 \sqrt{\frac{2}{N_1'}} \tag{6.122a, b}$$

for $n = 1, 2, \ldots, N_1$, where N_1 denotes the actual (user defined) number of harmonic functions and

$$N_1' = \left\lceil \frac{N_1}{\frac{2}{\pi}\arcsin(\kappa_0)} \right\rceil \tag{6.123}$$

is the virtual number of harmonic functions.

For the Doppler phases $\theta_{1,n}$, it is assumed that they are realizations of a random variable uniformly distributed within the interval $(0, 2\pi]$.

The calculation of the discrete Doppler frequencies $f_{3,n}$ of the deterministic Gaussian process $\tilde{\nu}_3(t)$ is performed exactly according to (6.75a) and (6.75b). Accordingly, for $c_{3,n}$ again the formula $c_{3,n} = \sqrt{2/N_3}$ for all $n = 1, 2, \ldots, N_3$ is used. The remaining parameters of the simulation model ($\rho, \theta_\rho, m_3, \sigma_3$) are identical to those of the reference model.

With (6.122a) and (6.122b), the characteristic quantities of the simulation model (6.121a)–(6.121f) can now be evaluated. A comparison with the corresponding quantities of the reference model (6.88a)–(6.88f) then gives us the desired information on the precision of the simulation model. As an example, the convergence behaviour of the normalized quantities $\ddot{\tilde{\psi}}_0/f_{max}^2$ and $\dot{\tilde{\phi}}_0/f_{max}$ is depicted in Figures 6.24(a) and 6.24(b), respectively. Just as in Figures 6.10(a) and 6.10(b), one can here as well see that the deviations of the depicted quantities of the simulation model are negligible compared to the reference model for all cases relevant in practice (i.e., $N_1 \geq 7$).

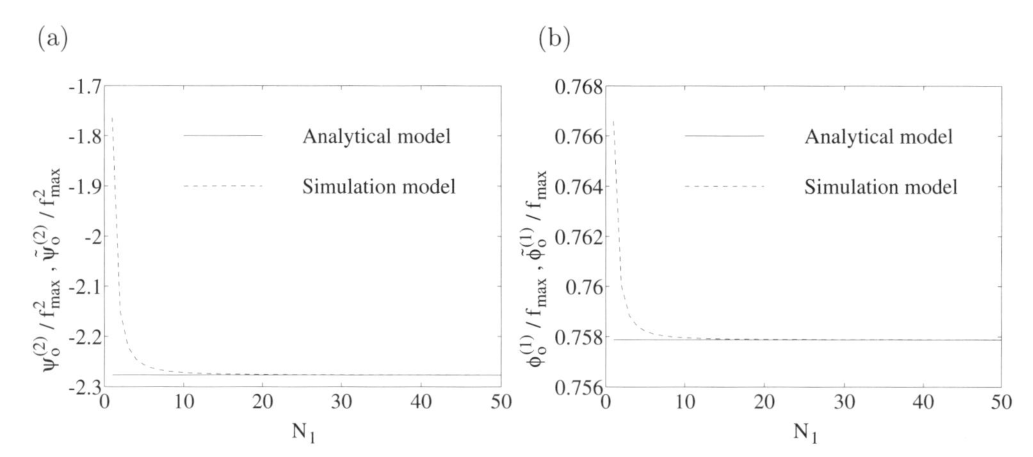

Figure 6.24: Illustration of (a) $\ddot{\psi}_0/f_{\max}^2$ and $\ddot{\tilde{\psi}}_0/f_{\max}^2$ as well as (b) $\dot{\phi}_0/f_{max}$ and $\dot{\tilde{\phi}}_0/f_{max}$ in terms of the N_1 (MEDS, $\sigma_0^2 = 2$, $\kappa_0 = 1/2$, $\theta_0 = 45°$).

For the level-crossing rate $\tilde{N}_\eta(r)$ and the average duration of fades $\tilde{T}_{\eta_-}(r)$ of the simulation model the statements made in Subsection 6.1.4 are also valid in the present case.

6.2.4 Applications and Simulation Results

This subsection intends to show us how the statistical properties of stochastic and deterministic extended Suzuki processes of Type II can be brought into accordance with those of measured channels. This is again performed by optimizing the primary model parameters. The basis for this is provided by the measurement results found in the literature [But83] ($F_{\eta+}^\star(r), N_\eta^\star(r), T_{\eta_-}^\star(r)$), which we have already introduced in Subsection 6.1.5. Only in this way, is a fair performance-comparison between extended Suzuki processes of Type I and those of Type II possible.

In the present case, the parameter vector $\mathbf{\Omega}$ is defined by

$$\mathbf{\Omega} := (\sigma_0, \kappa_0, \theta_0, \rho, \theta_\rho, \sigma_3, m_3, \kappa_c)\,. \tag{6.124}$$

This time, the vector $\mathbf{\Omega}$ contains all primary model parameters of the extended Suzuki process (Type II), also including κ_c, although exactly this parameter has no influence worth mentioning on the first and second order statistics of the process $\eta(t)$, if κ_c exceeds the value 10. It will be left to the optimization procedure to find a suitable value for this quantity.

Since the error function $E_2(\mathbf{\Omega})$ [see (6.76)] has turned out to be useful in our previous applications, we will also make use of it in the present minimization problem, where we again apply the Fletcher-Powell method [Fle63]. Of course, concerning the evaluation of (6.76), it has to be taken into account that now the complementary cumulative distribution function $F_{\eta+}(r/\rho) = 1 - F_{\eta-}(r/\rho)$ has to be calculated by means of (6.111) and that the level-crossing rate $N_\eta(r/\rho)$ is defined by (6.110). Table 6.3 shows the results obtained for the components of the parameter vector $\mathbf{\Omega}$ after the numerical minimization of the error function $E_2(\mathbf{\Omega})$.

Table 6.3: The optimized primary parameters of the reference channel model for areas with light and heavy shadowing.

Shadowing	σ_0	κ_0	θ_0	ρ	θ_ρ	σ_3	m_3	κ_c
heavy	0.2774	0.506	30°	0.269	45°	0.0905	0.0439	119.9
light	0.7697	0.4045	164°	1.567	127°	0.0062	-0.3861	1.735

With the results shown in Table 6.3 for the parameters σ_0, κ_0, and ρ, the Rice factor c_R [see (3.18)] of the extended Suzuki model (Type II), given by

$$c_R = \frac{\rho^2}{2\psi_0} = \frac{\pi}{4} \cdot \frac{\rho^2}{\sigma_0^2 \arcsin(\kappa_0)}, \tag{6.125}$$

takes on the values $c_R = 1.43\,\text{dB}$ (heavy shadowing) and $c_R = 8.93\,\text{dB}$ (light shadowing).

Figure 6.25(a) shows the complementary cumulative distribution function $F_{\eta+}(r/\rho)$ of the reference model in comparison with that of the real-world channel $F^\star_{\eta+}(r/\rho)$. At heavy shadowing, we obtain minor deviations at low (normalized to ρ) levels r/ρ. The deviations almost disappear as soon as r/ρ takes on medium or even large values. At light shadowing, on the other hand, the deviations are largest at medium levels, whereas they can be ignored at low levels.

Figure 6.25(b) shows the normalized level-crossing rate $N_\eta(r/\rho)/f_{max}$ of the reference model and that of the measured channel $N^\star_\eta(r/\rho)/f_{max}$. One can see that the two level-crossing rates match each other astonishingly well over the whole depicted amplitude range.

A comparison between the corresponding normalized average duration of fades is shown in Figure 6.25(c). The results presented there are quite good already, but it seems a likely supposition that there is still room for further improvement, which we can indeed achieve by a further model extension, as we will see in the next section.

At this point, a comparison of the performance between the two extended Suzuki processes (Type I and Type II) suggests itself. With regard to the complementary cumulative distribution function, both model types provide the same good results to a certain extent (compare Figure 6.25(a) with Figure 6.12(a)). However, the flexibility of the level-crossing rate of the extended Suzuki process of Type II seems to be higher

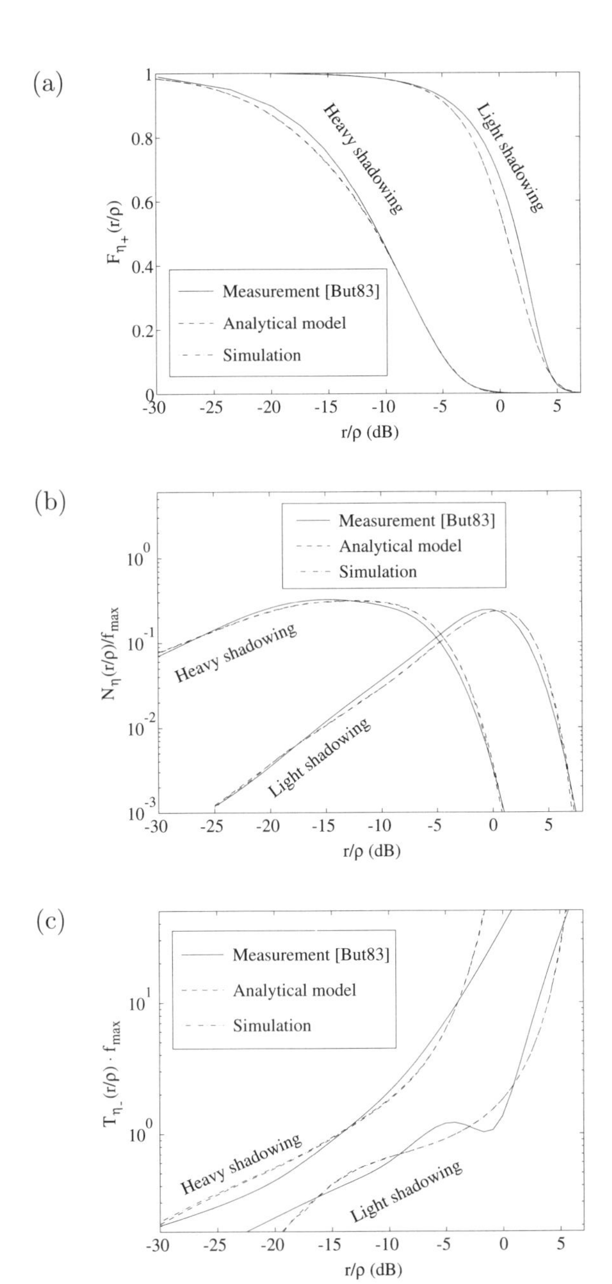

Figure 6.25: (a) Complementary cumulative distribution function $F_{\eta_+}(r/\rho)$, (b) normalized level-crossing rate $N_\eta(r/\rho)/f_{max}$, and (c) normalized average duration of fades $T_{\eta_-}(r/\rho) \cdot f_{max}$ for areas with heavy and light shadowing.

than that of Type I, which would explain the definitely better results of Figure 6.25(b) compared to those of Figure 6.12(b). To be fair though, we have to add that the higher flexibility comes along with a greater complexity of the reference model. Since the achievable improvements can only be reached with a higher numerical computation expenditure, the user himself has to decide from case to case, i.e., in our terminology "from channel to channel", whether the achievable improvements justify a higher analytical and numerical computation expenditure or not.

However, if the parameters of the reference model have been determined, then the determination of the parameters of the corresponding deterministic simulation model can be regarded as trivial due to the closed-form formulas derived here.

If we once again study the simulation models depicted in Figures 6.9 and 6.23, it becomes clear that the structure corresponding to the model Type II is, generally speaking, the more efficient one, and that the structure of Type I can only keep up with it, if N_2 is equal to zero, which is equivalent to the assumption that $\kappa_0 = 0$ holds.

Finally, the verification of the analytical results by means of simulation remains. Therefore, we design the deterministic processes $\tilde{\nu}_0(t)$ and $\tilde{\nu}_3(t)$ by applying the techniques described in the preceding Subsection 6.2.3 (modified MEDS with $N_1 = 25$ and $N_3 = 15$). The measurement of the functions $\tilde{F}_{\eta+}(r/\rho)$, $\tilde{N}_\eta(r/\rho)/f_{max}$, and $\tilde{T}_{\eta-}(r/\rho) \cdot f_{max}$ $N_3 = 15$) from a discrete-time simulation of the deterministic extended Suzuki process (Type II) $\tilde{\eta}(t)$ leads to the curves also depicted in Figures 6.25(a)–6.25(c). Again, there is a nearly complete correspondence between the reference model and the simulation model, so that the graphs corresponding to these models can hardly be distinguished from each other.

A small part of the sequence of the simulated deterministic process $\tilde{\eta}(t)$ is depicted in Figure 6.26(a) for an area with heavy shadowing and in Figure 6.26(b) for an area with light shadowing.

6.3 THE GENERALIZED RICE PROCESS

The extended Suzuki processes of Type I and Type II represent two classes of stochastic processes with different statistical properties. Both models are identical, however, if $\kappa_0 = 0$ holds in the former and if in the latter, the parameters κ_0 and θ_0 are given by $\kappa_0 = 1$ and $\theta_0 = \pi/2$, respectively. But in general, we can say that neither the extended Suzuki process of Type I is completely covered by that of Type II nor that the reverse is true. In [Pae96b] it has been pointed out, and shown later in [Pae97c], that both models can be combined in a single model. This so-called *generalized Suzuki model* contains the extended Suzuki processes of Type I and of Type II as special cases. The mathematical expenditure required to describe the generalized model is considerable, however, not much higher than that of Type II. Without the lognormal process, the *generalized Rice process* follows from the generalized Suzuki process. The generalized Rice process is considerably easier to describe and is in many cases sufficient for modelling frequency-nonselective mobile radio channels.

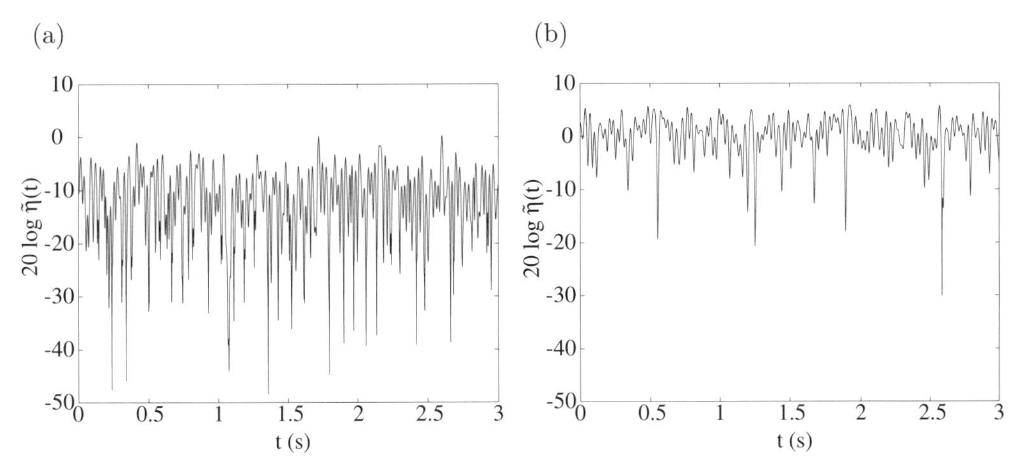

Figure 6.26: Simulation of deterministic extended Suzuki processes $\tilde{\eta}(t)$ of Type II for areas with (a) heavy shadowing and (b) light shadowing (MEDS, $N_1 = 25$, $N_3 = 15$, $f_{max} = 91\,\text{Hz}$).

This section deals with the description and the analysis of stochastic generalized Rice processes. Here, just as in previous sections, we will generally be concerned with the probability density function of the amplitude, the level-crossing rate, and the average duration of fades. Since the derivation of these quantities is again performed analogously to the procedure described in Subsection 6.1.1, we will be considerably briefer here. However, the comprehensibility of the derived results will still be maintained for the reader. Starting from the stochastic generalized Rice model, it then follows the derivation of the corresponding deterministic simulation model. Finally, the section closes with the fitting of the stochastic reference model and the deterministic simulation model to a real-world channel.

6.3.1 The Stochastic Generalized Rice Process

Let us study the reference model for a generalized Rice process $\xi(t)$ as depicted in Figure 6.27. The directly visible parameters of this model are θ_0, ρ, and θ_ρ, which are already known to us. We demand from the coloured real-valued Gaussian random processes $\nu_1(t)$ and $\nu_2(t)$ that they are zero-mean and statistically independent. For the Doppler power spectral density $S_{\nu_i\nu_i}(f)$ of the Gaussian random processes $\nu_i(t)$ $(i = 1, 2)$ it holds

$$S_{\nu_i\nu_i}(f) = \begin{cases} \dfrac{\sigma_i^2}{2\pi f_{max}\sqrt{1-(f/f_{max})^2}}, & |f| \leq \kappa_i f_{max}, \\ 0, & |f| > \kappa_i f_{max}, \end{cases} \tag{6.126}$$

where f_{max} again denotes the maximum Doppler frequency, and κ_i is a positive constant determining the Doppler bandwidth. Note that κ_i, together with the quantity σ_i^2, determines the variance of $\nu_i(t)$. In order to ensure that the chosen notation remains homogeneous, we make the following agreements: $\kappa_1 = 1$ and $\kappa_2 = \kappa_0$ with $\kappa_0 \in [0, 1]$, so that $S_{\nu_1\nu_1}(f)$ corresponds to the classical Jakes power spectral density (6.9a), and $S_{\nu_2\nu_2}(f)$ is identical to the restricted Jakes power spectral density (6.9c).

The reference model shown in Figure 6.27 includes two special cases:

$$\text{(i)} \quad \sigma_1^2 = \sigma_2^2 = \sigma_0^2 \quad \text{and} \quad \theta_0 = \pi/2\,, \tag{6.127a}$$

$$\text{(ii)} \quad \sigma_1^2 = 0 \quad \text{and} \quad \sigma_2^2 = 2\sigma_0^2\,. \tag{6.127b}$$

In case (i), exactly the Rice process depicted in Figure 6.1, whose underlying complex-valued Gaussian process is described by the left-hand side restricted Jakes power spectral density (6.2), follows from the generalized Rice process. If we leave the missing minus sign in the lower branch of the structure shown in Figure 6.14 aside,[3] then in case (ii) the extended Rice process (Figure 6.14) follows from the generalized Rice process (Figure 6.27).

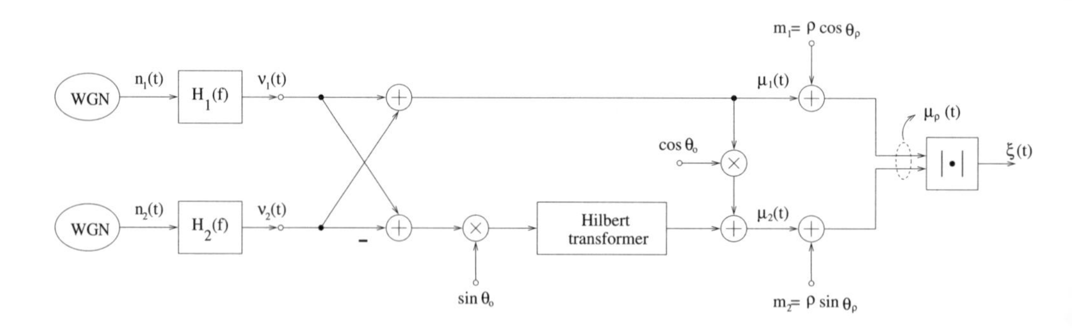

Figure 6.27: Reference model for generalized Rice processes $\xi(t)$.

Next, we are interested in the autocorrelation function $r_{\mu\mu}(\tau)$ and the Doppler power spectral density $S_{\mu\mu}(f)$ of the complex-valued process $\mu(t) = \mu_1(t) + j\mu_2(t)$. From Figure 6.27, we first read the relations

$$\mu_1(t) = \nu_1(t) + \nu_2(t) \tag{6.128}$$

and

$$\mu_2(t) = [\nu_1(t) + \nu_2(t)]\cos\theta_0 + [\check{\nu}_1(t) - \check{\nu}_2(t)]\sin\theta_0\,. \tag{6.129}$$

From these equations, we obtain the following relations for the autocorrelation functions $r_{\mu_1\mu_1}(\tau)$ and $r_{\mu_2\mu_2}(\tau)$, as well as for the cross-correlation functions $r_{\mu_1\mu_2}(\tau)$ and $r_{\mu_2\mu_1}(\tau)$

$$r_{\mu_1\mu_1}(\tau) = r_{\mu_2\mu_2}(\tau) = r_{\nu_1\nu_1}(\tau) + r_{\nu_2\nu_2}(\tau)\,, \tag{6.130a}$$

[3] It should be noted that the minus sign has no influence on the statistics of $\xi(t)$. Also, it can easily be obtained by substituting θ_0 with $-\theta_0$.

$$r_{\mu_1\mu_2}(\tau) = [r_{\nu_1\nu_1}(\tau) + r_{\nu_2\nu_2}(\tau)]\cos\theta_0 + [r_{\nu_1\breve{\nu}_1}(\tau) - r_{\nu_2\breve{\nu}_2}(\tau)]\sin\theta_0\,, \tag{6.130b}$$

$$r_{\mu_2\mu_1}(\tau) = [r_{\nu_1\nu_1}(\tau) + r_{\nu_2\nu_2}(\tau)]\cos\theta_0 - [r_{\nu_1\breve{\nu}_1}(\tau) - r_{\nu_2\breve{\nu}_2}(\tau)]\sin\theta_0\,, \tag{6.130c}$$

where $r_{\nu_i\nu_i}(\tau)$ $(i = 1, 2)$ denotes the inverse Fourier transform of (6.126), i.e,

$$r_{\nu_i\nu_i}(\tau) = \sigma_i^2 \frac{2}{\pi} \int_0^{\arcsin(\kappa_i)} \cos(2\pi f_{max}\tau \sin\varphi)\, d\varphi\,, \tag{6.131}$$

and $r_{\nu_i\breve{\nu}_i}(\tau)$, due to (2.56a), denotes the Hilbert transform of $r_{\nu_i\nu_i}(\tau)$, so that

$$r_{\nu_i\breve{\nu}_i}(\tau) = \sigma_i^2 \frac{2}{\pi} \int_0^{\arcsin(\kappa_i)} \sin(2\pi f_{max}\tau \sin\varphi)\, d\varphi \tag{6.132}$$

holds. By using the relation (6.5), the desired autocorrelation function $r_{\mu\mu}(\tau)$ can now be written as

$$r_{\mu\mu}(\tau) = 2[r_{\nu_1\nu_1}(\tau) + r_{\nu_2\nu_2}(\tau)] + j2[r_{\nu_1\breve{\nu}_1}(\tau) - r_{\nu_2\breve{\nu}_2}(\tau)]\sin\theta_0\,. \tag{6.133}$$

After performing the Fourier transform of (6.133) and taking the relation $S_{\nu_i\breve{\nu}_i}(f) = -j\,\mathrm{sgn}\,(f) S_{\nu_i\nu_i}(f)$ into account, we can then express the Doppler power spectral density $S_{\mu\mu}(f)$ in terms of $S_{\nu_i\nu_i}(f)$ [cf. (6.126)] as follows

$$\begin{aligned} S_{\mu\mu}(f) &= 2[1 + \mathrm{sgn}\,(f)\sin\theta_0] \cdot S_{\nu_1\nu_1}(f) \\ &\quad + 2[1 - \mathrm{sgn}\,(f)\sin\theta_0] \cdot S_{\nu_2\nu_2}(f)\,. \end{aligned} \tag{6.134}$$

An example of this in general unsymmetrical Doppler power spectral density is depicted in Figure 6.28.

It is obvious that for the two special cases (6.127a) and (6.127b) Figure 6.28 converts to Figures 6.2(c) and 6.15(b), respectively. Also, the Doppler power spectral density (6.134) contains the classical Jakes power spectral density according to (3.7) as further special case, because we obtain the latter with the parameter constellation $\sigma_1^2 = \sigma_2^2 = \sigma_0^2$, $\kappa_1 = \kappa_2 = 1$, and $\theta_0 = \pi/2$.

Next follows the derivation of the characteristic quantities $\psi_0^{(n)}$ and $\phi_0^{(n)}$ $(n = 0, 1, 2)$. Therefore, we substitute (6.130a) into (6.11a) and (6.130b) into (6.11b), which leads to the following expressions:

$$\psi_0^{(0)} = \psi_0 = \frac{\sigma_2^2}{2}\left[\left(\frac{\sigma_1}{\sigma_2}\right)^2 + \frac{2}{\pi}\arcsin(\kappa_0)\right], \tag{6.135a}$$

$$\psi_0^{(1)} = \dot{\psi}_0 = 0\,, \tag{6.135b}$$

$$\psi_0^{(2)} = \ddot{\psi}_0 = -(\pi\sigma_2 f_{max})^2\left\{\left(\frac{\sigma_1}{\sigma_2}\right)^2 + \frac{2}{\pi}\left[\arcsin(\kappa_0) - \frac{1}{2}\sin(2\arcsin(\kappa_0))\right]\right\}, \tag{6.135c}$$

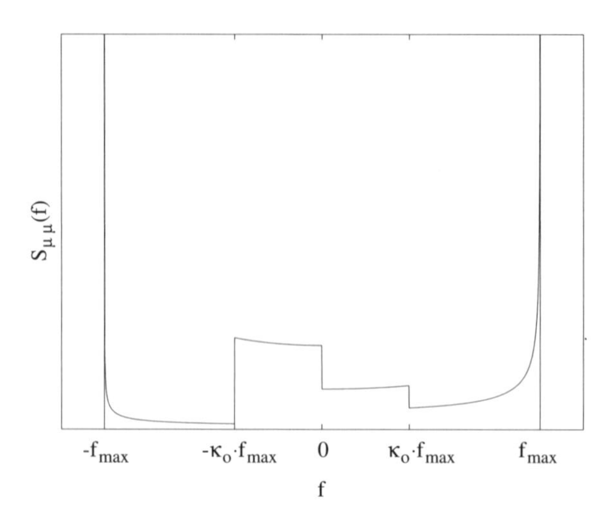

Figure 6.28: Unsymmetrical Doppler power spectral density $S_{\mu\mu}(f)$ ($\sigma_1^2 = 0.25$, $\sigma_2^2 = 1$, $\theta_0 = 15°$, $\kappa_0 = 0.4$).

$$\phi_0^{(0)} = \phi_0 = \psi_0 \cdot \cos\theta_0 \, , \tag{6.135d}$$

$$\phi_0^{(1)} = \dot{\phi}_0 = 2\sigma_2^2 f_{max} \left[\left(\frac{\sigma_1}{\sigma_2} \right)^2 - \left(1 - \sqrt{1-\kappa_0^2} \right) \right] \cdot \sin\theta_0 \, , \tag{6.135e}$$

$$\phi_0^{(2)} = \ddot{\phi}_0 = \ddot{\psi}_0 \cdot \cos\theta_0 \, , \tag{6.135f}$$

where $0 \leq \kappa_0 \leq 1$ and $-\pi \leq \theta_0 < \pi$. One may take into account that in the special case (i) described by (6.127a), the quantities presented above exactly result in the equations (6.12a)–(6.12f). On the other hand, the special case (ii) [cf. (6.127b)] leads to the formulae[4] (6.88a)–(6.88f).

With the characteristic quantities (6.135a)–(6.135f) the covariance matrix $\boldsymbol{C}_{\mu_\rho}(\tau)$ of the vector process $\boldsymbol{\mu}_\rho(t) = (\mu_{\rho_1}(t),\ \mu_{\rho_2}(t),\ \dot{\mu}_{\rho_1}(t),\ \dot{\mu}_{\rho_2}(t))$ at the same time t, i.e., $\tau = 0$, is completely determined. It holds

$$\boldsymbol{C}_{\mu_\rho}(0) = \boldsymbol{R}_\mu(0) = \begin{pmatrix} \psi_0 & \psi_0\cos\theta_0 & 0 & \dot{\phi}_0 \\ \psi_0\cos\theta_0 & \psi_0 & -\dot{\phi}_0 & 0 \\ 0 & -\dot{\phi}_0 & -\ddot{\psi}_0 & -\ddot{\psi}_0\cos\theta_0 \\ \dot{\phi}_0 & 0 & -\ddot{\psi}_0\cos\theta_0 & -\ddot{\psi}_0 \end{pmatrix} . \tag{6.136}$$

For us it is important now to realize that the covariance matrix (6.136) has the same form as (6.91). As a consequence of this, we again obtain the joint probability density function $p_{\xi\dot{\xi}\vartheta\dot{\vartheta}}(z, \dot{z}, \theta, \dot{\theta})$ described by (6.92), where we have to substitute the quantities ψ_0, $\ddot{\psi}_0$, and $\dot{\phi}_0$ with the equations derived above, i.e., (6.135a), (6.135c), and

[4] Due to the minus sign in the lower part of the sinal flow diagram shown in Figure 6.27, it has to be taken into account that the equations (6.88e) and (6.135e) have different signs.

(6.135e), respectively. Consequently, all relations derivable from these, as for example those for $p_\xi(z)$, $p_\vartheta(\theta)$, $N_\xi(r)$, and $T_{\xi_-}(r)$, exactly lead to the results already found in Subsection 6.2.1. In the formulae given there, we merely have to replace ψ_0, $\ddot{\psi}_0$, and $\dot{\phi}_0$ with (6.135a), (6.135c), and (6.135e), respectively. Therefore, all further calculations for the description of generalized Rice processes can at this point be omitted.

This fact, however, should not mislead us into concluding that extended Rice processes and generalized Rice processes are two different ways of describing one and the same stochastic process. The flexibility of generalized Rice processes is definitely higher than that of extended Rice processes. The reason for this lies in the additional primary model parameter σ_1^2, which is zero per definition for the extended Rice process and which contributes to a further de-coupling of the secondary model parameters $(\psi_0, \dot{\psi}_0, \ddot{\psi}_0, \phi_0, \dot{\phi}_0, \ddot{\phi}_0)$ of the generalized Rice process. In order to make this clear with the help of an example, we consider (6.88e). There, in the interval $(0, 1]$ no real number exists for the parameter κ_0, so that $\dot{\phi}_0 = 0$ holds. On the other hand, the quantity $\dot{\phi}_0$ according to (6.135e) behaves differently. Let $\sigma_1^2 \in [\sigma_2^2,\ 2\sigma_2^2)$, then a real-valued number

$$\kappa_0 = \frac{\sigma_1}{\sigma_2}\sqrt{2 - \left(\frac{\sigma_1}{\sigma_2}\right)^2} \tag{6.137}$$

always exists in the interval $(0, 1]$, so that $\dot{\phi}_0 = 0$ holds.

The multiplication of the generalized Rice process with a lognormal process results in the so-called *generalized Suzuki process* suggested in [Pae97c]. The generalized Suzuki process contains the classical Suzuki process [Suz77], the modified Suzuki process [Kra90b], as well as the two extended Suzuki processes of Type I [Pae98d] and of Type II [Pae97a] as special cases. This product process is described by the probability density function (6.107c), where we have to use the equation (6.135a) for ψ_0. Similarly, for the level-crossing rate one finds the expression (6.110). Now, however, it has to be emphasized that the entries $(\psi_0, \dot{\psi}_0, \ddot{\psi}_0, \phi_0, \dot{\phi}_0, \ddot{\phi}_0)$ of the covariance matrix $\boldsymbol{C}_{\mu_\rho}$ are defined by (6.135a)–(6.135f).

A detailed discussion of generalized Rice respectively Suzuki processes is not necessary for our purposes. Instead, we will continue with the design of deterministic generalized Rice processes.

6.3.2 The Deterministic Generalized Rice Process

We again proceed by at first replacing the coloured zero-mean Gaussian random processes $\nu_1(t)$ and $\nu_2(t)$ by a finite sum of N_i weighted harmonic functions of the form

$$\tilde{\nu}_i(t) = \sum_{n=1}^{N_i} c_{i,n} \cos(2\pi f_{i,n} t + \theta_{i,n}), \quad i = 1, 2. \tag{6.138}$$

When designing the deterministic processes (6.138), it has to be taken into account that $\tilde{\nu}_1(t)$ and $\tilde{\nu}_2(t)$ have to be uncorrelated, i.e., $f_{1,n} \neq f_{2,m}$ must hold for all

$n = 1, 2, \ldots, N_1$ and $m = 1, 2, \ldots, N_2$. With the deterministic processes designed in this way and the corresponding Hilbert transforms, i.e.,

$$\check{\tilde{\nu}}_i(t) = \sum_{n=1}^{N_i} c_{i,n} \sin(2\pi f_{i,n} t + \theta_{i,n}), \quad i = 1, 2, \tag{6.139}$$

we can directly replace the stochastic processes $\mu_1(t)$ and $\mu_2(t)$ [cf. (6.128) and (6.129), respectively] with the corresponding deterministic processes $\tilde{\mu}_1(t)$ and $\tilde{\mu}_2(t)$. Thus, the latter processes can be expressed as follows:

$$\begin{aligned} \tilde{\mu}_1(t) &= \sum_{n=1}^{N_1} c_{1,n} \cos(2\pi f_{1,n} t + \theta_{1,n}) \\ &\quad + \sum_{n=1}^{N_2} c_{2,n} \cos(2\pi f_{2,n} t + \theta_{2,n}), \end{aligned} \tag{6.140}$$

$$\begin{aligned} \tilde{\mu}_2(t) &= \sum_{n=1}^{N_1} c_{1,n} \cos(2\pi f_{1,n} t + \theta_{1,n} - \theta_0) \\ &\quad + \sum_{n=1}^{N_2} c_{2,n} \cos(2\pi f_{2,n} t + \theta_{2,n} + \theta_0). \end{aligned} \tag{6.141}$$

As a result, the *deterministic generalized Rice process* is completely determined, and we obtain the simulation system in the continuous-time representation form depicted in Figure 6.29.

Now, let $\theta_0 = \pi/2$, then the structure of the deterministic Rice process with cross-correlated underlying components (cf. Figure 6.9) follows from Figure 6.29. Moreover, in the special case $\sigma_2^2 = 0$, i.e., $N_2 = 0$, we obtain the deterministic extended Rice process depicted in the top part of Figure 6.23.

In the following, we will derive the characteristic quantities of the simulation model, i.e., $\tilde{\psi}_0^{(n)} = \tilde{r}_{\mu_1\mu_1}^{(n)}(0) = \tilde{r}_{\mu_2\mu_2}^{(n)}(0)$ and $\tilde{\phi}_0^{(n)} = \tilde{r}_{\mu_1\mu_2}^{(n)}(0)$ for $n = 0, 1, 2$. The autocorrelation functions $\tilde{r}_{\mu_1\mu_1}(\tau)$ and $\tilde{r}_{\mu_2\mu_2}(\tau)$ necessary for this can be expressed as

$$\begin{aligned} \tilde{r}_{\mu_1\mu_1}(\tau) &= \tilde{r}_{\mu_2\mu_2}(\tau) \\ &= \sum_{n=1}^{N_1} \frac{c_{1,n}^2}{2} \cos(2\pi f_{1,n} \tau) + \sum_{n=1}^{N_2} \frac{c_{2,n}^2}{2} \cos(2\pi f_{2,n} \tau), \end{aligned} \tag{6.142}$$

and for the cross-correlation function $\tilde{r}_{\mu_1\mu_2}(\tau)$ calculated according to (4.13) it holds

$$\begin{aligned} \tilde{r}_{\mu_1\mu_2}(\tau) &= \tilde{r}_{\mu_2\mu_1}(-\tau) \\ &= \sum_{n=1}^{N_1} \frac{c_{1,n}^2}{2} \cos(2\pi f_{1,n} \tau - \theta_0) + \sum_{n=1}^{N_2} \frac{c_{2,n}^2}{2} \cos(2\pi f_{2,n} \tau + \theta_0). \end{aligned} \tag{6.143}$$

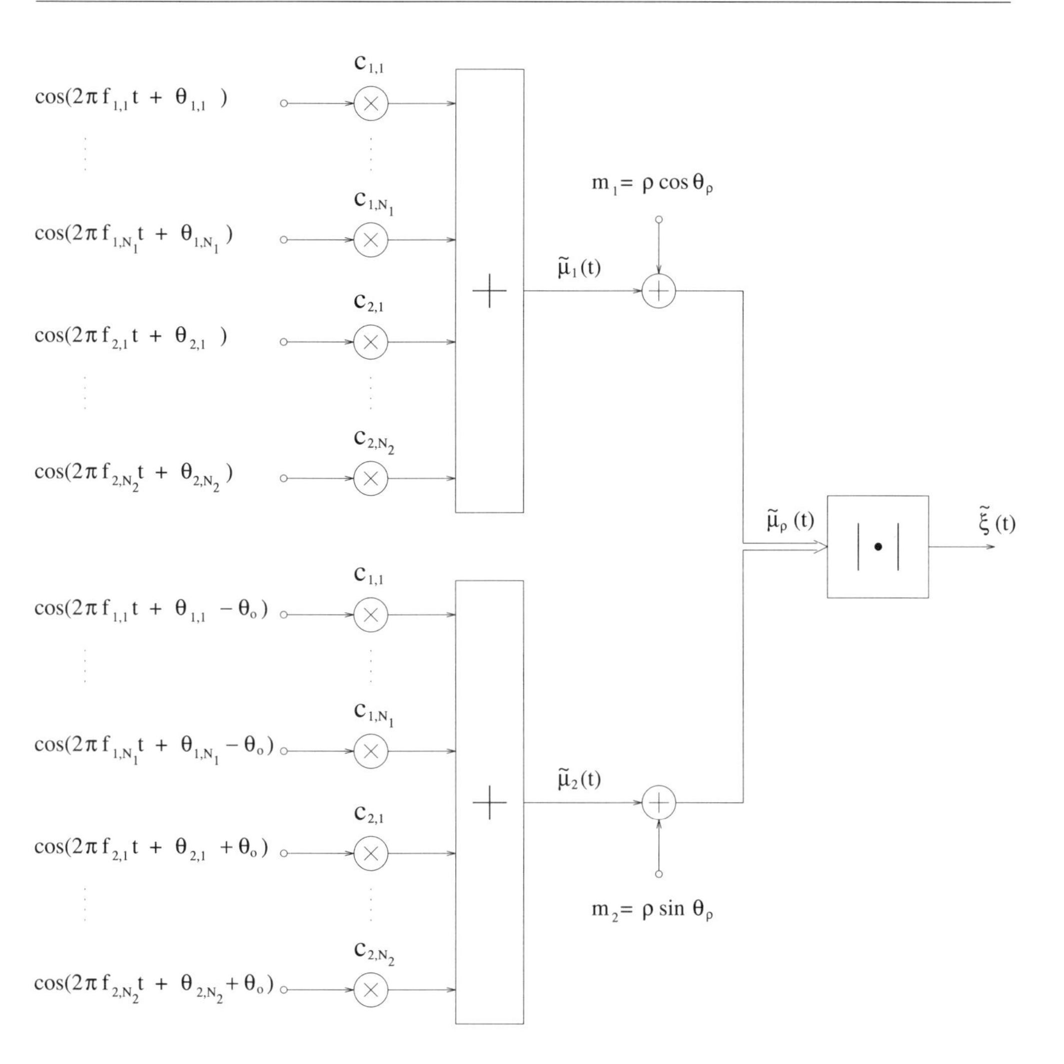

Figure 6.29: Deterministic simulation model for generalized Rice processes.

Thus, we obtain the following expressions for the characteristic quantities of the deterministic simulation model:

$$\tilde{\psi}_0^{(0)} = \tilde{\psi}_0 = \sum_{n=1}^{N_1} \frac{c_{1,n}^2}{2} + \sum_{n=1}^{N_2} \frac{c_{2,n}^2}{2} \,, \tag{6.144a}$$

$$\tilde{\psi}_0^{(1)} = \dot{\tilde{\psi}}_0 = 0 \,, \tag{6.144b}$$

$$\tilde{\psi}_0^{(2)} = \ddot{\tilde{\psi}}_0 = -2\pi^2 \left[\sum_{n=1}^{N_1} (c_{1,n} f_{1,n})^2 + \sum_{n=1}^{N_2} (c_{2,n} f_{2,n})^2 \right] , \tag{6.144c}$$

$$\tilde{\phi}_0^{(0)} = \tilde{\phi}_0 = \tilde{\psi}_0 \cdot \cos\theta_0 \,, \tag{6.144d}$$

$$\tilde{\phi}_0^{(1)} = \dot{\tilde{\phi}}_0 = \pi \left[\sum_{n=1}^{N_1} (c_{1,n}^2 f_{1,n}) - \sum_{n=1}^{N_2} (c_{2,n}^2 f_{2,n}) \right] \cdot \sin \theta_0 \,, \tag{6.144e}$$

$$\tilde{\phi}_0^{(2)} = \ddot{\tilde{\phi}}_0 = \ddot{\tilde{\psi}}_0 \cdot \cos \theta_0 \,. \tag{6.144f}$$

With these quantities, $\tilde{\beta}$ is also determined, because

$$\tilde{\beta} = -\ddot{\tilde{\psi}}_0 - \frac{\dot{\tilde{\phi}}_0^2}{\tilde{\psi}_0 \sin^2 \theta_0} \tag{6.145}$$

holds.

The calculation of the model parameters $f_{i,n}$ and $c_{i,n}$ is performed according to the method of exact Doppler spread. As described in Subsection 6.1.4, however, this procedure must be slightly modified due to $\kappa_2 = \kappa_0 \in (0, 1]$. Therefore, the formula (6.72) is also valid for the discrete Doppler frequencies $f_{i,n}$ in the present case, where we have to take into account that (6.73) holds. Similarly, the calculation of the Doppler coefficients $c_{i,n}$ is performed by using (6.74), where σ_0 has to be replaced by σ_i. Finally, for the Doppler phases $\theta_{i,n}$ it is assumed that these quantities are realizations (outcomes) of a random variable uniformly distributed within $(0, 2\pi]$.

Analysing the characteristic quantities of the simulation model, we restrict ourselves to $\tilde{\psi}_0$ and $\tilde{\beta}/f_{max}^2$. If these quantities are calculated according to (6.144a) and (6.145), respectively, by means of (6.72) and (6.74), then the convergence behaviour in terms of N_i $(N_1 = N_2)$ appears as depicted in Figures 6.30(a) and 6.30(b). The results shown are based on the primary model parameters σ_1^2, σ_2^2, and κ_0, as they are listed in the following subsection in Table 6.4.

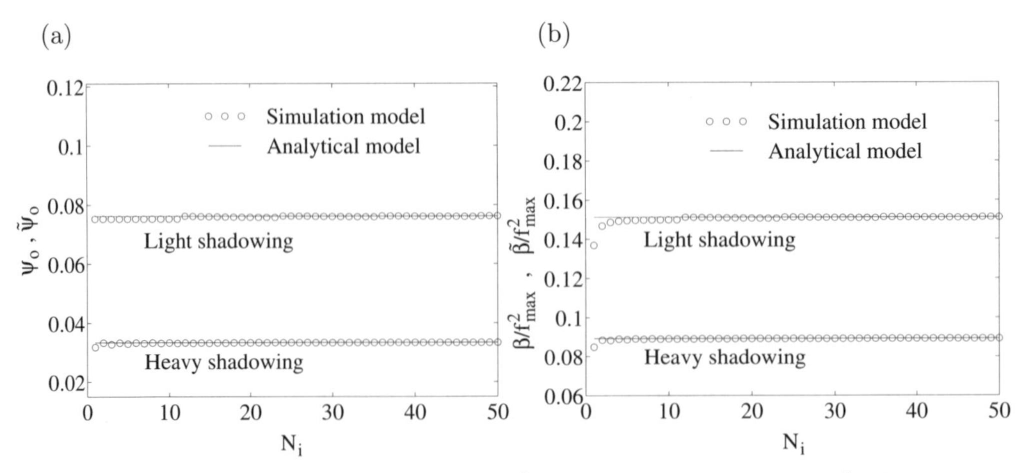

Figure 6.30: Illustration of (a) ψ_0 and $\tilde{\psi}_0$ as well as (b) β/f_{max}^2 and $\tilde{\beta}/f_{max}^2$ (MEDS, σ_i^2 and κ_2 according to Table 6.4).

Since the deviations between $\tilde{\psi}_0$ and ψ_0 as well as between $\tilde{\beta}/f_{max}^2$ and β/f_{max}^2 are negligible for all cases relevant in practice ($N_i \geq 7$), it follows that the probability density function $\tilde{p}_\xi(z)$, the level-crossing rate $\tilde{N}_\xi(r)$, and the average duration of fades $\tilde{T}_{\xi_-}(r)$ of the simulation model are extremely close to the corresponding quantities of the reference model.

6.3.3 Applications and Simulation Results

In this subsection, it will be shown that the statistical properties of stochastic and deterministic generalized Rice processes can be brought into astonishingly good agreement with real-world measurement results, even without multiplying the Rice process with a lognormal process. Since a fair comparison of the performance between different channel models is intended, we here again use the measurement results for $F_{\xi_+}^\star(r)$, $N_\xi^\star(r)$, and $T_{\xi_-}^\star(r)$ from [But83], which were also the basis for the experiments described in Subsections 6.1.5 and 6.2.4.

In the present case, the parameter vector $\mathbf{\Omega}$ contains all six primary model parameters. Thus, $\mathbf{\Omega}$ is defined by

$$\mathbf{\Omega} := (\sigma_1,\ \sigma_2,\ \kappa_0,\ \theta_0\ \rho,\ \theta_\rho)\,. \tag{6.146}$$

The optimization of the components of $\mathbf{\Omega}$ is again performed as described in Subsection 6.1.5 by minimizing the error function $E_2(\mathbf{\Omega})$ [cf. (6.76)] by means of the Fletcher-Powell algorithm [Fle63]. The optimization results found are presented in Table 6.4.

Table 6.4: The optimized primary model parameters of the reference model for areas with heavy and light shadowing.

Shadowing	σ_1	σ_2	κ_0	θ_0	ρ	θ_ρ
heavy	0.0894	0.7468	0.1651	0.3988	0.2626	30.3°
light	0.1030	0.9159	0.2624	0.3492	1.057	53.1°

The Rice factor c_R [see (3.18)], i.e.,

$$c_R = \frac{\rho^2}{2\psi_0} = \frac{\rho^2}{\sigma_2^2\left[\left(\frac{\sigma_1}{\sigma_2}\right)^2 + \frac{2}{\pi}\arcsin(\kappa_2)\right]}\,, \tag{6.147}$$

of the present model takes on the values $c_R = 0.134\,\text{dB}$ (heavy shadowing) and $c_R = 8.65\,\text{dB}$ (light shadowing), which are about as large as the Rice factors determined for the extended Suzuki model of Type II (cf. Subsection 6.2.4).

Figure 6.31(a) shows us the complementary cumulative distribution function $F_{\xi_+}(r/\rho)$ of the reference model and that of the measured channel $F_{\xi_+}^\star(r/\rho)$. Clearly visible deviations from the results depicted in Figure 6.25(a) do not seem to be apparent.

On the other hand, especially when considering the channel with heavy shadowing, we are able to achieve further improvement in view of fitting the normalized level-crossing rate of the reference model $N_\xi(r/\rho)/f_{max}$ to that of the measured channel $N_\xi^\star(r/\rho)/f_{max}$. This immediately becomes obvious if we compare Figure 6.31(b) with Figure 6.25(b). Especially with regard to the level-crossing rate, it seems as if the higher flexibility of the generalized Rice model has a positive effect.

Clearly visible are also the improvements achieved for the adapting of the normalized average duration of fades $T_{\xi_-}(r/\rho)\cdot f_{max}$ to $T_{\xi_-}^\star(r/\rho)\cdot f_{max}$. Concerning this statement, one may compare the two Figures 6.31(c) and 6.25(c). Now the present model is in very good agreement with the measurements, even at low levels.

Finally, it should be pointed out that the corresponding simulation results are also depicted in Figures 6.31(a)–6.31(c). For the realization and the simulation of the channel with heavy (light) shadowing, we have used $N_1 = N_2 = 7$ ($N_1 = N_2 = 15$) harmonic functions. In each of the two situations, a channel output sequence $\tilde{\xi}(kT_s)$ ($k = 1, 2, \ldots, N_s$) with $N_s = 3 \cdot 10^6$ sampling values was generated and used for the evaluation of the statistics. For the maximum Doppler frequency f_{max}, the value 91 Hz was chosen here, and the sampling interval T_s was prescribed by $T_s = 0.3\,\text{ms}$.

6.4 THE MODIFIED LOO MODEL

Loo developed a stochastic model for the modelling of frequency-nonselective terrestrial mobile radio channels on the basis of measurements in [Loo85]. This model was also the topic of further investigations in [Loo87, Loo90, Loo91, Loo96], which were summarized in [Loo98] at a later point. Loo's model is based on the physically plausibly reasoned assumption that the line-of-sight component underlies slow amplitude fluctuations caused by shadowing effects. In this model, it is assumed that the slow amplitude fluctuations of the line-of-sight component are lognormally distributed, while the fast fading, caused by the multipath propagation, behaves like a Rayleigh process.

In this section, we will combine Loo's stochastic model and the Rice process with cross-correlated in-phase and quadrature components to a superordinate model. The resulting model, which we will call the *modified Loo model*, then contains the original model suggested by Loo and the extended Rice process as respective special case.

6.4.1 The Stochastic Modified Loo Model

The model with which we will deal with in this subsection is depicted in Figure 6.32. It is a matter of the modified Loo model for which $\nu_1(t)$, $\nu_2(t)$, and $\nu_3(t)$ are uncorrelated zero-mean real-valued Gaussian random processes. Let the Doppler power spectral density $S_{\nu_i\nu_i}(f)$ of the Gaussian random processes $\nu_i(t)$ for $i = 1, 2$ be given by the restricted Jakes power spectral density (6.126) with $\kappa_i \in [0, 1]$, whereas we again use the Gaussian power spectral density according to (6.43) for $S_{\nu_3\nu_3}(f)$.

(a)

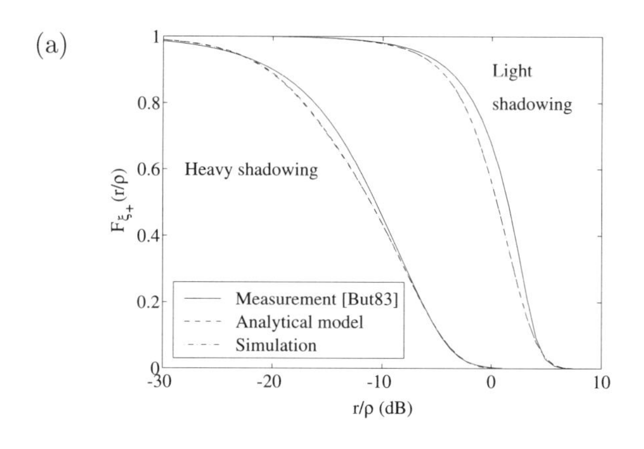

(b)

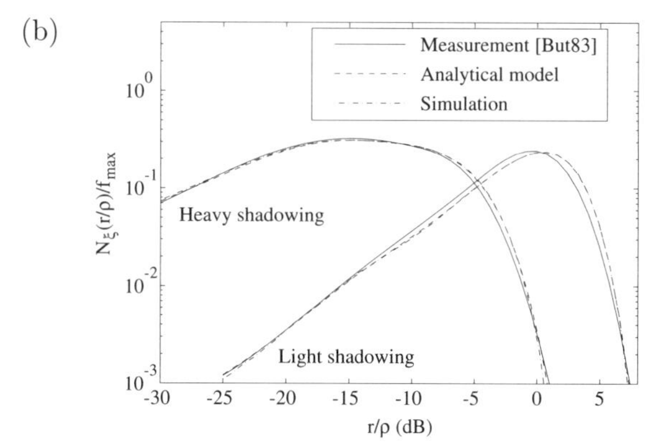

(c)

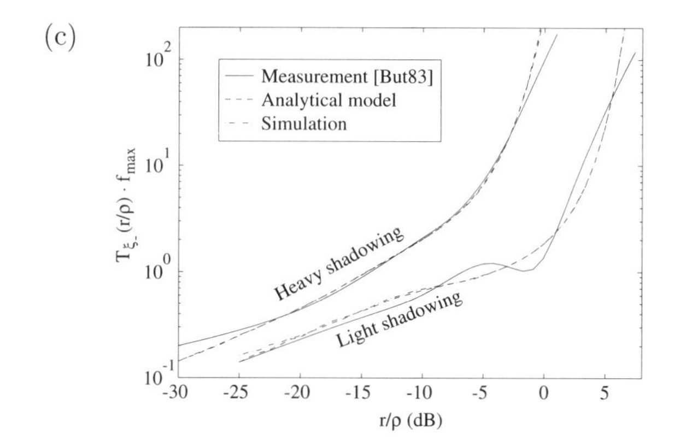

Figure 6.31: (a) Complementary cumulative distribution function $F_{\xi_+}(r/\rho)$, (b) normalized level-crossing rate $N_\xi(r/\rho)/f_{max}$, and (c) normalized average duration of fades $T_{\xi_-}(r/\rho) \cdot f_{max}$ for areas with heavy and light shadowing.

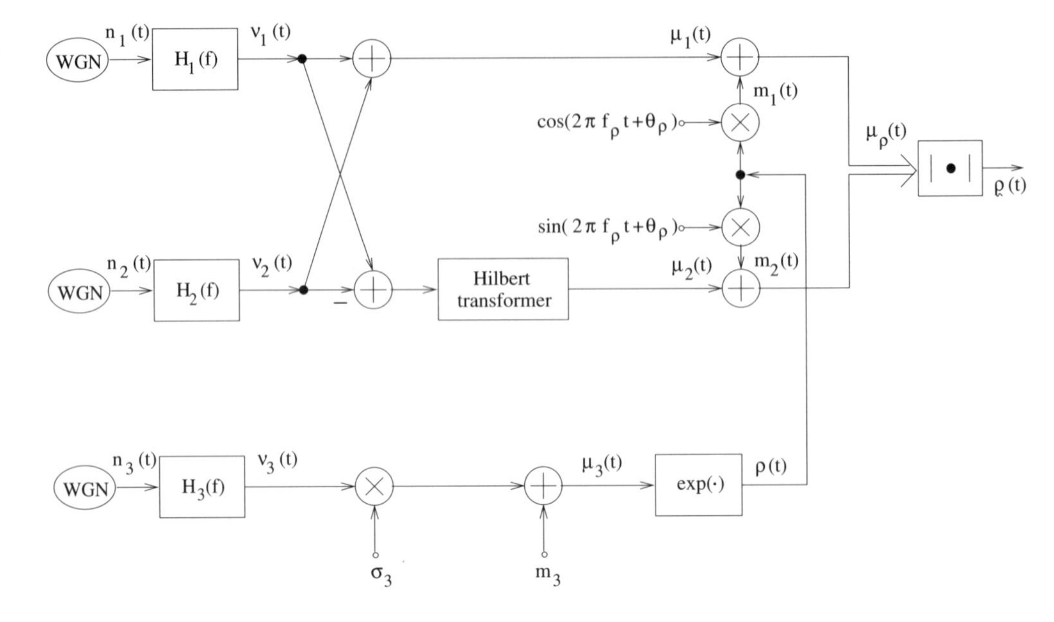

Figure 6.32: The modified Loo model (reference model).

In this model, the fast signal fluctuations caused by the multipath propagation are modelled in the equivalent complex baseband by a complex-valued Gaussian random process

$$\mu(t) = \mu_1(t) + j\mu_2(t)\,, \tag{6.148}$$

where its real and imaginary part

$$\mu_1(t) = \nu_1(t) + \nu_2(t)\,, \tag{6.149a}$$

$$\mu_2(t) = \check{\nu}_1(t) - \check{\nu}_2(t)\,, \tag{6.149b}$$

are statistically uncorrelated. Here, $\check{\nu}_i(t)$ $(i = 1, 2)$ again denotes the Hilbert transform of $\nu_i(t)$.

For the line-of-sight component $m(t) = m_1(t) + jm_2(t)$, we read from Figure 6.32 that

$$m(t) = \rho(t) \cdot e^{j(2\pi f_\rho t + \theta_\rho)} \tag{6.150}$$

holds, where f_ρ and θ_ρ again denote the Doppler frequency and the Doppler phase of the line-of-sight component, respectively, and

$$\rho(t) = e^{\sigma_3 \nu_3(t) + m_3} \tag{6.151}$$

designates a lognormal process with which the slow amplitude fluctuations of the line-of-sight component are modelled. For the spectral and statistical properties of the lognormal process (6.151), the statements made in Subsection 6.1.2 hold. Let us assume that the bandwidth of the Gaussian random process $\nu_3(t)$ is very small in comparison with the bandwidth of $\mu(t)$, so that, consequently, the amplitude $\rho(t)$ of the line-of-sight component (6.150) only varies relatively slowly compared to the fast signal fading.

The sum of the scattered component and the line-of-sight component results in the complex-valued Gaussian random process

$$\mu_\rho(t) = \mu(t) + m(t)\,, \tag{6.152}$$

whose real and imaginary part can be expressed — by using (6.149a), (6.149b), and (6.150) — as follows:

$$\mu_{\rho_1}(t) = \nu_1(t) + \nu_2(t) + \rho(t) \cdot \cos(2\pi f_\rho t + \theta_\rho)\,, \tag{6.153a}$$

$$\mu_{\rho_2}(t) = \check{\nu}_1(t) - \check{\nu}_2(t) + \rho(t) \cdot \sin(2\pi f_\rho t + \theta_\rho)\,. \tag{6.153b}$$

The absolute value of (6.152) finally results in a new stochastic process

$$\varrho(t) = \sqrt{\left[\mu_1(t) + \rho(t)\cos(2\pi f_\rho t + \theta_\rho)\right]^2 + \left[\mu_2(t) + \rho(t)\sin(2\pi f_\rho t + \theta_\rho)\right]^2}\,, \tag{6.154}$$

which is called the *modified Loo process.* This process will in the following be used as a stochastic model to describe the fading behaviour of frequency-nonselective satellite mobile radio channels.

The modified Loo model introduced here contains the following three special cases:

$$\text{(i)} \quad \sigma_1^2 = \sigma_2^2 = \sigma_0^2\,, \quad \kappa_1 = \kappa_2 = 1\,, \quad \text{and} \quad f_\rho = 0\,, \tag{6.155a}$$

$$\text{(ii)} \quad \sigma_2^2 = 0 \quad \text{or} \quad \kappa_2 = 0\,, \tag{6.155b}$$

$$\text{(iii)} \quad \sigma_1^2 = \sigma_2^2 = \sigma_0^2\,, \quad \kappa_1 = 1\,,\ \kappa_2 = \kappa_0\,, \quad \text{and} \quad \sigma_3^2 = 0\,. \tag{6.155c}$$

Further on, we will see that in the special case (i), the power spectral density $S_{\mu\mu}(f)$ of the complex-valued Gaussian random process $\mu(t)$ [see (6.148)] is equal to the Jakes power spectral density. Since the Gaussian random processes $\mu_1(t)$ and $\mu_2(t)$ are uncorrelated due to the symmetry of the Jakes power spectral density, the modified Loo model (Figure 6.32) can be reduced to the classical Loo model [Loo85, Loo91] depicted in Figure 6.33. One should take into account that also $f_\rho = 0$ holds, so that the power spectral density of the line-of-sight component does not experience a frequency shift (Doppler shift) in this model. For the second special case (ii), where σ_2^2 or κ_2 are equal to zero, the coloured Gaussian random process $\nu_2(t)$ can just as well be removed and one obtains the channel model proposed in [Pae98c], which stands out against the general variant due to its considerably smaller realization expense. Finally, the third special case (iii) leads to the Rice process depicted in Figure 6.1, for which the underlying Gaussian random processes $\mu_1(t)$ and $\mu_2(t)$ are, admittedly, also correlated, but for which the absolute value of the line-of-sight component $m(t)$ is time-independent, i.e., it then holds $|m(t)| = \rho(t) = \rho = e^{m_3}$.

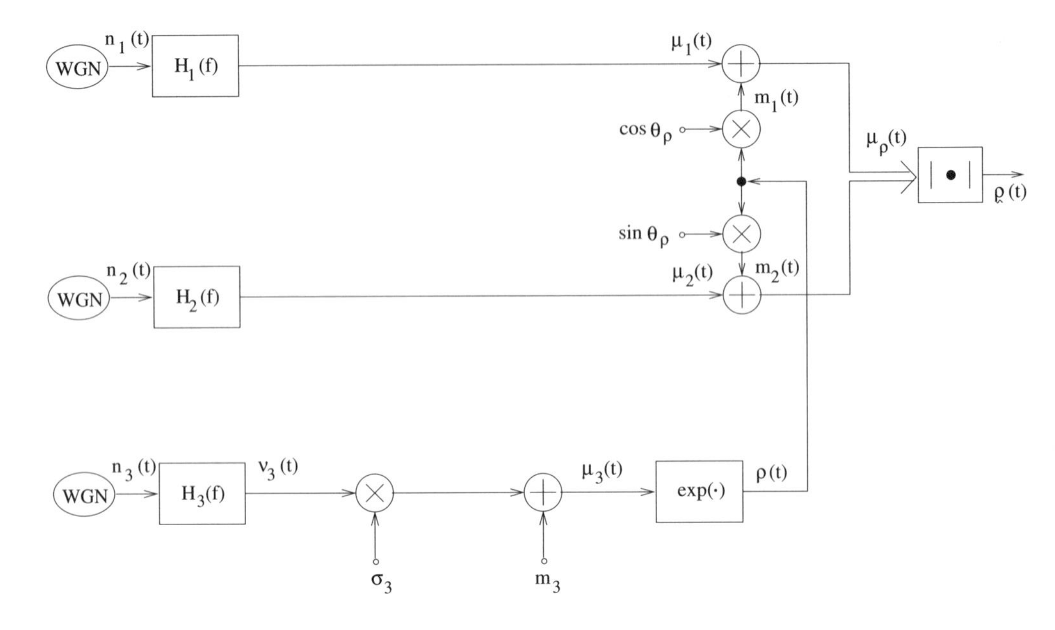

Figure 6.33: The classical Loo model (reference model).

6.4.1.1 Autocorrelation Function and Doppler Power Spectral Density

We are now interested in the autocorrelation function $r_{\mu_\rho\mu_\rho}(\tau)$ and in the corresponding Doppler power spectral density $S_{\mu_\rho\mu_\rho}(f)$ of the complex-valued random process $\mu_\rho(t)$ introduced by (6.152). Therefore, we at first calculate the autocorrelation function $r_{\mu_{\rho_i}\mu_{\rho_i}}(\tau)$ $(i = 1, 2)$ of the processes $\mu_{\rho_i}(t)$ as well as the cross-correlation function $r_{\mu_{\rho_1}\mu_{\rho_2}}(\tau)$ of the processes $\mu_{\rho_1}(t)$ and $\mu_{\rho_2}(t)$. By using (6.153a) and (6.153b), we obtain the following relations for these correlation functions:

$$r_{\mu_{\rho_1}\mu_{\rho_1}}(\tau) = r_{\mu_{\rho_2}\mu_{\rho_2}}(\tau) = r_{\nu_1\nu_1}(\tau) + r_{\nu_2\nu_2}(\tau) + \frac{1}{2} r_{\rho\rho}(\tau) \cdot \cos(2\pi f_\rho \tau), \quad (6.156a)$$

$$r_{\mu_{\rho_1}\mu_{\rho_2}}(\tau) = r^*_{\mu_{\rho_2}\mu_{\rho_1}}(-\tau) = r_{\nu_1\check{\nu}_1}(\tau) - r_{\nu_2\check{\nu}_2}(\tau) + \frac{1}{2} r_{\rho\rho}(\tau) \cdot \sin(2\pi f_\rho \tau), \quad (6.156b)$$

where $r_{\nu_i\nu_i}(\tau)$ $(i = 1, 2)$ describes the autocorrelation function of the Gaussian random process $\nu_i(t)$, and with $r_{\nu_i\check{\nu}_i}(\tau)$ $(i = 1, 2)$ exactly the cross-correlation function of $\nu_i(t)$ and $\check{\nu}_i(t)$ is meant. Recall that $r_{\nu_i\nu_i}(\tau)$ and $r_{\nu_i\check{\nu}_i}(\tau)$ are already known to us due to (6.131) and (6.132), respectively. Furthermore, $r_{\rho\rho}(\tau)$ describes the autocorrelation function of $\rho(t)$ [cf. (6.151)] in (6.156a) and in (6.156b). One may take into account that $\rho(t)$ has been introduced as lognormal process in this section. That is why the autocorrelation function $r_{\rho\rho}(\tau)$ of $\rho(t)$ can be directly identified with the right-hand side of (6.47). Hence, we can therefore directly write

$$r_{\rho\rho}(\tau) = e^{2m_3 + \sigma_3^2(1 + r_{\nu_3\nu_3}(\tau))}. \quad (6.157)$$

The autocorrelation function $r_{\mu_\rho\mu_\rho}(\tau)$ of the complex-valued process $\mu_\rho(t) = \mu_{\rho_1}(t) + j\mu_{\rho_2}(t)$ will in imitation of (6.5) be expressed in terms of the autocorrelation functions and the cross-correlation functions of $\mu_{\rho_1}(t)$ and $\mu_{\rho_2}(t)$ as follows

$$r_{\mu_\rho \mu_\rho}(\tau) = r_{\mu_{\rho 1} \mu_{\rho 1}}(\tau) + r_{\mu_{\rho 2} \mu_{\rho 2}}(\tau) + j\big(r_{\mu_{\rho 1} \mu_{\rho 2}}(\tau) - r_{\mu_{\rho 2} \mu_{\rho 1}}(\tau)\big) . \tag{6.158}$$

When studying (6.157) and taking (6.44) into account, we notice that $r_{\rho\rho}(\tau)$ is a real and even function in τ. From the relation (6.132), we can on the other hand conclude that $r_{\nu_i \check{\nu}_i}(\tau)$ is real and odd, so that from (6.156b) the relation $r_{\mu_{\rho 1} \mu_{\rho 2}}(\tau) = r^*_{\mu_{\rho 2} \mu_{\rho 1}}(-\tau) = -r_{\mu_{\rho 2} \mu_{\rho 1}}(\tau)$ follows. If we also take into account that $r_{\mu_{\rho 1} \mu_{\rho 1}}(\tau) = r_{\mu_{\rho 2} \mu_{\rho 2}}(\tau)$ holds, then (6.158) simplifies to

$$r_{\mu_\rho \mu_\rho}(\tau) = 2\big(r_{\mu_{\rho 1} \mu_{\rho 1}}(\tau) + j r_{\mu_{\rho 1} \mu_{\rho 2}}(\tau)\big) . \tag{6.159}$$

In this relation, we also substitute (6.156a) and (6.156b), so that we finally find the following expression for the desired autocorrelation function $r_{\mu_\rho \mu_\rho}(\tau)$

$$\begin{aligned} r_{\mu_\rho \mu_\rho}(\tau) &= 2\big(r_{\nu_1 \nu_1}(\tau) + j r_{\nu_1 \check{\nu}_1}(\tau)\big) \\ &\quad + 2\big(r_{\nu_2 \nu_2}(\tau) - j r_{\nu_2 \check{\nu}_2}(\tau)\big) + r_{\rho\rho}(\tau) e^{j2\pi f_\rho \tau} . \end{aligned} \tag{6.160}$$

After performing the Fourier transform of (6.160) and using the relation $S_{\nu_i \check{\nu}_i}(f) = -j \operatorname{sgn}(f) \cdot S_{\nu_i \nu_i}(f)$, we obtain the Doppler power spectral density $S_{\mu_\rho \mu_\rho}(f)$, which can be presented as follows

$$\begin{aligned} S_{\mu_\rho \mu_\rho}(f) &= 2\big(1 + \operatorname{sgn}(f)\big) S_{\nu_1 \nu_1}(f) \\ &\quad + 2\big(1 - \operatorname{sgn}(f)\big) S_{\nu_2 \nu_2}(f) + S_{\rho\rho}(f - f_\rho) , \end{aligned} \tag{6.161}$$

where $S_{\nu_i \nu_i}(f)$ $(i = 1, 2)$ is again given by (6.126), and $S_{\rho\rho}(f - f_\rho)$ can be identified with the right-hand side of (6.48) if the frequency variable f is substituted by $f - f_\rho$ there, i.e.,

$$S_{\rho\rho}(f - f_\rho) = e^{2m_3 + \sigma_3^2} \cdot \left[\delta(f - f_\rho) + \sum_{n=1}^{\infty} \frac{\sigma_3^{2n}}{n!} \cdot \frac{S_{\nu_3 \nu_3}\left(\frac{f - f_\rho}{\sqrt{n}}\right)}{\sqrt{n}} \right] , \tag{6.162}$$

where $S_{\nu_3 \nu_3}(f)$ denotes the Gaussian power spectral density according to (6.43).

Figures 6.34(a)–6.34(f) symbolically show how the generally unsymmetrical Doppler power spectral density $S_{\mu_\rho \mu_\rho}(f)$ is composed of the individual power spectral densities $S_{\nu_1 \nu_1}(f)$, $S_{\nu_2 \nu_2}(f)$, and $S_{\nu_3 \nu_3}(f)$. The spectra shown are valid for the following parameters: $\sigma_1^2 = \sigma_2^2 = 1$, $\kappa_1 = 0.8$, $\kappa_2 = 0.6$, $\sigma_3^2 = 0.01$, $m_3 = 0$, $f_\rho = 0.4 f_{max}$, $f_c = 0.13 f_{max}$, and $\sigma_c^2 = 100$.

From the general representation (6.161), we can easily derive the power spectral densities determined by the special cases (i)–(iii) according to (6.155a)–(6.155c), respectively. For example, on condition that (6.155a) holds, the Doppler power spectral

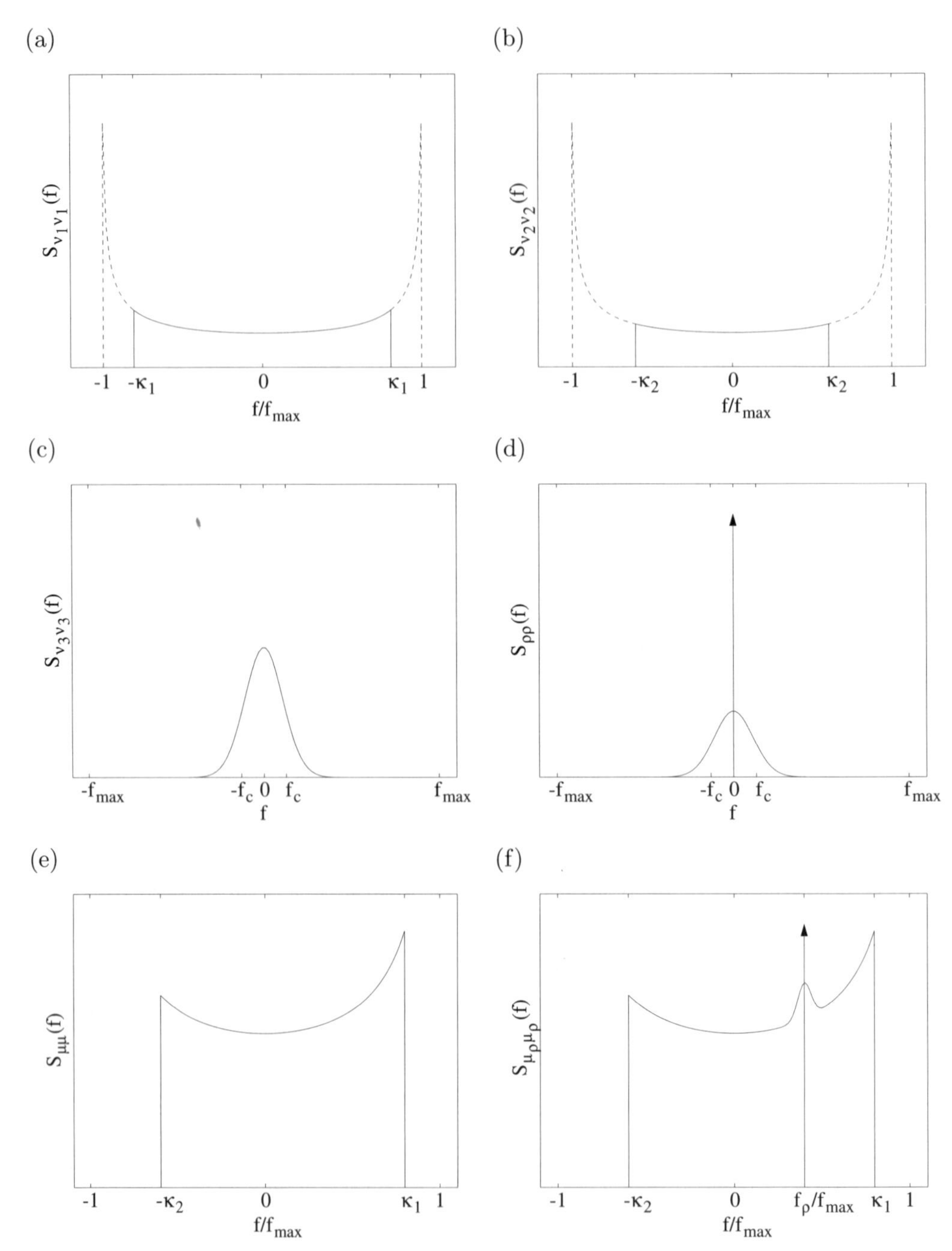

Figure 6.34: Various power spectral densities: restricted Jakes power spectral density (a) $S_{\nu_1\nu_1}(f)$ and (b) $S_{\nu_2\nu_2}(f)$, (c) Gaussian power spectral density $S_{\nu_3\nu_3}(f)$, (d) power spectral density $S_{\rho\rho}(f)$ of the lognormal process $\rho(t)$, (e) power spectral density $S_{\mu\mu}(f)$, and (f) resulting unsymmetrical Doppler power spectral density $S_{\mu_\rho\mu_\rho}(f)$.

density $S_{\mu_\rho\mu_\rho}(f)$ of the classical Loo model (see Figure 6.33) can be derived from (6.161) in the form

$$S_{\mu_\rho\mu_\rho}(f) = S_{\mu\mu}(f) + S_{\rho\rho}(f)\,, \tag{6.163}$$

where $S_{\mu\mu}(f)$ denotes the Jakes power spectral density according to (3.7), and $S_{\rho\rho}(f)$ represents the power spectral density of the lognormal process $\rho(t)$. An example of the shape of $S_{\rho\rho}(f)$ is depicted in Figure 6.34(d). For the special case (ii) determined by (6.155b), the Doppler power spectral density $S_{\mu_\rho\mu_\rho}(f)$ disappears for negative frequencies in Figure 6.34(f). Finally, in the special case (iii), $S_{\rho\rho}(f-f_\rho)$ only delivers a contribution to $S_{\mu_\rho\mu_\rho}(f)$ according to (6.161), which is characterized by a weighted Delta function at the point $f = f_\rho$.

Next, we will calculate the characteristic quantities $\psi_0^{(n)}$ and $\phi_0^{(n)}$ $(n = 0, 1, 2)$ valid for the modified Loo model. Therefore, we substitute $r_{\mu_1\mu_1}(\tau) = r_{\nu_1\nu_1}(\tau) + r_{\nu_2\nu_2}(\tau)$ and $r_{\mu_1\mu_2}(\tau) = r_{\nu_1\check{\nu}_1}(\tau) - r_{\nu_2\check{\nu}_2}(\tau)$ into (6.11a) and (6.11b), respectively, and obtain the following expressions by using (6.131) as well as (6.132):

$$\psi_0^{(0)} = \psi_0 = \frac{1}{\pi}\sum_{i=1}^{2}\sigma_i^2 \arcsin(\kappa_i)\,, \tag{6.164a}$$

$$\psi_0^{(1)} = \dot{\psi}_0 = 0\,, \tag{6.164b}$$

$$\psi_0^{(2)} = \ddot{\psi}_0 = -(\pi f_{max})^2\left[2\psi_0 - \frac{1}{\pi}\sum_{i=1}^{2}\sigma_i^2 \sin\left(2\arcsin(\kappa_i)\right)\right]\,, \tag{6.164c}$$

$$\phi_0^{(0)} = \phi_0 = 0\,, \tag{6.164d}$$

$$\phi_0^{(1)} = \dot{\phi}_0 = -2f_{max}\sum_{i=1}^{2}(-1)^i\sigma_i^2\left(1-\sqrt{1-\kappa_i^2}\right)\,, \tag{6.164e}$$

$$\phi_0^{(2)} = \ddot{\phi}_0 = 0\,, \tag{6.164f}$$

where $0 \leq \kappa_i \leq 1$ holds for $i = 1, 2$. In the special case (iii) described by (6.155c), one can easily convince oneself that the characteristic quantities (6.164a)–(6.164f) are identical to those described by (6.12a)–(6.12f), respectively.

6.4.1.2 Probability Density Function of the Amplitude and the Phase

In principle, the statistical properties of the modified Loo process $\varrho(t) = |\mu_\rho(t)|$ can again be calculated by means of the joint probability density function $p_{\mu_{\rho_1}\mu_{\rho_2}\dot{\mu}_{\rho_1}\dot{\mu}_{\rho_2}}(x_1, x_2, \dot{x}_1, \dot{x}_2)$ or $p_{\varrho\dot{\varrho}\vartheta\dot{\vartheta}}(z, \dot{z}, \theta, \dot{\theta})$, as in the previous cases. Due to the time variability of $\rho(t)$, the mathematical computation expenditure is in this case much higher than for the models analysed before, where $\rho(t) = \rho$ was always a constant quantity. Therefore, we will choose a more elegant alternative way, which leads us to our goal faster and which, furthermore, lets us profit from the results found in Section 6.1. Considering that the reference model depicted in Figure 6.1 is basically a special case of the modified Loo model shown in Figure 6.32 on condition that $\rho(t) = \rho$ holds, then the conditional probability density function $p_\varrho(z|\rho(t) = \rho)$ of the stochastic process $\varrho(t)$, which is defined by (6.154), has to be identical to (6.30). Therefore, we can write

$$p_\varrho(z|\rho(t)=\rho) = p_\xi(z) = \begin{cases} \frac{z}{\psi_0}\, e^{-\frac{z^2+\rho^2}{2\psi_0}}\, I_0\left(\frac{z\rho}{\psi_0}\right), & z \geq 0\,, \\ 0\,, & z < 0\,, \end{cases} \tag{6.165}$$

where ψ_0 describes the mean power of $\mu_i(t)$ $(i = 1, 2)$ according to (6.164a). Since the amplitude $\rho(t)$ of the line-of-sight component is lognormally distributed in the Loo model, i.e., the density $p_\rho(y)$ of $\rho(t)$ is given by the lognormal distribution [cf. (2.28)]

$$p_\rho(y) = \begin{cases} \frac{1}{\sqrt{2\pi}\sigma_3 y}\, e^{-\frac{(\ln y - m_3)^2}{2\sigma_3^2}}, & y \geq 0\,, \\ 0\,, & y < 0\,, \end{cases} \tag{6.166}$$

the probability density function $p_\varrho(z)$ of the modified Loo process $\varrho(t)$ can be derived from the joint probability density function $p_{\varrho\rho}(z, y)$ of the stochastic processes $\varrho(t)$ and $\rho(t)$ as follows:

$$\begin{aligned} p_\varrho(z) &= \int_0^\infty p_{\varrho\rho}(z, y)\, dy \\ &= \int_0^\infty p_\varrho(z|\rho(t)=y) \cdot p_\rho(y)\, dy \\ &= \int_0^\infty p_\xi(z;\rho=y) \cdot p_\rho(y)\, dy\,, \quad z \geq 0\,. \end{aligned} \tag{6.167}$$

If we now substitute (6.30) (or (6.165)) and (6.166) into (6.167), then we obtain the following expression for the probability density function $p_\varrho(z)$ of the modified Loo process $\varrho(t)$

$$p_\varrho(z) = \frac{z}{\sqrt{2\pi}\psi_0\sigma_3} \int_0^\infty \frac{1}{y}\, e^{-\frac{z^2+y^2}{2\psi_0}}\, I_0\left(\frac{zy}{\psi_0}\right) e^{-\frac{(\ln y - m_3)^2}{2\sigma_3^2}}\, dy\,, \quad z \geq 0\,, \tag{6.168}$$

where ψ_0 is given by (6.164a). We notice that the probability density function $p_\varrho(z)$ depends on three parameters, namely ψ_0, σ_3, and m_3. In connection with (6.155a), it now becomes apparent that (6.168) also holds for the classical Loo model, if we leave the influences of the parameters σ_i^2 and κ_i on ψ_0 aside. The same statement also holds for the special case (ii) introduced by (6.155b). Therefore, it is not surprising if one also finds the probability density function $p_\varrho(z)$ in the form (6.168), e.g., in [Loo85, Loo91, Loo98] and [Pae98c]. Differences, however, do occur for the level-crossing rate and the average duration of fades, as we will see in the following Subsection 6.4.1.3. For completeness, we will also briefly study the effects of the special case (iii) [see (6.155c)]. In the limit $\sigma_3^2 \to 0$, the lognormal distribution (6.166) converges to $p_\rho(y) = \delta(y - \rho)$, where $\rho = e^{m_3}$. In this case, the Rice distribution (6.165) follows directly from (6.167), where it has to be taken into account that ρ is equal to e^{m_3}.

In order to illustrate the probability density function $p_\varrho(z)$ of the modified Loo process $\varrho(t)$, we study Figures 6.35(a) and 6.35(b), which allow the influence of the parameters σ_3 and m_3, respectively, to stand out.

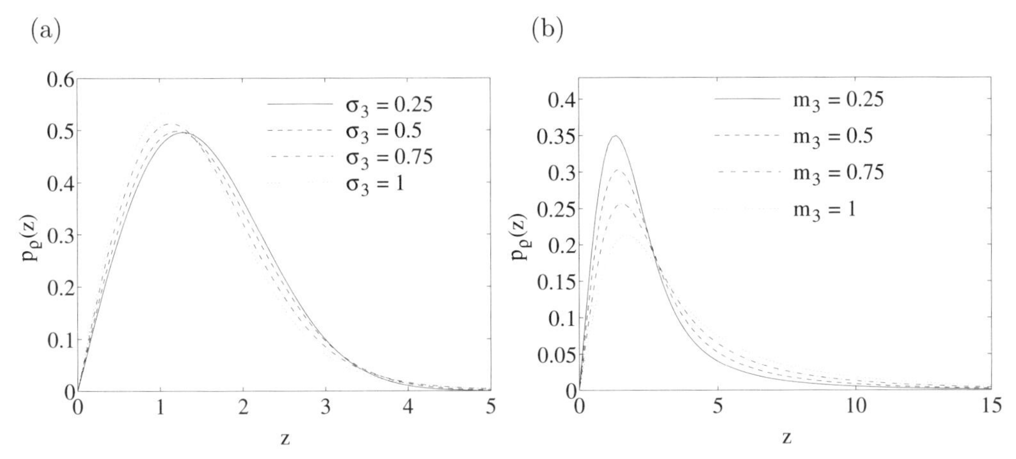

Figure 6.35: Probability density function $p_\varrho(z)$ of the amplitude $\varrho(t)$ of modified and classical Loo processes depending on: (a) σ_3 ($\psi_0 = 1, m_3 = -\sigma_3^2$) and (b) m_3 ($\psi_0 = 1, \sigma_3^2 = 1$).

Next, we will analyse the probability density function $p_\vartheta(\theta)$ of the phase $\vartheta(t) = \arg\{\mu_\rho(t)\}$ of the modified Loo model. Here, we proceed in a similar way as for the computation of $p_\varrho(z)$. In consideration of the present situation this means that we exploit the fact that the probability density function $p_\vartheta(\theta)$ for $\rho(t) = \rho = const.$ is identical to the right-hand side of (6.32). Hence, we have

$$
\begin{aligned}
p_\vartheta(\theta; t|\rho(t) = \rho) \;&=\; \frac{e^{-\frac{\rho^2}{2\psi_0}}}{2\pi} \left\{ 1 + \sqrt{\frac{\pi}{2\psi_0}}\, \rho \cos(\theta - 2\pi f_\rho t - \theta_\rho)\, e^{\frac{\rho^2 \cos^2(\theta - 2\pi f_\rho t - \theta_\rho)}{2\psi_0}} \right. \\
&\left. \left[1 + \operatorname{erf}\left(\frac{\rho \cos(\theta - 2\pi f_\rho t - \theta_\rho)}{\sqrt{2\psi_0}} \right) \right] \right\}, \quad -\pi \leq \theta \leq \pi \,. \quad (6.169)
\end{aligned}
$$

Since the conditional probability density function of the phase $\vartheta(t)$ for $f_\rho \neq 0$ is always a function of the time t according to this equation, we first perform an averaging of the expression above with respect to the time t. This leads to the uniform distribution

$$
\begin{aligned}
p_\vartheta(\theta|\rho(t) = \rho) \;&=\; \lim_{T\to\infty} \frac{1}{2T} \int_{-T}^{T} p_\vartheta(\theta; t|\rho(t) = \rho)\, dt \\
&=\; \frac{1}{2\pi}\,, \qquad -\pi \leq \theta \leq \pi \,. \qquad (6.170)
\end{aligned}
$$

The desired probability density function $p_\vartheta(\theta)$ of the phase $\vartheta(t) = \arg\{\mu_\rho(t)\}$ can now be determined by means of the joint probability density function $p_{\vartheta\rho}(\theta, y)$ of $\vartheta(t)$ and

$\rho(t)$ as follows:

$$\begin{aligned} p_\vartheta(\theta) &= \int_0^\infty p_{\vartheta\rho}(\theta, y)\, dy \\ &= \int_0^\infty p_\vartheta(\theta|\rho(t) = y) \cdot p_\rho(y)\, dy \\ &= \frac{1}{2\pi} \int_0^\infty p_\rho(y)\, dy \\ &= \frac{1}{2\pi}\,, \qquad -\pi \leq \theta \leq \pi\,. \end{aligned} \tag{6.171}$$

Thus, it is proven that the phase $\vartheta(t)$ of $\mu_\rho(t)$ is uniformly distributed in the interval $[-\pi, \pi]$, if the Doppler frequency f_ρ of the line-of-sight component $m(t)$ is not equal to zero. Similarly, an expression for $p_\vartheta(\theta)$ can be derived for the case $f_\rho = 0$. We will, however, refrain from a presentation of the resulting formula at this point.

6.4.1.3 Level-Crossing Rate and Average Duration of Fades

The derivation of the level-crossing rate $N_\varrho(r)$ of the modified Loo process $\varrho(t)$ is performed by means of the fundamental relation (6.33). Since the knowledge of the joint probability density function $p_{\varrho\dot{\varrho}}(z, \dot{z})$ of the processes $\varrho(t)$ and $\dot{\varrho}(t)$ at the same time t is necessary again, we will at first derive this. Therefore, we write

$$\begin{aligned} p_{\varrho\dot{\varrho}}(z, \dot{z}) &= \int_0^\infty \int_{-\infty}^\infty p_{\varrho\dot{\varrho}\rho\dot{\rho}}(z, \dot{z}, y, \dot{y})\, d\dot{y}\, dy \\ &= \int_0^\infty \int_{-\infty}^\infty p_{\varrho\dot{\varrho}}(z, \dot{z}|\rho(t) = y,\ \dot{\rho}(t) = \dot{y}) \cdot p_{\rho\dot{\rho}}(y, \dot{y})\, d\dot{y}\, dy\,. \end{aligned} \tag{6.172}$$

In the latter expression, $p_{\rho\dot{\rho}}(y, \dot{y})$ denotes the joint probability density function of $\rho(t)$ and $\dot{\rho}(t)$ at the same time t. Since the process $\rho(t)$ is lognormally distributed in the modified Loo model, we can directly identify $p_{\rho\dot{\rho}}(y, \dot{y})$ with the relation (6.53), i.e., we can write

$$p_{\rho\dot{\rho}}(y, \dot{y}) = \frac{e^{-\frac{(\ln y - m_3)^2}{2\sigma_3^2}}}{\sqrt{2\pi}\,\sigma_3 y} \cdot \frac{e^{-\frac{\dot{y}^2}{2\gamma(\sigma_3 y)^2}}}{\sqrt{2\pi\gamma}\,\sigma_3 y}\,, \tag{6.173}$$

where $\gamma = -\ddot{r}_{\nu_3\nu_3}(0) = (2\pi\sigma_c)^2$. At the beginning of Subsection 6.4.1, we assumed that the amplitude $\rho(t)$ of the line-of-sight component will itself only vary very slowly. Therefore, $\dot{\rho}(t) \approx 0$ must approximately hold, so that the probability density function $p_{\dot{\rho}}(\dot{y})$ of $\dot{\rho}(t)$ can be approximated by $p_{\dot{\rho}}(\dot{y}) \approx \delta(\dot{y})$. Since this always holds if γ is sufficiently small or if the frequency ratio $\kappa_c = f_{max}/f_c$ is sufficiently large, we can in this case replace (6.173) by the approximation

$$p_{\rho\dot{\rho}}(y, \dot{y}) \approx p_\rho(y) \cdot \delta(\dot{y})\,, \tag{6.174}$$

where $p_\rho(y)$ again denotes the lognormal distribution according to (6.166). Regarding the sifting property of the Delta function, we now substitute (6.174) into (6.172) and obtain the approximation

$$
\begin{aligned}
p_{\varrho\dot{\varrho}}(z,\dot{z}) &\approx \int_0^{\infty}\int_{-\infty}^{\infty} p_{\varrho\dot{\varrho}}(z,\dot{z}|\rho(t)=y,\ \dot{\rho}(t)=\dot{y})\cdot p_{\rho}(y)\ \delta(\dot{y})\ d\dot{y}\ dy \\
&= \int_0^{\infty} p_{\varrho\dot{\varrho}}(z,\dot{z}|\rho(t)=y,\dot{\rho}(t)=0)\cdot p_{\rho}(y)\ dy\,. \qquad (6.175)
\end{aligned}
$$

With this relation, we can now approximate the level-crossing rate $N_{\varrho}(r)$ [cf. (6.33)] as follows:

$$
\begin{aligned}
N_{\varrho}(r) &= \int_0^{\infty} \dot{z}\, p_{\varrho\dot{\varrho}}(r,\dot{z})\ d\dot{z} \\
&\approx \int_0^{\infty}\int_0^{\infty} \dot{z}\, p_{\varrho\dot{\varrho}}(z,\dot{z}|\rho(t)=y,\dot{\rho}(t)=0)\cdot p_{\rho}(y)\ dy\ d\dot{z} \\
&= \int_0^{\infty} N_{\varrho}(r|\rho(t)=y,\dot{\rho}(t)=0)\cdot p_{\rho}(y)\ dy\,. \qquad (6.176)
\end{aligned}
$$

Here, we have to take into account that the level-crossing rate $N_{\varrho}(r|\rho(t)=\rho,\dot{\rho}(t)=0)$ appearing under the integral of (6.176) exactly corresponds to the relation (6.37) derived in Subsection 6.1.1.2. If we now substitute this equation together with (6.166) into (6.176), then we obtain the following approximation for the level-crossing rate of the modified Loo process

$$
\begin{aligned}
N_{\varrho}(r) &\approx \int_0^{\infty} \frac{e^{-\frac{(\ln y-m_3)^2}{2\sigma_3^2}}}{\sqrt{2\pi}\sigma_3 y}\cdot\frac{r\sqrt{2\beta}}{\pi^{3/2}\psi_0}\,e^{-\frac{r^2+y^2}{2\psi_0}}\cdot\int_0^{\pi/2}\cosh\left(\frac{ry}{\psi_0}\cos\theta\right) \\
&\quad\cdot\left[e^{-(\alpha y\sin\theta)^2}+\sqrt{\pi}\alpha y\sin(\theta)\,\mathrm{erf}\,(\alpha y\sin\theta)\right]\ d\theta\ dy\,, \qquad (6.177)
\end{aligned}
$$

where the relations (6.27) and (6.28) hold for α and β, respectively, if there the formulae (6.164a), (6.164c), and (6.164e) are used for the characteristic quantities ψ_0, $\ddot{\psi}_0$, and $\dot{\phi}_0$, respectively.

The investigation of the special case (i) [see (6.155a)] at first provides $\alpha=0$. This leads to the fact that the approximation (6.177) can be simplified considerably. Thus, on condition that $\alpha=0$ holds, the level-crossing rate $N_{\varrho}(r)$ of the modified Loo model simplifies to that of the classical Loo model, which can be approximated as follows:

$$
\begin{aligned}
N_{\varrho}(r)|_{\alpha=0} &\approx \sqrt{\frac{\beta}{2\pi}}\cdot\frac{r}{\psi_0}\int_0^{\infty}\frac{e^{-\frac{(\ln y-m_3)^2}{2\sigma_3^2}}}{\sqrt{2\pi}\sigma_3 y}\cdot e^{-\frac{r^2+y^2}{2\psi_0}}\,I_0\left(\frac{ry}{\psi_0}\right)\ dy \\
&= \sqrt{\frac{\beta}{2\pi}}\int_0^{\infty} p_{\xi}(r;\rho=y)\cdot p_{\rho}(y)\ dy\,, \qquad (6.178)
\end{aligned}
$$

where the quantities β and ψ_0 are in this case given by $\beta=-2(\pi\sigma_0 f_{max})^2$ and $\psi_0=\sigma_0^2$, respectively, and $p_{\xi}(r;\rho=y)$ denotes the Rice distribution (2.26), if ρ is replaced by y there. Studying (6.178) and (6.167) it becomes clear that on condition that $\alpha = 0$ holds, the level-crossing rate $N_\varrho(r)$ is again proportional to the probability density function $p_\varrho(r)$. This is always the case if the Doppler power spectral density is symmetrical, which often does not correspond with reality. The special case (ii) [see (6.155b)] does not lead to a simplification of (6.177). Here, however, the characteristic quantities ψ_0, $\ddot{\psi}_0$, and $\dot{\phi}_0$ are coupled stronger to each other, so that the flexibility of $N_\varrho(r)$ suffers in the end. Finally, we will investigate the consequences which the special case (iii) [see (6.155c)] has on the level-crossing rate $N_\varrho(r)$. In the limit $\sigma_3^2 \to 0$, we obtain $p_\rho(y) = \delta(y - \rho)$ with $\rho = e^{m_3}$, so that (6.37) again follows from the right-hand side of (6.176). By the way, (6.174) is then exactly fulfilled, so that we can replace the approximations sign by an equals sign in (6.176) without hesitation.

In order to be able to calculate the average duration of fades

$$T_{\varrho_-}(r) = \frac{F_{\varrho_-}(r)}{N_\varrho(r)} \tag{6.179}$$

of the modified Loo process, we still need an expression for the cumulative distribution function $F_{\varrho_-}(r) = P(\varrho(t) \leq r)$ of the stochastic process $\varrho(t)$. For the derivation of $F_{\varrho_-}(r)$, we use (6.168) and obtain

$$\begin{aligned} F_{\varrho_-}(r) &= \int_0^r p_\varrho(z)\, dz \\ &= \frac{1}{\sqrt{2\pi}\psi_0\sigma_3} \int_0^r \int_0^\infty \frac{z}{y}\, e^{-\frac{z^2+y^2}{2\psi_0}} I_0\left(\frac{zy}{\psi_0}\right) e^{-\frac{(\ln y - m_3)^2}{2\sigma_3^2}}\, dy\, dz \\ &= 1 - \int_0^\infty Q_1\left(\frac{y}{\sqrt{\psi_0}}, \frac{r}{\sqrt{\psi_0}}\right) p_\rho(y)\, dy\,, \end{aligned} \tag{6.180}$$

where $Q_1(\cdot,\cdot)$ is the generalized Marcum's Q-function defined by (6.67).

In order to illustrate the results found for the level-crossing rate $N_\varrho(r)$ and the average duration of fades $T_{\varrho_-}(r)$, we study the graphs depicted in Figures 6.36(a)–6.36(d). In Figures 6.36(a) and 6.36(b), the normalized level-crossing rate $N_\varrho(r)/f_{max}$, calculated according to (6.177), is presented for various values of the parameters m_3 and σ_3, respectively. The figures below, Figures 6.36(c) and 6.36(d), each show the behaviour of the corresponding normalized average duration of fades $T_{\varrho_-}(r) \cdot f_{max}$.

6.4.2 The Deterministic Modified Loo Model

For the derivation of a proper simulation model for modified Loo processes, we proceed as in Subsection 6.1.4. That means, we replace the three stochastic Gaussian random processes $\nu_i(t)$ $(i = 1, 2, 3)$ by deterministic Gaussian processes $\tilde{\nu}_i(t)$ of the form (6.68). When constructing the sets $\{f_{1,n}\}$, $\{f_{2,n}\}$, and $\{f_{3,n}\}$, one has to take care that they are mutually disjoint (mutually exclusive), which leads to the fact that the resulting deterministic Gaussian processes $\tilde{\nu}_1(t)$, $\tilde{\nu}_2(t)$, and $\tilde{\nu}_3(t)$ are in pairs uncorrelated. The substitution $\nu_i(t) \to \tilde{\nu}_i(t)$ leads to $\mu_i(t) \to \tilde{\mu}_i(t)$, where the deterministic Gaussian processes $\tilde{\mu}_i(t)$ $(i = 1, 2, 3)$ can be expressed after a short side calculation as follows:

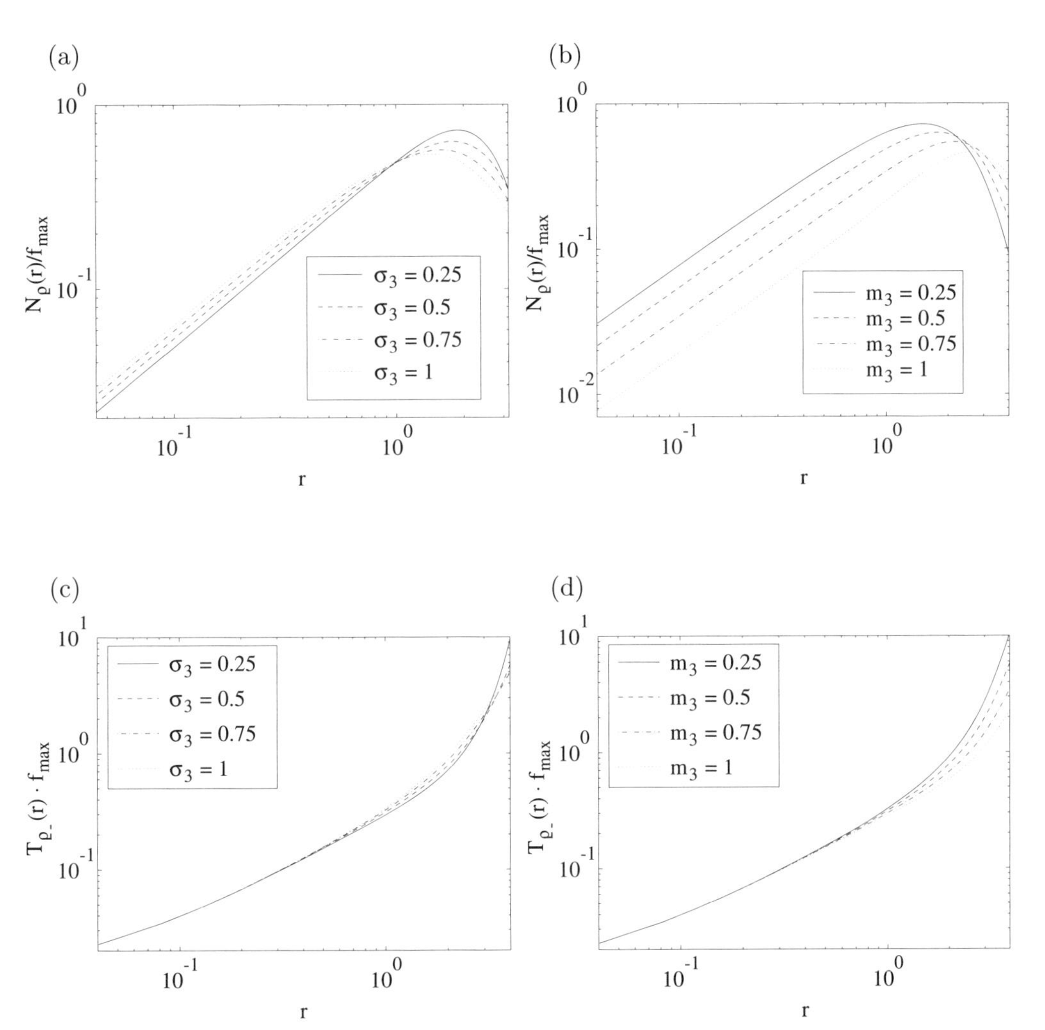

Figure 6.36: Normalized level-crossing rate $N_\varrho(r)/f_{max}$ of the modified Loo model for various values of (a) m_3 ($\sigma_3 = 1/2$) and (b) σ_3 ($m_3 = 1/2$) as well as (c) and (d) the corresponding normalized average duration of fades $T_{\varrho_-}(r) \cdot f_{max}$ ($\kappa_1 = \kappa_2 = 1,\ \psi_0 = 1,\ f_\rho = 0$).

$$\tilde{\mu}_1(t) = \sum_{n=1}^{N_1} c_{1,n} \cos(2\pi f_{1,n} t + \theta_{1,n}) + \sum_{n=1}^{N_2} c_{2,n} \cos(2\pi f_{2,n} t + \theta_{2,n}), \quad (6.181a)$$

$$\tilde{\mu}_2(t) = \sum_{n=1}^{N_1} c_{1,n} \sin(2\pi f_{1,n} t + \theta_{1,n}) - \sum_{n=1}^{N_2} c_{2,n} \sin(2\pi f_{2,n} t + \theta_{2,n}), \quad (6.181b)$$

$$\tilde{\mu}_3(t) = \sigma_3 \sum_{n=1}^{N_3} c_{3,n} \cos(2\pi f_{3,n} t + \theta_{3,n}) + m_3 . \quad (6.181c)$$

With these relations, the stochastic reference model (see Figure 6.32) can be directly transformed into the *deterministic Loo model*, shown in Figure 6.37. The output process $\tilde{\varrho}(t)$ of this model is mnemonically named *deterministic modified Loo process*.

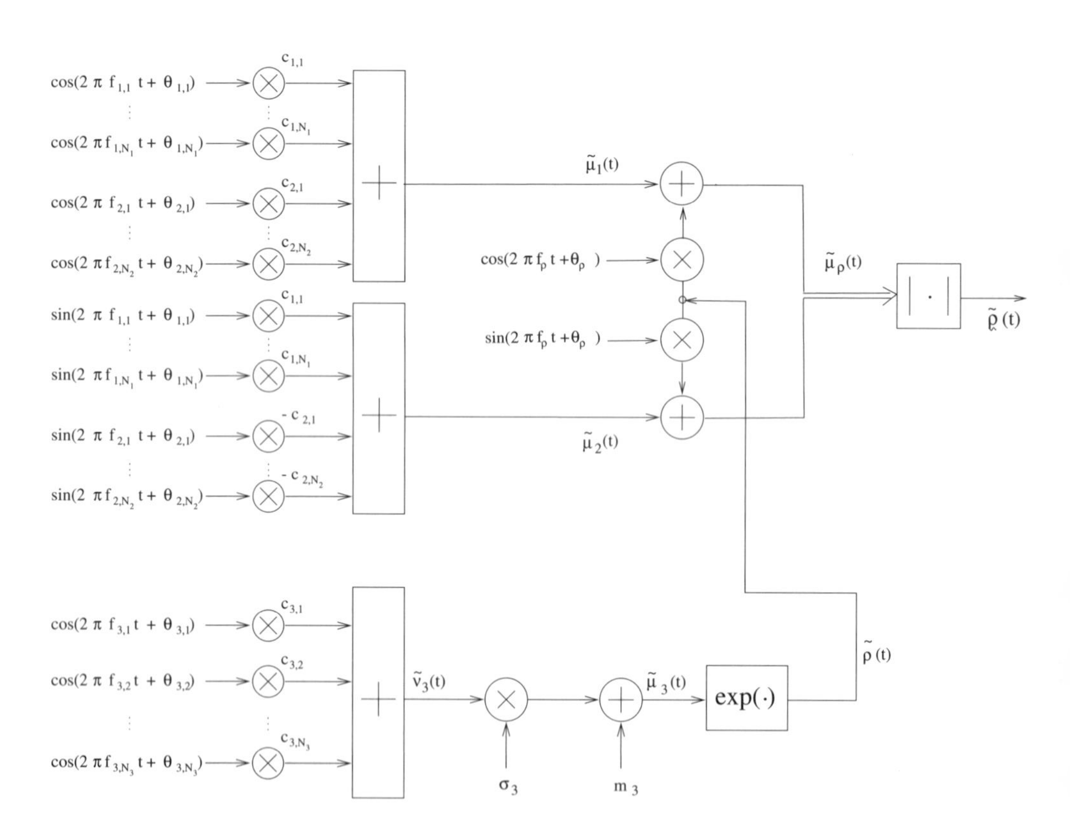

Figure 6.37: The deterministic modified Loo model (simulation model).

For the special case (i), introduced in (6.155a), the structure of the so-called deterministic classical Loo model follows from Figure 6.37. Here, each of the two deterministic processes $\tilde{\mu}_1(t)$ and $\tilde{\mu}_2(t)$, given by (6.181a) and (6.181b), respectively, can be replaced by the fundamental relation (4.4), as a result of which the realization expenditure of the model reduces considerably. The special case (ii) [see (6.155b)] also leads to a simplification of the structure of the simulation model, because $\sigma_2^2 = 0$ is equivalent to $N_2 = 0$. It should be noted that, for this case, one obtains the simulation model introduced in [Pae98c]. Finally, we want to point out that in the special case (6.155c), the structure of the deterministic Rice process with cross-correlated components follows from Figure 6.37, as we have already discovered from the top part of Figure 6.9.

On condition that $N_i \geq 7$ holds, the equations (6.168), (6.177), and (6.179) derived for the reference model in Subsection 6.4.1 approximately also hold for deterministic modified Loo processes $\tilde{\varrho}(t)$, if the substitutions $\psi_0 \to \tilde{\psi}_0$, $\ddot{\psi}_0 \to \ddot{\tilde{\psi}}_0$, and $\dot{\phi}_0 \to \dot{\tilde{\phi}}_0$ are performed in the formulae concerned. Here, the characteristic quantities $\tilde{\psi}_0$, $\ddot{\tilde{\psi}}_0$, and $\dot{\tilde{\phi}}_0$ of the simulation model are given by the relations (6.71a), (6.71b), and (6.71c) derived in Subsection 6.1.4, respectively. This is not particularly surprising because here as well as in Subsection 6.1.4, the deterministic Gaussian processes $\tilde{\mu}_1(t)$ and $\tilde{\mu}_2(t)$ are based on the same expressions. Differences, however, only occur in the calculation of the model parameters $f_{i,n}$ and $c_{i,n}$ for $i = 1, 2$. In the present case, we have to take into account that the Jakes power spectral density is in general left-hand side restricted as well as right-hand side restricted, due to $\kappa_1 \in [0, 1]$ and $\kappa_2 \in [0, 1]$. If we take this fact into account, when calculating the model parameters $f_{i,n}$ and $c_{i,n}$, by means of the method of exact Doppler spread (MEDS), then the following expressions hold for the deterministic modified Loo model:

$$f_{i,n} = f_{max} \sin\left[\frac{\pi}{2N_i'}\left(n - \frac{1}{2}\right)\right], \quad n = 1, 2, \ldots, N_i \quad (i = 1, 2), \tag{6.182a}$$

$$c_{i,n} = \frac{\sigma_i}{\sqrt{N_i'}}, \quad n = 1, 2, \ldots, N_i \quad (i = 1, 2), \tag{6.182b}$$

where

$$N_i' = \left\lceil \frac{N_i}{\frac{2}{\pi}\arcsin(\kappa_i)} \right\rceil, \quad i = 1, 2, \tag{6.183}$$

describes the virtual number of harmonic functions, and N_i again denotes the actual number, i.e., the number of harmonic functions set by the user. For the Doppler phases $\theta_{i,n}$, we assume as usual that these quantities are outcomes (realizations) of a random generator uniformly distributed in the interval $(0, 2\pi]$.

The design of the third deterministic Gaussian process $\tilde{\nu}_3(t)$ is performed exactly according to the method described in Subsection 6.1.4. In particular, the calculation of the discrete Doppler frequencies $f_{3,n}$ is carried out by means of the relation (6.75a) in connection with (6.75b), and for the Doppler coefficients $c_{i,n}$ the formula $c_{3,n} = \sqrt{2/N_3}$ again holds for all $n = 1, 2, \ldots, N_3$. The remaining parameters of the simulation model $(f_\rho, \theta_\rho, m_3, \sigma_3)$ of course correspond to those of the reference model, so that all parameters are now determined.

With the characteristic quantities $\tilde{\psi}_0$, $\ddot{\tilde{\psi}}_0$, and $\dot{\tilde{\phi}}_0$, the secondary model parameters of the simulation model

$$\tilde{\alpha} = \left(2\pi f_\rho - \frac{\dot{\tilde{\phi}}_0}{\tilde{\psi}_0}\right) \Big/ \sqrt{2\tilde{\beta}} \tag{6.184}$$

and

$$\tilde{\beta} = -\ddot{\tilde{\psi}}_0 - \dot{\tilde{\phi}}_0^2/\tilde{\psi}_0 \tag{6.185}$$

can be explicitly calculated similarly to (6.27) and (6.28). The convergence behaviour of $\tilde{\alpha}$ and $\tilde{\beta}/f_{max}^2$ is depicted in terms of N_i in Figure 6.38(a) and 6.38(b), respectively. The graphs shown hold for the primary model parameters σ_1, σ_2, κ_1, κ_2, and f_ρ presented in Table 6.5.

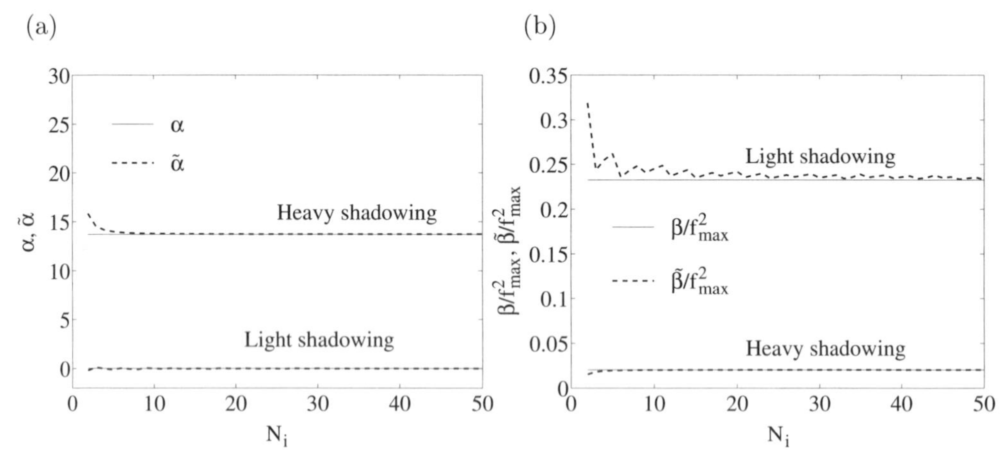

Figure 6.38: Illustration of (a) α and $\tilde{\alpha}$ as well as (b) β/f_{max}^2 and $\tilde{\beta}/f_{max}^2$ when using the MEDS with $N_1 = N_2$ but $N_1' \neq N_2'$ (σ_i, κ_i, and f_ρ according to Table 6.5).

Figure 6.39 is an example of the time behaviour of the deterministic Loo process $\tilde{\varrho}(t)$ (continuous line), where the values $N_1 = N_2 = N_3 = 13$ were chosen for the number of harmonic functions N_i ($i = 1, 2, 3$), and the maximum Doppler frequency f_{max} was again fixed by $f_{max} = 91$ Hz. This figure also illustrates the behaviour of the deterministic lognormal process $\tilde{\rho}(t)$ (dotted line).

A comparison between the statistical properties of the reference model and those of the simulation model is shown in Figures 6.40(a)–6.40(c). Except for the parameter $\kappa_c = f_{max}/f_c$, whose influence will be investigated here, all parameters of the simulation model and of the reference model were chosen exactly as in the previous example. The sampling interval T_s of the discrete deterministic Loo process $\tilde{\varrho}(kT_s)$ ($k = 1, 2, \ldots, K$) was given by $T_s = 1/(36.63 f_{max})$. Altogether $K = 3 \cdot 10^7$ sampling values of the process

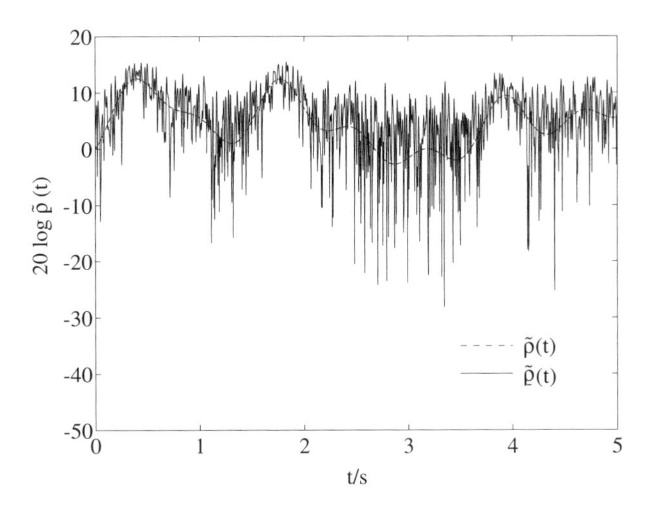

Figure 6.39: The deterministic processes $\tilde{\varrho}(t)$ and $\tilde{\rho}(t)$ ($\sigma_1^2 = \sigma_2^2 = 1$, $\kappa_1 = 0.8$, $\kappa_2 = 0.5$, $\sigma_3 = 0.5$, $m_3 = 0.25$, $f_\rho = 0.2 f_{max}$, $\theta_\rho = 0$, $\kappa_c = 50$, and $f_{max} = 91\,\text{Hz}$).

$\tilde{\varrho}(kT_s)$ ($k = 1, 2, \ldots, K$) have been simulated and used for the determination of the probability density function $\tilde{p}_\varrho(z)$ [see Figure 6.40(a)], the normalized level-crossing rate $\tilde{N}_\varrho(r)/f_{max}$ [see Figure 6.40(b)], and the normalized average duration of fades $\tilde{T}_{\varrho_-}(r) \cdot f_{max}$ [see Figure 6.40(c)] of the simulation model.

Figure 6.40(a) makes us recognize that the behaviour of the probability density function $\tilde{p}_\varrho(z)$ is not influenced by the quantity κ_c. This result was to be expected, because $\tilde{p}_\varrho(z)$ is, according to (6.168), independent of the bandwidth of the process $\nu_3(t)$, which completely explains the missing influence of the frequency ratio $\kappa_c = f_{max}/f_c$. The minor differences that can be observed between $p_\varrho(z)$ and $\tilde{p}_\varrho(z)$, are due to the limited numbers of harmonic functions, which were here equal to $N_i = 13$ for $i = 1, 2, 3$. It does not need to be explicitly emphasized that these deviations decrease if N_i increases, and that they converge against zero as $N_i \to \infty$.

The results of Figure 6.40(b) show us that the deviations between the level-crossing rate of the reference model and that of the simulation model are only relatively high for unrealistically small values of κ_c, i.e., $\kappa_c \leq 5$. On the contrary, for $\kappa_c \geq 20$ the differences between the analytical approximate solution (6.177) and the corresponding simulation results can be ignored.

Studying Figure 6.40(c), we notice that the same statements also hold for the average duration of fades. Consequently, the approximate solutions derived for this model for the level-crossing rate and the average duration of fades are very exact, provided that the frequency ratio $\kappa_c = f_{max}/f_c$ is greater than or equal to 20, i.e., if the amplitude of the line-of-sight component changes relatively slowly compared to the amplitude variations of the scattered component. We do not need to be afraid of a restriction connected with the boundary condition $\kappa_c \geq 20$ for practically relevant cases, because for real-world channels, $\kappa_c \gg 1$ holds anyway.

6.4.3 Applications and Simulation Results

In this subsection, we want to fit the statistic properties of the modified Loo model to the statistics of real-world channels. Just as for the extended Suzuki process of Type I and Type II as well as for the generalized Rice process, we here also use the measurement results presented in [But83] as a basis for the complementary cumulative distribution function, the level-crossing rate, and the average duration of fades.

In the following, we will choose the realistic value $\kappa_c = 20$ for the frequency ratio $\kappa_c = f_{max}/f_c$, so that the level-crossing rate $N_\varrho(r)$ of the Loo model is approximated by (6.177) very well. Without restriction of generality, we will also set the phase θ_ρ of the line-of-sight component to the arbitrary value $\theta_\rho = 0$.

The remaining free model parameters of the modified Loo model are the quantities σ_1, σ_2, κ_1, κ_2, σ_3, m_3, and f_ρ, which are set for the model fitting procedure. With these primary model parameters, we defined the parameter vector

$$\mathbf{\Omega} := \left(\sigma_1,\ \sigma_2,\ \kappa_1,\ \kappa_2,\ \sigma_3,\ m_3,\ f_\rho\right), \tag{6.186}$$

whose components are to be optimized according to the scheme described in Subsection 6.1.5. In order to minimize the error function $E_2(\mathbf{\Omega})$ [cf. (6.76)], we again make use of the Fletcher-Powell algorithm [Fle63]. The optimized components of the parameter vector $\mathbf{\Omega}$ obtained in this way are presented in Table 6.5 for areas with light and heavy shadowing.

Table 6.5: The optimized primary model parameters of the modified Loo model for areas with light and heavy shadowing.

Shadowing	σ_1	σ_2	κ_1	κ_2	σ_3	m_3	f_ρ/f_{max}
heavy	0	0.3856	0	0.499	0.5349	-1.593	0.1857
light	0.404	0.4785	0.6223	0.4007	0.2628	-0.0584	0.0795

For the modified Loo model, the Rice factor c_R is calculated as follows:

$$\begin{aligned} c_R &= \frac{E\left\{|m(t)|^2\right\}}{E\left\{|\mu(t)|^2\right\}} = \frac{E\left\{\varrho^2(t)\right\}}{2E\left\{\mu_i^2(t)\right\}} \qquad (i = 1, 2) \\ &= \frac{r_{\varrho\varrho}(0)}{2\psi_0} = \frac{\pi}{2} \cdot \frac{e^{2(m_3+\sigma_3^2)}}{\sum_{i=1}^{2} \sigma_i^2 \arcsin(\kappa_i)} \, . \end{aligned} \tag{6.187}$$

Thus, with the parameters taken from Table 6.5, the Rice factor c_R is $c_R = 1.7\,\text{dB}$ for heavy shadowing and $c_R = 8.96\,\text{dB}$ for light shadowing.

(a)

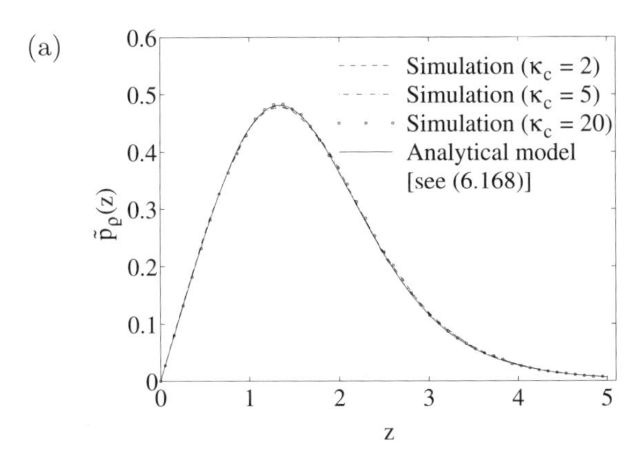

(b)

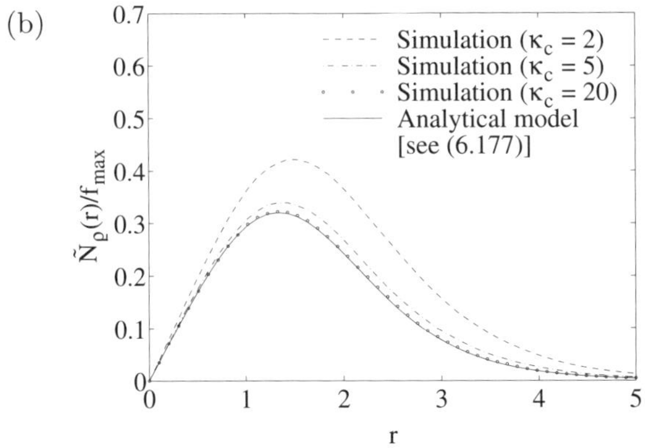

(c)

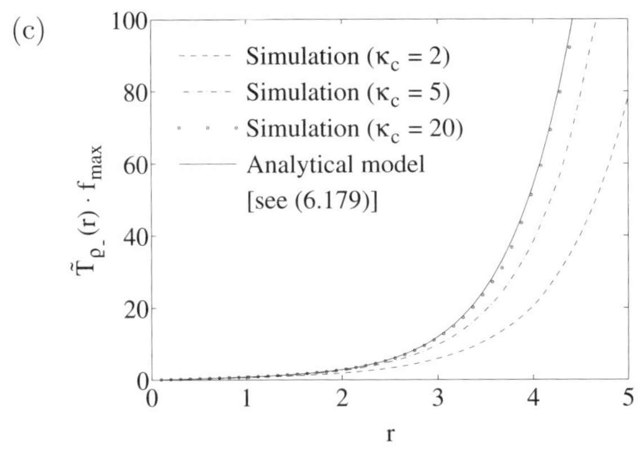

Figure 6.40: Comparisons between: (a) $p_\varrho(z)$ and $\tilde{p}_\varrho(z)$, (b) $N_\varrho(r)/f_{max}$ and $\tilde{N}_\varrho(r)/f_{max}$, as well as (c) $T_{\varrho_-}(r) \cdot f_{max}$ and $\tilde{T}_{\varrho_-}(r) \cdot f_{max}$ ($\sigma_1^2 = \sigma_2^2 = 1$, $\kappa_1 = 0.8$, $\kappa_2 = 0.5$, $\sigma_3 = 0.5$, $m_3 = 0.25$, $f_\rho = 0.2 f_{max}$, $\theta_\rho = 0$, and $f_{max} = 91\,\text{Hz}$).

In Figure 6.41(a), the complementary cumulative distribution function $F_{\varrho+}(r) = 1 - F_{\varrho-}(r)$ of the modified Loo model is depicted together with that of the measured channel for the areas with light shadowing and with heavy shadowing. Figure 6.41(b) makes it clear that the differences between the normalized level-crossing rate $N_{\varrho}(r)/f_{max}$ of the modified Loo model and the measured normalized level-crossing rate used here are acceptable.

Finally, Figure 6.41(c) shows the corresponding normalized average duration of fades. Here, an excellent agreement is again observable between the reference model and the measured channel.

For the verification of the analytical results, the corresponding simulation results are also depicted in Figures 6.41(a)–6.41(c). Therefore, the deterministic Gaussian processes $\tilde{\nu}_1(t)$, $\tilde{\nu}_2(t)$, and $\tilde{\nu}_3(t)$ were designed according to the method described in Subsection 6.4.2 by using $N_1 = N_2 = N_3 = 15$ cosine functions.

Finally, the deterministic modified Loo process $\tilde{\varrho}(t)$ is depicted in Figures 6.42(a) and 6.42(b) for areas with light shadowing and for areas with heavy shadowing, respectively.

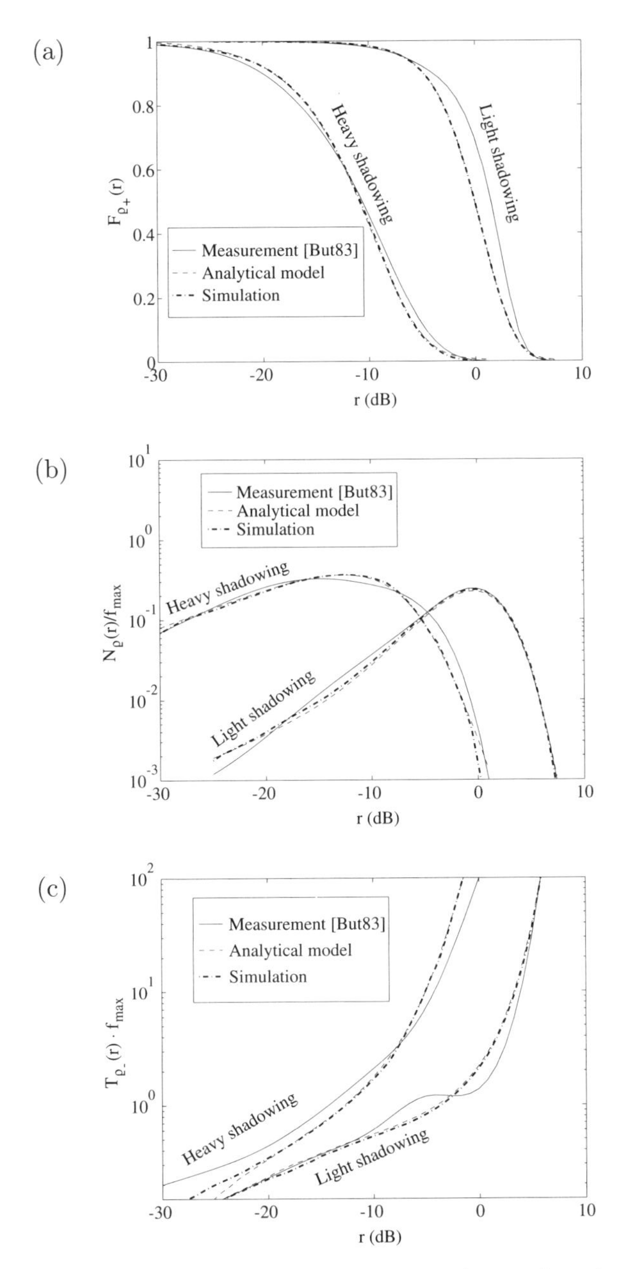

Figure 6.41: (a) Complementary cumulative distribution function $F_{\varrho_+}(r)$, (b) normalized level-crossing rate $N_{\varrho}(r)/f_{max}$, and (c) normalized average duration of fades $T_{\varrho_-}(r/\rho) \cdot f_{max}$ for areas with light and heavy shadowing.

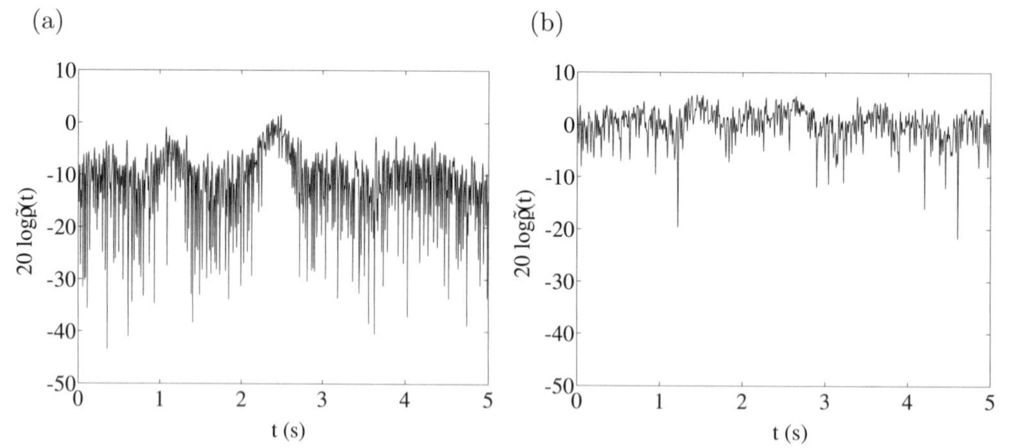

Figure 6.42: Simulation of deterministic modified Loo processes $\tilde{\varrho}(t)$ for areas with (a) heavy shadowing and (b) light shadowing (MEDS, $N_1 = N_2 = N_3 = 15$, $f_{max} = 91\,\text{Hz}$, $\kappa_c = 20$).

7

FREQUENCY-SELECTIVE STOCHASTIC AND DETERMINISTIC CHANNEL MODELS

So far, we have been exclusively concerned with the description of frequency-nonselective mobile radio channels, which are characterized by the fact that the differences between the propagation delays of the received electromagnetic waves can be ignored compared to the symbol interval. Obviously, this assumption is less and less justified the shorter the symbol interval or the higher the data rate becomes. Channels for which the propagation delay differences cannot be ignored in comparison with the symbol interval, therefore, represent a further important class of channels, namely the class of *frequency-selective channels*. The statistic and deterministic modelling of this class of channels is the topic of this chapter.

In the literature, the number of publications merely dealing with frequency-selective mobile radio channels has grown so much that even scientists mainly involved with this subject are running the risk of losing the overview. Therefore, it is here impossible to mention every author who has made a contribution to this subject, particularly since this introduction can only present a small selection of publications. In order to organize this group of themes systematically, it is sensible to make a rough classification of the publications, dividing them into the following categories: theory, measurement, and simulation.

In the first category belong works in which the description and analysis of mobile radio channels are mainly treated theoretically. The most important article in this category is indisputably [Bel63]. In this fundamental work on stochastic time-variant linear systems, Bello introduces the WSSUS[1] model that is employed almost exclusively for the description of frequency-selective mobile radio channels. With this model, the input-output behaviour of mobile radio channels can be described in the equivalent complex baseband in a relatively simple manner, since the channel

---

[1] WSSUS: wide-sense stationary uncorrelated scattering

is assumed to be quasi-stationary during the observation time interval. Empirically grounded statements have shown [Par82] that the assumption of quasi-stationarity is justified, when the mobile unit covers a distance in the order of less than a few tens of the wavelength of the carrier signal. The articles [Par82, Ste87, Lor85] give a good overview of the most important characteristics of time-variant channels, to which naturally mobile radio channels also belong. Books discussing this subject are, for example, [Ste94, Stu96, Pro95] and [Jun97]. A deep insight into the theory of WSSUS models can be obtained by studying [Fle90]. A detailed analysis of WSSUS models can also be found in [Sad98] for instance, where correlation functions as well as scattering functions are derived on the assumption of non-uniformly distributed angles of arrival. The article [Fle96] gives an overview of the state of research in the field of channel modelling, carried out until 1996 by European research projects such as COST 207, RACE CODIT, and RACE ATDMA. In the meantime, intensive research on spatial-temporal channel models for future mobile radio systems with adaptive antennas has been carried out [Lib99]. Detailed articles giving an overview with many references concerning this subject are the publications [Ert98] and [Mar99].

In the second category belong publications reporting on experimental measurement results of mobile radio channels as well as works treating the technology of mobile radio channel measurement. Certainly, the works by Young [You52], NyLund [Nyl68], Cox [Cox72, Cox73], Nielson [Nie78], as well as by Bajwa and Parsons [Baj82] belong to the pioneering works in the field of channel measurements. The subject of channel measuring is treated in an easily understandable manner in the overview article by Andersen et al. [And95]. In this publication, mobile radio channels are divided into classes depending on the environment, and typical measured characteristic quantities for different propagation scenarios are given. Particularly in connection with the measurement of system functions of mobile radio channels, the papers [Wer91] and [Kat95] are interesting. For the measurement of the propagation properties of mobile radio channels, special measuring devices are required. They are denoted as channel sounders. At the Telecommunications Institute of the University of Erlangen–Nuremberg, Germany, the three channel sounders RUSK 400, RUSK 5000, and RUSK X have been developed [Mar94a] as part of research projects in association with the Deutsche Telekom in Darmstadt, Germany. Detailed information on the principle of the applied measurement methods of the channel sounders can be found in [Mar92, Mar94a, Mar94b]. Results of measurement campaigns performed with the device type RUSK 5000 are, for example, reported in [Kad91, Goe92a, Goe92b]. The channel sounders RUSK 400 and RUSK 5000 have been produced merely as prototypes, whereas RUSK X and the succeeding models RUSK SX and RUSK WLL have for some time already been commercially sold by the company MEDAV GmbH. Recently, the device RUSK ATM also became part of the RUSK channel sounders' family. This device arose within the framework of the project ATMmobil, which was promoted by the BMFT and supported by the company MEDAV [Tho99]. With this vector channel sounder, in particular directional resolved measurements of mobile radio channels can be carried out in the frequency band of 5 GHz up to 6 GHz. A further channel sounder named SIMOCS 2000 has been produced by the Siemens AG in Munich, Germany. The principle of the functionality of SIMOCS 2000 is described in [Fel94] and [Jun97]. Furthermore, it should be mentioned that Zollinger [Zol93] has developed a channel sounder at the ETH (Swiss Federal Institute of Technology Zurich) in Zurich, Switzerland. The channel sounder ECHO 24 (**ETH Ch**annel **So**under operating at **24** GHz) also traces its descent from the ETH. With this channel sounder, complex channel impulse responses can be measured in buildings with a temporal resolution of 2 ns [Hed99].

In the third and last category finally belong works stressing the development of simulation models of frequency-selective mobile radio channels. Concerning the method of realization of these so-called channel simulators, one distinguishes between hardware realization and software realization. Hardware realizations can in addition be divided into analog and digital channel simulators. Analog channel simulators (e.g., [Cap80, Ber86]) model the channel in the high-frequency band or in the intermediate-frequency band, where surface acoustic wave (SAW) filters are applied to realize different multipath propagation delays. Digital channel simulators in general perform all arithmetic operations that become necessary in the complex baseband in real-time using digital signal processors [Sch89] or vector processors (e.g., [Ehr82, Sch90]). In most applications, however, the channel simulation does not take place under real-time conditions, but is performed on a workstation or a PC. As design methods for the required algorithms, in principle both the filter method (e.g., [Fec93a, Lau94]) and the Rice method (e.g., [Schu89, Hoe90, Hoe92, Cre95, Yip95, Pae95b]) are possible. Incidentally, both of these methods are eligible for the design of channel simulators for mobile communication systems with frequency-hopping capabilities, which has been shown by using the filter method in [Lam97] and by using the Rice method in [Pae97b].

The present Chapter 7 is structured as follows. In order to illustrate the path geometry for multipath fading channels with different propagation delays, we will at first describe the ellipses model introduced by Parsons and Bajwa [Par82] in Section 7.1. In Section 7.2, we will be concerned with the system theoretical description of frequency-selective channels. In this context, we will discuss four system functions introduced by Bello [Bel63]. It will be pointed out how to get various insights with these system functions into the input-output behaviour of linear time-variant systems. Section 7.3 contains a description of the theory of frequency-selective stochastic channel models, which is also going back to Bello [Bel63]. Here, the WSSUS channel model is of central importance. In particular, stochastic system functions as well as the characteristic quantities derivable from these will be introduced to characterize the statistical properties of WSSUS channel models. These models are also suitable for modelling channels specified by the European working group COST 207 [COS86, COS89]. The description of the COST 207 channel models is the topic of Subsection 7.3.3. Section 7.4 is dedicated to the mathematical description of frequency-selective deterministic channel models. The mathematical methods applied in this section are an extension of the theory of deterministic processes introduced in Chapter 4. Finally, Chapter 7 ends with the derivation of deterministic simulation models for the channel models according to COST 207.

7.1 THE ELLIPSES MODEL OF PARSONS AND BAJWA

During the transmission of data, the emitted electromagnetic waves are influenced by a multitude of various obstacles. Depending on the geometric dimensions and the electromagnetic properties of these obstacles, one can distinguish between reflected waves, scattered waves, diffracted waves, and absorbed waves. For our purposes, a strict distinction between reflection, scattering, and diffraction is not as useful as the exact knowledge of the location and the consistency of each individual obstacle. Here, it is sufficient to merely speak of scattering, and — for the sake of simplicity — to introduce elliptical scattering zones, which lead us to the ellipses model of Parsons and Bajwa [Par82] (see also [Par89] and [Par92]) shown in Figure 7.1. All ellipses are confocal, i.e., they have common focal points Tx and Rx, which in our case coincide with the position of the transmitter (Tx) and the receiver (Rx). As is well known, the ellipse is the set of all points for which the sum of the distances to the focal points Tx and Rx is equal. Referring to Figure 7.1, this means that the propagation paths $Tx - A - Rx$ and $Tx - C - Rx$ have the same path length. However, the respective angles of arrival are different, and, consequently, the corresponding Doppler frequencies, caused by the movement of the transmitter (receiver), are also different. The exact opposite occurs for the (multipath) propagation paths $Tx - A - Rx$ and $Tx - B - Rx$, where the path lengths are different, but the angles of arrival are equal, and, thus, the Doppler frequencies are equal too.

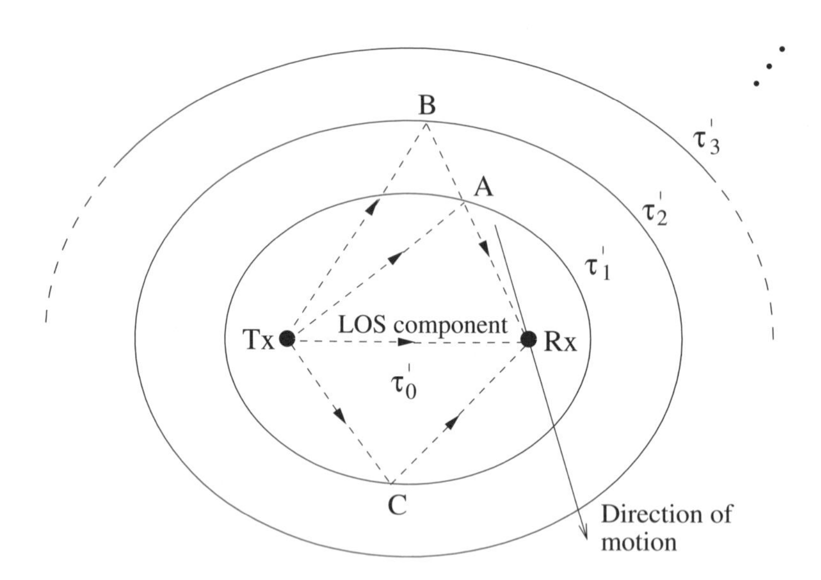

Figure 7.1: The ellipses model describing the path geometry according to Parsons and Bajwa [Par82].

The path length of each wave determines the propagation delay and essentially also the average power of the wave at the antenna of the receiver. Every wave in the scattering zone characterized by the ℓth ellipses undergoes the same *discrete propagation delay*

$$\tau'_\ell = \tau'_0 + \ell\,\Delta\tau', \quad \ell = 0, 1, \ldots, \mathcal{L} - 1, \tag{7.1}$$

where τ'_0 is the propagation delay of the line-of-sight (LOS) component, $\Delta\tau'$ is an infinitesimal propagation delay, and $\mathcal{L}$ denotes the number of paths with different propagation delays. It is evident that the ellipses model increases in precision if $\mathcal{L}$ increases and $\Delta\tau'$ becomes smaller. In the limit $\mathcal{L} \to \infty$ and $\Delta\tau' \to 0$, the discrete propagation delay τ'_ℓ results in the *continuous propagation delay* τ' restricted to the interval $[\tau'_0, \tau'_{max}]$. Here, τ'_{max} characterizes the maximum propagation delay, which depends on the environment. The maximum propagation delay τ'_{max} is chosen in such a way that the contributions of the scattered components with propagation delays τ' greater than τ'_{max} can be ignored.

In the following discussion, we will see that the ellipses model forms to a certain extent the physical basis for the modelling of frequency-selective channels. In particular, the number of paths $\mathcal{L}$ with different propagation delays exactly corresponds to the number of delay elements required for the tapped-delay-line structure of the time-variant filter used for modelling frequency-selective channels. In order to achieve an economical realization, $\mathcal{L}$ should be kept as small as possible.

7.2 SYSTEM THEORETICAL DESCRIPTION OF FREQUENCY-SELECTIVE CHANNELS

Using the system functions introduced by Bello [Bel63], the input and output signals of frequency-selective channels can be related to each other in different ways. The starting point for the derivation of the system functions is based on the assumption that the channel is a linear time-variant system in the equivalent complex baseband. In time-variant systems, the impulse response — denoted by $h_0(t_0, t)$ — is a function of the time t_0 at which the channel has been excited by the impulse $\delta(t - t_0)$, and the time t at which the effect of the impulse is observed at the output of the channel. The relation between the impulse $\delta(t - t_0)$ and the corresponding impulse response $h_0(t_0, t)$ can therefore be expressed by

$$\delta(t - t_0) \to h_0(t_0, t)\,. \tag{7.2}$$

Since every physical channel is causal, the impulse cannot produce an effect before the impulse has excited the channel. This is the so-called *law of causality* that can be expressed by

$$h_0(t_0, t) = 0 \quad \text{for} \quad t < t_0\,. \tag{7.3}$$

Using the impulse response $h_0(t_0, t)$, we now want to compute the output signal $y(t)$ of the channel for an arbitrary input signal $x(t)$. For this purpose, we at first represent $x(t)$ as an infinite densely superposition of weighted delta functions. By applying the sifting property of delta functions, this leads us to

$$x(t) = \int_{-\infty}^{\infty} x(t_0)\,\delta(t - t_0)\,dt_0\,. \tag{7.4}$$

Alternatively, we can also use the expression

$$x(t) = \lim_{\Delta t_0 \to 0} \sum_{t_0} x(t_0)\, \delta(t - t_0)\, \Delta t_0 \,. \tag{7.5}$$

Since the channel was assumed to be linear, we may employ the principle of superposition [Lue90]. Hence, by using (7.2), the response to the sum in (7.5) can be written as

$$\sum_{t_0} x(t_0)\, \delta(t - t_0)\, \Delta t_0 \to \sum_{t_0} x(t_0)\, h_0(t_0, t)\, \Delta t_0 \,. \tag{7.6}$$

For the desired relation

$$x(t) \to y(t) \,, \tag{7.7}$$

we obtain the following result from (7.6) in the limit $\Delta t_0 \to 0$

$$\int_{-\infty}^{\infty} x(t_0)\, \delta(t - t_0)\, dt_0 \to \int_{-\infty}^{\infty} x(t_0)\, h_0(t_0, t)\, dt_0 \,. \tag{7.8}$$

If we now make use of the causality property (7.3), then the output signal is given by

$$y(t) = \int_{-\infty}^{t} x(t_0)\, h_0(t_0, t)\, dt_0 \,. \tag{7.9}$$

Next, we substitute the variable t_0 by the propagation delay

$$\tau' = t - t_0 \,, \tag{7.10}$$

which defines the time elapsed from the moment at which the channel was excited by the impulse to the moment at which the response was observed at the output of the channel. Substituting t_0 by $t - \tau'$ in (7.9) results in

$$y(t) = \int_{0}^{\infty} x(t - \tau')\, h(\tau', t)\, d\tau' \,. \tag{7.11}$$

In order to simplify the notation, the time-variant impulse response $h_0(t - \tau', t)$ has been replaced by $h(\tau', t) := h_0(t - \tau', t)$. Physically, the time-variant impulse response $h(\tau', t)$ can be interpreted as the response of the channel at the time t to a delta impulse that stimulated the channel at the time $t - \tau'$. Considering (7.10), the causality property (7.3) can be expressed by

$$h(\tau', t) = 0 \quad \text{for} \quad \tau' < 0 \,. \tag{7.12}$$

From (7.11), we now directly obtain the tapped-delay-line model shown in Figure 7.2 of a frequency-selective channel with the time-variant impulse response $h(\tau', t)$. Note that the tapped-delay-line model can be interpreted as transversal filter with time-variant coefficients.

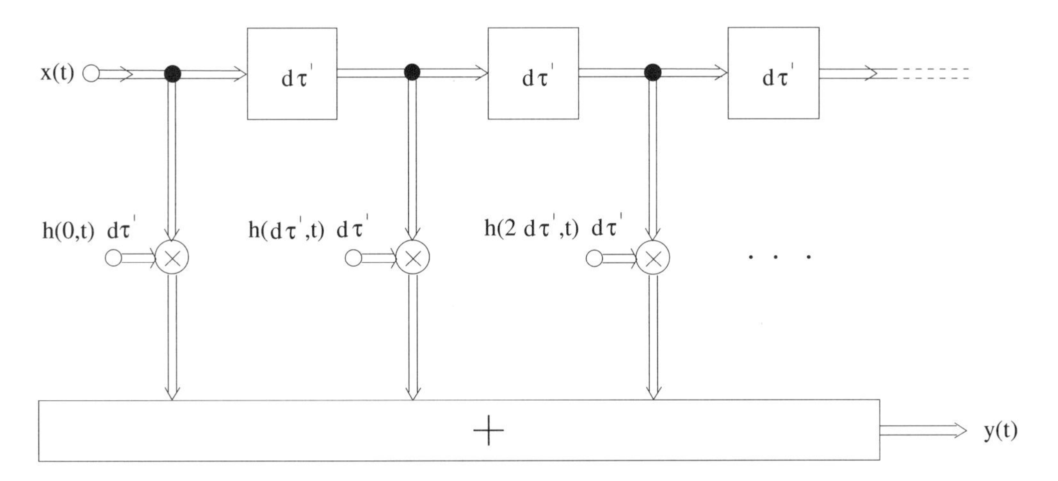

Figure 7.2: Tapped-delay-line representation of a frequency-selective and time-variant channel in the equivalent complex baseband.

Modelling mobile radio channels by using a tapped-delay-line structure with time-variant coefficients gives a deep insight into the channel distortions caused by scattering components with different propagation delays. Thus, it is recognizable that the received signal, for example, is composed of an infinite number of delayed and weighted replicas of the transmitted signal. In digital data transmission this causes intersymbol interferences (ISI) which have to be eliminated as best as possible in the receiver, e.g., by using equalizers. Moreover, the close relation between the tapped-delay-line structure of the channel model and the ellipses model, described in the previous section, becomes obvious.

The *time-variant transfer function* $H(f',t)$ of the channel is defined by the Fourier transform of the time-variant impulse response $h(\tau',t)$ with respect to the propagation delay variable τ', i.e.,

$$H(f',t) := \int_0^\infty h(\tau',t)\, e^{-j2\pi f'\tau'}\, d\tau', \tag{7.13}$$

which is expressed symbolically by $h(\tau',t) \overset{\tau' \quad f'}{\circ\!\!-\!\!\bullet} H(f',t)$. Here, we realize that $H(f',t)$ only fulfils the condition $H^*(f',t) = H(-f',t)$ if $h(\tau',t)$ is a real-valued function. Starting from (7.11), we can represent the input-output relation with $H(f',t)$ as follows

$$y(t) = \int_{-\infty}^{\infty} X(f')\, H(f',t)\, e^{j2\pi f't}\, df', \tag{7.14}$$

where $X(f')$ is the Fourier transform of the input signal $x(t)$ at $f = f'$.

Now, let $x(t)$ be a complex oscillation of the form

$$x(t) = A\, e^{j2\pi f't}, \tag{7.15}$$

where A denotes a complex-valued constant, then it follows from (7.11)

$$y(t) = A \int_0^\infty h(\tau', t)\, e^{j2\pi f'(t-\tau')}\, d\tau' \,. \tag{7.16}$$

Using (7.13), we may also write

$$y(t) = A\, H(f', t)\, e^{j2\pi f' t} \,. \tag{7.17}$$

Therefore, the response of the channel can in this case be represented by the input signal (7.15) weighted by the time-variant transfer function $H(f', t)$. The form (7.17) consequently makes clear that the time-variant transfer function $H(f', t)$ can be measured directly in the relevant frequency band by sinusoidal excitation.

Neither the time-variant impulse response $h(\tau', t)$ nor the corresponding transfer function $H(f', t)$ allow an insight into the phenomena caused by the Doppler effect. In order to eliminate this disadvantage, we apply the Fourier transform on $h(\tau', t)$ with respect to the time variable t. In this way, we obtain a further system function

$$s(\tau', f) := \int_{-\infty}^{\infty} h(\tau', t)\, e^{-j2\pi f t}\, dt \,, \tag{7.18}$$

which is called the *Doppler-variant impulse response.*

Instead of (7.18), we may also write $h(\tau', t) \overset{t \;\; f}{\circ\!\!-\!\!\!-\!\!\bullet} s(\tau', f)$. Expressing the time-variant impulse response $h(\tau', t)$ by the inverse Fourier transform of $s(\tau', f)$, allows the representation of (7.11) in the form

$$y(t) = \int_0^\infty \int_{-\infty}^{\infty} x(t - \tau')\, s(\tau', f)\, e^{j2\pi f t}\, df\, d\tau' \,. \tag{7.19}$$

This relation shows that the output signal $y(t)$ can be represented by an infinite sum of delayed, weighted, and Doppler shifted replicas of the input signal $x(t)$. Signals delayed during transmission in the range of $[\tau', \tau' + d\tau')$ and affected by a Doppler shift within $[f, f + df)$ are weighted by the differential part $s(\tau', f)\, df\, d\tau'$. The Doppler-variant impulse response $s(\tau', f)$, therefore, explicitly describes the dispersive behaviour of the channel as a function of both the propagation delays τ' and the Doppler frequencies f. Consequently, the physical interpretation of $s(\tau', f)$ directly leads to the ellipses model shown in Figure 7.1.

A further system function, the so-called *Doppler-variant transfer function* $T(f', f)$, is defined by the two-dimensional Fourier transform of the time-variant impulse response $h(\tau', t)$ according to

$$T(f', f) := \int_{-\infty}^{\infty} \int_0^\infty h(\tau', t)\, e^{-j2\pi(ft + f'\tau')}\, d\tau'\, dt \,. \tag{7.20}$$

Due to (7.13) and (7.18), we may also write $T(f', f) \overset{f \;\; t}{\bullet\!\!-\!\!\!-\!\!\circ} H(f', t)$ or $T(f', f) \overset{f' \;\; \tau'}{\bullet\!\!-\!\!\!-\!\!\circ} s(\tau', f)$ for (7.20).

The computation of the Fourier transform of (7.11) with respect to the time variable t allows the representation of the spectrum $Y(f)$ of the output signal $y(t)$ in the form

$$Y(f) = \int_{-\infty}^{\infty} X(f - f') \, T(f - f', f') \, df' \, . \tag{7.21}$$

Finally, we exchange the frequency variables f and f' and obtain

$$Y(f') = \int_{-\infty}^{\infty} X(f' - f) \, T(f' - f, f) \, df \, . \tag{7.22}$$

This equation shows how a relation between the spectrum of the output-signal and the input-signal can be established by making use of the Doppler-variant transfer function $T(f', f)$. Regarding (7.22), it becomes obvious that the spectrum of the output signal can be interpreted as a superposition of an infinite number of Doppler shifted and filtered replicas of the spectrum of the input signal.

At the end of this section, we keep in mind that the four system functions $h(\tau', t)$, $H(f', t)$, $s(\tau', f)$, and $T(f', f)$ are related in pairs by the Fourier transform. The Fourier transform relationships established above are illustrated in Figure 7.3.

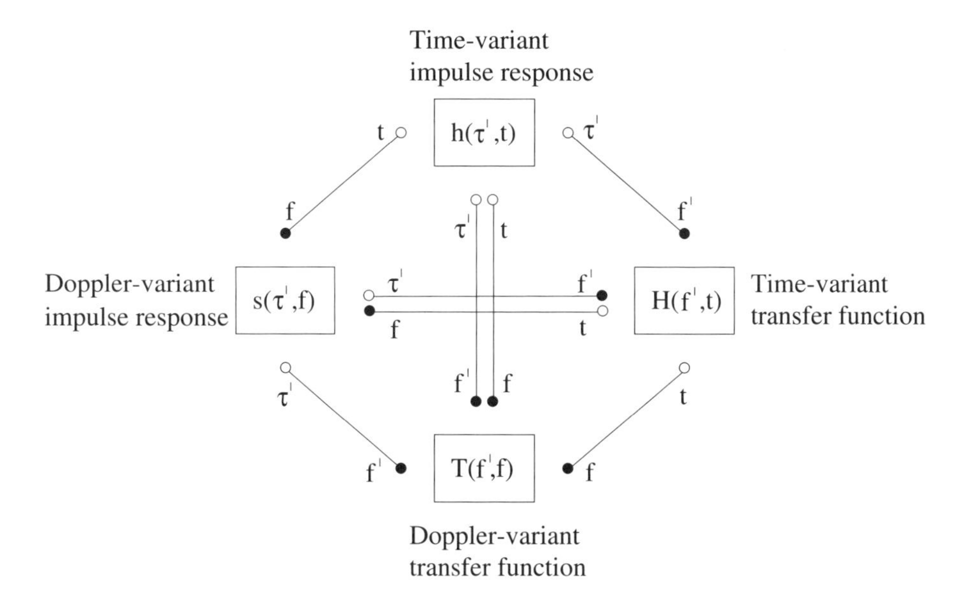

Figure 7.3: Fourier transform relationships between the system functions according to Bello [Bel63].

7.3 FREQUENCY-SELECTIVE STOCHASTIC CHANNEL MODELS

7.3.1 Correlation Functions

In the following, we consider the channel as a stochastic system. In this case, the four functions $h(\tau', t)$, $H(f', t)$, $s(\tau', f)$, and $T(f', f)$ are stochastic system functions. Generally, these stochastic system functions can be described by the following autocorrelation functions:

$$r_{hh}(\tau_1', \tau_2'; t_1, t_2) := E\{h^*(\tau_1', t_1)\, h(\tau_2', t_2)\}\,, \tag{7.23a}$$
$$r_{HH}(f_1', f_2'; t_1, t_2) := E\{H^*(f_1', t_1)\, H(f_2', t_2)\}\,, \tag{7.23b}$$
$$r_{ss}(\tau_1', \tau_2'; f_1, f_2) := E\{s^*(\tau_1', f_1)\, s(\tau_2', f_2)\}\,, \tag{7.23c}$$
$$r_{TT}(f_1', f_2'; f_1, f_2) := E\{T^*(f_1', f_1)\, T(f_2', f_2)\}\,. \tag{7.23d}$$

Since the system functions are related by the Fourier transform, it is not surprising that analog relations are also valid for the autocorrelation functions. For example, (7.23a) and (7.23b) are related by

$$\begin{aligned} r_{HH}(f_1', f_2'; t_1, t_2) &:= E\{H^*(f_1', t_1)\, H(f_2', t_2)\} \\ &= E\left\{\int_{-\infty}^{\infty} h^*(\tau_1', t_1)\, e^{j2\pi f_1' \tau_1'}\, d\tau_1' \int_{-\infty}^{\infty} h(\tau_2', t_2)\, e^{-j2\pi f_2' \tau_2'}\, d\tau_2'\right\} \\ &= \int_{-\infty}^{\infty}\int_{-\infty}^{\infty} E\{h^*(\tau_1', t_1)\, h(\tau_2', t_2)\}\, e^{j2\pi(f_1'\tau_1' - f_2'\tau_2')}\, d\tau_1'\, d\tau_2' \\ &= \int_{-\infty}^{\infty}\int_{-\infty}^{\infty} r_{hh}(\tau_1', \tau_2'; t_1, t_2)\, e^{j2\pi(f_1'\tau_1' - f_2'\tau_2')}\, d\tau_1'\, d\tau_2'\,. \end{aligned} \tag{7.24}$$

Finally, we replace the variable f_1' by $-f_1'$ on both sides of the last equation, to make clear that $r_{HH}(-f_1', f_2'; t_1, t_2)$ is the two-dimensional Fourier transform of $r_{hh}(\tau_1', \tau_2'; t_1, t_2)$ with respect to the two propagation delay variables τ_1' and τ_2'. This can be expressed symbolically by the notation $r_{hh}(\tau_1', \tau_2'; t_1, t_2) \overset{\tau_1', \tau_2' \quad f_1', f_2'}{\circ\!\!-\!\!\!-\!\!\bullet} r_{HH}(-f_1', f_2'; t_1, t_2)$. The Fourier transform relationships between all the other pairs of (7.23a)–(7.23d) can be derived in a similar way. As a result, one finds the relationships between the autocorrelation functions of the stochastic system functions shown in Figure 7.4.

In order to describe the input-output relation of the stochastic channel, we assume that the input signal $x(t)$ is a stochastic process with the known autocorrelation function $r_{xx}(t_1, t_2) := E\{x^*(t_1)\, x(t_2)\}$. As (7.11) is valid for deterministic systems as well as for stochastic systems, we can express the autocorrelation function $r_{yy}(t_1, t_2)$ of the output signal $y(t)$ as a function of $r_{xx}(t_1, t_2)$ and $r_{hh}(\tau_1', \tau_2'; t_1, t_2)$ as follows:

$$\begin{aligned} r_{yy}(t_1, t_2) &:= E\{y^*(t_1)\, y(t_2)\} \\ &= E\left\{\int_0^{\infty}\int_0^{\infty} x^*(t_1 - \tau_1')\, x(t_2 - \tau_2')\, h^*(\tau_1', t_1)\, h(\tau_2', t_2)\, d\tau_1'\, d\tau_2'\right\} \end{aligned}$$

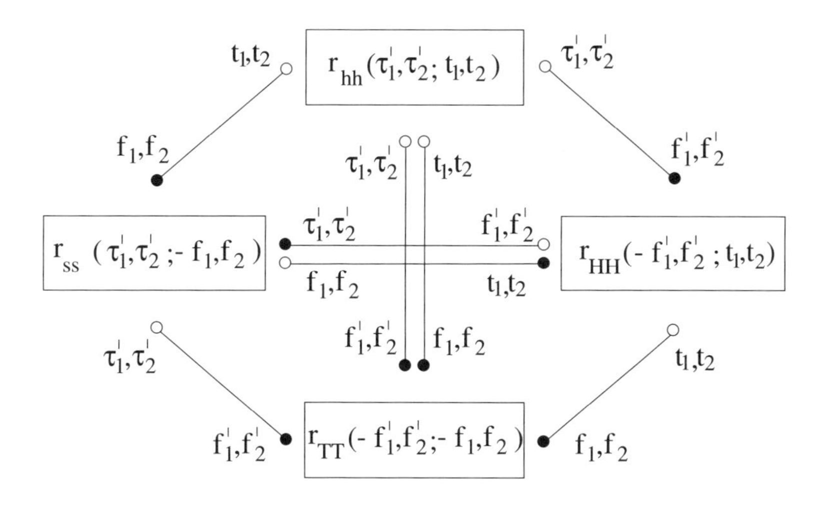

Figure 7.4: Relationships between the autocorrelation functions of the stochastic system functions.

$$= \int_0^\infty \int_0^\infty E\{x^*(t_1 - \tau_1')\, x(t_2 - \tau_2')\}\, E\{h^*(\tau_1', t_1)\, h(\tau_2', t_2)\}\, d\tau_1'\, d\tau_2'$$

$$= \int_0^\infty \int_0^\infty r_{xx}(t_1 - \tau_1'; t_2 - \tau_2')\, r_{hh}(\tau_1', \tau_2'; t_1, t_2)\, d\tau_1'\, d\tau_2' \,. \tag{7.25}$$

In the derivation above, we have implicitly assumed that the time-variant impulse response $h(\tau', t)$ of the channel and the input signal $x(t)$ are statistically independent.

Significant simplifications can be made by assuming that the time-variant impulse response $h(\tau', t)$ is stationary in the wide sense with respect to t, and that the scattering components with different propagation delays are statistically uncorrelated. Basing on these assumptions, Bello introduced the so-called WSSUS model in his essential work [Bel63] on stochastic time-variant linear systems. The description of the WSSUS model is the topic of the following subsection.

7.3.2 The WSSUS Model According to Bello

The WSSUS model enables the statistical description of the input-output relation of mobile radio channels for the transmission of bandpass signals in the equivalent complex baseband for observation periods in which the stationarity of the channel is ensured in the wide sense. According to empirical studies [Par82], the channel can be considered as wide-sense stationary as long as the mobile unit covers a distance in the dimension of a few tens of the wavelength of the carrier signal.

7.3.2.1 WSS Models

A channel model with a wide-sense stationary impulse response is called *WSS channel model* (WSS, wide-sense stationary). Instead of the term WSS channel model, we also use the short form *WSS model*, since it is evident that this model is exclusively used for modelling channels. The assumption of wide-sense stationarity leads to the fact that the two autocorrelation functions in (7.23a) and (7.23b) are invariant with respect to a translation of time, i.e., the autocorrelation functions $r_{hh}(\tau_1', \tau_2'; t_1, t_2)$ and $r_{HH}(f_1', f_2'; t_1, t_2)$ merely depend on the time difference $\tau := t_2 - t_1$. With $t_1 = t$ and $t_2 = t + \tau$, we can therefore write in case of WSS models:

$$r_{hh}(\tau_1', \tau_2'; t, t+\tau) = r_{hh}(\tau_1', \tau_2'; \tau), \tag{7.26a}$$

$$r_{HH}(f_1', f_2'; t, t+\tau) = r_{HH}(f_1', f_2'; \tau). \tag{7.26b}$$

The restricted behaviour of these two autocorrelation functions certainly has consequences for the remaining autocorrelation functions (7.23c) and (7.23d). To clarify this, we look at the Fourier transform relation between $r_{hh}(\tau_1', \tau_2'; t_1, t_2)$ and $r_{ss}(\tau_1', \tau_2'; f_1, f_2)$ which can, by considering Figure 7.4, be formulated as follows

$$r_{ss}(\tau_1', \tau_2'; f_1, f_2) = \int_{-\infty}^{\infty} \int_{-\infty}^{\infty} r_{hh}(\tau_1', \tau_2'; t_1, t_2)\, e^{j2\pi(t_1 f_1 - t_2 f_2)}\, dt_1\, dt_2\,. \tag{7.27}$$

The substitutions of the variables $t_1 \rightarrow t$ and $t_2 \rightarrow t + \tau$, in connection with (7.26a), result in

$$r_{ss}(\tau_1', \tau_2'; f_1, f_2) = \int_{-\infty}^{\infty} e^{-j2\pi(f_2 - f_1)t}\, dt \int_{-\infty}^{\infty} r_{hh}(\tau_1', \tau_2'; \tau)\, e^{-j2\pi f_2 \tau}\, d\tau\,. \tag{7.28}$$

The first integral on the right-hand side of (7.28) can be identified with the delta function $\delta(f_2 - f_1)$. Consequently, $r_{ss}(\tau_1', \tau_2'; f_1, f_2)$ can be expressed by

$$r_{ss}(\tau_1', \tau_2'; f_1, f_2) = \delta(f_2 - f_1)\, S_{ss}(\tau_1', \tau_2'; f_1), \tag{7.29}$$

where $S_{ss}(\tau_1', \tau_2'; f_1)$ is the Fourier transform of the autocorrelation function $r_{hh}(\tau_1', \tau_2'; \tau)$ with respect to the time separation variable τ. The assumption that the time-variant impulse response $h(\tau', t)$ is wide-sense stationary therefore leads to the fact that the system functions $s(\tau_1', f_1)$ and $s(\tau_2', f_2)$ are statistically uncorrelated if the Doppler frequencies f_1 and f_2 are different.

It can be shown in a similar way that (7.23d) can be represented in the form

$$r_{TT}(f_1', f_2'; f_1, f_2) = \delta(f_2 - f_1)\, S_{TT}(f_1', f_2'; f_1), \tag{7.30}$$

where $S_{TT}(f_1', f_2'; f_1)$ denotes the Fourier transform of the autocorrelation function $r_{HH}(f_1', f_2'; \tau)$ with respect to τ. From (7.30), it can be realized that the system functions $T(f_1', f_1)$ and $T(f_2', f_2)$ are statistically uncorrelated for different Doppler frequencies f_1 and f_2.

Since the time-variant impulse response $h(\tau', t)$ results from a superposition of a multitude of scattered components, it can generally be stated that the WSS assumption leads to the fact that scattering components with different Doppler frequencies or different angles of arrival are statistically uncorrelated.

7.3.2.2 US Models

A second important class of channel models is obtained by assuming that scattering components with different propagation delays are statistically uncorrelated. These channel models are called *US channel models* or *US models* (US, **u**ncorrelated **s**cattering). The autocorrelation functions $r_{hh}(\tau_1', \tau_2'; t_1, t_2)$ and $r_{ss}(\tau_1', \tau_2'; f_1, f_2)$ of US models can first of all be described formally by

$$r_{hh}(\tau_1', \tau_2'; t_1, t_2) = \delta(\tau_2' - \tau_1')\, S_{hh}(\tau_1'; t_1, t_2)\,, \tag{7.31a}$$

$$r_{ss}(\tau_1', \tau_2'; f_1, f_2) = \delta(\tau_2' - \tau_1')\, S_{ss}(\tau_1'; f_1, f_2)\,. \tag{7.31b}$$

The singular behaviour of the autocorrelation function (7.31a) has significant consequences on the tapped-delay-line model shown in Figure 7.2, because the time-variant coefficients of this model are now uncorrelated as a result of the US assumption. In practice, the coefficients of the tapped-delay-line model are realized almost exclusively by coloured Gaussian random processes. It should be taken into account that uncorrelated Gaussian random processes are also statistically independent.

Formal expressions for the autocorrelation functions of the remaining stochastic system functions $H(f', t)$ and $T(f', f)$ can easily be determined by using the relations shown in Figure 7.4. With the substitutions $f_1' = f'$ and $f_2' = f' + v'$, the following equations can be found:

$$r_{HH}(f', f' + v'; t_1, t_2) = r_{HH}(v'; t_1, t_2)\,, \tag{7.32a}$$

$$r_{TT}(f', f' + v'; f_1, f_2) = r_{TT}(v'; f_1, f_2)\,. \tag{7.32b}$$

Obviously, the autocorrelation functions of the system functions $H(f', t)$ and $T(f', f)$ only depend on the frequency difference $v' := f_2' - f_1'$. As a consequence, US models are wide-sense stationary with respect to the frequency f'.

If we now compare the above mentioned autocorrelation functions of US models with those derived for WSS models, then we notice that they are dual to each other. Therefore, we can say that the class of US models stands in a duality relationship to the class of WSS models.

7.3.2.3 WSSUS Models

The most important class of stochastic time-variant linear channel models is represented by models belonging to the class of WSS models as well as to the class of US models. These channel models with wide-sense stationary impulse responses and uncorrelated scattering components are called WSSUS channel models or simply WSSUS models (WSSUS, **w**ide-**s**ense **s**tationary **u**ncorrelated **s**cattering). Due to their simplicity, they are of great practical importance and are nowadays almost exclusively employed for modelling frequency-selective mobile radio channels.

In the case of the WSSUS assumption, the autocorrelation function of the time-variant impulse response $h(\tau', t)$ has to be describable by (7.26a) as well as by (7.31a). Hence, we may formally write

$$r_{hh}(\tau_1', \tau_2'; t, t+\tau) = \delta(\tau_2' - \tau_1')\, S_{hh}(\tau_1', \tau)\,, \tag{7.33}$$

where $S_{hh}(\tau_1', \tau)$ is called the *delay cross-power spectral density.* With this representation it becomes obvious that the time-variant impulse response $h(\tau', t)$ of WSSUS models has the characteristic properties of non-stationary white noise with respect to the propagation delay τ', on the one hand, and is also wide-sense stationary with respect to the time t, on the other hand.

By analogy, we can directly obtain the autocorrelation function of $T(f', f)$ by bringing (7.30) and (7.32b) together. Thus, for WSSUS models, it holds

$$r_{TT}(f', f' + \upsilon'; f_1, f_2) = \delta(f_2 - f_1)\, S_{TT}(\upsilon', f_1)\,, \tag{7.34}$$

where $S_{TT}(\upsilon', f_1)$ is called *Doppler cross-power spectral density.* This result shows that the system function $T(f', f)$ of WSSUS models behaves like non-stationary white noise with respect to the Doppler frequency f and like a wide-sense stationary stochastic process with respect to the frequency f'.

Furthermore, we can combine the relations (7.29) and (7.31b) and obtain the autocorrelation function of $s(\tau', f)$ in the form

$$r_{ss}(\tau_1', \tau_2'; f_1, f_2) = \delta(f_2 - f_1)\, \delta(\tau_2' - \tau_1')\, S(\tau_1', f_1)\,. \tag{7.35}$$

From this, we conclude that the system function $s(\tau', f)$ of WSSUS models has the character of non-stationary white noise with respect to τ' as well as with respect to f. In [Bel63], Bello called the function $S(\tau_1', f_1)$ appearing in (7.35) the *scattering function.*

Finally, by combining (7.26b) and (7.32a), it follows for the autocorrelation function of $H(f', t)$ the relation

$$r_{HH}(f', f' + \upsilon'; t, t+\tau) = r_{HH}(\upsilon', \tau)\,. \tag{7.36}$$

The autocorrelation function $r_{HH}(\upsilon', \tau)$ is called the *time-frequency correlation function.* Regarding (7.36) it becomes obvious that the system function $H(f', t)$ of WSSUS models has the properties of wide-sense stationary stochastic processes with respect to f' and t.

Figure 7.4 shows the universally valid relationships between the autocorrelation functions of the four system functions. With the expressions (7.33)–(7.36), it is now possible to derive the specific relations valid for WSSUS models. One may therefore study Figure 7.5, where the relationships between the delay cross-power spectral density $S_{hh}(\tau', \tau)$, the time-frequency correlation function $r_{HH}(\upsilon', \tau)$, the Doppler cross-power spectral density $S_{TT}(\upsilon', f)$, and the scattering function $S(\tau', f)$ are shown.

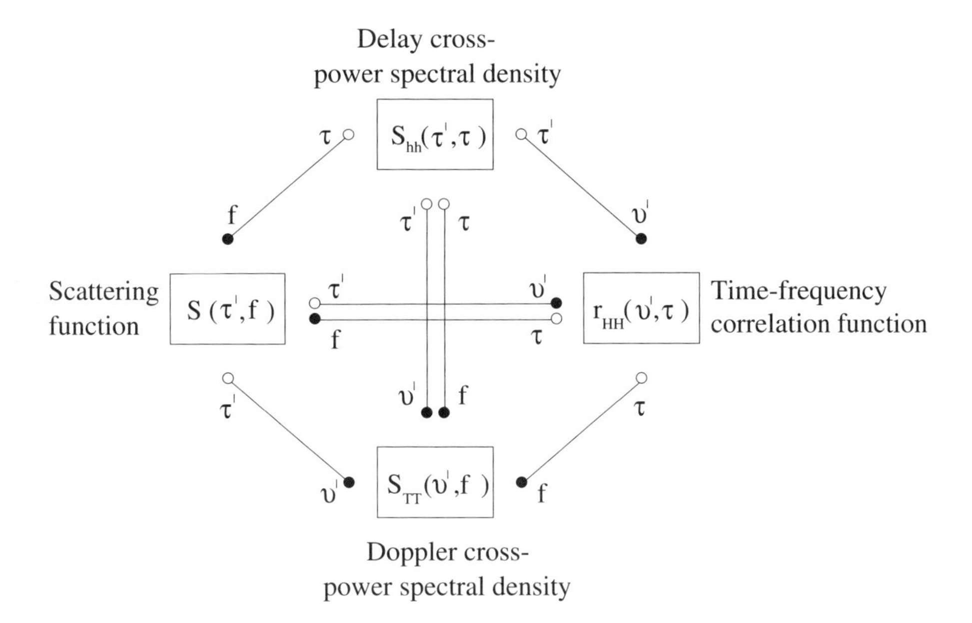

Figure 7.5: Relationships between the delay cross-power spectral density $S_{hh}(\tau', \tau)$, the time-frequency correlation function $r_{HH}(\upsilon', \tau)$, the Doppler cross-power spectral density $S_{TT}(\upsilon', f)$, and the scattering function $S(\tau', f)$ of WSSUS models.

Here, it should be noted that the substitutions $f_1 \to f$ and $\tau_1' \to \tau'$ have been carried out for reasons of simplifying the notation.

Figure 7.5 makes clear that the knowledge of one of the four depicted functions is sufficient to calculate the remaining three. For example, from the scattering function $S(\tau', f)$, we can directly obtain the delay cross-power spectral density $S_{hh}(\tau', \tau)$ by computing the inverse Fourier transform with respect to the Doppler frequency f, i.e.,

$$S_{hh}(\tau', \tau) = \int_{-\infty}^{\infty} S(\tau', f)\, e^{j2\pi f \tau}\, df\,, \tag{7.37}$$

where $\tau = t_2 - t_1$.

The delay cross-power spectral density $S_{hh}(\tau', \tau)$ at $\tau = 0$ defines the so-called *delay power spectral density* $S_{\tau'\tau'}(\tau')$, which is due to (7.37) related to the scattering function $S(\tau', f)$ according to

$$S_{\tau'\tau'}(\tau') := S_{hh}(\tau', 0) = \int_{-\infty}^{\infty} S(\tau', f)\, df\,. \tag{7.38}$$

The delay power spectral density $S_{\tau'\tau'}(\tau')$ determines the average power of scattering components occurring with the propagation delay τ'. It can easily be shown that $S_{\tau'\tau'}(\tau')$ is proportional to the probability density function of the propagation delays τ'. From the delay power spectral density $S_{\tau'\tau'}(\tau')$, two important characteristic quantities for the characterization of WSSUS models can be derived: the *average delay* and the *delay spread*.

Average delay: The average delay $B^{(1)}_{\tau'\tau'}$ is defined by the first moment of $S_{\tau'\tau'}(\tau')$, i.e.,

$$B^{(1)}_{\tau'\tau'} := \frac{\int_{-\infty}^{\infty} \tau' \, S_{\tau'\tau'}(\tau') \, d\tau'}{\int_{-\infty}^{\infty} S_{\tau'\tau'}(\tau') \, d\tau'} \, . \tag{7.39}$$

It corresponds to the centre of gravity of the delay power spectral density $S_{\tau'\tau'}(\tau')$. The average delay $B^{(1)}_{\tau'\tau'}$ is the statistical mean delay that a carrier signal experiences during the transmission over a multipath fading channel.

Delay spread: The delay spread $B^{(2)}_{\tau'\tau'}$ is defined by the square root of the second central moment of $S_{\tau'\tau'}(\tau')$, i.e.,

$$B^{(2)}_{\tau'\tau'} := \sqrt{\frac{\int_{-\infty}^{\infty} \left(\tau' - B^{(1)}_{\tau'\tau'}\right)^2 S_{\tau'\tau'}(\tau') \, d\tau'}{\int_{-\infty}^{\infty} S_{\tau'\tau'}(\tau') \, d\tau'}} \, . \tag{7.40}$$

The delay spread $B^{(2)}_{\tau'\tau'}$ provides us with a measure of the time spread of an impulse passed through a multipath fading channel.

From Figure 7.5, we realize that the Doppler cross-power spectral density $S_{TT}(\upsilon', f)$ is the Fourier transform of the scattering function $S(\tau', f)$ with respect to the delay τ', i.e., the relation

$$S_{TT}(\upsilon', f) = \int_{-\infty}^{\infty} S(\tau', f) \, e^{-j2\pi\upsilon'\tau'} \, d\tau' \tag{7.41}$$

holds, where $\upsilon' = f_2' - f_1'$.

For $\upsilon' = 0$, the already known Doppler power spectral density $S_{\mu\mu}(f)$ follows from the Doppler cross-power spectral density $S_{TT}(\upsilon', f)$, because

$$S_{\mu\mu}(f) := S_{TT}(0, f) = \int_{-\infty}^{\infty} S(\tau', f) \, d\tau' \tag{7.42}$$

holds. The Doppler power spectral density $S_{\mu\mu}(f)$ gives the average power of the scattering components occurring with the Doppler frequency f. In Appendix A, it is shown that $S_{\mu\mu}(f)$ is proportional to the probability density function of the Doppler frequencies f. Remember that two important characteristic quantities can be derived from the Doppler power spectral density $S_{\mu\mu}(f)$, namely, the average Doppler shift $B^{(1)}_{\mu\mu}$ [cf. (3.13a)] and the Doppler spread $B^{(2)}_{\mu\mu}$ [cf. (3.13b)].

According to Figure 7.5, the time-frequency correlation function $r_{HH}(\upsilon', \tau)$ can be calculated from the scattering function $S(\tau', f)$ as follows:

$$r_{HH}(\upsilon', \tau) = \int_{-\infty}^{\infty} \int_{-\infty}^{\infty} S(\tau', f) \, e^{-j2\pi(\upsilon'\tau' - f\tau)} \, d\tau' \, df \, , \tag{7.43}$$

where $\upsilon' = f_2' - f_1'$ and $\tau = t_2 - t_1$. Alternatively, we could have calculated $r_{HH}(\upsilon', \tau)$ by applying the Fourier transform on $S_{hh}(\tau', \tau)$ with respect to the propagation delays τ' or via the inverse Fourier transform of $S_{TT}(\upsilon', f)$ with respect to the Doppler frequency f.

From the time-frequency correlation function $r_{HH}(\upsilon', \tau)$, two further correlation functions can be derived. They are called *frequency correlation function* and *time correlation function*. From each of these functions, a further important characteristic quantity can be derived: the *coherence bandwidth* and the *coherence time*.

Frequency correlation function: The *frequency correlation function* $r_{\tau'\tau'}(\upsilon')$ is defined by the time-frequency correlation function $r_{HH}(\upsilon', \tau)$ at $\tau = t_2 - t_1 = 0$, i.e.,

$$\begin{aligned} r_{\tau'\tau'}(\upsilon') &:= r_{HH}(\upsilon', 0) \\ &= \int_{-\infty}^{\infty} \int_{-\infty}^{\infty} S(\tau', f)\, e^{-j2\pi\upsilon'\tau'}\, d\tau'\, df \\ &= \int_{-\infty}^{\infty} S_{\tau'\tau'}(\tau')\, e^{-j2\pi\upsilon'\tau'}\, d\tau' . \end{aligned} \tag{7.44}$$

Obviously, the frequency correlation function $r_{\tau'\tau'}(\upsilon')$ is the Fourier transform of the delay power spectral density $S_{\tau'\tau'}(\tau')$. The frequency correlation function characterizes the similarity of the time-variant transfer functions $H(f_1', t)$ and $H(f_2', t)$ as a function of the frequency separation variable $\upsilon' = f_2' - f_1'$.

Coherence bandwidth: The frequency separation variable $\upsilon' = B_C$ that fulfils the condition

$$|r_{\tau'\tau'}(B_C)| = \frac{1}{2} |r_{\tau'\tau'}(0)| \tag{7.45}$$

is called the *coherence bandwidth*.

Since, referring to (7.44), the frequency correlation function $r_{\tau'\tau'}(\upsilon')$ and the delay power spectral density $S_{\tau'\tau'}(\tau')$ form a Fourier transform pair, the coherence bandwidth B_C is, according to the uncertainty principle of communications engineering [Lue90], approximately reciprocally proportional to the delay spread $B_{\tau'\tau'}^{(2)}$. With an increasing ratio of signal bandwidth to coherence bandwidth, the expenditure of signal equalization in the receiver grows. An important special case occurs if the coherence bandwidth B_C is much greater than the symbol rate f_{sym}, i.e., if

$$B_C \gg f_{sym} \quad \text{or} \quad B_{\tau'\tau'}^{(2)} \ll T_{sym} \tag{7.46a, b}$$

holds, where $T_{sym} = 1/f_{sym}$ denotes the symbol interval. In this case, the effect of the impulse dispersion can be ignored and the time-variant impulse response $h(\tau', t)$ of the channel can approximately be represented by

$$h(\tau', t) = \delta(\tau') \cdot \mu(t) , \tag{7.47}$$

where $\mu(t)$ is a proper complex stochastic process. Using (7.11), the output signal $y(t)$ may be expressed as

$$y(t) = \mu(t) \cdot x(t) \,. \tag{7.48}$$

Due to the multiplicative relation between $\mu(t)$ and $x(t)$, we speak of *multiplicative fading* in this context. After substituting (7.47) in (7.13), we obtain the following expression for the time-variant transfer function $H(f', t)$

$$H(f', t) = \mu(t) \,, \tag{7.49}$$

In this case, the time-variant transfer function is obviously independent of the frequency f'. Thus, the channel is said to be *frequency-nonselective*, because all frequency components of the transmitted signal are subjected to the same variations. A frequency-nonselective modelling of the mobile radio channel is always adequate if the delay spread $B_{\tau'\tau'}^{(2)}$ does not exceed 10 per cent to 20 per cent of the symbol interval T_{sym}.

Time correlation function: The *time correlation function* $r_{\mu\mu}(\tau)$ is defined by the time-frequency correlation function $r_{HH}(\upsilon', \tau)$ at $\upsilon' = f_2' - f_1' = 0$, i.e.,

$$\begin{aligned} r_{\mu\mu}(\tau) &:= r_{HH}(0, \tau) \\ &= \int_{-\infty}^{\infty} \int_{-\infty}^{\infty} S(\tau', f)\, e^{j2\pi f \tau}\, d\tau'\, df \\ &= \int_{-\infty}^{\infty} S_{\mu\mu}(f)\, e^{j2\pi f \tau}\, df \,. \end{aligned} \tag{7.50}$$

This correlation function describes the correlation properties of the received scattered components as a function of the time difference $\tau = t_2 - t_1$.

Coherence time: The time interval $\tau = T_C$ that fulfils the condition

$$|r_{\mu\mu}(T_C)| = \frac{1}{2} |r_{\mu\mu}(0)| \tag{7.51}$$

is called the *coherence time.*

According to (7.50), the time correlation function $r_{HH}(0, \tau)$ and the Doppler power spectral density $S_{\mu\mu}(f)$ form a Fourier transform pair. Consequently, the coherence time T_C behaves approximately reciprocally proportional to the Doppler spread $B_{\mu\mu}^{(2)}$. The smaller the ratio of the coherence time T_C and the symbol interval T_{sym} is, the higher are the demands on the tracing performance of the channel estimator in the receiver. If the coherence time T_C is much larger than the symbol interval T_{sym}, i.e.,

$$T_C \gg T_{sym} \quad \text{or} \quad B_{\mu\mu}^{(2)} \ll f_{sym} \,, \tag{7.52a, b}$$

then the impulse response of the channel may be regarded as approximately constant for the duration of one symbol. In this case, we speak of slow fading.

Figure 7.6 once more shows the relationships between the correlation functions and the power spectral densities introduced in this subsection in conjunction with the characteristic quantities of WSSUS models derived from these.

This figure vividly shows us that for WSSUS models the knowledge of the scattering function $S(\tau', f)$ is sufficient to determine all correlation functions and power spectral densities as well as their characteristic quantities such as the delay spread and Doppler spread.

7.3.3 The Channel Models According to COST 207

In 1984, the European working group COST[2] 207 was established by CEPT.[3] At that time, this working group developed suitable channel models for typical propagation environments, in view of the planned pan-European mobile communication system GSM. The typical propagation environments are classifiable into areas with rural character (RA, Rural Area), areas typical for cities and suburbs (TU, Typical Urban), densely built urban areas with bad propagation conditions (BU, Bad Urban), and hilly terrains (HT, Hilly Terrain). Basing on the WSSUS assumption, the working group COST 207 developed specifications for the delay power spectral density and the Doppler power spectral density for these four classes of propagation environments [COS86, COS89]. The main results will be presented subsequently.

The specification of typical delay power spectral densities $S_{\tau'\tau'}(\tau')$ is based on the assumption that the corresponding probability density function $p_{\tau'}(\tau')$, which is proportional to $S_{\tau'\tau'}(\tau')$, can be represented by one or more negative exponential functions. The delay power spectral density functions $S_{\tau'\tau'}(\tau')$ of the channel models according to COST 207 are shown in Table 7.1 and in Figure 7.7. The real-valued constant quantities c_{RA}, c_{TU}, c_{BU}, and c_{HT} introduced there can in principle be chosen arbitrarily. Hence, they can be determined in such a way that the average delay power is equal to one for example, i.e., $\int_0^\infty S_{\tau'\tau'}(\tau')\,d\tau' = 1$. In this case, it holds:

$$c_{RA} = \frac{9.2}{1-e^{-6.44}}\,, \qquad c_{TU} = \frac{1}{1-e^{-7}}\,, \tag{7.53a, b}$$

$$c_{BU} = \frac{2}{3(1-e^{-5})}\,, \qquad c_{HT} = \frac{1}{(1-e^{-7})/3.5+(1-e^{-5})/10}\,. \tag{7.53c, d}$$

In the GSM system, the symbol interval T_{sym} is defined by $T_{sym} = 3.7\mu s$. If we

[2] COST: European Cooperation in the Field of Scientific and Technical Research.
[3] CEPT: Conference of European Posts and Telecommunications Administrations.

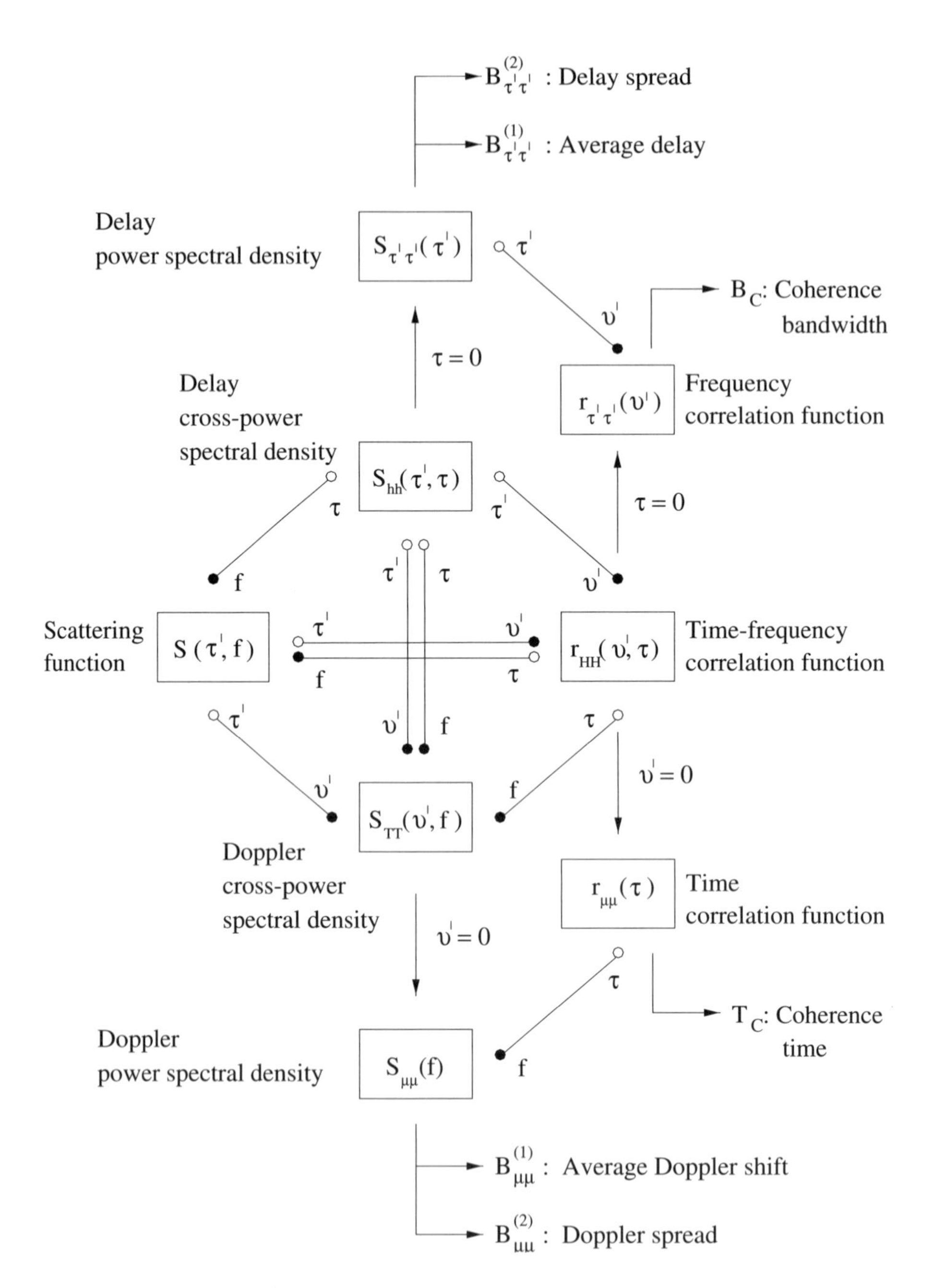

Figure 7.6: Relationships between the correlation functions, power spectral densities, and characteristic quantities of WSSUS models.

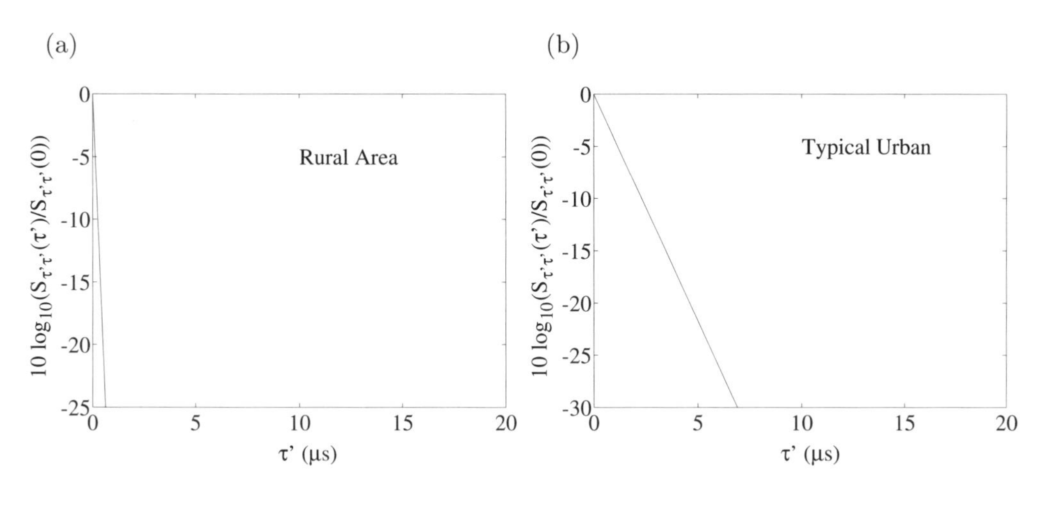

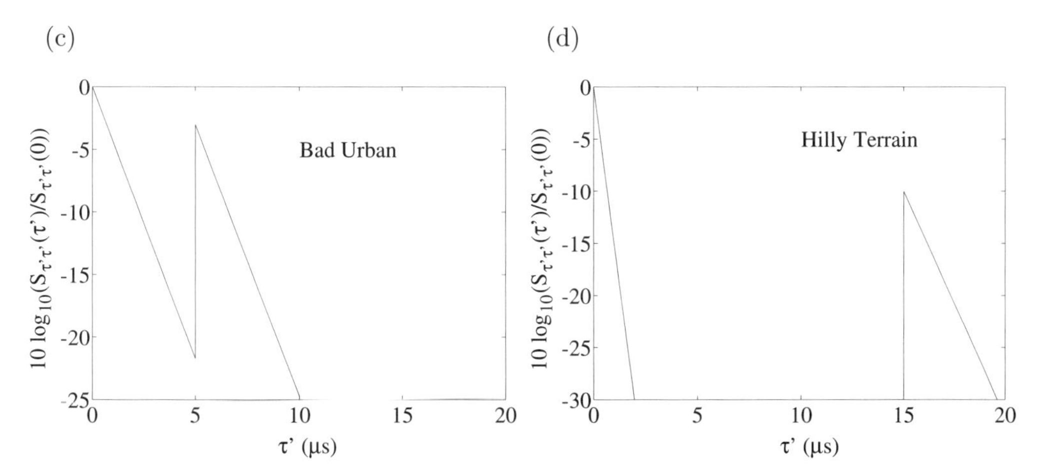

Figure 7.7: Delay power spectral densities $S_{\tau'\tau'}(\tau')$ of the channel models according to COST 207 [COS89].

Table 7.1: Specification of typical delay power spectral densities $S_{\tau'\tau'}(\tau')$ according to COST 207 [COS89].

Propagation area	Delay power spectral density $S_{\tau'\tau'}(\tau')$		Delay spread $B^{(2)}_{\tau'\tau'}$
Rural Area (RA)	$c_{RA}\, e^{-9.2\tau'/\mu s}$, 0,	$0 \le \tau' < 0.7\mu s$ else	$0.1\ \mu s$
Typical Urban (TU)	$c_{TU}\, e^{-\tau'/\mu s}$, 0,	$0 \le \tau' < 7\mu s$ else	$0.98\ \mu s$
Bad Urban (BU)	$c_{BU}\, e^{-\tau'/\mu s}$, $c_{BU}\, \frac{1}{2} e^{(5-\tau'/\mu s)}$, 0,	$0 \le \tau' < 5\mu s$ $5\mu s \le \tau' < 10\mu s$ else	$2.53\ \mu s$
Hilly Terrain (HT)	$c_{HT}\, e^{-3.5\tau'/\mu s}$, $c_{HT}\, 0.1 e^{(15-\tau'/\mu s)}$, 0,	$0 \le \tau' < 2\mu s$ $15\mu s \le \tau' < 20\mu s$ else	$6.88\ \mu s$

bring T_{sym} in relation to the delay spread $B^{(2)}_{\tau'\tau'}$, which is listed in the last column of Table 7.1, then we realize that (7.46b) is only fulfilled for the Rural Area (RA). Consequently, the RA channel belongs to the class of frequency-nonselective channels, whereas the other channels (TU, BU, HT) are frequency-selective.

Table 7.2 shows the four types of Doppler power spectral densities $S_{\mu\mu}(f)$ specified by COST 207. They are also presented graphically in Figure 7.8 for better illustration. For the real-valued constants A_1 and A_2, preferably the values $A_1 = 50/\left(\sqrt{2\pi}3f_{max}\right)$ and $A_2 = 10^{1.5}/\left[\sqrt{2\pi}\left(\sqrt{10}+0.15\right)f_{max}\right]$ are chosen, since it is then ensured that $\int_{-\infty}^{\infty} S_{\mu\mu}(f)\,df$ is equal to one. The classical Jakes power spectral density only occurs in the case of very short propagation delays ($\tau' \le 0.5\,\mu$s) [see Figures 7.8(a) and 7.8(d)]. Only in this case, the assumptions that the amplitudes of the scattering components are homogeneous and the angles of arrival are uniformly distributed between 0 and 2π are justified. For scattering components with medium and long propagation delays τ', however, it is assumed that the corresponding Doppler frequencies are normally distributed, resulting in a Doppler power spectral density with a Gaussian shape [see Figures 7.8(b) and 7.8(c)]. This had already been pointed out by Cox [Cox73] at a very early stage after performing extensive empirical investigations.

From Tables 7.1 and 7.2, it can be seen that the delay power spectral density $S_{\tau'\tau'}(\tau')$

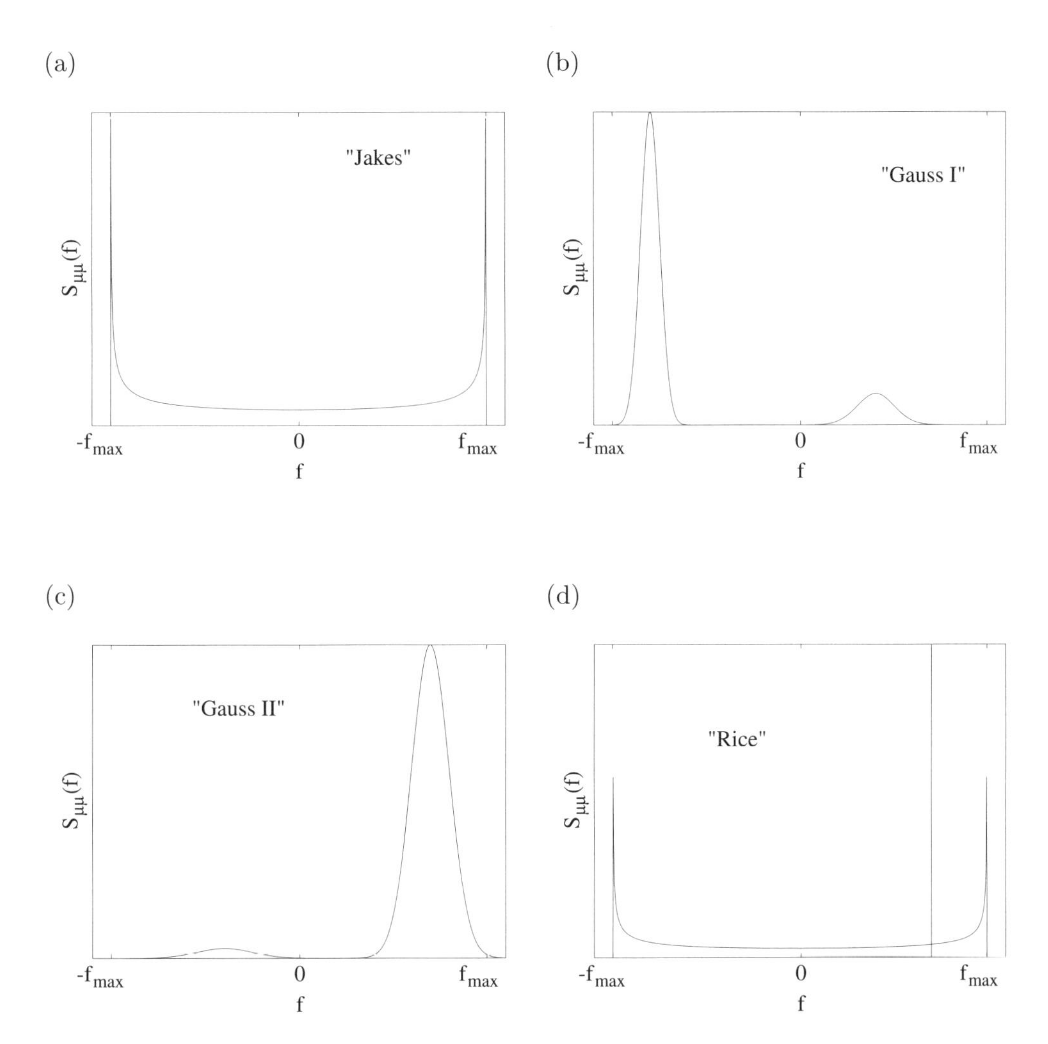

Figure 7.8: Doppler power spectral densities $S_{\mu\mu}(f)$ of the channel models according to COST 207 [COS89].

Table 7.2: Specification of typical Doppler power spectral densities $S_{\mu\mu}(f)$ according to COST 207 [COS89], where $G(A_i, f_i, s_i)$ is defined by $G(A_i, f_i, s_i) := A_i \exp\{-\frac{(f-f_i)^2}{2s_i^2}\}$.

Type	Doppler power spectral density $S_{\mu\mu}(f)$	Propagation delay τ'	Doppler spread $B_{\mu\mu}^{(2)}$
"Jakes"	$\frac{1}{\pi f_{max}\sqrt{1-(f/f_{max})^2}}$	$0 \leq \tau' \leq 0.5\mu s$	$f_{max}/\sqrt{2}$
"Gauss I"	$G(A_1, -0.8f_{max}, 0.05f_{max})$ $+G(A_1/10, 0.4f_{max}, 0.1f_{max})$	$0.5\mu s \leq \tau' \leq 2\mu s$	$0.45f_{max}$
"Gauss II"	$G(A_2, 0.7f_{max}, 0.1f_{max})$ $+G(A_2/10^{1.5}, -0.4f_{max}, 0.15f_{max})$	$\tau' \geq 2\mu s$	$0.25f_{max}$
"Rice"	$\frac{0.41^2}{\pi f_{max}\sqrt{1-(f/f_{max})^2}}$ $+0.91^2\ \delta(f - 0.7f_{max})$	$\tau' = 0\mu s$	$0.39f_{max}$

is independent of the Doppler frequencies f, but the propagation delays τ' have a decisive influence on the shape of the Doppler power spectral density $S_{\mu\mu}(f)$. However, this is not valid for rural areas, where only the classical Jakes power spectral density is used. In this special case, the scattering function $S(\tau', f)$ can be represented by the product of the delay power spectral density and the Doppler power spectral density, i.e.,

$$S(\tau', f) = S_{\tau'\tau'}(\tau') \cdot S_{\mu\mu}(f). \tag{7.54}$$

Channels with a scattering function of the form (7.54) are called *independent time dispersive and frequency dispersive channels.* For this class of channels, the physical mechanism causing the propagation delays is independent from that which is responsible for the Doppler effect [Fle90].

Regarding the design of hardware or software simulation models for frequency-selective channels, a discretization of the delay power spectral density $S_{\tau'\tau'}(\tau')$ has to be performed. In particular, the propagation delays τ' have to be made discrete and adapted to the sampling interval. This is the reason why discrete $\mathcal{L}$-path channel models have been specified in [COS89] for the four propagation areas (RA, TU, BU, HT). Some of these specified $\mathcal{L}$-path channel models are listed in Table 7.3 for $\mathcal{L} = 4$ and $\mathcal{L} = 6$. The resulting scattering functions $S(\tau', f)$ are shown in Figures 7.9(a)–(d). In [COS89], moreover, alternative 6-path channel models as well as more complex, but therefore more exact, 12-path channel models have been specified. They are presented in Appendix E for the sake of completeness.

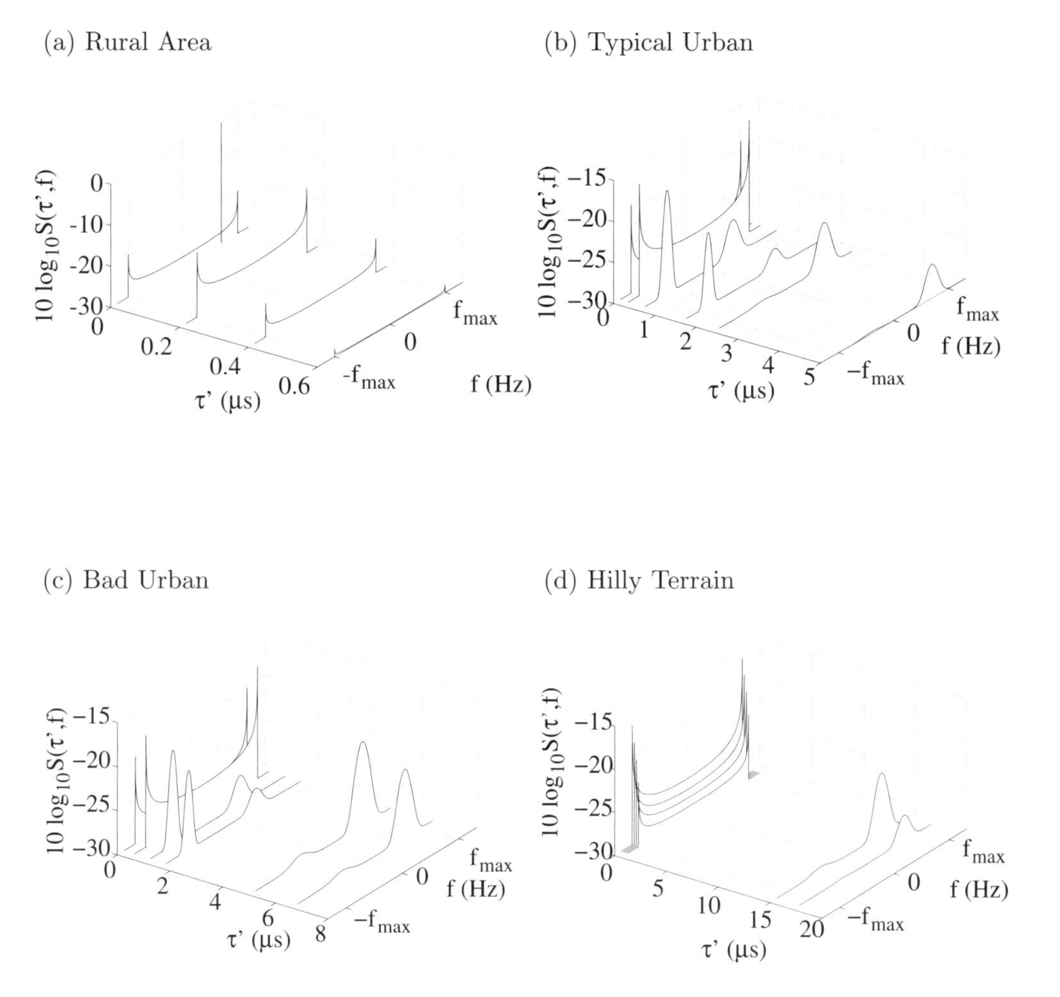

Figure 7.9: Scattering functions $S(\tau', f)$ of the $\mathcal{L}$-path channel models according to COST 207 [COS89].

Table 7.3: Specification of the $\mathcal{L}$-path channel models according to COST 207 [COS89], where $\mathcal{L} = 4$ (RA) and $\mathcal{L} = 6$ (TU, BU, HT).

Path no. ℓ	**Propagation delay** τ'_ℓ	**Path power (lin.)**	**Path power (dB)**	**Category of the Doppler power spectral density**	**Delay spread** $B^{(2)}_{\tau'\tau'}$
(a) Rural Area					
0	0.0 μs	1	0	"Rice"	0.1 μs
1	0.2 μs	0.63	-2	"Jakes"	
2	0.4 μs	0.1	-10	"Jakes"	
3	0.6 μs	0.01	-20	"Jakes"	
(b) Typical Urban					
0	0.0 μs	0.5	-3	"Jakes"	1.1 μs
1	0.2 μs	1	0	"Jakes"	
2	0.6 μs	0.63	-2	"Gauss I"	
3	1.6 μs	0.25	-6	"Gauss I"	
4	2.4 μs	0.16	-8	"Gauss II"	
5	5.0 μs	0.1	-10	"Gauss II"	
(c) Bad Urban					
0	0.0 μs	0.5	-3	"Jakes"	2.4 μs
1	0.4 μs	1	0	"Jakes"	
2	1.0 μs	0.5	-3	"Gauss I"	
3	1.6 μs	0.32	-5	"Gauss I"	
4	5.0 μs	0.63	-2	"Gauss II"	
5	6.6 μs	0.4	-4	"Gauss II"	
(d) Hilly Terrain					
0	0.0 μs	1	0	"Jakes"	5.0 μs
1	0.2 μs	0.63	-2	"Jakes"	
2	0.4 μs	0.4	-4	"Jakes"	
3	0.6 μs	0.2	-7	"Jakes"	
4	15.0 μs	0.25	-6	"Gauss II"	
5	17.2 μs	0.06	-12	"Gauss II"	

7.4 FREQUENCY-SELECTIVE DETERMINISTIC CHANNEL MODELS

In this section, we will deal with the derivation and the analysis of frequency-selective deterministic channel models. For this purpose, we again apply the principle of deterministic channel modelling described in Section 4.1.

7.4.1 System Functions of Frequency-Selective Deterministic Channel Models

Starting point for the derivation of the system functions of frequency-selective deterministic channel models is the time-variant impulse response consisting of a sum of $\mathcal{L}$ discrete propagation paths

$$\tilde{h}(\tau', t) = \sum_{\ell=0}^{\mathcal{L}-1} \tilde{a}_\ell \, \tilde{\mu}_\ell(t) \, \delta(\tau' - \tilde{\tau}'_\ell) \,. \tag{7.55}$$

The quantities $\tilde{a}_\ell$ in (7.55) are real-valued and they are called the *delay coefficients*. As we will see later on, both the delay coefficients $\tilde{a}_\ell$ and the discrete propagation delays $\tilde{\tau}'_\ell$ [see (7.1)] determine the delay power spectral density of frequency-selective deterministic channel models. Strictly speaking, the delay coefficient $\tilde{a}_\ell$ is a measure of the square root of the average delay power which is assigned to the ℓth discrete propagation path. In general, one can say that the delay coefficients $\tilde{a}_\ell$ and the discrete propagation delays $\tilde{\tau}'_\ell$ determine the frequency-selective behaviour of the channel, which can be attributed to the effect of multipath propagation. In the present case, it is assumed that elliptical scattering zones with different discrete axes are the reason for multipath propagation. The disturbances of the channel caused by the Doppler effect, i.e., the disturbances caused by the motion of the receiver (transmitter), are modelled in (7.55), according to the principle of deterministic channel modelling, by complex deterministic Gaussian processes

$$\tilde{\mu}_\ell(t) = \tilde{\mu}_{1,\ell}(t) + j\tilde{\mu}_{2,\ell}(t) \,, \quad \ell = 0, 1, \ldots, \mathcal{L}-1 \,, \tag{7.56a}$$

where

$$\tilde{\mu}_{i,\ell}(t) = \sum_{n=1}^{N_{i,\ell}} c_{i,n,\ell} \cos(2\pi f_{i,n,\ell} t + \theta_{i,n,\ell}) \,, \quad i = 1, 2 \,. \tag{7.56b}$$

Here, $N_{i,\ell}$ denotes the number of harmonic functions belonging to the real part ($i = 1$) or the imaginary part ($i = 2$) of the ℓth propagation path. In (7.56b), $c_{i,n,\ell}$ is the Doppler coefficient of the nth component of the ℓth propagation path, and the remaining model parameters $f_{i,n,\ell}$ and $\theta_{i,n,\ell}$ are, as stated before, called the Doppler frequencies and the Doppler phases, respectively.

Figure 7.10 shows the structure of the complex Gaussian random process $\tilde{\mu}_\ell(t)$ in the continuous-time representation. To ensure that the simulation model derived below has the same striking properties as a US model, the complex deterministic Gaussian processes $\tilde{\mu}_\ell(t)$ must be uncorrelated for different propagation paths. Therefore, it is inevitable that the deterministic Gaussian processes $\tilde{\mu}_\ell(t)$ and $\tilde{\mu}_\lambda(t)$ are designed in such a way that they are uncorrelated for $\ell \neq \lambda$, where $\ell, \lambda = 0, 1, \ldots, \mathcal{L}-1$. This demand can easily be fulfilled. One merely has to ensure that the discrete Doppler frequencies $f_{i,n,\ell}$ are designed so that the resulting sets $\{f_{i,n,\ell}\}$ are disjoint (mutually exclusive) for different propagation paths. For the simulation model, the demand for uncorrelated scattering (US) propagation can therefore be formulated as follows:

$$\text{US} \quad \Longleftrightarrow \quad f_{i,n,\ell} \neq f_{j,m,\lambda} \quad \text{for} \quad \ell \neq \lambda\,, \tag{7.57}$$

where $i, j = 1, 2$, $n = 1, 2, \ldots, N_{i,\ell}$, $m = 1, 2, \ldots, N_{j,\lambda}$, and $\ell, \lambda = 0, 1, \ldots, \mathcal{L}-1$.

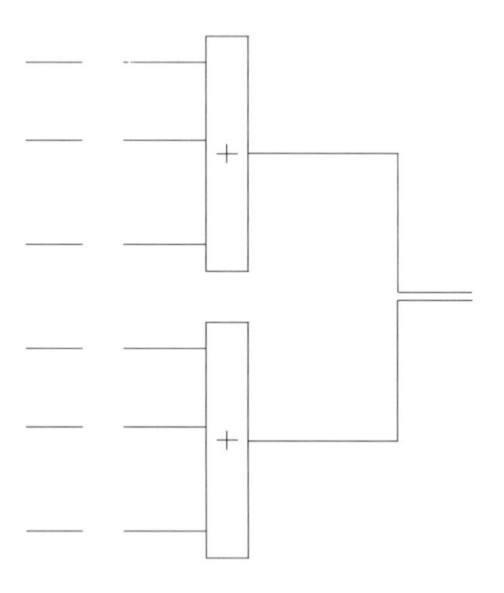

Figure 7.10: Simulation model for complex deterministic Gaussian processes $\tilde{\mu}_\ell(t)$.

In the following, we assume that the US condition (7.57) is always fulfilled. In this case, the correlation properties of the complex deterministic Gaussian processes $\tilde{\mu}_\ell(t)$ introduced by (7.56a) can be described by

$$\lim_{T\to\infty} \frac{1}{2T} \int_{-T}^{T} \tilde{\mu}_\ell^*(t)\, \tilde{\mu}_\lambda(t+\tau)\, dt = \begin{cases} \tilde{r}_{\mu_\ell \mu_\ell}(\tau)\,, & \text{if} \quad \ell = \lambda\,, \\ 0\,, & \text{if} \quad \ell \neq \lambda\,, \end{cases} \tag{7.58}$$

where

$$\tilde{r}_{\mu_\ell \mu_\ell}(\tau) = \sum_{i=1}^{2} \tilde{r}_{\mu_{i,\ell}\mu_{i,\ell}}(\tau)\,, \tag{7.59a}$$

$$\tilde{r}_{\mu_{i,\ell}\mu_{i,\ell}}(\tau) = \sum_{n=1}^{N_{i,\ell}} \frac{c_{i,n,\ell}^2}{2} \cos(2\pi f_{i,n,\ell}\tau) \tag{7.59b}$$

holds for $i = 1, 2$ and $\ell, \lambda = 0, 1, \ldots, \mathcal{L} - 1$.

At this stage, it should be mentioned that all parameters determining the statistical behaviour of the time-variant impulse response $\tilde{h}(\tau', t)$ can be calculated in such a way that the scattering function of the deterministic system approximates a given specified or measured scattering function. A procedure for this will be introduced in Subsection 7.4.4. We may therefore assume that the parameters mentioned above are not only known, but also constant quantities, which will not be changed during the channel simulation run. In this case, the time-variant impulse response $\tilde{h}(\tau', t)$ is a deterministic function (sample function) which will consequently be called the *time-variant deterministic impulse response.* It defines a further important class of channel models. In the following, channel models with an impulse response according to (7.55) will be called *DGUS*[4] *models.*

Since the discrete propagation delays $\tilde{\tau}'_\ell$ in (7.55) cannot become negative, $\tilde{h}(\tau', t)$ fulfils the causality condition, i.e., it holds

$$\tilde{h}(\tau', t) = 0 \quad \text{for} \quad \tau' < 0 \,. \tag{7.60}$$

By analogy to (7.11), we can compute the output signal $y(t)$ for any given input signal $x(t)$ by applying

$$y(t) = \int_0^\infty x(t - \tau')\, \tilde{h}(\tau', t)\, d\tau' \,. \tag{7.61}$$

If we now employ the expression (7.55) for the time-variant deterministic impulse response $\tilde{h}(\tau', t)$, we obtain

$$y(t) = \sum_{\ell=0}^{\mathcal{L}-1} \tilde{a}_\ell\, \tilde{\mu}_\ell(t)\, x(t - \tilde{\tau}'_\ell) \,. \tag{7.62}$$

Hence, the output signal $y(t)$ of the channel can be interpreted as a superposition of $\mathcal{L}$ delayed versions of the input signal $x(t - \tilde{\tau}'_\ell)$, where each of the delayed versions is weighted by a constant delay coefficient $\tilde{a}_\ell$ and a time-variant complex deterministic Gaussian process $\tilde{\mu}_\ell(t)$. Without restriction of generality, we may ignore the propagation delay of the line-of-sight component in this model. To simplify matters, we define $\tilde{\tau}'_0 := 0$. This does not cause any problem, because only the propagation delay differences $\Delta\tilde{\tau}'_\ell = \tilde{\tau}'_\ell - \tilde{\tau}'_{\ell-1}$ ($\ell = 1, 2, \ldots, \mathcal{L} - 1$) are relevant for the system behaviour. From (7.62) follows the tapped-delay-line structure shown in Figure 7.11 of a deterministic simulation model for a frequency-selective mobile radio channel in the continuous-time representation.

The discrete-time simulation model, required for computer simulations, can be obtained from the continuous-time structure, e.g., by substituting $\tilde{\tau}'_\ell \rightarrow \ell T'_s$, $x(t) \rightarrow x(kT'_s)$, $y(t) \rightarrow y(kT'_s)$ and $\tilde{\mu}_\ell(t) \rightarrow \tilde{\mu}_\ell(kT_s)$, where T_s and T'_s denote sampling

[4] DGUS is introduced here as an abbreviation for "deterministic Gaussian uncorrelated scattering".

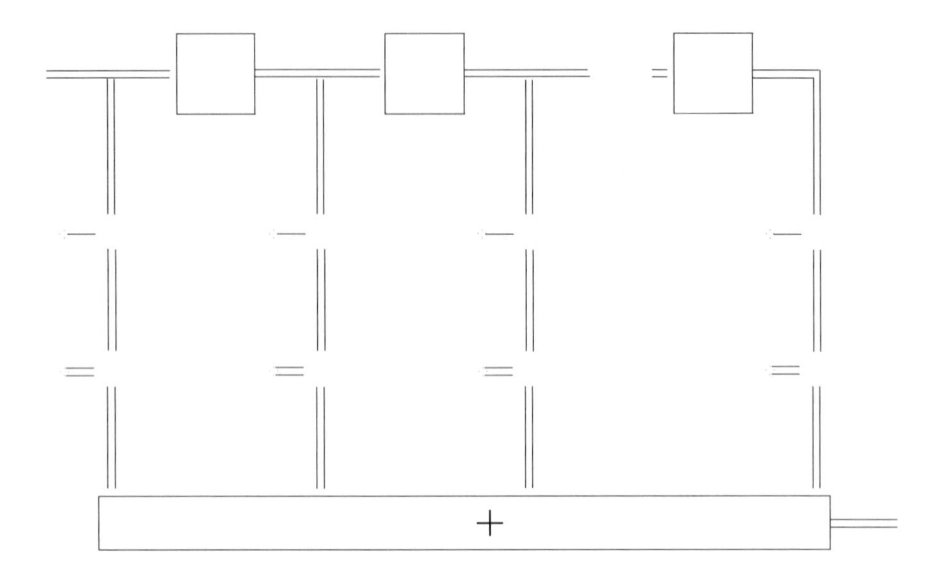

Figure 7.11: Deterministic simulation model for a frequency-selective mobile radio channel in the equivalent complex baseband.

intervals, k is an integer, and ℓ refers to the ℓth propagation path ($\ell = 0, 1, \ldots, \mathcal{L}-1$). For the propagation delay differences $\Delta\tilde{\tau}'_\ell = \tilde{\tau}'_\ell - \tilde{\tau}'_{\ell-1}$, we in this case obtain $\Delta\tilde{\tau}'_\ell \to T'_s$ for all $\ell = 1, 2, \ldots, \mathcal{L}-1$. The sampling intervals T_s and T'_s have to be sufficiently small, but must not necessarily be identical. Between T_s and T'_s, we can therefore establish the general relation $T_s = m'_s T'_s$, where $m'_s \in \mathbb{N}$ is in the following called the *sampling rate ratio*. The larger (smaller) the sampling rate ratio m'_s is chosen, the higher (lower) the simulation speed of the channel simulator is and the larger (smaller) the error occurring due to the discretization of $\tilde{\mu}_\ell(t)$ is. The sampling rate ratio m'_s enables the user to find a good compromise between the simulation speed and the precision of the channel model. As a guideline, m'_s should be chosen so that the sampling interval T_s satisfies the condition $T'_s \leq T_s \leq T_{sym}$ for any given symbol interval T_{sym}. The upper limit $T_s = T_{sym}$ corresponds to the often made assumption that the impulse response is constant for the duration of one data symbol. However, this assumption is only justified if the product $f_{max} T_{sym}$ is very small.

From the general relation (7.62), two important special cases can be derived. These are characterized by

$$\text{(i)} \quad \tilde{a}_0 \neq 0\,, \quad \tilde{a}_1 = \tilde{a}_2 = \ldots = \tilde{a}_{\mathcal{L}-1} = 0 \tag{7.63a}$$

and

$$\text{(ii)} \quad \tilde{\mu}_\ell(t) = \tilde{\mu}_\ell = const.\,, \quad \forall\, \ell = 0, 1, \ldots, \mathcal{L}-1\,. \tag{7.63b}$$

The first special case (i) describes a channel for which all scattering components caused by obstacles situated relatively far away from the receiver can be ignored. Using $\tilde{\mu}(t) := \tilde{a}_0 \tilde{\mu}_0(t)$, we can in this case represent the time-variant deterministic impulse response by

$$\tilde{h}(\tau', t) = \delta(\tau') \cdot \tilde{\mu}(t) . \tag{7.64}$$

A comparison with (7.47) shows that we are here dealing with a frequency-nonselective channel model. This also explains the fact that the multiplicative relation

$$y(t) = \tilde{\mu}(t) \cdot x(t) \tag{7.65}$$

follows from (7.61).

The second special case (ii) always occurs if both the transmitter and the receiver are not moving. In this case, the Doppler effect disappears and the deterministic Gaussian processes $\tilde{\mu}_\ell(t)$ become complex-valued constants $\tilde{\mu}_\ell$ for all discrete paths $\ell = 0, 1, \ldots, \mathcal{L} - 1$. From (7.55) then follows the impulse response of a time-invariant finite impulse response (FIR) filter with $\mathcal{L}$ complex-valued coefficients

$$\tilde{h}(\tau') = \sum_{\ell=0}^{\mathcal{L}-1} a_\ell \, \delta(\tau' - \tilde{\tau}'_\ell) , \tag{7.66}$$

where $a_\ell := \tilde{a}_\ell \, \tilde{\mu}_\ell$ for $\ell = 0, 1, \ldots, \mathcal{L} - 1$.

Now, we consider the general case more detailed. By analogy to (7.13), we define the time-variant transfer function $\tilde{H}(f', t)$ by the Fourier transform of the time-variant deterministic impulse response $\tilde{h}(\tau', t)$ with respect to the propagation delay τ', i.e., we may write $\tilde{h}(\tau', t) \overset{\tau' \;\; f'}{\circ\!\!-\!\!\!-\!\!\bullet} \tilde{H}(f', t)$. If we replace the impulse response $h(\tau', t)$ by the deterministic impulse response $\tilde{h}(\tau', t)$ in (7.13), and take (7.55) into account, then we can easily derive the following closed-form solution for the *time-variant transfer function* $\tilde{H}(f', t)$ *of DGUS models*

$$\tilde{H}(f', t) = \sum_{\ell=0}^{\mathcal{L}-1} \tilde{a}_\ell \, \tilde{\mu}_\ell(t) \, e^{-j2\pi f' \tilde{\tau}'_\ell} . \tag{7.67}$$

It is obvious that $\tilde{H}(f', t)$ is deterministic, because the Fourier transform of a deterministic function again results in a deterministic function. For the description of the input-output relationship of DGUS models, we may refer to (7.14), where of course the time-variant transfer function $H(f', t)$ has to be replaced by $\tilde{H}(f', t)$. Moreover, (7.62) can directly be derived from (7.14). Therefore, $H(f', t)$ has to be substituted by $\tilde{H}(f', t)$ in (7.14), where $\tilde{H}(f', t)$ is given by (7.67).

An insight into the phenomena of the Doppler effect can be obtained from the Doppler-variant impulse response $\tilde{s}(\tau', f)$. This function is defined by the Fourier transform of $\tilde{h}(\tau', t)$ with respect to the time variable t, i.e., $\tilde{h}(\tau', t) \overset{t \;\; f}{\circ\!\!-\!\!\!-\!\!\bullet} \tilde{s}(\tau', f)$. Using the expression (7.55), we obtain the following closed-form solution for the *Doppler-variant impulse response* $\tilde{s}(\tau', f)$ of DGUS models

$$\tilde{s}(\tau', f) = \sum_{\ell=0}^{\mathcal{L}-1} \tilde{a}_\ell \, \tilde{\Xi}_\ell(f) \, \delta(\tau' - \tilde{\tau}'_\ell) , \tag{7.68}$$

where $\tilde{\Xi}_\ell(f)$ denotes the Fourier transform of $\tilde{\mu}_\ell(t)$, i.e.,

$$\tilde{\Xi}_\ell(f) = \tilde{\Xi}_{1,\ell}(f) + j\tilde{\Xi}_{2,\ell}(f)\,, \quad \ell = 0, 1, \ldots, \mathcal{L} - 1\,, \tag{7.69a}$$

$$\tilde{\Xi}_{i,\ell}(f) = \sum_{n=1}^{N_{i,\ell}} \frac{c_{i,n,\ell}}{2} \left[\delta(f - f_{i,n,\ell})\, e^{j\theta_{i,n,\ell}} + \delta(f + f_{i,n,\ell})\, e^{-j\theta_{i,n,\ell}}\right], \quad i = 1, 2\,. \tag{7.69b}$$

Thus, $\tilde{s}(\tau', f)$ is a two-dimensional discrete line spectrum, where the spectral lines are located at the discrete positions $(\tau', f) = (\tilde{\tau}'_\ell, \pm f_{i,n,\ell})$ and weighted by the complex-valued factors $\frac{1}{2}\tilde{a}_\ell c_{i,n,\ell} e^{\pm j\theta_{i,n,\ell}}$. For the description of the input-output behaviour, the relation (7.19) is useful, if the Doppler-variant impulse response $s(\tau', f)$ is substituted by $\tilde{s}(\tau', f)$ there. It should also be observed that (7.62) follows from (7.19), if in the latter equation $s(\tau', f)$ is replaced by (7.68).

Finally, we consider the *Doppler-variant transfer function* $\tilde{T}(f', f)$ *of DGUS models* which is defined by the two-dimensional Fourier transform of the time-variant deterministic impulse response $\tilde{h}(\tau', t)$, i.e., $\tilde{h}(\tau', t) \overset{\tau', t \quad f', f}{\circ\!\!-\!\!\!-\!\!\bullet} \tilde{T}(f', f)$. Due to $\tilde{h}(\tau', t) \overset{\tau' \quad f'}{\circ\!\!-\!\!\!-\!\!\bullet} \tilde{H}(f', t)$ and $\tilde{h}(\tau', t) \overset{t \quad f}{\circ\!\!-\!\!\!-\!\!\bullet} \tilde{s}(\tau', f)$, the computation of an expression for $\tilde{T}(f', f)$ can also be carried out via the one-dimensional Fourier transform $\tilde{H}(f', t) \overset{t \quad f}{\circ\!\!-\!\!\!-\!\!\bullet} \tilde{T}(f', f)$ or $\tilde{s}(\tau', f) \overset{\tau' \quad f'}{\circ\!\!-\!\!\!-\!\!\bullet} \tilde{T}(f', f)$. No matter which procedure we decide upon, we in any case obtain the following closed-form expression for the Doppler-variant transfer function $\tilde{T}(f', f)$ of the deterministic system

$$\tilde{T}(f', f) = \sum_{\ell=0}^{\mathcal{L}-1} \tilde{a}_\ell\, \tilde{\Xi}_\ell(f)\, e^{-j2\pi f' \tilde{\tau}'_\ell}\,. \tag{7.70}$$

Summarizing, we should keep in mind that when the model parameters $\{c_{i,n,\ell}\}$, $\{f_{i,n,\ell}\}$, $\{\theta_{i,n,\ell}\}$, $\{\tilde{a}_\ell\}$, $\{\tilde{\tau}'_\ell\}$, $\{N_{i,\ell}\}$, and $\mathcal{L}$ are known and constant quantities, the four system functions $\tilde{h}(\tau', t)$, $\tilde{H}(f', t)$, $\tilde{s}(\tau', f)$, and $\tilde{T}(f', f)$ can be calculated explicitly. By analogy to Figure 7.3, the system functions of deterministic channel models are related in pairs by the Fourier transform. The resulting relationships are illustrated in Figure 7.12.

7.4.2 Correlation Functions and Power Spectral Densities of DGUS Models

With reference to the WSSUS model, analog relations can be established in the general sense for the correlation functions and power spectral densities of the frequency-selective deterministic channel model (DGUS model). In particular, the correlation functions of the four system functions $\tilde{h}(\tau', t)$, $\tilde{H}(f', t)$, $\tilde{s}(\tau', f)$, and $\tilde{T}(f', f)$ of the deterministic system can be represented by the following relations:

$$\tilde{r}_{hh}(\tau'_1, \tau'_2; t, t+\tau) \quad = \quad \delta(\tau'_2 - \tau'_1)\, \tilde{S}_{hh}(\tau'_1, \tau)\,, \tag{7.71a}$$

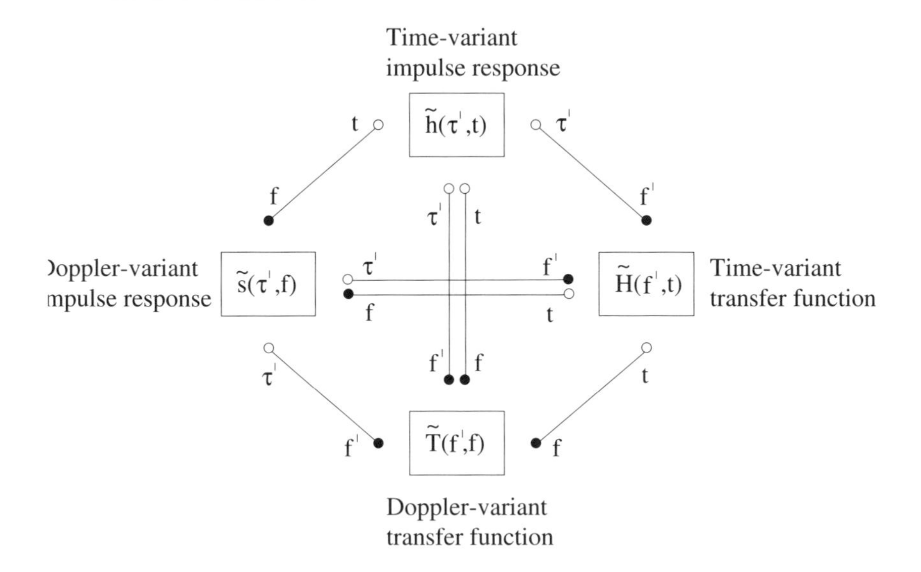

Figure 7.12: Relationships between the system functions of frequency-selective deterministic channel models.

$$\tilde{r}_{HH}(f', f' + \upsilon'; t, t + \tau) = \tilde{r}_{HH}(\upsilon', \tau), \tag{7.71b}$$

$$\tilde{r}_{ss}(\tau_1', \tau_2'; f_1, f_2) = \delta(f_2 - f_1)\, \delta(\tau_2' - \tau_1')\, \tilde{S}(\tau_1', f_1), \tag{7.71c}$$

$$\tilde{r}_{TT}(f', f' + \upsilon'; f_1, f_2) = \delta(f_2 - f_1)\, \tilde{S}_{TT}(\upsilon', f_1). \tag{7.71d}$$

In these equations, $\tilde{S}_{hh}(\tau_1', \tau)$ denotes the delay cross-power spectral density, $\tilde{r}_{HH}(\upsilon', \tau)$ is the time-frequency correlation function, $\tilde{S}(\tau_1', f_1)$ is the scattering function, and $\tilde{S}_{TT}(\upsilon', f_1)$ is the Doppler cross-power spectral density of the deterministic system. Two of these quantities at a time form a Fourier transform pair in the same way as for the WSSUS model. By analogy to Figure 7.5, we obtain the relations depicted in Figure 7.13 for frequency-selective deterministic channel models. In order to simplify the notation, the variables τ_1' and f_1 have again been replaced by τ' and f, respectively.

The interpretation of $\tilde{h}(\tau', t)$ as time-variant deterministic function, enables us to derive closed-form solutions for the correlation functions (7.71a)–(7.71d), and, hence, also for the functions shown in Figure 7.13. This provides the basis for analysing the statistical properties of the deterministic channel model analytically. We will deal with this task in the following.

Therefore, we at first define the autocorrelation function of the time-variant deterministic impulse response $\tilde{h}(\tau', t)$ as follows

$$\tilde{r}_{hh}(\tau_1', \tau_2'; t, t + \tau) := \lim_{T \to \infty} \frac{1}{2T} \int_{-T}^{T} \tilde{h}^*(\tau_1', t)\, \tilde{h}(\tau_2', t + \tau)\, dt\,. \tag{7.72}$$

It should be taken into account that the time averaging, which has to be carried out here, is in contrast to (7.23a), whereas the computation of the autocorrelation function

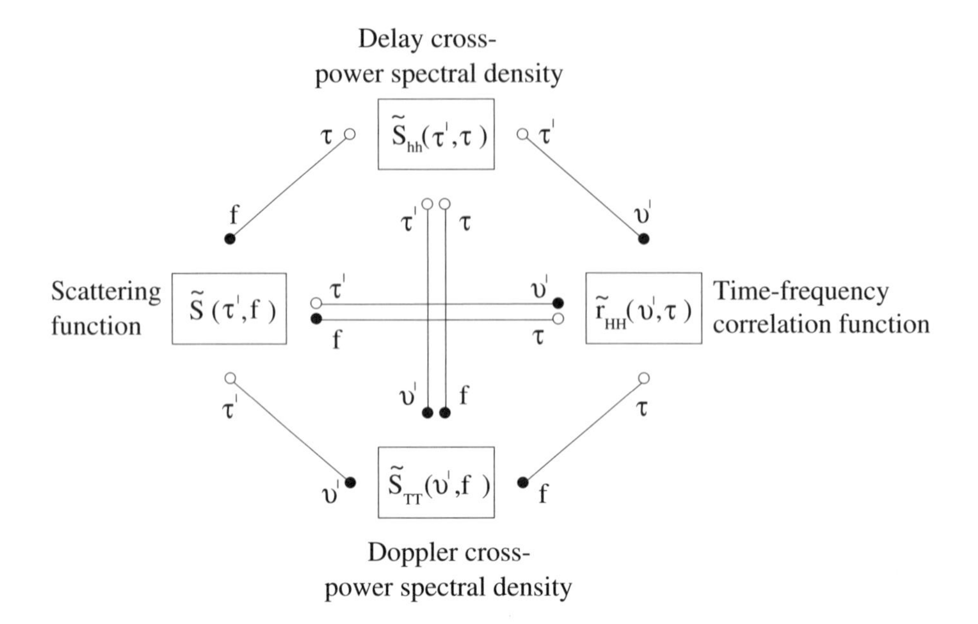

Figure 7.13: Relationships between the delay cross-power spectral density $\tilde{S}_{hh}(\tau', \tau)$, the time-frequency correlation function $\tilde{r}_{HH}(\upsilon', \tau)$, the scattering function $\tilde{S}(\tau', f)$, and the Doppler cross-power spectral density $\tilde{S}_{TT}(\upsilon', f)$ of DGUS models.

of the stochastic impulse response $h(\tau', t)$ requires statistical averaging (ensemble average). In the equation above, we use the expression (7.55) for $\tilde{h}(\tau', t)$, so that we may write

$$\begin{aligned}
\tilde{r}_{hh}(\tau_1', \tau_2'; t, t+\tau) &= \lim_{T\to\infty} \frac{1}{2T} \int_{-T}^{T} \left[\sum_{\ell=0}^{\mathcal{L}-1} \tilde{a}_\ell \, \tilde{\mu}_\ell^*(t) \, \delta(\tau_1' - \tilde{\tau}_\ell') \right] \\
&\qquad \cdot \left[\sum_{\lambda=0}^{\mathcal{L}-1} \tilde{a}_\lambda \, \tilde{\mu}_\lambda(t+\tau) \, \delta(\tau_2' - \tilde{\tau}_\lambda') \right] dt \\
&= \lim_{T\to\infty} \sum_{\ell=0}^{\mathcal{L}-1} \sum_{\lambda=0}^{\mathcal{L}-1} \tilde{a}_\ell \, \tilde{a}_\lambda \, \delta(\tau_1' - \tilde{\tau}_\ell') \, \delta(\tau_2' - \tilde{\tau}_\lambda') \\
&\qquad \cdot \frac{1}{2T} \int_{-T}^{T} \tilde{\mu}_\ell^*(t) \, \tilde{\mu}_\lambda(t+\tau) \, dt \, .
\end{aligned} \tag{7.73}$$

Using (7.58), it follows

$$\tilde{r}_{hh}(\tau_1', \tau_2'; t, t+\tau) = \sum_{\ell=0}^{\mathcal{L}-1} \tilde{a}_\ell^2 \, \tilde{r}_{\mu_\ell \mu_\ell}(\tau) \, \delta(\tau_1' - \tilde{\tau}_\ell') \, \delta(\tau_2' - \tilde{\tau}_\ell') \, . \tag{7.74}$$

Generally, the product of two delta functions is not defined. But in the present case, however, the first delta function appearing in (7.74) depends on the variable τ_1' and the second one on τ_2'. Since τ_1' and τ_2' are independent variables, the product will not cause any problems in the two-dimensional (τ_1', τ_2')-plane. Furthermore, $\delta(\tau_1' - \tilde{\tau}_\ell')\,\delta(\tau_2' - \tilde{\tau}_\ell')$ is equivalent to $\delta(\tau_1' - \tilde{\tau}_\ell')\,\delta(\tau_2' - \tilde{\tau}_1')$, so that (7.74) can be represented by

$$\tilde{r}_{hh}(\tau_1', \tau_2'; t, t+\tau) = \delta(\tau_2' - \tilde{\tau}_1')\,\tilde{S}_{hh}(\tau_1', \tau)\,, \tag{7.75}$$

where

$$\tilde{S}_{hh}(\tau', \tau) = \sum_{\ell=0}^{\mathcal{L}-1} \tilde{a}_\ell^2\,\tilde{r}_{\mu_\ell\mu_\ell}(\tau)\,\delta(\tau' - \tilde{\tau}_\ell') \tag{7.76}$$

denotes the *delay cross-power spectral density* of frequency-selective deterministic channel models. Note that (7.75) has the same form as (7.33). The delay cross-power spectral density $\tilde{S}_{hh}(\tau', \tau)$ can be computed explicitly in connection with the autocorrelation functions (7.59a) and (7.59b) if all model parameters $\{c_{i,n,\ell}\}$, $\{f_{i,n,\ell}\}$, $\{\tilde{a}_\ell\}$, $\{\tilde{\tau}_\ell'\}$, $\{N_{i,\ell}\}$, and $\mathcal{L}$ are known.

The Fourier transform of the delay cross-power spectral density $\tilde{S}_{hh}(\tau', \tau)$ with respect to the propagation delay τ' results in the *time-frequency correlation function*

$$\tilde{r}_{HH}(\upsilon', \tau) = \sum_{\ell=0}^{\mathcal{L}-1} \tilde{a}_\ell^2\,\tilde{r}_{\mu_\ell\mu_\ell}(\tau)\,e^{-j2\pi\upsilon'\tilde{\tau}_\ell'} \tag{7.77}$$

of the deterministic system.

Preferably, we also refer to the delay cross-power spectral density $\tilde{S}_{hh}(\tau', \tau)$ in order to calculate an analytical expression for the scattering function. The Fourier transform of (7.76) with respect to τ immediately leads to the expression

$$\tilde{S}(\tau', f) = \sum_{\ell=0}^{\mathcal{L}-1} \tilde{a}_\ell^2\,\tilde{S}_{\mu_\ell\mu_\ell}(f)\,\delta(\tau' - \tilde{\tau}_\ell')\,, \tag{7.78}$$

which describes the *scattering function* of frequency-selective deterministic channel models. In this equation,

$$\tilde{S}_{\mu_\ell\mu_\ell}(f) = \sum_{i=1}^{2}\sum_{n=1}^{N_{i,\ell}} \frac{c_{i,n,\ell}^2}{4}\left[\delta(f - f_{i,n,\ell}) + \delta(f + f_{i,n,\ell})\right]\,,\ \ell = 0, 1, \ldots, \mathcal{L}-1\,, \tag{7.79}$$

represents the Doppler power spectral density of the ℓth scattering component, which is defined by the Fourier transform of the autocorrelation function $\tilde{r}_{\mu_\ell\mu_\ell}(\tau)$ according to (7.59a). Now it becomes obvious that the scattering function $\tilde{S}(\tau', f)$ of deterministic channel models can be represented by a finite sum of weighted delta functions. The delta functions are located in the two-dimensional (τ', f)-plain at the positions $(\tilde{\tau}_\ell', \pm f_{i,n,\ell})$, and are weighted by the constants $(\tilde{a}_\ell c_{i,n,\ell})^2/4$. Without restriction of generality, we assume in the following that the scattering function $\tilde{S}(\tau', f)$ is normalized to unity, so that the volume under $\tilde{S}(\tau', f)$ is equal to one, i.e.,

$$\int_{-\infty}^{\infty}\int_{0}^{\infty} \tilde{S}(\tau', f)\,d\tau'\,df = 1\,. \tag{7.80}$$

To ensure that (7.80) is definitely fulfilled, the Doppler coefficients $c_{i,n,\ell}$ and the delay coefficients $\tilde{a}_\ell$ have to fulfil the boundary conditions

$$\sum_{n=1}^{N_{i,\ell}} c_{i,n,\ell}^2 = 1 \quad \text{and} \quad \sum_{\ell=0}^{\mathcal{L}-1} \tilde{a}_\ell^2 = 1\,. \tag{7.81a, b}$$

Finally, we determine the Fourier transform of the scattering function $\tilde{S}(\tau', f)$ with respect to τ' in order to obtain the *Doppler cross-power spectral density*

$$\tilde{S}_{TT}(\upsilon', f) = \sum_{\ell=0}^{\mathcal{L}-1} \tilde{a}_\ell^2 \, \tilde{S}_{\mu_\ell\mu_\ell}(f) \, e^{-j2\pi\upsilon'\tilde{\tau}'_\ell} \tag{7.82}$$

of frequency-selective deterministic channel models. We can easily assure ourselves of the fact that one also obtains the Doppler cross-power spectral density $\tilde{S}_{TT}(\upsilon', f)$ in the presented form (7.82), if the alternative possibility — via the Fourier transform of the time-frequency correlation function $\tilde{r}_{HH}(\upsilon', \tau)$ with respect to τ — is made use of, where in this case the relation (7.77) has to be used for $\tilde{r}_{HH}(\upsilon', \tau)$.

Thus, it has been shown that the four functions $\tilde{S}_{hh}(\tau', \tau)$, $\tilde{r}_{HH}(\upsilon', \tau)$, $\tilde{S}(\tau, f)$, and $\tilde{S}_{TT}(\upsilon', f)$ characterizing the deterministic system can be calculated analytically, if the relevant model parameters $\{c_{i,n,\ell}\}$, $\{f_{i,n,\ell}\}$, $\{\tilde{a}_\ell\}$, $\{\tilde{\tau}'_\ell\}$, $\{N_{i,\ell}\}$, and $\mathcal{L}$ are known.

7.4.3 Delay Power Spectral Density, Doppler Power Spectral Density, and Characteristic Quantities of DGUS Models

In this subsection, simple closed-form solutions will be derived for the fundamental characteristic functions and quantities of DGUS models, such as the delay power spectral density, Doppler power spectral density, and delay spread. For this purpose, we will here discuss the terms introduced for stochastic models (WSSUS models) in Subsection 7.3.2.3 for deterministic systems.

Delay power spectral density: Let $\tilde{S}(\tau', f)$ be the scattering function of a deterministic channel model, then, by analogy to (7.38), the corresponding *Delay power spectral density* $\tilde{S}_{\tau'\tau'}(\tau')$ is defined by

$$\begin{aligned} \tilde{S}_{\tau'\tau'}(\tau') &:= \tilde{S}_{hh}(\tau', 0) \\ &= \int_{-\infty}^{\infty} \tilde{S}(\tau', f) \, df\,. \end{aligned} \tag{7.83}$$

After employing (7.78) and considering the boundary condition (7.81a), it follows

$$\tilde{S}_{\tau'\tau'}(\tau') = \sum_{\ell=0}^{\mathcal{L}-1} \tilde{a}_\ell^2 \, \delta(\tau' - \tilde{\tau}'_\ell)\,. \tag{7.84}$$

Hence, the delay power spectral density $\tilde{S}_{\tau'\tau'}(\tau')$ is a discrete line spectrum, where the spectral lines are located at the discrete positions $\tau' = \tilde{\tau}'_\ell$ and weighted by the constants $\tilde{a}_\ell^2$. Consequently, the behaviour of $\tilde{S}_{\tau'\tau'}(\tau')$ is completely determined by the model parameters $\tilde{a}_\ell$, $\tilde{\tau}'_\ell$, and $\mathcal{L}$. It should be pointed out that the area under the Delay power spectral density $\tilde{S}_{\tau'\tau'}(\tau')$ is equal to one due to (7.81b), i.e., $\int_0^\infty \tilde{S}_{\tau'\tau'}(\tau')\,d\tau' = 1$.

Average delay: Let $\tilde{S}_{\tau'\tau'}(\tau')$ be the delay power spectral density of a deterministic channel model, then the first moment of $\tilde{S}_{\tau'\tau'}(\tau')$ is called the *average delay* $\tilde{B}^{(1)}_{\tau'\tau'}$. Thus, by analogy to (7.39), the definition

$$\tilde{B}^{(1)}_{\tau'\tau'} := \frac{\int_{-\infty}^{\infty} \tau' \, \tilde{S}_{\tau'\tau'}(\tau')\,d\tau'}{\int_{-\infty}^{\infty} \tilde{S}_{\tau'\tau'}(\tau')\,d\tau'} \tag{7.85}$$

holds. Putting (7.84) in (7.85) and taking the boundary condition (7.81b) into account, then the average delay $\tilde{B}^{(1)}_{\tau'\tau'}$ can be expressed in closed form as follows

$$\tilde{B}^{(1)}_{\tau'\tau'} = \sum_{\ell=0}^{\mathcal{L}-1} \tilde{\tau}'_\ell \, \tilde{a}_\ell^2 \,. \tag{7.86}$$

Delay spread: The square root of the second central moment of $\tilde{S}_{\tau'\tau'}(\tau')$ is called the *Delay spread* $\tilde{B}^{(2)}_{\tau'\tau'}$, which is, by analogy to (7.40), defined by

$$\tilde{B}^{(2)}_{\tau'\tau'} := \sqrt{\frac{\int_{-\infty}^{\infty} \left(\tau' - \tilde{B}^{(1)}_{\tau'\tau'}\right)^2 \tilde{S}_{\tau'\tau'}(\tau')\,d\tau'}{\int_{-\infty}^{\infty} \tilde{S}_{\tau'\tau'}(\tau')\,d\tau'}} \,. \tag{7.87}$$

With (7.84) and (7.81b), the closed-form expression

$$\tilde{B}^{(2)}_{\tau'\tau'} = \sqrt{\sum_{\ell=0}^{\mathcal{L}-1} (\tilde{\tau}'_\ell \, \tilde{a}_\ell)^2 - \left(\tilde{B}^{(1)}_{\tau'\tau'}\right)^2} \tag{7.88}$$

can be derived, where $\tilde{B}^{(1)}_{\tau'\tau'}$ is the average delay according to (7.86).

Doppler power spectral density: Let $\tilde{S}(\tau', f)$ be the scattering function of a deterministic channel model, then — by analogy to (7.42) — the corresponding *Doppler power spectral density* $\tilde{S}_{\mu\mu}(f)$ can be determined via the relation

$$\begin{aligned} \tilde{S}_{\mu\mu}(f) &:= \tilde{S}_{TT}(0, f) \\ &= \int_{-\infty}^{\infty} \tilde{S}(\tau', f)\,d\tau' \,. \end{aligned} \tag{7.89}$$

With the scattering function $\tilde{S}(\tau', f)$ given by (7.78), we can now derive a closed-form solution for the Doppler power spectral density $\tilde{S}_{\mu\mu}(f)$ of the deterministic system.

Thus, we obtain

$$\tilde{S}_{\mu\mu}(f) = \sum_{\ell=0}^{\mathcal{L}-1} \tilde{a}_\ell^2 \tilde{S}_{\mu_\ell\mu_\ell}(f)\,, \tag{7.90}$$

where $\tilde{S}_{\mu_\ell\mu_\ell}(f)$ denotes the Doppler power spectral density of the ℓth scattering component determined by (7.79). This result shows that the Doppler power spectral density $\tilde{S}_{\mu\mu}(f)$ of frequency-selective deterministic channel models is given by the sum of the Doppler power spectral densities $\tilde{S}_{\mu_\ell\mu_\ell}(f)$ of all propagation paths $\ell = 0, 1, \ldots, \mathcal{L}$, where each individual Doppler power spectral density $\tilde{S}_{\mu_\ell\mu_\ell}(f)$ has to be weighted by the square of the corresponding delay coefficient. Here, the square of the delay coefficient $\tilde{a}_\ell^2$ represents the *path power*, that is the mean (average) power of the ℓth scattering component.

With knowledge of the Doppler power spectral density $\tilde{S}_{\mu\mu}(f)$ or $\tilde{S}_{\mu_\ell\mu_\ell}(f)$, the *average Doppler shift* and the *Doppler spread* can be computed. The definition, derivation, and discussion of these characteristic quantities have already been performed in Section 4.2. We will refrain from a recapitulation of these results at this place.

Frequency correlation function: Let $\tilde{r}_{HH}(\upsilon', \tau)$ be the time-frequency correlation function of a deterministic channel model. Then, by analogy to (7.44), the *frequency correlation function* $\tilde{r}_{\tau'\tau'}(\upsilon')$ is defined by the time-frequency correlation function $\tilde{r}_{HH}(\upsilon', \tau)$ at $\tau = t_2 - t_1 = 0$, i.e.,

$$\begin{aligned}\tilde{r}_{\tau'\tau'}(\upsilon') &:= \tilde{r}_{HH}(\upsilon', 0)\\ &= \int_{-\infty}^{\infty}\int_{-\infty}^{\infty} \tilde{S}(\tau', f)\, e^{-j2\pi\upsilon'\tau'}\, d\tau'\, df\\ &= \int_{-\infty}^{\infty} \tilde{S}_{\tau'\tau'}(\tau')\, e^{-j2\pi\upsilon'\tau'}\, d\tau'\,.\end{aligned} \tag{7.91}$$

A closed-form expression for the frequency correlation function $\tilde{r}_{\tau'\tau'}(\upsilon')$ is obtained in a simple way by setting $\tau = 0$ in (7.77). Taking the boundary condition (7.81a) into consideration, which implies that $\tilde{r}_{\mu_\ell\mu_\ell}(0) = 1$ holds for all $\ell = 0, 1, \ldots, \mathcal{L}-1$, we then obtain

$$\tilde{r}_{\tau'\tau'}(\upsilon') = \sum_{\ell=0}^{\mathcal{L}-1} \tilde{a}_\ell^2\, e^{-j2\pi\upsilon'\tilde{\tau}'_\ell}\,. \tag{7.92}$$

Coherence bandwidth: Let $\tilde{r}_{\tau'\tau'}(\upsilon')$ be the frequency correlation function given by (7.92), then the frequency separation variable $\upsilon' = \tilde{B}_C$ for which

$$|\tilde{r}_{\tau'\tau'}(\tilde{B}_C)| = \frac{1}{2}|\tilde{r}_{\tau'\tau'}(0)| \tag{7.93}$$

holds, is called the *coherence bandwidth* of deterministic channel models. With (7.92) and taking the boundary condition (7.81b) into consideration, we obtain the transcendental equation

$$\left| \sum_{\ell=0}^{\mathcal{L}-1} \tilde{a}_\ell^2 \, e^{-j2\pi \tilde{B}_C \tau'_\ell} \right| - \frac{1}{2} = 0 \,. \tag{7.94}$$

The smallest positive value for $\tilde{B}_C$ which fulfils the equation above is the coherence bandwidth. Apart from simple special cases, (7.94) has generally to be solved by means of numerical *root-finding techniques.* The *Newton-Raphson method* is one of the most powerful and well-known numerical methods for solving root-finding problems.

Time correlation function: Let $\tilde{r}_{HH}(\upsilon', \tau)$ be the time-frequency correlation function of a deterministic channel model. Then, by analogy to (7.50), the *time correlation function* $\tilde{r}_{\mu\mu}(\tau)$ is defined by the time-frequency correlation function $\tilde{r}_{HH}(\upsilon', \tau)$ at $\upsilon' = f'_2 - f'_1 = 0$, i.e.,

$$\begin{aligned} \tilde{r}_{\mu\mu}(\tau) &:= \tilde{r}_{HH}(0, \tau) \\ &= \int_{-\infty}^{\infty} \int_{-\infty}^{\infty} \tilde{S}(\tau', f) \, e^{j2\pi f \tau} \, d\tau' \, df \\ &= \int_{-\infty}^{\infty} \tilde{S}_{\mu\mu}(f) \, e^{j2\pi f \tau} \, df \,. \end{aligned} \tag{7.95}$$

We consider (7.77) at $\upsilon' = 0$ and, thus, obtain

$$\tilde{r}_{\mu\mu}(\tau) = \sum_{\ell=0}^{\mathcal{L}-1} \tilde{a}_\ell^2 \, \tilde{r}_{\mu_\ell \mu_\ell}(\tau) \,. \tag{7.96}$$

Coherence time: Let $\tilde{r}_{\mu\mu}(\tau)$ be the time correlation function given by (7.96), then the time interval $\tau = \tilde{T}_C$ for which

$$|\tilde{r}_{\mu\mu}(\tilde{T}_C)| = \frac{1}{2} |\tilde{r}_{\mu\mu}(0)| \tag{7.97}$$

holds, is called the *coherence time* of deterministic channel models. Substituting (7.96) in (7.97), and taking (7.59a) and (7.59b) into account, results in the transcendental equation

$$\left| \sum_{i=1}^{2} \sum_{\ell=0}^{\mathcal{L}-1} \sum_{n=1}^{N_{i,\ell}} \frac{(\tilde{a}_\ell c_{i,n,\ell})^2}{2} \cos(2\pi f_{i,n,\ell} \tilde{T}_C) \right| - \frac{1}{2} = 0 \,, \tag{7.98}$$

from which the coherence time $\tilde{T}_C$ can be computed by applying numerical zero finding techniques. The smallest positive value for $\tilde{T}_C$ which solves (7.98) is the coherence time.

In order to facilitate an overview, the above derived relationships between the correlation functions and the power spectral densities as well as the characteristic quantities of frequency-selective deterministic channel models derivable from these are depicted in Figure 7.14.

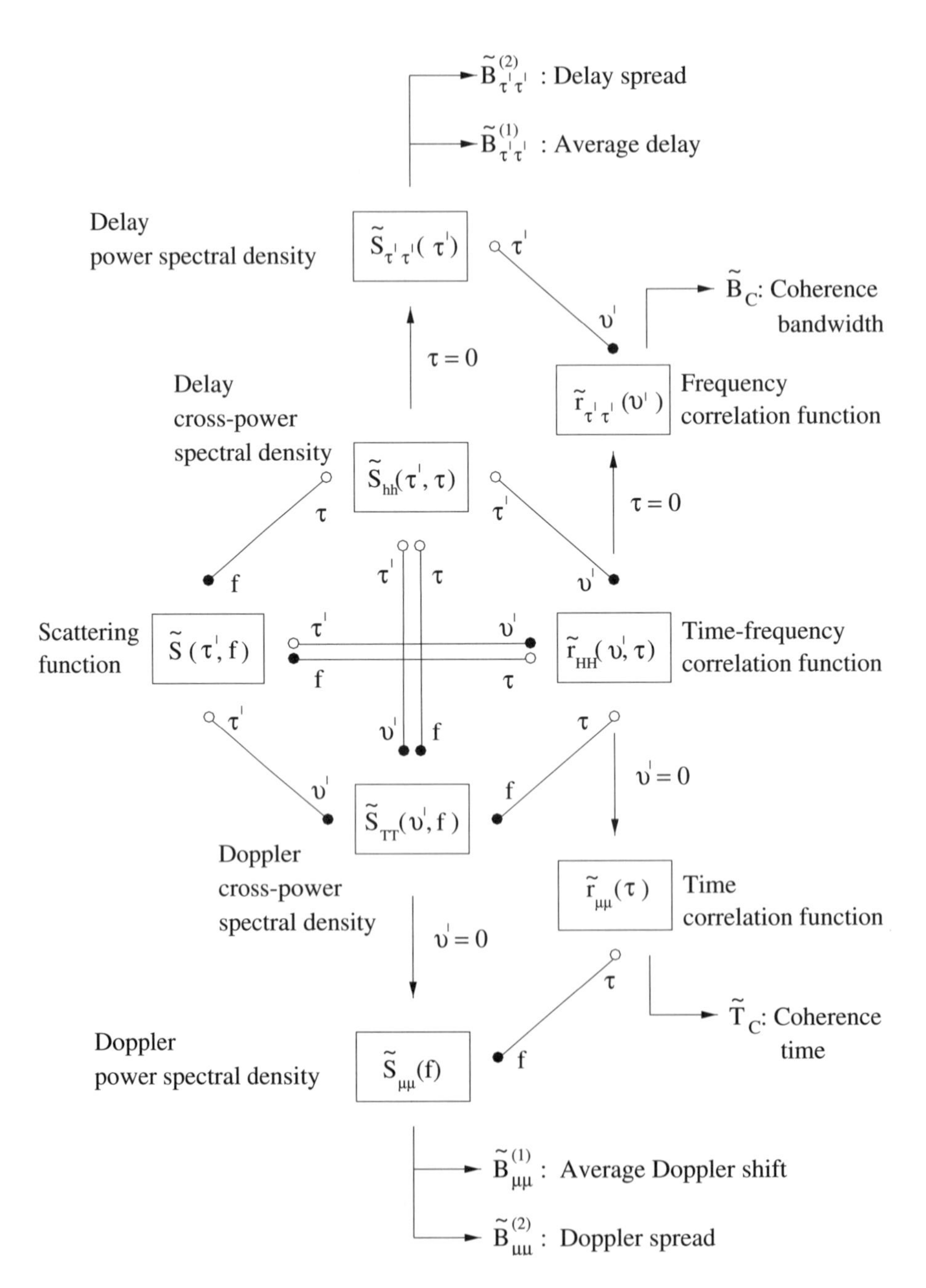

Figure 7.14: Relationships between the correlation functions, the power spectral densities, and the characteristic quantities of DGUS models.

7.4.4 Determination of the Model Parameters of DGUS Models

In this subsection, we are concerned with the determination of the model parameters $\tilde{\tau}'_\ell$, $\tilde{a}_\ell$, $f_{i,n,\ell}$, $c_{i,n\ell}$, and $\theta_{i,n,\ell}$ of the simulation model shown in Figure 7.11, and, thus, of the DGUS model determined by (7.55). The starting point of the procedure described here is the scattering function $S(\tau', f)$ of a given stochastic channel model. Since the procedure is universally valid, $S(\tau', f)$ can, for example, be any specified scattering function. The method may just as well be applied if $S(\tau', f)$ is the result of an evaluation of a single snapshot measurement obtained from a real-world channel.

From the scattering function $S(\tau', f)$, which is assumed to be known henceforth, the corresponding delay power spectral density $S_{\tau'\tau'}(\tau')$ and the Doppler power spectral density $S_{\mu\mu}(f)$ are determined first. For this purpose, we use the relations

$$S_{\tau'\tau'}(\tau') = \int_{-\infty}^{\infty} S(\tau', f)\, df \quad \text{and} \quad S_{\mu\mu}(f) = \int_{-\infty}^{\infty} S(\tau', f)\, d\tau' , \qquad \text{(7.99a, b)}$$

which are defined by (7.38) and (7.42), respectively. The causality condition (7.12) leads to $S_{\tau'\tau'}(\tau') = 0$ if $\tau' < 0$. Furthermore, we assume that all scattering components with propagation delays $\tau' > \tau'_{max}$ can be ignored. For the Delay power spectral density, we then may generally write

$$S_{\tau'\tau'}(\tau') = 0 \quad \text{for} \quad \tau' \notin I = [0, \tau'_{max}] \, . \qquad (7.100)$$

Next, we perform a partition of the interval $I = [0, \tau'_{max}]$ into a number of $\mathcal{L}$ disjoint subintervals I_ℓ according to $I = \bigcup_{\ell=0}^{\mathcal{L}-1} I_\ell$. This partition is realized in a way that allows us to consider the Delay power spectral density $S_{\tau'\tau'}(\tau')$ and the Doppler power spectral density $S_{\mu_\ell\mu_\ell}(f)$ appertaining to I_ℓ as independent within each subinterval I_ℓ. From this, it follows that the scattering function $S(\tau', f)$ can be expressed by means of $S_{\tau'\tau'}(\tau')$ and $S_{\mu_\ell\mu_\ell}(f)$ as

$$S(\tau', f) = \sum_{\ell=0}^{\mathcal{L}-1} S_{\mu_\ell\mu_\ell}(f)\, S_{\tau'\tau'}(\tau') \Big|_{\tau' \in I_\ell} \, . \qquad (7.101)$$

Continuing from this form, we will now determine the model parameters of the deterministic system.

7.4.4.1 Determination of the discrete propagation delays and delay coefficients

The discrete propagation delays $\tilde{\tau}'_\ell$ are integer multiples of the sampling interval T'_s, i.e.,

$$\tilde{\tau}'_\ell = \ell \cdot T'_s \, , \quad \ell = 0, 1, \ldots, \mathcal{L} - 1 \, , \qquad (7.102)$$

where the number of discrete paths $\mathcal{L}$ with different propagation delays is given by

$$\mathcal{L} = \left\lfloor \frac{\tau'_{max}}{T'_s} \right\rfloor + 1 \, . \qquad (7.103)$$

Thus, the ratio τ'_{max}/T'_s determines the number of delay elements shown in Figure 7.11. Note that $\mathcal{L} \to \infty$ as $T'_s \to 0$.

With the discrete propagation delays $\tilde{\tau}'_\ell$ given by (7.102) and the sampling interval T'_s, the subintervals I_ℓ required for the partition of the interval $I = [0, \tau'_{max}] = \bigcup_{\ell=0}^{\mathcal{L}-1} I_\ell$ can be defined as follows:

$$I_\ell := \begin{cases} [0,\, T'_s/2) & \text{for} \quad \ell = 0\,, \\ [\tilde{\tau}'_\ell - T'_s/2,\, \tilde{\tau}'_\ell + T'_s/2) & \text{for} \quad \ell = 1, 2, \ldots, \mathcal{L}-2\,, \\ [\tilde{\tau}'_\ell - T'_s/2,\, \tau'_{max}] & \text{for} \quad \ell = \mathcal{L}-1\,. \end{cases} \tag{7.104}$$

Next, we demand that the areas under the delay power spectral densities $S_{\tau'\tau'}(\tau')$ and $\tilde{S}_{\tau'\tau'}(\tau')$ are identical within each subinterval I_ℓ, i.e., we demand that

$$\int_{\tau' \in I_\ell} S_{\tau'\tau'}(\tau')\, d\tau' = \int_{\tau' \in I_\ell} \tilde{S}_{\tau'\tau'}(\tau')\, d\tau' \tag{7.105}$$

holds for all $\ell = 0, 1, \ldots, \mathcal{L}-1$. Substituting $\tilde{S}_{\tau'\tau'}(\tau')$ by the expression (7.84) in the right-hand side of the equation above and applying the sifting property of delta functions leads directly to the following explicit formula for the delay coefficients

$$\tilde{a}_\ell = \sqrt{\int_{\tau' \in I_\ell} S_{\tau'\tau'}(\tau')\, d\tau'}\,, \quad \ell = 0, 1, \ldots, \mathcal{L}-1\,, \tag{7.106}$$

where I_ℓ are the subintervals defined by (7.104). This result shows that the delay coefficient $\tilde{a}_\ell$ of the ℓth propagation path is the square root of the average path power within the subinterval I_ℓ.

Next, we will consider the limit of the delay power spectral density $\tilde{S}_{\tau'\tau'}(\tau')$ for $\mathcal{L} \to \infty$ and $T'_s \to 0$. For this purpose, we substitute (7.106) into (7.84) and obtain [Pae95b]

$$\begin{aligned} \lim_{\substack{\mathcal{L}\to\infty \\ T'_s\to 0}} \tilde{S}_{\tau'\tau'}(\tau') &= \lim_{\substack{\mathcal{L}\to\infty \\ T'_s\to 0}} \sum_{\ell=0}^{\mathcal{L}-1} \left[\int_{\tau' \in I_\ell} S_{\tau'\tau'}(\tau')\, d\tau' \right] \delta(\tau' - \tilde{\tau}'_\ell) \\ &= \lim_{\mathcal{L}\to\infty} \sum_{\ell=0}^{\mathcal{L}-1} S_{\tau'\tau'}(\tilde{\tau}'_\ell)\, \delta(\tau' - \tilde{\tau}'_\ell)\; \Delta\tilde{\tau}'_\ell \\ &= \int_0^\infty S_{\tau'\tau'}(\tilde{\tau}'_\ell)\, \delta(\tau' - \tilde{\tau}'_\ell)\; d\tilde{\tau}'_\ell \\ &= S_{\tau'\tau'}(\tau')\,. \end{aligned} \tag{7.107}$$

Thus, it becomes obvious that $\tilde{S}_{\tau'\tau'}(\tau')$ converges to $S_{\tau'\tau'}(\tau')$ if the number of discrete propagation paths $\mathcal{L}$ tends to infinity. Consequently, this also holds for the average delay $\tilde{B}^{(1)}_{\tau'\tau'}$ and the delay spread $\tilde{B}^{(2)}_{\tau'\tau'}$ of the simulation model, i.e., we obtain $\tilde{B}^{(1)}_{\tau'\tau'} \to B^{(1)}_{\tau'\tau'}$ as $\mathcal{L} \to \infty$ ($T'_s \to 0$). For $\mathcal{L} < \infty$ ($T'_s > 0$), however, we generally have to write $\tilde{B}^{(1)}_{\tau'\tau'} \approx B^{(1)}_{\tau'\tau'}$ and $\tilde{B}^{(2)}_{\tau'\tau'} \approx B^{(2)}_{\tau'\tau'}$. Especially for the delay power spectral densities of the channel models according to COST 207, which are depicted in Figure 7.7 (see also Table 7.1), the quality of the approximation $\tilde{B}^{(i)}_{\tau'\tau'} \approx B^{(i)}_{\tau'\tau'}$ is shown for $i = 1, 2$ in Figures 7.15(a)–7.15(d) as a function of the number of discrete propagation paths $\mathcal{L}$.

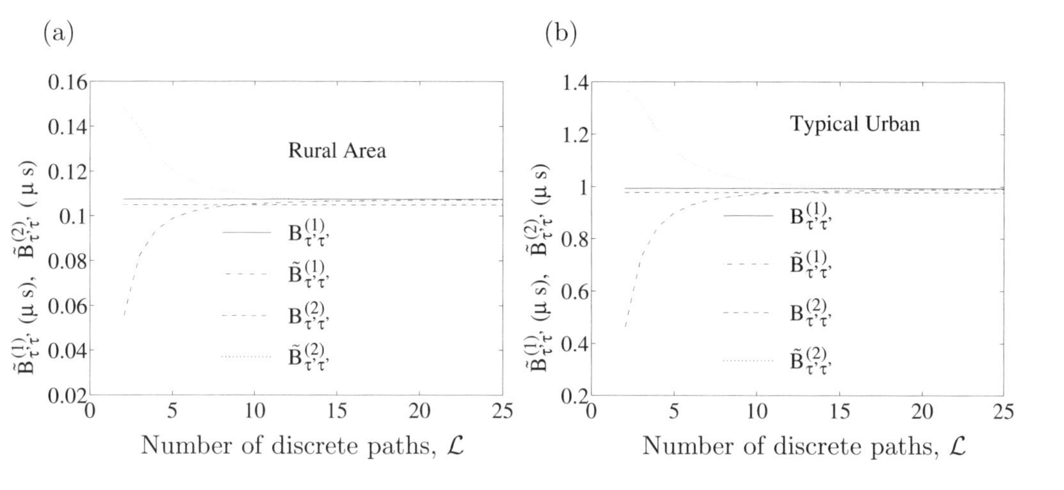

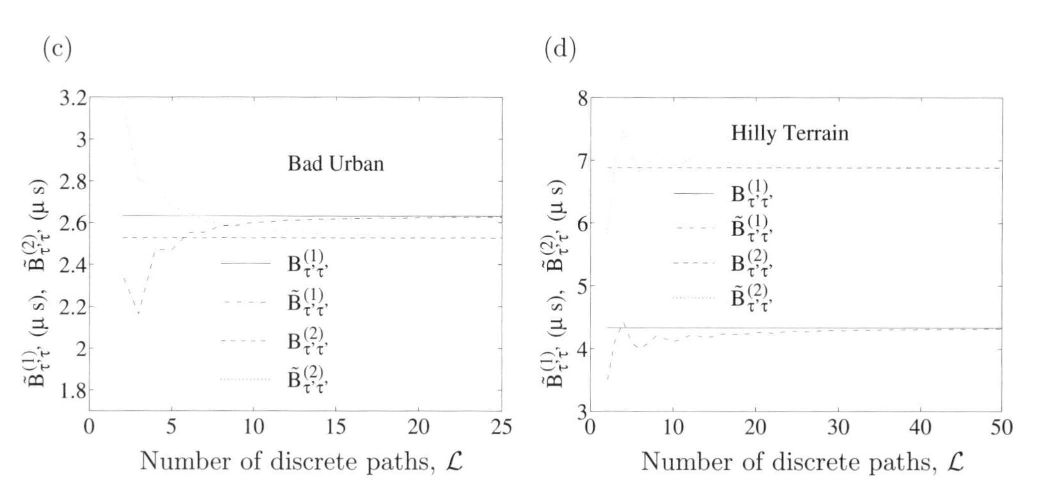

Figure 7.15: Average delay $\tilde{B}^{(1)}_{\tau'\tau'}$ and delay spread $\tilde{B}^{(2)}_{\tau'\tau'}$ of the delay power spectral densities according to COST 207 [COS89]: (a) Rural Area, (b) Typical Urban, (c) Bad Urban, and (d) Hilly Terrain.

7.4.4.2 Determination of the discrete Doppler frequencies and Doppler coefficients

The discrete Doppler frequencies $f_{i,n,\ell}$ and the Doppler coefficients $c_{i,n,\ell}$ can be determined by applying the methods described in Section 5.1. As well as the method of exact Doppler spread (MEDS), we preferably also use the L_p-norm method (LPNM). The first method mentioned is especially recommended for the Jakes power spectral density. However, it has to be taken into account that the complex deterministic Gaussian processes $\tilde{\mu}_\ell(t)$ are designed in such a way that $\tilde{\mu}_\ell(t)$ and $\tilde{\mu}_\lambda(t)$ are uncorrelated for $\ell \neq \lambda$ ($\ell, \lambda = 0, 1, \ldots, \mathcal{L}-1$). This is always the case if the discrete Doppler frequencies $f_{i,n,\ell}$ fulfil the condition (7.57). Using the MEDS, this condition is always fulfilled in case the number of harmonic functions $N_{i,\ell}$ are chosen in such a way that $N_{i,\ell} \neq N_{j,\lambda}$ holds if $\ell \neq \lambda$, where $i, j = 1, 2$ and $\ell, \lambda = 0, 1, \ldots, \mathcal{L}-1$. However, when using the LPNM method, we do not have to take the inequality $N_{i,\ell} \neq N_{j,\lambda}$ into account, because even for $N_{i,\ell} = N_{j,\lambda}$, disjoint sets $\{f_{i,n,\ell}\}$ and $\{f_{j,m,\lambda}\}$ with $\ell \neq \lambda$ can easily be found, so that the resulting deterministic processes $\tilde{\mu}_\ell(t)$ and $\tilde{\mu}_\lambda(t)$ are uncorrelated for $\ell \neq \lambda$. For this purpose, it is sufficient to either minimize the L_p-norm (5.61) by using different values for the parameter p, or by performing the optimization of each set $\{f_{i,n,\ell}\}$ of discrete Doppler frequencies with different values for the quantity τ_{max} defining the upper limit of the integral in (5.61). Having this in mind, the numerical optimization of the autocorrelation function $\tilde{r}_{\mu_{i,\ell}\mu_{i,\ell}}(\tau)$ [see (7.59b)] guarantees that the desired property

$$\{f_{i,n,\ell}\} \bigcap \{f_{j,m,\lambda}\} = \emptyset \quad \Longleftrightarrow \quad \ell \neq \lambda \tag{7.108}$$

is usually fulfilled for all $i, j = 1, 2$, $n = 1, 2, \ldots, N_{i,\ell}$, $m = 1, 2, \ldots, N_{j,\lambda}$, and $\ell, \lambda = 0, 1, \ldots, \mathcal{L}-1$.

7.4.4.3 Determination of the Doppler phases

In Subsections 7.4.2 and 7.4.3, it was pointed out that the Doppler phases $\theta_{i,n,\ell}$ have no influence on the system functions shown in Figure 7.14. Hence, we may conclude that the fundamental statistical properties of DGUS models are independent of the choice of the Doppler phases $\theta_{i,n,\ell}$. The statements made in Section 5.2 are still valid for the frequency-selective case. Therefore, we once again may assume that the Doppler phases $\theta_{i,n,\ell}$ are realizations of a random variable, uniformly distributed in the interval $(0, 2\pi]$. Alternatively, $\theta_{i,n,\ell}$ can also be determined by applying the deterministic procedure described in Section 5.2. In both cases different events (sets) $\{\theta_{i,n,\ell}\}$ always result in different realizations (sample functions) for the time-variant impulse response $\tilde{h}(\tau', t)$ but, nevertheless, all impulse responses have the same statistical properties. In other words: every realization of the impulse response $\tilde{h}(\tau', t)$ contains the complete statistical information.

7.4.5 Deterministic Simulation Models for the Channel Models According to COST 207

At the end of this chapter, we will once more pick up the channel models according to COST 207 [COS89] and will show how to develop efficient simulation models for them. For this purpose, we restrict our attention to the 4-path and 6-path channel models (RA, TU, BU, HT) specified in Table 7.3. Due to the fact that these models are already presented in a discrete form with respect to τ', the discrete propagation delays $\tilde{\tau}'_\ell$ can directly be equated with the values for τ'_ℓ, presented in Table 7.3, i.e.,

$\tilde{\tau}'_\ell = \tau'_\ell$. Adaptation of the sampling interval T'_s to the discrete propagation delays $\tilde{\tau}'_\ell$ is achieved here by $\tilde{\tau}'_\ell = q_\ell \cdot T'_s$, where q_ℓ denotes an integer and T'_s is the greatest common divisor of $\tau'_1, \tau'_2, \ldots, \tau'_{\mathcal{L}-1}$, i.e., $T'_s = \gcd\{\tau'_\ell\}_{\ell=1}^{\mathcal{L}-1}$. The corresponding delay coefficients $\tilde{a}_\ell$ are identical to the square root of the path power as listed in Table 7.3.

The specifications for the Doppler power spectral density can be found in Table 7.2. In the case of the Jakes power spectral density, we determine the model parameters $f_{i,n,\ell}$ and $c_{i,n,\ell}$ by applying the L_p-norm method described in Subsection 5.1.5, taking into account that (7.108) is fulfilled. For the Gaussian power spectral densities (Gauss I and Gauss II), the third variant of the L_p-norm method (LPNM III) is of advantage. For the solution of the present problem, it is recommended to start with a Gaussian random process $\nu_{i,\ell}(t)$ having a symmetrical Gaussian power spectral density of the form

$$S_{\nu_{i,\ell}\nu_{i,\ell}}(f) = A_{i,\ell}\, e^{-\frac{f^2}{2s_{i,\ell}^2}}, \quad i = 1, 2, \tag{7.109}$$

and then perform a frequency shift of $f_{i,0,\ell}$, which finally results in

$$S_{\mu_\ell\mu_\ell}(f) = \sum_{i=1}^{2} S_{\nu_{i,\ell}\nu_{i,\ell}}(f - f_{i,0,\ell}), \tag{7.110}$$

where $A_{i,\ell}$, $s_{i,\ell}$, and $f_{i,0,\ell}$ denote the quantities specified in Table 7.2. The autocorrelation function required for the minimization of the error function (5.65) is in the present case given by the inverse Fourier transform of (7.109), i.e.,

$$r_{\nu_{i,\ell}\nu_{i,\ell}}(\tau) = \sigma_{i,\ell}^2\, e^{-2(\pi s_{i,\ell}\tau)^2}, \tag{7.111}$$

where $\sigma_{i,\ell}^2 = \sqrt{2\pi}\, A_{i,\ell}\, s_{i,\ell}$ describes the variance of the Gaussian random process $\nu_{i,\ell}(t)$. For the simulation model this means that we first have to determine the model parameters $f_{i,n,\ell}$ and $c_{i,n,\ell}$ of the deterministic process

$$\tilde{\nu}_{i,\ell}(t) = \sum_{n=1}^{N_{i,\ell}} c_{i,n,\ell} \cos(2\pi f_{i,n,\ell} + \theta_{i,n,\ell}) \tag{7.112}$$

by using the LPNM III. The application of the frequency translation theorem of the Fourier transform then provides the demanded complex deterministic Gaussian process in the form

$$\begin{aligned} \tilde{\mu}_\ell(t) &= \sum_{i=1}^{2} \tilde{\nu}_{i,\ell}(t)\, e^{-j2\pi f_{i,0,\ell} t} \\ &= \sum_{i=1}^{2} \tilde{\nu}_{i,\ell}(t) \cos(2\pi f_{i,0,\ell} t) - j \sum_{i=1}^{2} \tilde{\nu}_{i,\ell}(t) \sin(2\pi f_{i,0,\ell} t)\,. \end{aligned} \tag{7.113}$$

The resulting simulation model for the complex deterministic process $\tilde{\mu}_\ell(t)$ is shown in Figure 7.16.

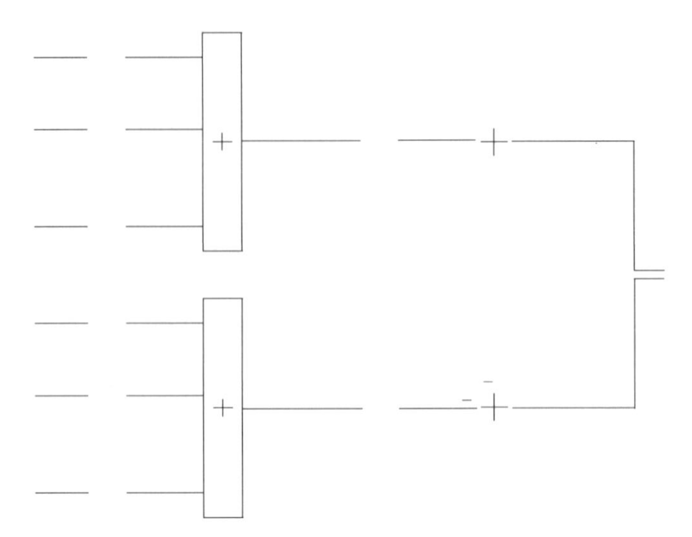

Figure 7.16: Simulation model for complex deterministic Gaussian processes $\tilde{\mu}_\ell(t)$ by using the frequency-shifted Gaussian power spectral densities according to COST 207 [see Table 7.2].

Using the L_p-norm method, we have the chance to choose an equal number of harmonic functions $N_{i,\ell}$ not only for all propagation paths, but also for the corresponding real and imaginary parts, without violating the condition (7.108). As an example, we fix $N_{i,\ell}$ for the $\mathcal{L}$-path channel models according to COST 207 by $N_{i,\ell} = 10$ ($\forall\, i = 1, 2$, and $\ell = 0, 1, \ldots, \mathcal{L} - 1$). For the maximum Doppler frequency f_{max}, we choose the value 91 Hz. Now, the remaining model parameters can be computed by using the method described above. Knowing the model parameters, not only the scattering function $\tilde{S}(\tau', f)$ [see (7.78)], but also all other correlation functions, power spectral densities, and characteristic quantities shown in Figure 7.14 can be determined analytically. For example, the resulting scattering functions $\tilde{S}(\tau', f)$ of the deterministic simulation models are shown in Figures 7.17(a)–7.17(d) for the $\mathcal{L}$-path channel models listed in Table 7.3.

At the end of this chapter it should be mentioned that the processing of the discrete input signal $x(kT_s')$ and the corresponding output signal $y(kT_s')$ is performed with the sampling rate $f_s' = 1/T_s'$, whereas the sampling of the complex deterministic Gaussian process $\tilde{\mu}_\ell(t)$ ($\ell = 0, 1, \ldots, \mathcal{L} - 1$) takes place at the discrete time instants $t = k\,T_s = k\,m_s'\,T_s'$. It should be noted that the statements made in Subsection 7.4.1 have to be taken into account for the choice of the sampling rate ratio m_s'.

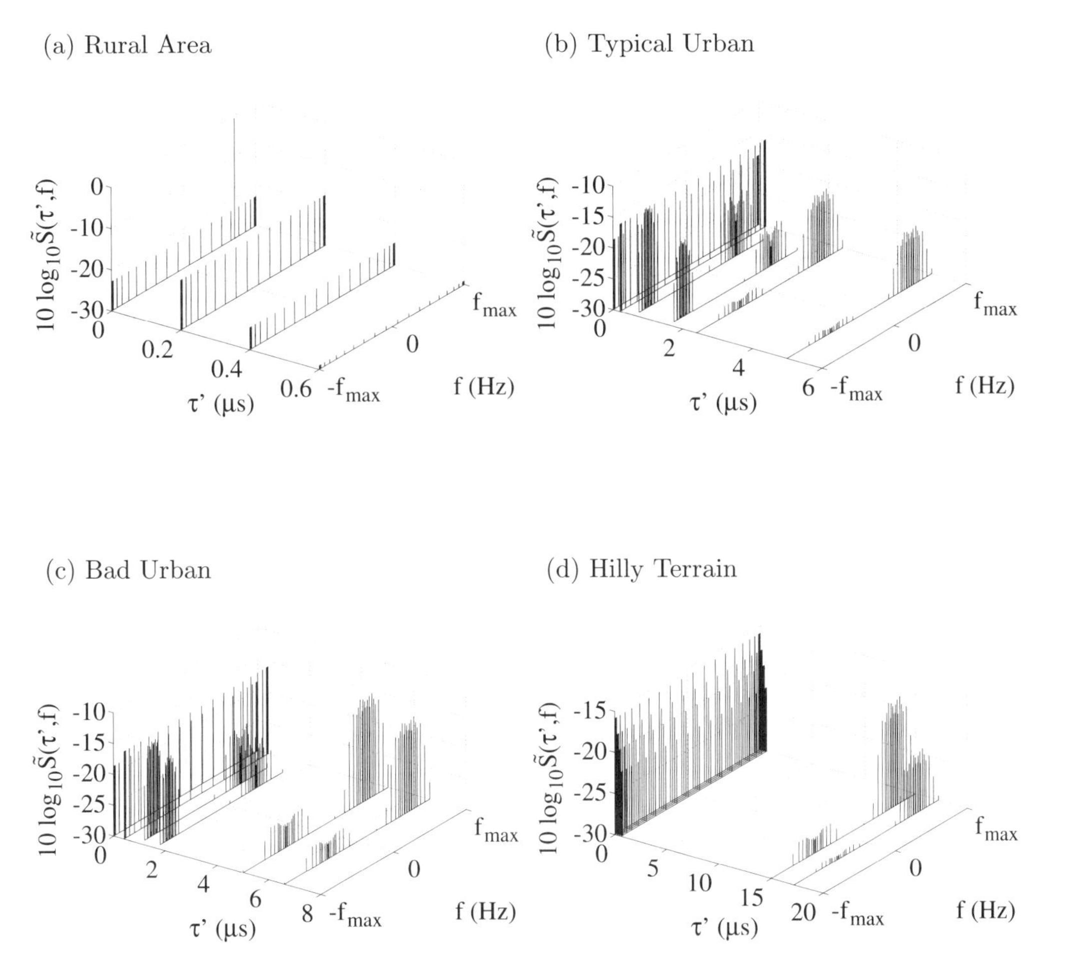

Figure 7.17: Scattering function $\tilde{S}(\tau', f)$ of deterministic channel models on the basis of the $\mathcal{L}$-path channel models according to COST 207 [COS89]: (a) Rural Area, (b) Typical Urban, (c) Bad Urban, and (d) Hilly Terrain.

8

FAST CHANNEL SIMULATORS

The description of the channel simulators considered up to now has always been performed by using the continuous-time representation. In Section 4.1, it was stated that a discrete-time simulation model, which is required for computer simulations, can directly be obtained from the continuous-time simulation model by substituting in the latter the time variable t by $t = kT_s$, where T_s denotes the sampling interval. This way of implementation will henceforth be denoted by *direct realization* and the corresponding simulation model will be called *direct system*. In order to realize a real-valued deterministic Gaussian process by using the direct realization, N_i harmonic functions as well as several multiplications and additions have to be computed at each instant k. Since the number of harmonic functions N_i is the decisive quantity determining the computation time, the efficiency can only be increased essentially by reducing N_i. On the other hand, we know from our investigations in Chapter 5 that a natural lower limit at $N_i = 7$ exists, and, consequently, choosing $N_i < 7$ will result in heavy losses in quality. Thus, the possibilities for a further increase of the speed of direct systems with $N_i = 7$ are exhausted to a large extent. A speed-up of the simulator without accepting losses in precision can only be attained with indirect realization forms.

In this chapter, several ways of indirect implementation forms will be investigated. The basic idea which enables the derivation of new structures for the simulation of deterministic processes is based on taking advantage of the periodicity of harmonic functions. During the set-up phase, each of the N_i harmonic functions is sampled only once within its basic period. The samples are then stored in N_i tables. During the simulation run, the registers of each table are read out cyclically and summed up.

In this manner, it is possible to realize simulation models for complex-valued Gaussian random processes by merely using adders, storage elements, and a simple address generator. Time-consuming trigonometric operations as well as the implementation of multiplications are then no longer required. This results in *fast channel simulators* [Pae98f, Pae00e, Pae99b], which are applicable for all frequency-nonselective and frequency-selective channel models that can be derived from (complex-valued) Gaussian random processes. Since the proposed principle can be generalized easily, we will restrict our attention in this chapter to the derivation of fast channel simulators for Rayleigh channels.

For that purpose, we will employ the discrete-time representation to describe so-called discrete-time deterministic processes in Section 8.1. This class of processes opens up new possibilities to establish indirect realization forms, where three of the most relevant will be presented in Section 8.2. The elementary and statistical properties of discrete deterministic processes will then be examined in Section 8.3. Section 8.4 deals with the analysis of the required realization expenditure as well as with the measurement of the speed of fast channel simulators. Finally, a comparison with a filter method based simulation model for Rayleigh processes will be carried out in Section 8.5.

8.1 DISCRETE DETERMINISTIC PROCESSES

Our starting point is the deterministic Gaussian process $\tilde{\mu}_i(t)$ introduced by (4.4). Sampling this process at $t = kT_s$ will result in a discrete-time signal (sequence)

$$\tilde{\mu}_i[k] := \tilde{\mu}_i(kT_s) = \sum_{n=1}^{N_i} c_{i,n} \cos(2\pi f_{i,n} kT_s + \theta_{i,n}) \,. \tag{8.1}$$

With respect to a preferably efficient realization, the range of values has to be limited for the discrete Doppler frequencies $f_{i,n}$ as well as for the Doppler phases $\theta_{i,n}$. Thus, for the reciprocal value of the discrete Doppler frequencies $1/f_{i,n}$, for example, only integer multiples of the sampling interval T_s are henceforth permissible. The Doppler phases $\theta_{i,n}$ are subject to a similar restriction. According to two mappings, defined below, we obtain from $f_{i,n} \rightarrow \bar{f}_{i,n}$ and $\theta_{i,n} \rightarrow \bar{\theta}_{i,n}$ *quantized Doppler frequencies* $\bar{f}_{i,n}$ and *quantized Doppler phases* $\bar{\theta}_{i,n}$, respectively. Provided that the deviations between $f_{i,n}$ and $\bar{f}_{i,n}$ are sufficiently small, and, consequently, $\bar{f}_{i,n} \approx f_{i,n}$ holds, then

$$\bar{\mu}_i[k] := \bar{\mu}_i(kT_s) = \sum_{n=1}^{N_i} c_{i,n} \cos(2\pi \bar{f}_{i,n} kT_s + \bar{\theta}_{i,n}) \tag{8.2}$$

describes a sequence, which is equivalent to (8.1) (with respect to the relevant statistical properties). In the following, the sequence $\bar{\mu}_i[k]$ is called *discrete deterministic Gaussian process.* Thereby, the Doppler coefficients in (8.2) are identical to those in (8.1), whereas the quantized Doppler frequencies $\bar{f}_{i,n}$ are related to the quantities $f_{i,n}$ and T_s according to

$$\bar{f}_{i,n} := \frac{1}{T_s \, \mathrm{round}\,\{1/(f_{i,n}T_s)\}} \tag{8.3}$$

for all $n = 1, 2, \ldots, N_i$.[1] We call

$$L_{i,n} = \frac{1}{\bar{f}_{i,n}T_s} = \mathrm{round}\left\{\frac{1}{f_{i,n}T_s}\right\} \tag{8.4}$$

the period of the individual discrete harmonic elementary function $\bar{\mu}_{i,n}[k] = c_{i,n} \cos(2\pi \bar{f}_{i,n} kT_s + \bar{\theta}_{i,n})$, i.e., it applies $\bar{\mu}_{i,n}[k] = \bar{\mu}_{i,n}[k + L_{i,n}]$. Note that the rounding

[1] The operator round$\{x\}$ in (8.3) rounds the real-valued number x to the nearest integer.

operation used in (8.4) always results in a natural number for the period $L_{i,n}$. In the next section, we will see that this will turn out to be a clear advantage for the realization.

The quantized Doppler phases $\bar{\theta}_{i,n}$ in (8.2) are calculated from the given quantities $\theta_{i,n}$ according to the expression

$$\bar{\theta}_{i,n} := \frac{2\pi}{L_{i,n}} \operatorname{round}\left\{\frac{L_{i,n}}{2\pi}\,\theta_{i,n}\right\} \tag{8.5}$$

for all $n = 1, 2, \ldots, N_i$. Remember that the Doppler phases $\theta_{i,n}$ are real-valued numbers within the interval $(0, 2\pi]$, whereas the quantized values $\bar{\theta}_{i,n}$ according to (8.5) are elements of the set

$$\bar{\Theta}_{i,n} = \left\{2\pi\frac{1}{L_{i,n}},\ 2\pi\frac{2}{L_{i,n}},\ \ldots\ ,\ 2\pi\frac{L_{i,n}-1}{L_{i,n}},\ 2\pi\right\}. \tag{8.6}$$

The mapping $\theta_{i,n} \rightarrow \bar{\theta}_{i,n}$ according to (8.5) has been chosen in such a way that $\bar{\theta}_{i,n} \in \bar{\Theta}_{i,n}$ is as close as possible to $\theta_{i,n}$.

By using $x - 1/2 \leq \operatorname{round}\{x\} \leq x + 1/2$, one can show that in the limit $T_s \rightarrow 0$ from (8.3) and (8.5) it follows $\bar{f}_{i,n} = f_{i,n}$ and $\bar{\theta}_{i,n} = \theta_{i,n}$, respectively. However, for sufficiently small sampling intervals T_s, we can write $\bar{f}_{i,n} \approx f_{i,n}$ and $\bar{\theta}_{i,n} \approx \theta_{i,n}$. At this point, we want to note that the quality of the approximation $\bar{\theta}_{i,n} \approx \theta_{i,n}$ under particular conditions, which will be discussed in detail in Section 8.3, does not affect the statistical properties of $\bar{\mu}_i[k]$. On the other hand, the deviations between $\bar{f}_{i,n}$ and $f_{i,n}$ determined by the sampling interval T_s cannot be ignored without hesitation, which will also be substantiated in Section 8.3. As an appropriate measure of the deviation between $\bar{f}_{i,n}$ and $f_{i,n}$, we consider the relative error

$$\varepsilon_{\bar{f}_{i,n}} = \frac{\bar{f}_{i,n} - f_{i,n}}{f_{i,n}} \tag{8.7}$$

represented in Figure 8.1. From this figure, it can be realized that the quality of the approximation $\bar{f}_{i,n} \approx f_{i,n}$ decreases if the sampling interval T_s increases. This result indicates that the statistical properties of $\bar{\mu}_i[k]$ depend on the size of the sampling interval T_s. However, for $T_s < 1/(10 f_{i,n})$ the absolute value of the relative error $|\varepsilon_{\bar{f}_{i,n}}|$ is below a limit of 5 per cent, which can be tolerated in most practical applications.

Obviously, the discrete deterministic Gaussian process $\bar{\mu}_i[k]$ introduced by (8.2) can be derived from the continuous-time deterministic process $\tilde{\mu}_i(t)$ by sampling the latter at time instants $t = kT_s$ and, furthermore, by replacing the quantities $f_{i,n}$ and $\theta_{i,n}$ by their quantized versions $\bar{f}_{i,n}$ and $\bar{\theta}_{i,n}$, respectively, i.e.,

$$\tilde{\mu}_i(t) \xrightarrow{\ t \rightarrow kT_s\ } \tilde{\mu}_i[k] := \tilde{\mu}_i(kT_s) \xrightarrow[\ ]{\ \substack{f_{i,n} \rightarrow \bar{f}_{i,n} \\ \theta_{i,n} \rightarrow \bar{\theta}_{i,n}}\ } \bar{\mu}_i[k] := \bar{\mu}_i(kT_s)\,. \tag{8.8}$$

From the fact that $\bar{f}_{i,n}$ and $\bar{\theta}_{i,n}$ converge to $f_{i,n}$ and $\theta_{i,n}$, respectively, as T_s tends to zero, it follows: $\bar{\mu}_i[k] \rightarrow \tilde{\mu}_i(t)$ as $T_s \rightarrow 0$. Considering the results of Chapter 4,

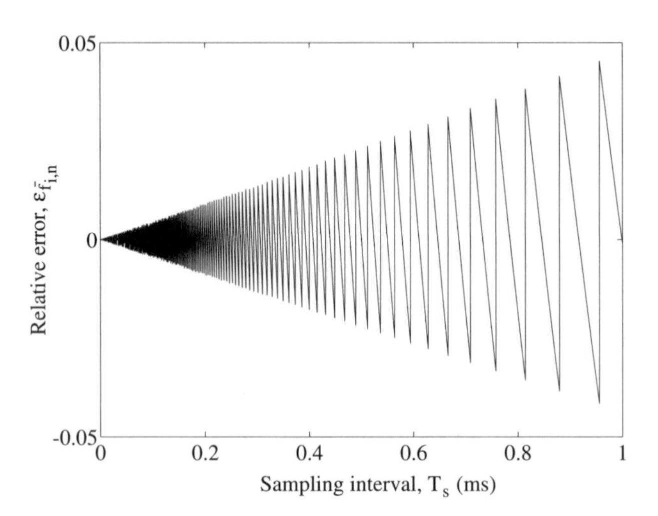

Figure 8.1: Relative error $\varepsilon_{\bar{f}_{i,n}}$ according to (8.7) for $f_{i,n} = 91\,\text{Hz}$ as a function of the sampling interval T_s.

it becomes obvious that the discrete deterministic Gaussian process $\bar{\mu}_i[k]$ tends to a sample function of the Gaussian random process $\mu_i(t)$ as $T_s \to 0$ and $N_i \to \infty$.

By analogy with (4.5), we here introduce the complex-valued sequence

$$\bar{\mu}[k] = \bar{\mu}_1[k] + j\bar{\mu}_2[k] \tag{8.9}$$

as *complex discrete deterministic Gaussian process* and we call its absolute value

$$\bar{\zeta}[k] = |\bar{\mu}[k]| = |\bar{\mu}_1[k] + j\bar{\mu}_2[k]| \tag{8.10}$$

discrete deterministic Rayleigh process. Moreover, we will in the following study the phase $\bar{\vartheta}[k] = \arg\{\bar{\mu}[k]\}$ defined by the discrete deterministic process

$$\bar{\vartheta}[k] = \arctan\left\{\frac{\bar{\mu}_2[k]}{\bar{\mu}_1[k]}\right\}. \tag{8.11}$$

8.2 REALIZATION OF DISCRETE DETERMINISTIC PROCESSES

The discrete deterministic processes introduced above open up new possibilities for the development of fast channel simulators. In the following, three procedures will be presented.

8.2.1 Tables System

The basic idea of the tables system is to store the samples of one period of the sequence $\bar{\mu}_{i,n}[k] = c_{i,n}\cos(2\pi\bar{f}_{i,n}kT_s + \bar{\theta}_{i,n})$ into a table and to read out the table entries cyclically during the simulation [Pae00e]. For the design of a simulation model of

Rayleigh channels, $N_1 + N_2$ tables instead of $N_1 + N_2$ harmonic functions are required. By means of an address generator, the values stored in the tables are accessed. At any discrete time $k = 0, 1, 2, \ldots$, the discrete sequence $\bar{\mu}[k] = \bar{\mu}_1[k] + j\bar{\mu}_2[k]$ can simply be reconstructed by summing up the selected entries of the table as shown in Figure 8.2. After taking the absolute value, the desired discrete deterministic Rayleigh process $\bar{\zeta}[k]$ is then available.

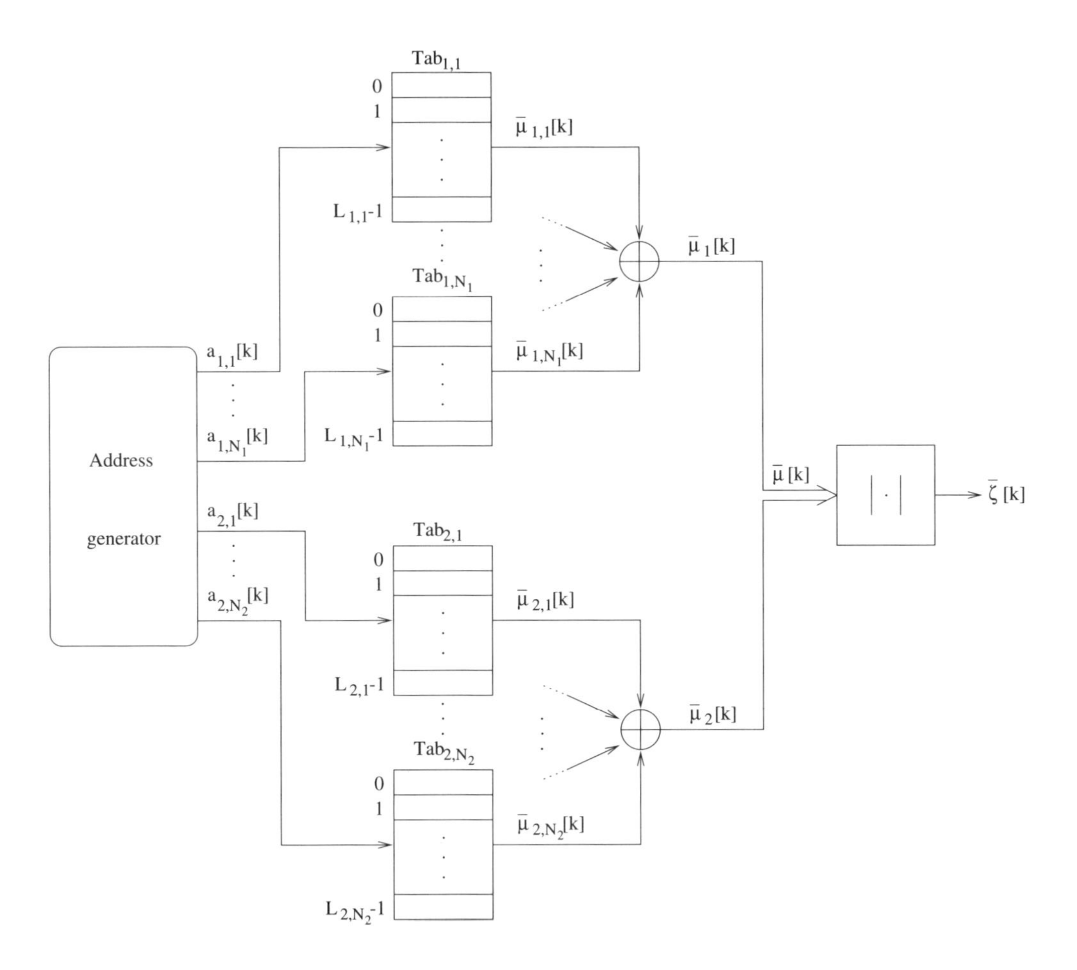

Figure 8.2: Tables system for the fast simulation of Rayleigh channels.

The table, in which the information of one period of a discrete harmonic elementary function $\bar{\mu}_{i,n}[k]$ is stored, will be denoted by $\text{Tab}_{i,n}$. The entry of the table $\text{Tab}_{i,n}$ at position $l \in \{0, 1, \ldots, L_{i,n} - 1\}$ corresponds to the value of $\bar{\mu}_{i,n}[k]$ at $k = l$, i.e., it holds

$$\bar{\mu}_{i,n}[l] = c_{i,n} \cos(2\pi \bar{f}_{i,n} l T_s + \bar{\theta}_{i,n}) \tag{8.12}$$

for all $n = 1, 2, \ldots, N_i$ $(i = 1, 2)$. Now, reading out the entries of the table $\text{Tab}_{i,n}$ cyclically, results in the sequence $\{\bar{\mu}_{i,n}[0], \bar{\mu}_{i,n}[1], \ldots, \bar{\mu}_{i,n}[L_{i,n} - 1], \bar{\mu}_{i,n}[L_{i,n}] = \bar{\mu}_{i,n}[0], \ldots\}$. Hence, by exploiting the periodicity, $\bar{\mu}_{i,n}[k]$ can be reconstructed completely for all $k = 0, 1, 2, \ldots$ The length of the table $\text{Tab}_{i,n}$ is identical to the period $L_{i,n}$ of $\bar{\mu}_{i,n}[k]$. In consequence, the total amount of storage elements required for the implementation of discrete deterministic processes $\bar{\mu}_i[k]$ is given by the sum $\sum_{n=1}^{N_i} L_{i,n}$. Due to (8.4), the total memory size is not only determined by the number of used tables N_i, but also by the value of the sampling interval T_s or, alternatively, the sampling frequency $f_s = 1/T_s$. In Figures 8.3(a) and 8.3(b), the table lengths $L_{i,n}$ as well as their resulting sums are depicted as a function of the normalized sampling frequency f_s/f_{max} for commonly used values of $N_1 = 7$ and $N_2 = 8$, respectively. Thereby, the MEDS has been applied.

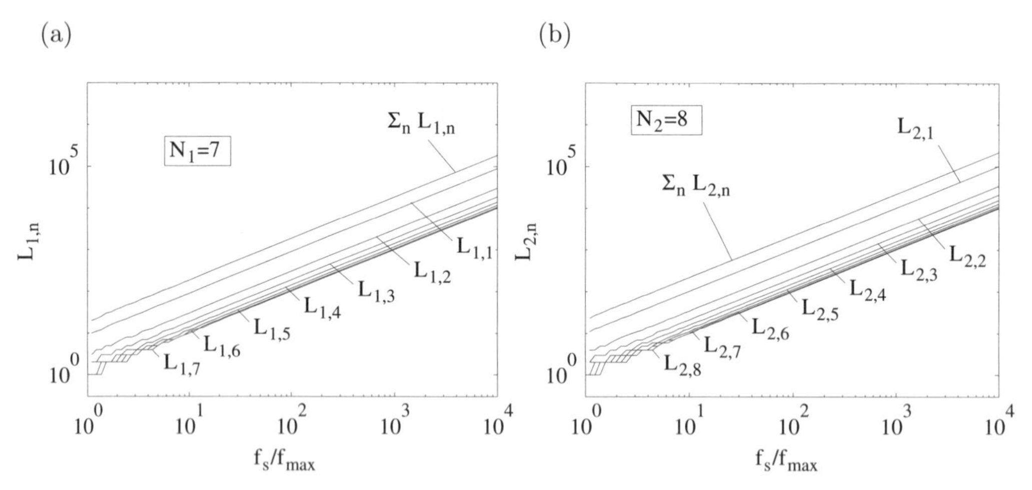

Figure 8.3: Table lengths $L_{i,n}$ as a function of the normalized sampling frequency f_s/f_{max}: (a) $L_{1,n}$ for $N_1 = 7$ and (b) $L_{2,n}$ for $N_2 = 8$ (MEDS, Jakes PSD, $f_{max} = 91\,\text{Hz}$, $\sigma_0^2 = 1$).

Viewing Figures 8.3(a) and 8.3(b), one realizes that within the range of small values of f_s/f_{max}, two or even more tables $\text{Tab}_{i,n}$ can have the same length. The problems associated with this phenomenon will be discussed later in Subsection 8.3.2.

The task of the address generator shown in Figure 8.2 is to find the correct position of the table entries required to reconstruct $\bar{\mu}[k] = \bar{\mu}_1[k] + j\bar{\mu}_2[k]$ for any instant $k = 0, 1, 2, \ldots$. Therefore, the address generator has to generate altogether $N_1 + N_2$ addresses for each discrete time k. As can be seen from Figure 8.2, $a_{i,n}[k]$ denotes the address of the table $\text{Tab}_{i,n}$ at the discrete time k. Figure 8.4 illustrates the mode of operation of the address generator.

At the instant $k = 0$, the address $a_{i,n}[0]$ points at the register $\bar{\mu}_{i,n}[0]$ of the table $\text{Tab}_{i,n}$. At the next instant $k = 1$, $a_{i,n}[1]$ refers to $\bar{\mu}_{i,n}[1]$, etc., up to the instant $k = L_{i,n} - 1$, where the address $a_{i,n}[L_{i,n} - 1]$ points to the last position of the table

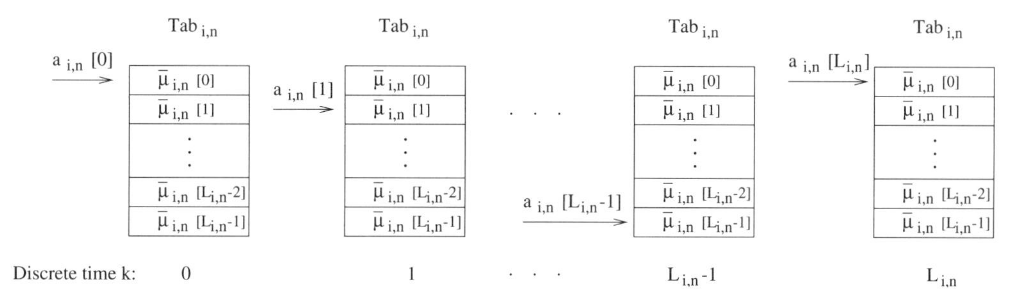

Figure 8.4: Mode of operation of the address generator.

with the entry $\bar{\mu}_{i,n}[L_{i,n}-1]$. At the following instant $k = L_{i,n}$, the address $a_{i,n}[L_{i,n}]$ is reset to $a_{i,n}[0]$, which points at the initial position $\bar{\mu}_{i,n}[0]$.

Starting with the initial addresses $a_{i,n}[0] = 0$ and applying the modulo operation, all addresses $a_{i,n}[k]$ can be found at any instant $k > 0$ by using the following recursive algorithm:

$$a_{i,n}[k] = (a_{i,n}[k-1]+1) \bmod L_{i,n}, \tag{8.13}$$

where $n = 1, 2, \ldots, N_i$ $(i = 1, 2)$. It should be mentioned that the modulo operation in (8.13) has been applied here only for mathematical convenience. For the realization of the algorithm on a computer, only one addition and a simple conditional control flow statement (*if-else* statement) are required for the computation of $a_{i,n}[k]$.

Thus, the entire tables system (see Figure 8.2) only consists of adders, storage elements, and simple conditional operators. Multiplications as well as trigonometric operations no longer have to be carried out for the computation of $\bar{\mu}[k] = \bar{\mu}_1[k] + j\bar{\mu}_2[k]$.

8.2.2 Matrix System

The matrix system combines the N_i tables to a *channel matrix* $\boldsymbol{M}_i$. The number of rows of the channel matrix $\boldsymbol{M}_i$ is identical to the number of tables N_i. Thereby, the nth row of $\boldsymbol{M}_i$ contains the entries of the table $\text{Tab}_{i,n}$. As a result, the length of the largest table, i.e., $L_{i,max} = \max\{L_{i,n}\}_{n=1}^{N_i}$, defines the number of columns of the channel matrix $\boldsymbol{M}_i$. Without loss of generality, we assume in the following that $L_{i,max} = L_{i,1}$ holds, which is actually always the case by using the MEDS (see Figure 8.3). The first $L_{i,n}$ entries of the nth row of $\boldsymbol{M}_i$ are exactly identical to the entries of the table $\text{Tab}_{i,n}$, whereas the rest of the row is filled up with zeros. Thus, the channel matrix $\boldsymbol{M}_i \in \mathbb{R}^{N_i \times L_{i,1}}$ can be represented as follows:

$$\boldsymbol{M}_i = \begin{pmatrix} \bar{\mu}_{i,1}[0] & \cdots\cdots\cdots\cdots\cdots\cdots & & & & \bar{\mu}_{i,1}[L_{i,1}-1] \\ \bar{\mu}_{i,2}[0] & \cdots\cdots\cdots & \bar{\mu}_{i,2}[L_{i,2}-1] & 0 & \cdots & 0 \\ \vdots & & & & \ddots & \vdots \\ \bar{\mu}_{i,N_i}[0] & \cdots \quad \bar{\mu}_{i,N_i}[L_{i,N_i}-1] & & 0 & \cdots & 0 \end{pmatrix}. \tag{8.14}$$

The channel matrix $\boldsymbol{M}_i$ contains the complete information needed for the reconstruction of $\bar{\mu}_i[k]$. In order to guarantee the correct reconstruction of $\bar{\mu}_i[k]$ for all values of $k = 0, 1, 2, \ldots$, it is necessary to select from each row of $\boldsymbol{M}_i$ one entry at the correct position. This can be achieved by introducing a further matrix $\boldsymbol{S}_i$, which will henceforth be called the *selection matrix*. The entries of the selection matrix $\boldsymbol{S}_i$ are time variant quantities, which can only take the values 0 or 1. There is a close relation between the address generator introduced in the previous subsection and the selection matrix $\boldsymbol{S}_i$. This becomes obvious by noting that the entries of $\boldsymbol{S}_i = (s_{l,n}) \in \{0,1\}^{L_{i,1} \times N_i}$ can be calculated at any instant k by using the addresses $a_{i,n}[k]$ (8.13) according to

$$s_{l,n} = s_{l,n}[k] = \begin{cases} 1 & \text{if} \quad l = a_{i,n}[k] \\ 0 & \text{if} \quad l \neq a_{i,n}[k] \end{cases} \tag{8.15}$$

for all $l = 0, 1, \ldots, L_{i,1} - 1$ and $n = 1, 2, \ldots, N_i$ $(i = 1, 2)$.

The discrete deterministic Gaussian process $\bar{\mu}_i[k]$ can now be obtained from the product of the channel matrix $\boldsymbol{M}_i$ and the selection matrix $\boldsymbol{S}_i$ as follows:

$$\bar{\mu}_i[k] = \operatorname{tr}(\boldsymbol{M}_i \cdot \boldsymbol{S}_i), \tag{8.16}$$

where $\operatorname{tr}(\cdot)$ denotes the trace[2] [Zur92, Hor85].

Using (8.16), we can thus also express the complex discrete deterministic Gaussian process (8.9) in an alternative form

$$\bar{\mu}[k] = \operatorname{tr}(\boldsymbol{M}_1 \cdot \boldsymbol{S}_1) + j \operatorname{tr}(\boldsymbol{M}_2 \cdot \boldsymbol{S}_2). \tag{8.17}$$

It is worth mentioning that the number of columns (rows) of the channel matrix $\boldsymbol{M}_i$ (selection matrix $\boldsymbol{S}_i$) tends to infinity as $T_s \to 0$ and, thus, $\bar{\mu}_i[k]$ converges to $\tilde{\mu}_i(t)$. In the limits $T_s \to 0$ and $N_i \to \infty$, the number of columns and the number of rows of both the channel matrix $\boldsymbol{M}_i$ and the selection matrix $\boldsymbol{S}_i$ tend to infinity. In this case, the complex discrete deterministic Gaussian process $\bar{\mu}[k]$ converges, as it was expected, to a sample function of the complex stochastic Gaussian random process $\mu(t)$.

An equivalent representation of the discrete deterministic Rayleigh process $\bar{\zeta}[k]$, introduced by (8.10), can be obtained by taking the absolute value of (8.17), i.e.,

$$\bar{\zeta}[k] = |\bar{\mu}[k]| = |\operatorname{tr}(\boldsymbol{M}_1 \cdot \boldsymbol{S}_1) + j \operatorname{tr}(\boldsymbol{M}_2 \cdot \boldsymbol{S}_2)|. \tag{8.18}$$

For the sake of completeness, we write the phase $\bar{\vartheta}[k]$ of $\bar{\mu}[k] = \bar{\mu}_1[k] + j\bar{\mu}_2[k]$ in the form

$$\bar{\vartheta}[k] = \arctan\left\{ \frac{\operatorname{tr}(\boldsymbol{M}_2 \cdot \boldsymbol{S}_2)}{\operatorname{tr}(\boldsymbol{M}_1 \cdot \boldsymbol{S}_1)} \right\}. \tag{8.19}$$

---

[2] The trace of a square matrix $\mathbf{A} = (a_{n,m}) \in \mathbb{R}^{N \times N}$ is defined by the sum of the main diagonal entries $a_{n,m}$, i.e., $\operatorname{tr}(\mathbf{A}) = \sum_{n=1}^{N} a_{n,n}$.

It is evident that in the limit $T_s \to 0$ it follows: $\bar{\zeta}[k] \to \tilde{\zeta}(t)$ and $\bar{\vartheta}[k] \to \tilde{\vartheta}(t)$. Furthermore, the sequences $\bar{\zeta}[k]$ and $\bar{\vartheta}[k]$ are converging for $T_s \to 0$ and $N_i \to \infty$ to a sample function of the corresponding stochastic processes $\zeta(t)$ and $\vartheta(t)$, respectively.

It should be mentioned that the computation of the discrete deterministic processes (8.16)–(8.19) by taking the trace of the product of two matrices is not a very efficient approach due to the large number of multiplications and additions that have to be carried out. However, considerable simplifications are possible if all unnecessary operations such as multiplications with zero and one are avoided at the beginning. In this case, the matrix system reduces to the tables system. In other words: the matrix system actually represents no genuine alternative realization form to the tables system, but provides some new aspects regarding the interpretation and representation of discrete-time deterministic processes.

8.2.3 Shift Register System

From the tables system (see Figure 8.2), we can derive the shift register system depicted in Figure 8.5 by replacing in the former the tables $\text{Tab}_{i,n}$ with feedback *shift registers* $\text{Reg}_{i,n}$. Instead of $N_1 + N_2$ tables, now $N_1 + N_2$ shift registers are required for the realization of $\bar{\mu}[k] = \bar{\mu}_1[k] + j\bar{\mu}_2[k]$. The length of the shift register $\text{Reg}_{i,n}$ is thereby identical to the length $L_{i,n}$ of the corresponding table $\text{Tab}_{i,n}$. During the simulation set-up phase, the shift registers $\text{Reg}_{i,n}$ are filled at the positions $l \in \{0, 1, \ldots, L_{i,n} - 1\}$ with the values $\bar{\mu}_{i,n}[l] = c_{i,n}\cos(2\pi \bar{f}_{i,n} l T_s + \bar{\theta}_{i,n})$, where $n = 1, 2, \ldots, N_i$ and $i = 1, 2$. Throughout the simulation run phase, the contents of the shift registers are shifted by one position to the right at every clock pulse (see Figure 8.5). Due to the links created between the shift register outputs (positions 0) with their respective inputs (positions $L_{i,n} - 1$), it is ensured that the discrete deterministic processes $\bar{\mu}_{i,n}[k]$ and, consequently, also $\bar{\mu}[k] = \bar{\mu}_1[k] + j\bar{\mu}_2[k]$ as well as $\bar{\zeta}[k] = |\bar{\mu}[k]|$ can be reconstructed for all $k = 0, 1, 2, \ldots$

Note that in comparison with the tables system, no address generator is needed, but instead of this, $\sum_{i=1}^{2}\sum_{n=1}^{N_i} L_{i,n}$ register entries have to be shifted at every clock pulse, which — especially for software realizations in connection with large register lengths — does not lead to a satisfying solution. For that reason, we will prefer the tables system to the shift register system and will turn our attention in the next section to the analysis of the properties of discrete deterministic processes.

8.3 PROPERTIES OF DISCRETE DETERMINISTIC PROCESSES

By analogy to the analysis of continuous-time deterministic processes (Chapter 4), we start in Subsection 8.3.1 with the investigation of the elementary properties of discrete deterministic processes, and then we will continue with the analysis of the statistical properties in the following Subsection 8.3.2.

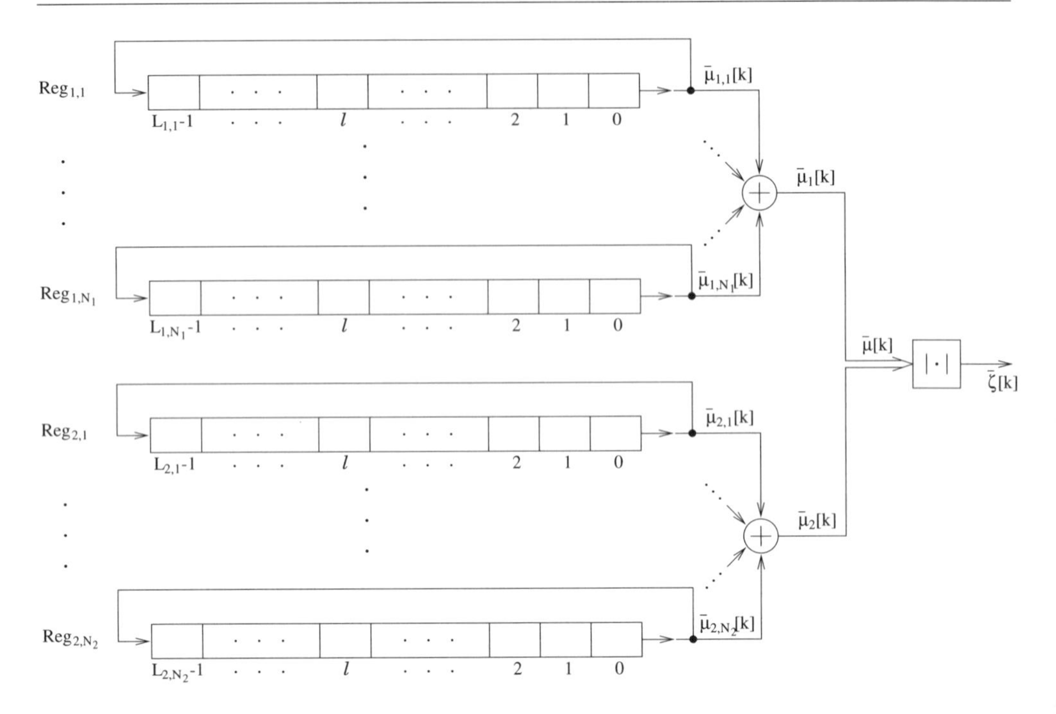

Figure 8.5: Realization of discrete deterministic Rayleigh processes $\bar{\zeta}[k]$ by using shift registers.

8.3.1 Elementary Properties of Discrete Deterministic Processes

The interpretation of $\bar{\mu}_i[k]$ as discrete deterministic process, i.e., as a mapping of the form

$$\bar{\mu}_i : \mathbb{Z} \to \mathbb{R} \, , \qquad k \mapsto \bar{\mu}_i[k] \, , \tag{8.20}$$

allows us to establish a close relationship with the investigations performed in Section 4.2. Therefore, we proceed analogously to Section 4.2 and derive simple closed solutions for the fundamental characteristic quantities of $\bar{\mu}_i[k]$ such as mean value, mean power, autocorrelation sequence, etc.

Mean value: Let $\bar{\mu}_i[k]$ be a discrete deterministic process with $\bar{f}_{i,n} \neq 0$ ($n = 1, 2, \ldots, N_i$). Then, by using (2.77) and (8.2), it can be shown that the mean value of $\bar{\mu}_i[k]$ is equal to

$$\bar{m}_{\mu_i} = \lim_{K \to \infty} \frac{1}{2K+1} \sum_{k=-K}^{K} \bar{\mu}_i[k] = 0 \, . \tag{8.21}$$

It is henceforth assumed that $\bar{f}_{i,n} \neq 0$ holds for all $n = 1, 2, \ldots, N_i$ and $i = 1, 2$.

Mean power: Let $\bar{\mu}_i[k]$ be a discrete deterministic process. Then, it follows by using (2.78) and (8.2) that its mean power is given by

$$\bar{\sigma}^2_{\mu_i} = \lim_{K\to\infty} \frac{1}{2K+1} \sum_{k=-K}^{K} \bar{\mu}_i^2[k] = \sum_{n=1}^{N_i} \frac{c_{i,n}^2}{2} . \tag{8.22}$$

In particular, by applying the MEDS, we obtain due to (5.73) the desired result $\bar{\sigma}^2_{\mu_i} = \sigma_0^2$.

Autocorrelation sequence: Let $\bar{\mu}_i[k]$ be a discrete deterministic process. Then, it follows from (2.79) and (8.2) that the autocorrelation sequence of $\bar{\mu}_i[k]$ can be expressed by

$$\begin{aligned}\bar{r}_{\mu_i\mu_i}[\kappa] &= \lim_{K\to\infty} \frac{1}{2K+1} \sum_{k=-K}^{K} \bar{\mu}_i[k]\,\bar{\mu}_i[k+\kappa] \\ &= \sum_{n=1}^{N_i} \frac{c_{i,n}^2}{2} \cos(2\pi \bar{f}_{i,n} T_s \kappa) .\end{aligned} \tag{8.23}$$

A comparison with (4.11) shows that $\bar{r}_{\mu_i\mu_i}[\kappa]$ can be obtained from $\tilde{r}_{\mu_i\mu_i}(\tau)$ if $\tilde{r}_{\mu_i\mu_i}(\tau)$ is sampled at $\tau = \kappa T_s$ and if additionally the quantities $f_{i,n}$ are substituted by $\bar{f}_{i,n}$. In addition, we realize that also in the discrete-time case, the quantized Doppler phases $\bar{\theta}_{i,n}$ have no influence on the behaviour of the autocorrelation sequence $\bar{r}_{\mu_i\mu_i}[\kappa]$. Observe that from (8.22) and (8.23) the relation $\bar{\sigma}^2_{\mu_i} = \bar{r}_{\mu_i\mu_i}[0]$ can directly be obtained.

The deviations between $\tilde{r}_{\mu_i\mu_i}[\kappa] := \tilde{r}_{\mu_i\mu_i}(\kappa T_s)$ and $\bar{r}_{\mu_i\mu_i}[\kappa]$, caused by the quantization of the discrete Doppler frequencies $f_{i,n}$, can be observed in Figure 8.6. Thereby, the MEDS has been applied by using $N_i = 8$ harmonic functions (tables). Figure 8.6(a) shows that for sufficiently small sampling intervals ($T_s = 0.1\,\text{ms}$) no significant differences occur between $\tilde{r}_{\mu_i\mu_i}[\kappa]$ and $\bar{r}_{\mu_i\mu_i}[\kappa]$ if $\tau = \kappa T_s$ is within its range of interest $\tau =\in [0,\, N_i/(2f_{max})]$. However, this does not apply for large values of T_s, as can be seen when considering Figure 8.6(b), where the corresponding ratios in case of $T_s = 1\,\text{ms}$ are shown.

Cross-correlation sequence: Let $\bar{\mu}_1[k]$ and $\bar{\mu}_2[k]$ be two discrete deterministic processes. Then, it follows from (2.80) in connection with (8.2) that the cross-correlation sequence is equal to

$$\bar{r}_{\mu_1\mu_2}[\kappa] = 0 , \tag{8.24}$$

if $\bar{f}_{1,n} \neq \pm\bar{f}_{2,m}$ is fulfilled for all $n = 1, 2, \ldots, N_1$ and $m = 1, 2, \ldots, N_2$, or

$$\bar{r}_{\mu_1\mu_2}[\kappa] = \sum_{\substack{n=1 \\ \bar{f}_{1,n}=\pm\bar{f}_{2,m}}}^{\max\{N_1,N_2\}} \frac{c_{1,n}c_{2,m}}{2} \cos(2\pi \bar{f}_{1,n} T_s \kappa - \bar{\theta}_{1,n} \pm \bar{\theta}_{2,m}) , \tag{8.25}$$

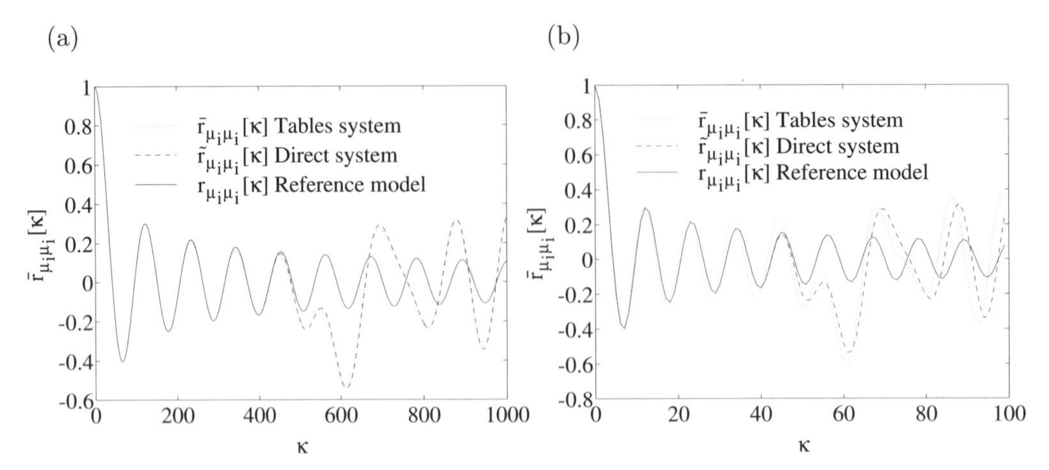

Figure 8.6: Autocorrelation sequence $\bar{r}_{\mu_i\mu_i}[\kappa]$ of discrete deterministic Gaussian processes $\bar{\mu}_i[k]$ for (a) $T_s = 0.1\,\text{ms}$ and (b) $T_s = 1\,\text{ms}$ (MEDS, Jakes PSD, $N_i = 8$, $f_{max} = 91\,\text{Hz}$, $\sigma_0^2 = 1$).

if $\bar{f}_{1,n} = \pm\bar{f}_{2,m}$ holds for one or several pairs (n, m). Notice that $\bar{r}_{\mu_1\mu_2}[\kappa]$ can immediately be derived from $\tilde{r}_{\mu_1\mu_2}(\tau)$ if in (4.12) and (4.13) the continuous variable τ is replaced by κT_s, and, additionally, the quantities $f_{i,n}$ and $\theta_{i,n}$ are substituted by their quantized quantities $\bar{f}_{i,n}$ and $\bar{\theta}_{i,n}$, respectively. The two cross-correlation sequences $\bar{r}_{\mu_1\mu_2}[\kappa]$ and $\bar{r}_{\mu_2\mu_1}[\kappa]$ are related by $\bar{r}_{\mu_2\mu_1}[\kappa] = \bar{r}^*_{\mu_1\mu_2}[-\kappa] = \bar{r}_{\mu_1\mu_2}[-\kappa]$.

Power spectral density: Let $\bar{\mu}_i[k]$ be a discrete deterministic process. Then, it follows for the power spectral density by applying the discrete Fourier transform (2.81) in connection with (8.23)

$$\bar{S}_{\mu_i\mu_i}(f) = \frac{1}{T_s} \sum_{\nu=-\infty}^{\infty} \sum_{n=1}^{N_i} \frac{c_{i,n}^2}{4} \left[\delta(f - \bar{f}_{i,n} - \nu f_s) + \delta(f + \bar{f}_{i,n} - \nu f_s)\right], \qquad (8.26)$$

where $f_s = 1/T_s$ denotes the sampling frequency. Thus, the power spectral density $\bar{S}_{\mu_i\mu_i}(f)$ is a symmetrical line spectrum, where the spectral lines are located at $f = \pm\bar{f}_{i,n} + \nu f_s$ and weighted by the factor $c_{i,n}^2/(4T_s)$. Using (2.82) and taking $\tilde{S}_{\mu_i\mu_i}(f)$ according to (4.14) into account, one can derive the following relation between $\bar{S}_{\mu_i\mu_i}(f)$ and $\tilde{S}_{\mu_i\mu_i}(f)$

$$\bar{S}_{\mu_i\mu_i}(f) = \frac{1}{T_s} \sum_{\nu=-\infty}^{\infty} \tilde{S}_{\mu_i\mu_i}(f - \nu f_s)\big|_{f_{i,n}=\bar{f}_{i,n}} . \qquad (8.27)$$

The equation above makes clear that the power spectral density $\bar{S}_{\mu_i\mu_i}(f)$ of the discrete deterministic process $\bar{\mu}_i[k]$ can be represented by an infinite sum of weighted and frequency-shifted versions of the power spectral density $\tilde{S}_{\mu_i\mu_i}(f)$ of the corresponding continuous-time deterministic process $\tilde{\mu}_i(t)$, where the weighting factor is equal to $1/T_s$ and the shift frequencies are integer multiples of the sampling frequency f_s. In addition, the quantities $f_{i,n}$ have to be replaced by $\bar{f}_{i,n}$.

Cross-power spectral density: Let $\bar{\mu}_1[k]$ and $\bar{\mu}_2[k]$ be two discrete deterministic processes. Then, it follows from (2.81) by using (8.24) and (8.25) that the cross-power spectral density of $\bar{\mu}_1[k]$ and $\bar{\mu}_2[k]$ can be expressed by

$$\bar{S}_{\mu_1\mu_2}(f) = 0\,, \tag{8.28}$$

if $\bar{f}_{1,n} \neq \pm\bar{f}_{2,m}$ holds for all $n = 1, 2, \ldots, N_1$ and $m = 1, 2, \ldots, N_2$, or

$$\begin{aligned}\bar{S}_{\mu_1\mu_2}(f) \;=\; & \frac{1}{4T_s} \sum_{\nu=-\infty}^{\infty} \sum_{\substack{n=1 \\ \bar{f}_{1,n}=\pm\bar{f}_{2,m}}}^{\max\{N_1,N_2\}} c_{1,n}c_{2,m} \Big[\delta(f - \bar{f}_{1,n} - \nu f_s) \cdot e^{-j(\bar{\theta}_{1,n} \mp \bar{\theta}_{2,m})} \\ & + \delta(f + \bar{f}_{1,n} - \nu f_s) \cdot e^{j(\bar{\theta}_{1,n} \mp \bar{\theta}_{2,m})}\Big]\,, \end{aligned} \tag{8.29}$$

if $\bar{f}_{1,n} = \pm\bar{f}_{2,m}$ is valid for one or several pairs (n, m). Employing (4.15), (4.16), and (2.82), the results of (8.28) and (8.29) can be combined as follows

$$\bar{S}_{\mu_1\mu_2}(f) = \frac{1}{T_s} \sum_{\nu=-\infty}^{\infty} \tilde{S}_{\mu_1\mu_2}(f - \nu f_s)\Big|_{\substack{f_{i,n}=\bar{f}_{i,n} \\ \theta_{i,n}=\bar{\theta}_{i,n}}}\,. \tag{8.30}$$

The cross-power spectral densities $\bar{S}_{\mu_1\mu_2}(f)$ and $\bar{S}_{\mu_2\mu_1}(f)$ are related by $\bar{S}_{\mu_2\mu_1}(f) = \bar{S}^*_{\mu_1\mu_2}(f)$.

Average Doppler shift: Let $\bar{\mu}_i[k]$ be a discrete deterministic process with the power spectral density $\bar{S}_{\mu_i\mu_i}(f)$ as given by (8.26). Then, the corresponding average Doppler shift $\bar{B}^{(1)}_{\mu_i\mu_i}$ is defined by

$$\bar{B}^{(1)}_{\mu_i\mu_i} := \frac{\int\limits_{-f_s/2}^{f_s/2} f\,\bar{S}_{\mu_i\mu_i}(f)\,df}{\int\limits_{-f_s/2}^{f_s/2} \bar{S}_{\mu_i\mu_i}(f)\,df} = \frac{1}{2\pi j} \cdot \frac{\dot{\bar{r}}_{\mu_i\mu_i}[0]}{\bar{r}_{\mu_i\mu_i}[0]}\,. \tag{8.31}$$

In contrast to (3.13a) and (4.17), where the integration is carried out over the entire frequency range, the limits of the integration in (8.31) are restricted to the Nyquist range defined by the frequency interval $[-f_s/2,\, f_s/2)$. In the special case that the Doppler power spectral density has a symmetrical shape, i.e., $\bar{S}_{\mu_i\mu_i}(f) = \bar{S}_{\mu_i\mu_i}(-f)$, it follows directly

$$\bar{B}^{(1)}_{\mu_i\mu_i} = B^{(1)}_{\mu_i\mu_i} = 0\,. \tag{8.32}$$

A comparison with (4.18) shows that neither the effect caused by the substitution of the time variable t by $t = kT_s$ nor the quantization of the discrete Doppler frequencies has an influence on the average Doppler shift.

Doppler spread: Let $\bar{\mu}_i[k]$ be a discrete deterministic process with power spectral density $\bar{S}_{\mu_i\mu_i}(f)$ as given by (8.26). Then, the corresponding Doppler spread $\bar{B}^{(2)}_{\mu_i\mu_i}$ is defined by

$$\bar{B}^{(2)}_{\mu_i\mu_i} := \sqrt{\frac{\int\limits_{-f_s/2}^{f_s/2} (f - \bar{B}^{(1)}_{\mu_i\mu_i})^2 \, \bar{S}_{\mu_i\mu_i}(f) \, df}{\int\limits_{-f_s/2}^{f_s/2} \bar{S}_{\mu_i\mu_i}(f) \, df}}$$

$$= \frac{1}{2\pi} \sqrt{\left(\frac{\dot{\bar{r}}_{\mu_i\mu_i}[0]}{\bar{r}_{\mu_i\mu_i}[0]} \right)^2 - \frac{\ddot{\bar{r}}_{\mu_i\mu_i}[0]}{\bar{r}_{\mu_i\mu_i}[0]}} \,. \tag{8.33}$$

Using (8.31), (8.32), and $\bar{\sigma}^2_{\mu_i} = \bar{r}_{\mu_i\mu_i}[0]$, we can — especially for symmetrical Doppler power spectral densities — express the last equation as follows

$$\bar{B}^{(2)}_{\mu_i\mu_i} = \frac{\sqrt{\bar{\beta}_i}}{2\pi\bar{\sigma}_{\mu_i}} \,, \tag{8.34}$$

where

$$\bar{\beta}_i = -\ddot{\bar{r}}_{\mu_i\mu_i}[0] = 2\pi^2 \sum_{n=1}^{N_i} (c_{i,n}\bar{f}_{i,n})^2 \,. \tag{8.35}$$

It should be remembered that the MEDS has been developed especially for the Jakes power spectral density. In Subsection 5.1.6, we learned that in this case the Doppler spread of the continuous-time simulation model is identical to the Doppler spread of the reference model, i.e., $\tilde{B}^{(2)}_{\mu_i\mu_i} = B^{(2)}_{\mu_i\mu_i}$. This relationship is now only approximately valid. The reason for this is that although $\bar{\sigma}^2_{\mu_i} = \tilde{\sigma}^2_{\mu_i} = \sigma_0^2$ holds, but due to $\bar{f}_{i,n} \approx f_{i,n}$ it follows $\bar{\beta}_i \approx \tilde{\beta}_i = \beta_i$ and, thus,

$$\bar{B}^{(2)}_{\mu_i\mu_i} \approx \tilde{B}^{(2)}_{\mu_i\mu_i} = B^{(2)}_{\mu_i\mu_i} \,. \tag{8.36}$$

The deviation between $\bar{B}^{(2)}_{\mu_i\mu_i}$ and $B^{(2)}_{\mu_i\mu_i}$ or between $\bar{\beta}_i$ and β_i is basically determined by the chosen value for the sampling interval T_s. We will find out more details about this by analysing the model error of discrete-time systems.

Model error: Let $\bar{\mu}_i[k]$ be a discrete deterministic process introduced by (8.2). Then, the model error $\Delta\bar{\beta}_i$ of the discrete-time system is defined by

$$\Delta\bar{\beta}_i := \bar{\beta}_i - \beta_i \,. \tag{8.37}$$

Using (3.29) and (8.35), the model error $\Delta\bar{\beta}_i$ can easily be evaluated for all parameter computation methods described in Chapter 5 as a function of N_i and T_s or alternatively $f_s = 1/T_s$. An example of the behaviour of the relative model error $\Delta\bar{\beta}_i/\beta_i$ of discrete-time systems is shown in Figure 8.7 as a function of the normalized sampling frequency f_s/f_{max}. Thereby, the MEDS has been applied with $N_i = 7$ (Jakes PSD, $f_{max} = 91\,\text{Hz}$, $\sigma_0^2 = 1$).

Figure 8.7 clearly illustrates that the model error $\Delta\bar{\beta}_i/\beta_i$ decreases if the sampling frequency f_s increases. In the limit $f_s \to \infty$ or $T_s \to 0$, we obtain $\Delta\bar{\beta}_i/\beta_i \to 0$

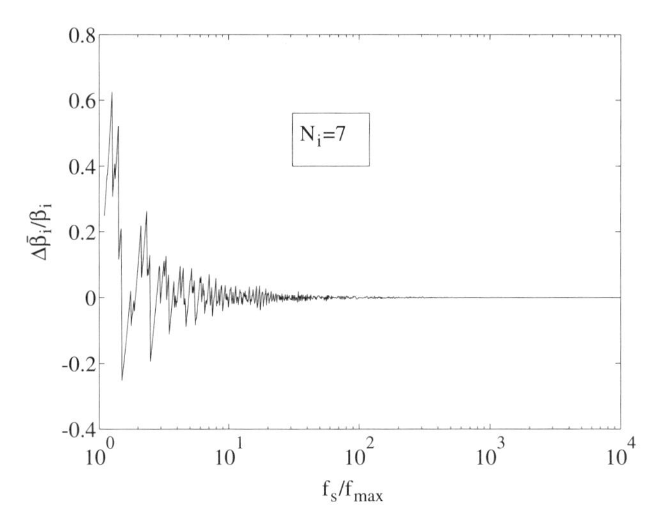

Figure 8.7: Relative model error $\Delta\bar{\beta}_i/\beta_i$ of the discrete-time system (MEDS, Jakes PSD, $N_i = 7$, $f_{max} = 91\,\text{Hz}$, $\sigma_0^2 = 1$).

as expected, since it is well known that the quantized Doppler frequencies $\bar{f}_{i,n}$ are approaching the quantities $f_{i,n}$ as $T_s \to 0$. In case of the MSEM, this directly results in $\bar{\beta}_i \to \tilde{\beta}_i = \beta_i$ and, thus, $\Delta\bar{\beta}_i \to 0$.

Periodicity: Let $\bar{\mu}_i[k]$ be a discrete deterministic process with arbitrary but nonzero parameters $c_{i,n}$, $\bar{f}_{i,n}$ (and $\bar{\theta}_{i,n}$). Then, $\bar{\mu}_i[k]$ is periodic with the least common multiple (lcm) of the set $\{L_{i,n}\}_{n=1}^{N_i}$, i.e., the period L_i of $\bar{\mu}_i[k]$ is equal to

$$L_i = \text{lcm}\,\{L_{i,n}\}_{n=1}^{N_i}\,. \tag{8.38}$$

In order to prove this theorem, we have to show that

$$\bar{\mu}_i[k] = \bar{\mu}_i[k + L_i] \tag{8.39}$$

is valid for all $k \in \mathbb{Z}$. Since L_i is the least common multiple of the set $\{L_{i,n}\}_{n=1}^{N_i}$, L_i must be an integer multiple of every table length $L_{i,n}$. Thus, we may write

$$L_{i,n} = \frac{L_i}{q_{i,n}}\,, \tag{8.40}$$

where $q_{i,n}$ is a natural number, which can be different for every $L_{i,n}$. Since the table length $L_{i,n}$ is identical to the period of $\bar{\mu}_{i,n}[k]$, the product $q_{i,n} \cdot L_{i,n}$ has to fulfil the relation

$$\bar{\mu}_{i,n}[k] = \bar{\mu}_{i,n}[k + q_{i,n}L_{i,n}] \qquad \forall\, k \in \mathbb{Z}\,. \tag{8.41}$$

Using the last two equations, we can prove the validity of (8.39) in the following way:

$$\bar{\mu}_i[k] \;=\; \sum_{n=1}^{N_i} \bar{\mu}_{i,n}[k]$$

$$
\begin{aligned}
&= \sum_{n=1}^{N_i} \bar{\mu}_{i,n}[k + q_{i,n} L_{i,n}] \\
&= \sum_{n=1}^{N_i} \bar{\mu}_{i,n}[k + L_i] \\
&= \bar{\mu}_i[k + L_i] \qquad \forall\, k \in \mathbb{Z}\,.
\end{aligned}
\tag{8.42}
$$

From L_i being the least common multiple of the set $\{L_{i,n}\}_{n=1}^{N_i}$, it follows that L_i is the smallest (positive) value for which (8.39) is valid. Consequently, L_i is said to be the period of the discrete deterministic process $\bar{\mu}_i[k]$.

We will point out here that an upper bound on the period L_i (8.38) is given by the product of all table lengths $L_{i,n}$, i.e.,

$$
\hat{L}_i = \prod_{n=1}^{N_i} L_{i,n}\,. \tag{8.43}
$$

Taking the above remarks into account, it can easily be shown that $\hat{L}_i$ also fulfils (8.39). However, the period L_i and its upper bound $\hat{L}_i$ are related by $\hat{L}_i \geq L_i$.

From the fact that the table length $L_{i,n}$ depends on the sampling frequency f_s, it follows that the period L_i depends on f_s as well. This dependency is illustrated in Figure 8.8, where the period L_i and its upper bound $\hat{L}_i$ are presented as a function of the normalized sampling frequency f_s/f_{max}. Thereby, the results are deliberately shown for a small, medium, and large number of tables ($N_i = 7,\ N_i = 14,\ N_i = 21$) in order to make clear that both N_i and f_s have a decisive influence on the period L_i. We can also note, especially for low values of N_i, that the period L_i is often close to its upper bound $\hat{L}_i$. The easily computable expression (8.43) therefore allows in general to estimate the period L_i with sufficient precision. Furthermore, it can be realized by considering Figure 8.8 that the period L_i is very large even for small values of f_s/f_{max}. For that reason, we may denote $\bar{\mu}_i[k]$ as quasi-nonperiodic discrete deterministic Gaussian process, provided that the sampling frequency f_s is sufficiently large, i.e., $f_s > 20 f_{max}$.

Next, we will examine the period of discrete deterministic Rayleigh processes $\bar{\zeta}[k]$. Therefore, we consider the following theorem:

Let $\bar{\mu}_1[k]$ and $\bar{\mu}_2[k]$ be two discrete deterministic Gaussian processes, which are periodic with L_1 and L_2, respectively. Then, the discrete deterministic Rayleigh process $\bar{\zeta}[k] = |\bar{\mu}_1[k] + j\bar{\mu}_2[k]|$ is periodic with the period

$$
L = \operatorname{lcm}\{L_1, L_2\}\,. \tag{8.44}
$$

The proof of this theorem is similar to the proof of (8.39) allowing us this time to present an abridged version. Due to (8.44), two natural numbers q_1 and q_2 exist, which fulfil the equations $L = q_1 L_1$ and $L = q_2 L_2$, respectively. Thus, it follows

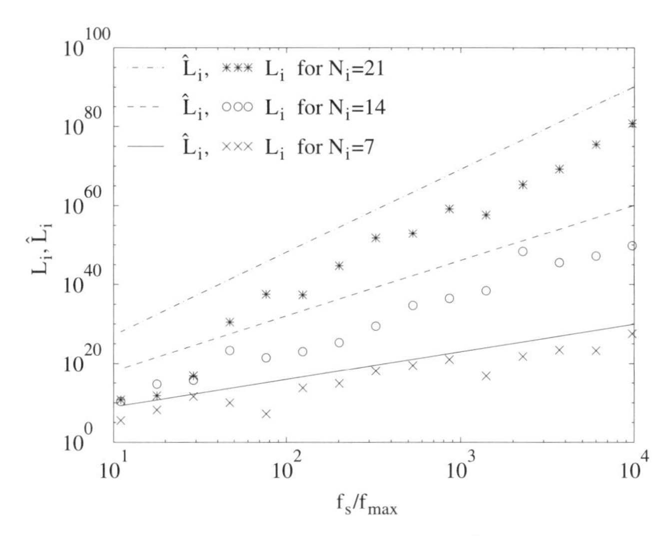

Figure 8.8: Period L_i of $\bar{\mu}_i[k]$ and its upper limit $\hat{L}_i$ as a function of the normalized sampling frequency f_s/f_{max} (MEDS, Jakes PSD, $f_{max} = 91\,\text{Hz}$, $\sigma_0^2 = 1$).

$$\begin{aligned}
\bar{\zeta}[k] &= |\bar{\mu}_1[k] + j\bar{\mu}_2[k]| \\
&= |\bar{\mu}_1[k + L_1] + j\bar{\mu}_2[k + L_2]| \\
&= |\bar{\mu}_1[k + q_1 L_1] + j\bar{\mu}_2[k + q_2 L_2]| \\
&= |\bar{\mu}_1[k + L] + j\bar{\mu}_2[k + L]| \\
&= \bar{\zeta}[k + L] \qquad \forall\, k \in \mathbb{Z}\,.
\end{aligned} \tag{8.45}$$

This shows that $\bar{\zeta}[k]$ is periodic with L. Since L is due to (8.44) the smallest integer number which fulfils (8.45), $L = \text{lcm}\,\{L_1, L_2\}$ must be the period of the discrete deterministic Rayleigh process $\bar{\zeta}[k]$. An upper limit on L is given by

$$\hat{L} = L_1 L_2 \geq L = \text{lcm}\,\{L_1, L_2\}\,. \tag{8.46}$$

8.3.2 Statistical Properties of Discrete Deterministic Processes

This subsection begins with the analysis of the probability density function and the cumulative distribution function of the amplitude and phase of complex discrete deterministic Gaussian processes $\bar{\mu}[k] = \bar{\mu}_1[k] + j\bar{\mu}_2[k]$. Subsequently, it follows the investigation of the level-crossing rate and the average duration of fades of discrete deterministic Rayleigh processes $\bar{\zeta}[k]$ introduced by (8.10). When analysing the statistical properties of discrete deterministic processes, we always assume that all model parameters ($c_{i,n}$, $\bar{f}_{i,n}$, and $\bar{\theta}_{i,n}$) are constant quantities. However, we get access to the analysis of the statistical properties by picking up the numbers (samples) of the discrete deterministic Gaussian process $\bar{\mu}_i[k]$ at random instants k, i.e., we assume in this subsection that k is a random variable, uniformly distributed in the interval $\mathbb{Z}$.

8.3.2.1 Probability Density Function and Cumulative Distribution Function of the Amplitude and the Phase

In this subsection, we will derive analytical expressions for the probability density function and cumulative distribution function of the amplitude as well as the phase of complex discrete deterministic Gaussian processes $\bar{\mu}[k]$. Let us start by considering one single discrete harmonic elementary sequence of the form

$$\bar{\mu}_{i,n}[k] = c_{i,n}\cos(2\pi \bar{f}_{i,n} kT_s + \bar{\theta}_{i,n}), \tag{8.47}$$

where the model parameters $c_{i,n}$, $\bar{f}_{i,n}$, and $\bar{\theta}_{i,n}$ are arbitrary but nonzero quantities and k is the uniformly distributed random variable mentioned above. Since $\bar{\mu}_{i,n}[k]$ is periodic with $L_{i,n}$, we can assume, without restriction of generality, that the random variable k is limited to the half-open interval $[0, L_{i,n})$. In this case, $\bar{\mu}_{i,n}[k]$ has no longer to be regarded as a deterministic sequence but as a random variable, whose possible elementary events (outcomes or realizations) are the elements of the set $\{\bar{\mu}_{i,n}[0], \bar{\mu}_{i,n}[1], \ldots, \bar{\mu}_{i,n}[L_{i,n}-1]\}$. Thereby, it should be noted that each elementary event occurs with the probability $1/L_{i,n}$. Consequently, the probability density function of $\bar{\mu}_{i,n}[k]$ can be written as

$$\bar{p}_{\mu_{i,n}}(x) = \frac{1}{L_{i,n}} \sum_{l=0}^{L_{i,n}-1} \delta(x - \bar{\mu}_{i,n}[l]), \tag{8.48}$$

where $n = 1, 2, \ldots, N_i$ $(i = 1, 2)$. Since the discrete harmonic elementary sequence $\bar{\mu}_{i,n}[k]$ converges to the corresponding harmonic elementary function $\tilde{\mu}_{i,n}(t)$ [see (4.27)] as the sampling interval T_s tends to zero, the discrete probability density function $\bar{p}_{\mu_{i,n}}(x)$ converges consequently to the continuous probability density function $\tilde{p}_{\mu_{i,n}}(x)$ defined by (4.28), i.e., in the limit $T_s \to 0$ it follows $\bar{p}_{\mu_{i,n}}(x) \to \tilde{p}_{\mu_{i,n}}(x)$. An example of the probability density function $\bar{p}_{\mu_{i,n}}(x)$ of $\bar{\mu}_{i,n}[k]$ is shown in Figure 8.9(a) for the case $T_s = 0.1\,\text{ms}$. As well as that, Figure 8.9(b) illustrates the results obtained after taking the limit $T_s \to 0$.

Following the approach described above, we proceed with the derivation of the probability density function $\bar{p}_{\mu_i}(x)$ of discrete deterministic Gaussian processes $\bar{\mu}_i[k]$. Due to the periodicity of $\bar{\mu}_i[k]$, we can restrict k to the half-open interval $[0, L_i)$. Therefore, let k be a random variable, uniformly distributed over $[0, L_i)$, then $\bar{\mu}_i[k]$ [see (8.2)] is also a random variable whose elementary events $\bar{\mu}_i[0]$, $\bar{\mu}_i[1], \ldots, \bar{\mu}_i[L_i-1]$ are uniformly distributed. By analogy to (8.48), the probability density function of discrete deterministic Gaussian processes $\bar{\mu}_i[k]$ can be expressed by

$$\bar{p}_{\mu_i}(x) = \frac{1}{L_i} \sum_{l=0}^{L_i-1} \delta(x - \bar{\mu}_i[l]). \tag{8.49}$$

This result shows that the density $\bar{p}_{\mu_i}(x)$ of $\bar{\mu}_i[k]$ can be represented as a weighted sum of delta functions. Thereby, the delta functions are located at $\bar{\mu}_i[0]$, $\bar{\mu}_i[1], \ldots, \bar{\mu}_i[L_i-1]$ and weighted by the reciprocal value of the period L_i. Notice that $\bar{p}_{\mu_i}(x)$ does not result from the convolution $\bar{p}_{\mu_{i,1}}(x) * \bar{p}_{\mu_{i,2}}(x) * \ldots * \bar{p}_{\mu_{i,N_i}}(x)$, because the random variables $\bar{\mu}_{i,1}[k], \bar{\mu}_{i,2}[k], \ldots, \bar{\mu}_{i,N_i}[k]$ are, strictly speaking, not statistically

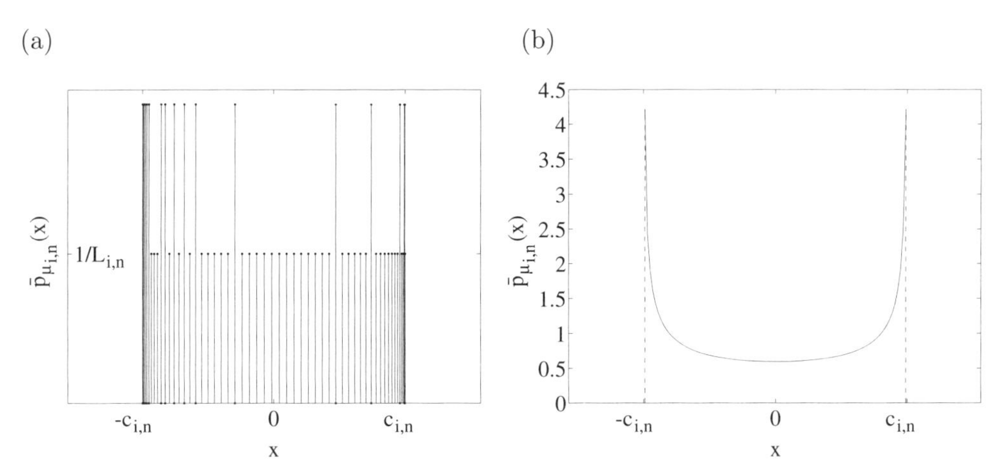

Figure 8.9: Probability density function $\bar{p}_{\mu_{i,n}}(x)$ of $\bar{\mu}_{i,n}(x)$ with (a) $T_s = 0.1$ ms and (b) $T_s \to 0$ (MEDS, Jakes PSD, $N_i = 7$, $n = 7$, $f_{max} = 91$ Hz, $\sigma_0^2 = 1$).

independent. Regarding the tables system, for example, the statistical dependency finds expression in the fact that the address generator in general does not produce the maximum number of different address combinations (states). This, in the ultimate analysis, is the reason why the actual period L_i and the maximum period $\hat{L}_i$ are related by the inequality $L_i \leq \hat{L}_i$. It should also be noted that in the limit $T_s \to 0$ it follows $\bar{p}_{\mu_i}(x) \to \tilde{p}_{\mu_i}(x)$, where $\tilde{p}_{\mu_i}(x)$ [see (4.34)] is the probability density function of $\tilde{\mu}_i(t)$. Moreover, $\bar{p}_{\mu_i}(x)$ approaches to the Gaussian probability density function $p_{\mu_i}(x)$ defined by (4.36) as $T_s \to 0$ and $N_i \to \infty$. For $T_s > 0$, it is not advisable to analyse the difference between the probability density functions $\bar{p}_{\mu_i}(x)$ and $\tilde{p}_{\mu_i}(x)$ directly, because the former density is a discrete function, and the latter is a continuous function. However, this problem can easily be avoided by considering the cumulative distribution function $\bar{F}_{\mu_i}(r)$ of the discrete deterministic Gaussian process $\bar{\mu}_i[k]$. From (8.49), we obtain immediately

$$\bar{F}_{\mu_i}(r) = \frac{1}{L_i} \sum_{l=0}^{L_i-1} \int_0^r \delta(x - \bar{\mu}_i[l])\, dx\,, \quad r \geq 0\,. \tag{8.50}$$

A comparison of $\bar{F}_{\mu_i}(r)$ with the cumulative distribution function $\tilde{F}_{\mu_i}(r)$ of the corresponding continuous-time deterministic Gaussian process $\tilde{\mu}_i(t)$

$$\tilde{F}_{\mu_i}(r) = \frac{1}{2} + 2r \int_0^{\infty} \left[\prod_{n=1}^{N_i} J_0(2\pi c_{i,n}\nu) \right] \text{sinc}\,(2\pi\nu r)\, d\nu\,, \quad r \geq 0 \tag{8.51}$$

is shown in Figure 8.10. The analytical expression for the cumulative distribution function $\tilde{F}_{\mu_i}(r)$ given above can directly be obtained after substituting the probability density function (4.34) in $\tilde{F}_{\mu_i}(r) = \int_{-\infty}^{r} \tilde{p}_{\mu_i}(x)\, dx$ and then solving the integral with respect to the independent variable x.

In addition, the cumulative distribution function

$$F_{\mu_i}(r) = \frac{1}{2}\left[1 + \text{erf}\left(\frac{r}{\sqrt{2}\sigma_0}\right)\right], \quad r \geq 0\,, \tag{8.52}$$

of the zero-mean Gaussian random process $\mu_i(t)$ represents in Figure 8.10 the behaviour of the reference model.

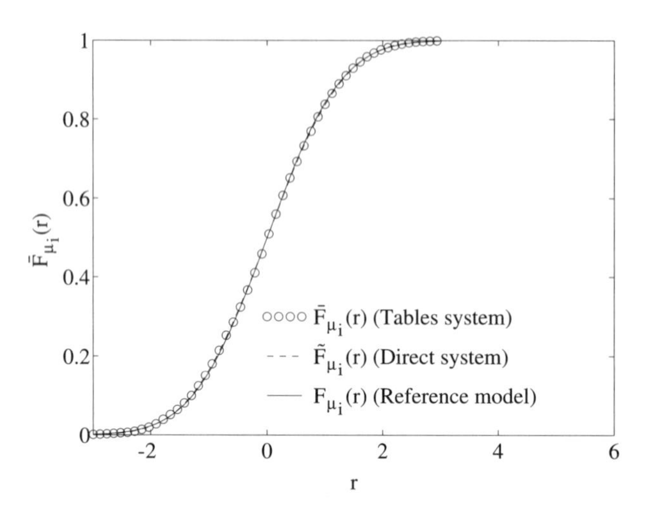

Figure 8.10: Cumulative distribution function $\bar{F}_{\mu_i}(x)$ of discrete deterministic Gaussian processes $\bar{\mu}_i[k]$ for $T_s = 0.1\,\text{ms}$ (MEDS, Jakes PSD, $N_i = 7$, $f_{max} = 91\,\text{Hz}$, $\sigma_0^2 = 1$).

For sufficiently small values of the sampling interval T_s, the period L_i becomes very large (see Figure 8.8) and, consequently, the sample space $\{\bar{\mu}_i[l]\}_{l=0}^{L_i-1}$ becomes very large as well. In such cases, it is not possible to evaluate the cumulative distribution function $\bar{F}_{\mu_i}(r)$ according to (8.50) exactly without exceeding any reasonably chosen time-out interval for the computer simulation. Fortunately, this problem can be avoided, because one even obtains excellent results by merely evaluating $K \ll L_i$ elements of the subset $\{\bar{\mu}_i[k]\}_{k=0}^{K-1}$, as demonstrated in Figure 8.10. This figure shows an almost perfect correspondence between $\bar{F}_{\mu_i}(r)$ and $\tilde{F}_{\mu_i}(r)$ or $F_{\mu_i}(x)$, although (8.50) has been evaluated by using only $K = 50 \cdot 10^3 \ll L_i$ samples $\bar{\mu}_i[k]$ $(k = 0, 1, \ldots, K-1)$.

Next, we will examine the probability density function and the cumulative distribution function of discrete deterministic Rayleigh processes $\bar{\zeta}[k]$. Thereby, we take into account that $\bar{\zeta}[k]$ is periodic with $L = \text{lcm}\{L_1, L_2\}$. Let us assume until further notice that k is a random variable, uniformly distributed in the interval $[0, L)$. Then, it follows that $\bar{\zeta}[k]$ defined by (8.10) is also a random variable, where each of the possible outcomes $\bar{\zeta}[0]$, $\bar{\zeta}[1], \ldots, \bar{\zeta}[L-1]$ occurs with the probability $1/L$. By analogy to (8.49), we can thus write for the probability density function $\bar{p}_\zeta(z)$ of discrete deterministic Rayleigh processes $\bar{\zeta}[k]$

$$\bar{p}_\zeta(z) = \frac{1}{L} \sum_{l=0}^{L-1} \delta(z - \bar{\zeta}[l]), \quad z \geq 0. \tag{8.53}$$

This result allows us to express the cumulative distribution function of $\bar{\zeta}[k]$ as

$$\bar{F}_{\zeta_-}(r) = \frac{1}{L}\sum_{l=0}^{L-1}\int_0^r \delta(z-\bar{\zeta}[l])\,dz\,, \quad r \geq 0\,. \tag{8.54}$$

Note that due to $\bar{\zeta}[k] \to \tilde{\zeta}(t)$ as $T_s \to 0$, it follows $\bar{p}_\zeta(z) \to \tilde{p}_\zeta(z)$ and $\bar{F}_{\zeta_-}(r) \to \tilde{F}_{\zeta_-}(r)$. Thereby, $\tilde{p}_\zeta(z)$ results from (4.47a) with $\rho = 0$, which allows us to present the cumulative distribution function $\tilde{F}_{\zeta_-}(r)$ of $\tilde{\zeta}(t)$ as

$$\begin{aligned}\tilde{F}_{\zeta_-}(r) &= \int_0^r \tilde{p}_\zeta(z)\,dz \\ &= 4r\int_0^\infty J_1(2\pi r z)\int_0^{\pi/2}\left[\prod_{n=1}^{N_1} J_0(2\pi c_{1,n} z\cos\theta)\right] \\ &\quad \left[\prod_{n=1}^{N_2} J_0(2\pi c_{2,n} z\sin\theta)\right]\,d\theta\,dz\,, \quad r \geq 0\,. \end{aligned} \tag{8.55}$$

Finally, it should be noted that after performing the limits $T_s \to 0$ and $N_i \to \infty$, the identity $\bar{F}_{\zeta_-}(r) = F_{\zeta_-}(r)$ is obtained, where

$$F_{\zeta_-}(r) = 1 - e^{-\frac{r^2}{2\sigma_0^2}}\,, \quad r \geq 0\,, \tag{8.56}$$

describes the cumulative distribution function of Rayleigh processes.

The cumulative distribution functions (8.54)–(8.56) are depicted in Figure 8.11. For the evaluation of $\bar{F}_{\zeta_-}(r)$ according to (8.54), $K = 50 \cdot 10^3 \ll L$ samples $\bar{\zeta}[k]$ $(k = 0, 1, \ldots, K-1)$ have been used. The sampling interval T_s has been chosen sufficiently small ($T_s = 0.1\,\text{ms}$).

Now, let us analyse in detail the influence of the sampling interval T_s on the statistics of $\bar{\zeta}[k]$. In particular, it is our intention to answer the following question: what is the maximum value of T_s for which $\bar{F}_{\zeta_-}(r)$ does not perceptibly differ from $\tilde{F}_{\zeta_-}(r)$? Up to now, we have in general assumed that T_s is sufficiently small without concretely saying what the phrase 'sufficiently small' really means. In the following, we want to make up for this by deriving a lower limit for T_s. To illustrate the problem that occurs when T_s exceeds a certain critical threshold, we consider the graphs presented in Figure 8.12. In contrast to the cumulative distribution function $\bar{F}_{\zeta_-}(r)$ shown in Figure 8.11, we have used in the present case $K = L = 9240$ samples $\bar{\zeta}[k]$ $(k = 0, 1, \ldots, K-1)$ for the computation of $\bar{F}_{\zeta_-}(r)$ by using (8.54). At the same time, the sampling interval T_s has been increased from $T_s = 0.1\,\text{ms}$ up to $T_s = 5\,\text{ms}$. Obviously, this seems to be problematical, because different realizations of the quantized Doppler phases $\{\bar{\theta}_{i,n}\}_{n=1}^{N_i}$ are now leading to different cumulative distribution functions $\bar{F}_{\zeta_-}(r)$, which may differ considerably (see Figure 8.12). It should be observed that in this example the sampling theorem for low-pass signals [Fet96] is still fulfilled, because the chosen values $T_s = 5\,\text{ms}$, i.e., $f_s = 200\,\text{Hz}$, and $f_{max} = 91\,\text{Hz}$ are sufficient for the sampling

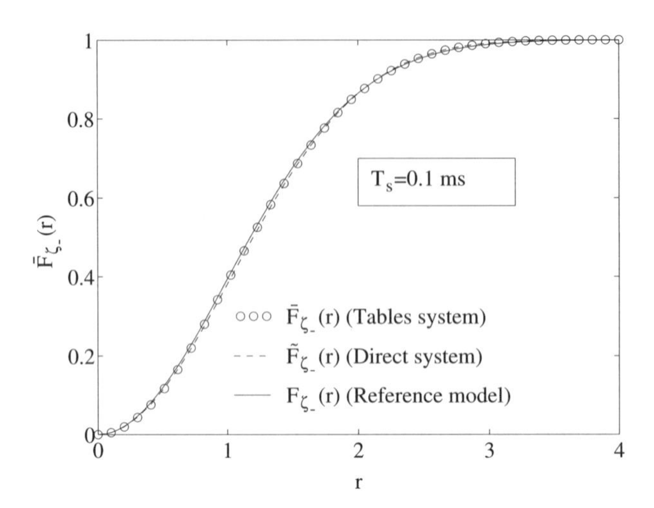

Figure 8.11: Cumulative distribution function $\bar{F}_{\zeta_-}(r)$ of discrete deterministic Rayleigh processes $\bar{\zeta}[k]$ for $T_s = 0.1\,\text{ms}$ (MEDS, Jakes PSD, $N_1 = 7$, $N_2 = 8$, $f_{max} = 91\,\text{Hz}$, $\sigma_0^2 = 1$).

theorem (2.85), i.e., it yields $f_s > 2f_{max}$. By fulfilling the sampling theorem, it is guaranteed that the continuous-time function $\bar{\zeta}(t)$ can be reconstructed completely from its samples $\tilde{\zeta}[k]$. But in addition to that, the sampling theorem provides no further information, for example, about the uniqueness of the cumulative distribution function $\bar{F}_{\zeta_-}(r)$ of $\bar{\zeta}[k]$.

The reason for the problem illustrated in Figure 8.12 can be put down to the fact that due to (8.3) the quantized Doppler frequencies $f_{i,n}$ are related to the sampling interval T_s. The consequence of this relation is that from the requirement

$$f_{i,n} \neq f_{j,m} \tag{8.57}$$

it does not inevitably follow that by increasing T_s the two inequalities

$$\bar{f}_{i,n} \neq \bar{f}_{j,m} \iff L_{i,n} \neq L_{j,m} \tag{8.58}$$

are also fulfilled, where $n = 1, 2, \ldots, N_i$ and $m = 1, 2, \ldots, N_j$ $(i, j = 1, 2)$. If T_s exceeds a certain threshold, then one or several pairs (n, m) exist for which $\bar{f}_{1,n} = \bar{f}_{2,m}$ and, thus, $L_{1,n} = L_{2,m}$ hold. In this case, the discrete harmonic elementary sequences $\bar{\mu}_{1,n}[k]$ and $\bar{\mu}_{2,n}[k]$ are identical apart from a phase shift. Hence, it follows that the discrete deterministic Gaussian processes $\bar{\mu}_1[k]$ and $\bar{\mu}_2[k]$ are correlated. Moreover, by increasing T_s it can also be the case that $\bar{f}_{i,n} = \bar{f}_{i,m} \Leftrightarrow L_{i,n} = L_{i,m}$ holds for $i = 1, 2$ and $n \neq m$. This, by the way, becomes obvious by examining the graphs shown in Figures 8.3(a) and 8.3(b) for $f_s/f_{max} < 10$.

For the derivation of a lower limit on the sampling frequency $f_{s,min}$, the auxiliary function

$$\Delta_{n,m}^{(i,j)} := L_{i,n} - L_{j,m} \tag{8.59}$$

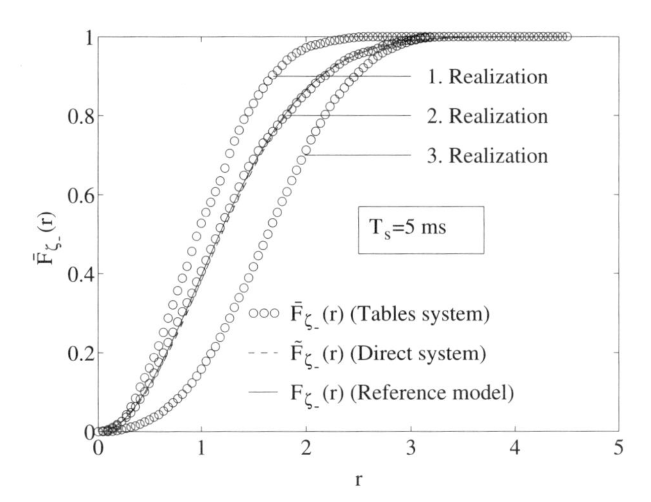

Figure 8.12: Cumulative distribution function $\bar{F}_{\zeta_-}(r)$ of discrete deterministic Rayleigh processes $\bar{\zeta}[k]$ for $T_s = 5\,\mathrm{ms}$ and different realizations of the discrete Doppler phases $\{\bar{\theta}_{i,n}\}_{n=1}^{N_i}$ (MEDS, Jakes PSD, $N_1 = 7$, $N_2 = 8$, $f_{max} = 91\,\mathrm{Hz}$, $\sigma_0^2 = 1$).

is useful. Using (8.4), we can also write

$$\Delta_{n,m}^{(i,j)} = \mathrm{round}\left\{\frac{f_s}{f_{i,n}}\right\} - \mathrm{round}\left\{\frac{f_s}{f_{j,m}}\right\}, \tag{8.60}$$

where $n = 1, 2, \ldots, N_i$ and $m = 1, 2, \ldots, N_j$ $(i, j = 1, 2)$. The lower limit on the sampling frequency $f_{s,min}$ is determined by those pairs (n, m) and (i, j) for which, by decreasing f_s, the auxiliary function (8.60) is zero for the first time. Hence,

$$f_{s,min} = \max\left\{f_s \,\middle|\, \Delta_{n,m}^{(i,j)} = 0 \quad \forall\, i, j = 1, 2\right\}_{n,m=1}^{N_i,N_j}. \tag{8.61}$$

This result can be summarized in the following statement: Let us assume that the elements of the two sets $\{f_{1,n}\}_{n=1}^{N_1}$ and $\{f_{2,m}\}_{m=1}^{N_2}$ fulfil the property $f_{i,n} \neq f_{j,m}$, then the corresponding elements of sets $\{\bar{f}_{1,n}\}_{n=1}^{N_1}$ and $\{\bar{f}_{2,m}\}_{m=1}^{N_2}$ fulfil the analogous property $\bar{f}_{i,n} \neq \bar{f}_{j,m}$ for all $n = 1, 2, \ldots, N_1$ and $m = 1, 2, \ldots, N_2$ $(i, j = 1, 2)$, if the sampling frequency f_s is above the threshold defined by (8.61), i.e., $f_s > f_{s,min}$. In this case, it follows that if the processes $\tilde{\mu}_1(t)$ and $\tilde{\mu}_2(t)$ are uncorrelated, then the corresponding sequences $\bar{\mu}_1[k]$ and $\bar{\mu}_2[k]$ are also uncorrelated.

Two examples showing the results of the evaluation of (8.61) by using the MEDS are presented in Figure 8.13. In particular by using the MEDS, the lower limit for the sampling frequency $f_{s,min}$ is determined by that value for f_s for which the auxiliary function $\Delta_{N_1,N_2}^{(1,2)}$ becomes zero for the first time. Problems caused by correlation (see Figure 8.12) can thus be avoided if the sampling frequency f_s is above the threshold shown in Figure 8.13, which is not the case for the negative examples shown in Figure 8.12.

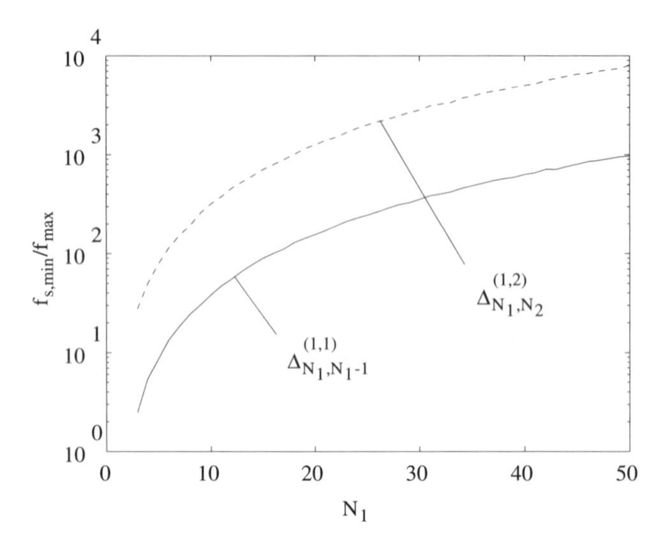

Figure 8.13: Lower limit on the sampling frequency $f_{s,min}$ as a function of N_1 (MEDS, Jakes PSD, $N_2 = N_1 + 1$, $f_{max} = 91\,\text{Hz}$, $\sigma_0^2 = 1$).

It should finally be mentioned that the lower limit $f_{s,min}$ according to (8.61) is sufficient, but not necessary in order to fulfil the condition (8.58), i.e., it cannot be excluded that there exist values for f_s which are below $f_{s,min}$ and even though the two inequalities in (8.58) are fulfilled. Here, it is not our intention to discuss this problem in detail. Instead of this, we consider $f_s > 20 f_{max}$ as a rule of thumb, which turned out to be very useful for most practical applications. Hence, we say that the sampling frequency f_s is sufficiently large, if f_s is larger than $20 f_{max}$.

We will proceed with the analysis of the probability density function $\bar{p}_\vartheta(\theta)$ of the phase of complex discrete deterministic Gaussian processes $\bar{\mu}[k]$. Of course, similar arguments to those used for the derivation of (8.53) can be applied here to achieve the present aim. However, we prefer a more simple and straightforward approach by substituting in (8.53) the amplitude $\bar{\zeta}[l]$ by the phase $\bar{\vartheta}[l]$ allowing us directly to express the probability density function $\bar{p}_\vartheta(\theta)$ of the phase as

$$\bar{p}_\vartheta(\theta) = \frac{1}{L} \sum_{l=0}^{L-1} \delta(\theta - \bar{\vartheta}[l]) \,, \quad |\theta| \leq \pi \,, \tag{8.62}$$

where $\bar{\vartheta}[l] = \arctan\{\bar{\mu}_2[l]/\bar{\mu}_1[l]\}$ denotes the phase of the complex deterministic Gaussian process $\bar{\mu}[k] = \bar{\mu}_1[k] + j\bar{\mu}_2[k]$ at instants $k = l \in \{0, 1, \ldots, L-1\}$. Using (8.62), the corresponding cumulative distribution function can be written as

$$\bar{F}_\vartheta(\varphi) = \frac{1}{L} \sum_{l=0}^{L-1} \int_{-\pi}^{\varphi} \delta(\theta - \bar{\vartheta}[l]) \, d\varphi \,, \quad |\varphi| \leq \pi \,. \tag{8.63}$$

It should be mentioned that due to $\bar{\vartheta}[k] \to \tilde{\vartheta}(t)$ as $T_s \to 0$, it also follows $\bar{p}_\vartheta(\theta) \to \tilde{p}_\vartheta(\theta)$ and, thus, $\bar{F}_\vartheta(\varphi) \to \tilde{F}_\vartheta(\varphi)$ as $T_s \to 0$, where $\tilde{p}_\vartheta(\theta)$ can be obtained from (4.47b) for $\rho = 0$, so that the cumulative distribution function $\tilde{F}_\vartheta(\varphi)$ of the phase $\tilde{\vartheta}(t)$ of $\tilde{\mu}(t) = \tilde{\mu}_1(t) + j\tilde{\mu}_2(t)$ can be expressed as

$$\begin{aligned}\tilde{F}_\vartheta(\varphi) &= \int_{-\pi}^{\varphi} \tilde{p}_\vartheta(\theta)\, d\theta \\ &= 4 \int_{-\pi}^{\varphi} \int_{0}^{\infty} z \left\{ \int_{0}^{\infty} \left[\prod_{n=1}^{N_1} J_0(2\pi c_{1,n}\nu_1) \right] \cos(2\pi\nu_1 z \cos\theta)\, d\nu_1 \right\} \\ &\quad \left\{ \int_{0}^{\infty} \left[\prod_{m=1}^{N_2} J_0(2\pi c_{2,m}\nu_2) \right] \cos(2\pi\nu_2 z \sin\theta)\, d\nu_2 \right\} dz\, d\theta\,, \quad |\varphi| \leq \pi\,. \end{aligned} \tag{8.64}$$

As $T_s \to 0$ and $N_i \to \infty$, it follows $\bar{F}_\vartheta(\varphi) \to F_\vartheta(\varphi)$, where

$$F_\vartheta(\varphi) = \frac{1}{2}\left(1 + \frac{\varphi}{\pi}\right), \quad |\varphi| \leq \pi\,, \tag{8.65}$$

is the cumulative distribution function of the uniformly distributed phase of zero-mean complex Gaussian random processes $\mu(t) = \mu_1(t) + j\mu_2(t)$.

Figure 8.14 illustrates the cumulative distribution functions (8.63)–(8.65). The evaluation of (8.63) has been performed by using $K = 50 \cdot 10^3 \ll L$ samples (outcomes) of the sample space $\{\bar{\vartheta}[l]\}_{l=0}^{L-1}$. For the sampling interval T_s, the value $T_s = 0.1$ ms has been chosen. Thus, by using the MEDS with the parameters specified in the figure caption of Figure 8.14, the relation f_s/f_{max} is close to the threshold $f_{s,min}/f_{max}$ (see Figure 8.13).

8.3.2.2 Level-Crossing Rate and Average Duration of Fades

In contrast to continuous-time deterministic Rayleigh processes for which analytical expressions for both the level-crossing rate and the average duration of fades have been derived (see Appendix C), up to now no comparable solutions for discrete-time deterministic Rayleigh processes exist. In the following, we restrict our investigation to the derivation of approximate formulas by assuming that the normalized sampling frequency f_s/f_{max} lies above the threshold shown in Figure 8.13. Thereby, the number of harmonic functions (tables) N_i is assumed to be sufficiently large, i.e., $N_i \geq 7$. Moreover, we assume that the relative model error $\Delta\bar{\beta}_i/\beta_i$ of the discrete-time system is small, which is in particular the case when the MEDS is applied on condition that $f_s > f_{s,min}$ is fulfilled (see Figure 8.7). Taking into account that the probability density function $\bar{p}_{\mu_i}(x)$ of discrete deterministic processes $\bar{\mu}_i[k]$ is asymptotically equal to the probability density function $\tilde{p}_{\mu_i}(x)$ of continuous-time deterministic processes $\tilde{\mu}_i(t)$, i.e., $\bar{p}_{\mu_i}(x) \sim \tilde{p}_{\mu_i}(x)$, then we can summarize the above mentioned statements and assumptions as follows:

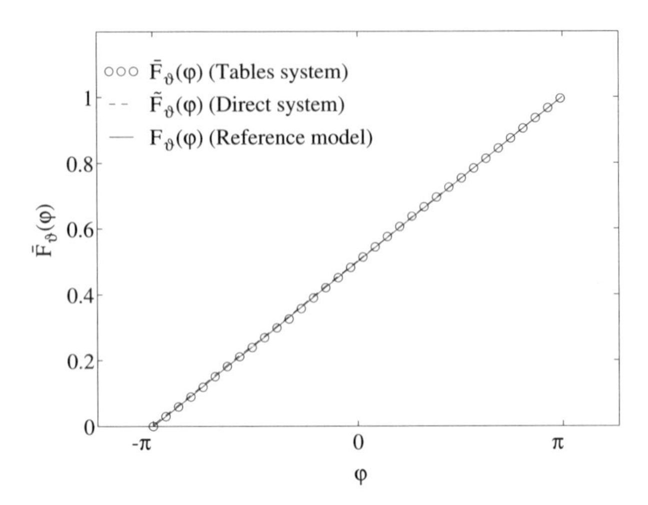

Figure 8.14: Cumulative distribution function $\bar{F}_\vartheta(\varphi)$ of the phase $\bar{\vartheta}[k]$ of complex discrete deterministic Gaussian processes $\bar{\mu}[k] = \bar{\mu}_1[k] + j\bar{\mu}_2[k]$ for $T_s =$ 0.1 ms (MEDS, Jakes PSD, $N_1 = 7$, $N_2 = 8$, $f_{max} = 91\,\text{Hz}$, $\sigma_0^2 = 1$).

$$\text{(i)} \qquad \bar{p}_{\mu_i}(x) \sim \tilde{p}_{\mu_i}(x) \approx p_{\mu_i}(x)\,, \tag{8.66a}$$

$$\text{(ii)} \qquad \bar{\beta} = (\bar{\beta}_1 + \bar{\beta}_2)/2 \approx \beta = \beta_1 = \beta_2\,. \tag{8.66b}$$

Taking these assumptions into account, the level-crossing rate $\bar{N}_\zeta(r)$ and the average duration of fades $\bar{T}_{\zeta_-}(r)$ of discrete deterministic Rayleigh processes $\bar{\zeta}[k]$ are in principle still given by the approximations (4.66) and (4.70), respectively. However, we only have to evaluate these equations for the case $\rho = 0$ and after that, we have to replace the model error $\Delta\beta$ by $\Delta\bar{\beta}$. This results in

$$\bar{N}_\zeta(r) \approx N_\zeta(r)\left(1 + \frac{\Delta\bar{\beta}}{2\beta}\right), \tag{8.67a}$$

$$\bar{T}_{\zeta_-}(r) \approx T_{\zeta_-}(r)\left(1 - \frac{\Delta\bar{\beta}}{2\beta}\right), \tag{8.67b}$$

where $\Delta\bar{\beta} = \bar{\beta} - \beta$. In (8.67a), $N_\zeta(r)$ denotes the level-crossing rate of Rayleigh processes as defined by (2.60) and in (8.67b), $T_{\zeta_-}(r)$ refers to the average duration of fades introduced by (2.65). As $T_s \to 0$ and $N_i \to \infty$ it follows $\bar{N}_\zeta(r) \to N_\zeta(r)$ and $\bar{T}_{\zeta_-} \to T_{\zeta_-}(r)$.

Figure 8.15(a) shows an example for the normalized level-crossing rate $\bar{N}_\zeta(r)/f_{max}$ of discrete deterministic Rayleigh processes $\bar{\zeta}[k]$. Just as in the previous examples, we computed here the model parameters by using the MEDS with $N_1 = 7$ and $N_2 = 8$. For the sampling interval T_s, again, the value $T_s = 0.1\,\text{ms}$ has been chosen. The corresponding normalized average duration of fades $\bar{T}_{\zeta_-}(r) \cdot f_{max}$ is illustrated in Figure 8.15(b).

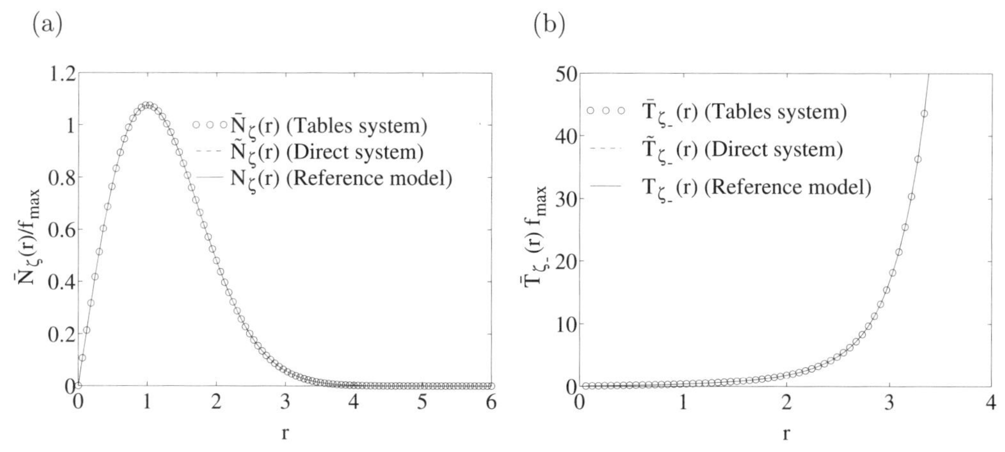

Figure 8.15: (a) Normalized level-crossing rate $\bar{N}_\zeta(r)/f_{max}$ and (b) normalized average duration of fades $\bar{T}_{\zeta_-}(r)\cdot f_{max}$ of discrete deterministic Rayleigh processes $\bar{\zeta}[k]$ for $T_s = 0.1$ ms (MEDS, Jakes PSD, $N_1 = 7$, $N_2 = 8$, $f_{max} = 91$ Hz, $\sigma_0^2 = 1$).

8.4 REALIZATION EXPENDITURE AND SIMULATION SPEED

In this section, we will examine in detail the efficiency of the tables system (see Figure 8.2). Moreover, the tables system's efficiency will be compared to that of the corresponding discrete-time direct system, which is obtained by replacing the continuous-time variable t by kT_s in Figure 4.3. For convenience, we ignore the influence of the line-of-sight component by choosing $\rho = 0$. Let us assume in the following that the set-up phase has been completed, so that we can restrict our investigations to the computation expenditure required for the generation of the respective complex-valued channel output sequence.

It can easily be seen from Figure 8.2 that the operations listed in Table 8.1 have to be carried out at each instant k in order to compute one sample of the complex discrete deterministic Gaussian process $\bar{\mu}[k] = \bar{\mu}_1[k] + j\bar{\mu}_2[k]$. One realizes that merely additions and simple conditional control flow statements (*if-else* statements) are required. The additions are needed for the generation of the addresses within the address generator as well as for adding up the tables outputs, whereas the conditional control flow statements are only required for the generation of the addresses within the address generator.

The number of operations required for the generation of the complex-valued sequence $\tilde{\mu}[k] = \tilde{\mu}_1[k] + j\tilde{\mu}_2[k]$ by employing the direct system is also listed in Table 8.1. Thereby, normalized Doppler frequencies $\Omega_{i,n} = 2\pi f_{i,n} T_s$ have been used in order to avoid unnecessary multiplications within the arguments of the harmonic functions.

The results shown in Table 8.1 can be summarized as follows: all multiplications can be avoided, the number of additions remains unchanged, and all trigonometric operations can be substituted by simple if-else statements, when the tables system is used instead of the direct system for the generation of complex-valued channel output sequences. It

Table 8.1: Number of operations required for the computation of $\bar{\mu}[k]$ (tables system) and $\tilde{\mu}[k]$ (direct system).

Number of operations	Tables system	Direct system
# Multiplications	0	$2(N_1 + N_2)$
# Additions	$2(N_1 + N_2) - 2$	$2(N_1 + N_2) - 2$
# Trig. operations	0	$N_1 + N_2$
# If-else operations	$N_1 + N_2$	0

is therefore not surprising when it turns out in the following that the tables system has clear advantages in comparison with the direct system with respect to the simulation speed.

As an appropriate measure of the simulation speed of channel simulators, we introduce the *iteration time* defined by

$$\Delta T_{sim} = \frac{T_{sim}}{K}, \tag{8.68}$$

where T_{sim} denotes the simulation time required for the computation of K samples of the complex-valued channel output sequence. Thus, the quantity ΔT_{sim} represents the average computation time per complex-valued channel output sample. Figure 8.16 shows the iteration time ΔT_{sim} for both the direct system and the tables system as function of the number of harmonic functions (tables) N_1. The model parameters $f_{i,n}$ and $c_{i,n}$ have been computed by applying the MEDS with $N_2 = N_1+1$ and by using the JM with $N_2 = N_1$. The algorithms of the channel simulators have been implemented on a computer by using MATLAB and the simulation results for T_{sim} are obtained by running the programs on a workstation (HP 730). For each run, the number of samples of the complex-valued channel output sequence was equal to $K = 10^4$.

The results illustrated in Figure 8.16 clearly show the difference in speed of the treated channel simulators. When using the MEDS, for example, the simulation speed of the tables system is approximately 3.8 times higher than that of the direct system. Applying the JM, we can exploit the fact that the discrete Doppler frequencies $f_{1,n}$ and $f_{2,n}$ are identical, whereas the corresponding Doppler phases $\theta_{1,n}$ and $\theta_{2,n}$ are zero for all $n = 1, 2, \ldots, N_1$ ($N_1 = N_2$). This enables a drastic reduction of the complexity of both simulation systems. The consequence for the direct system is that only N_1 instead of N_1+N_2 harmonic functions have to be evaluated at each instant k. The speed of the direct system can thus be increased by approximately a factor of two (see Figure 8.16). The properties of the JM ($f_{1,n} = f_{2,n}$, $c_{1,n} \neq c_{2,n}$, $\theta_{1,n} = \theta_{2,n} = 0$, $N_1 = N_2$) furthermore imply that the tables $\text{Tab}_{1,n}$ and $\text{Tab}_{2,n}$ of the tables system have the same length, i.e., it holds $L_{1,n} = L_{2,n}$ for all $n = 1, 2, \ldots, N_1$ ($N_1 = N_2$). Bearing this in mind and noticing that from $\theta_{i,n} = 0$ it follows immediately $\bar{\theta}_{i,n} = 0$, it is seen that the address generator only needs to compute half of the usually required number of addresses. This is the reason for the fact that the speed of the tables system increases approximately for another 40 per cent (see Figure 8.16).

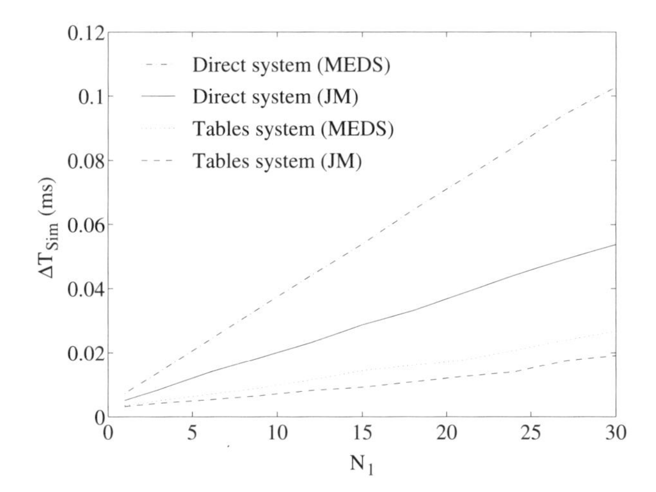

Figure 8.16: Iteration time ΔT_{sim} as a function of the number of harmonic functions (tables) N_1 (MEDS with $N_2 = N_1 + 1$, JM with $N_2 = N_1$, $f_{max} = 91\,\text{Hz}$, $\sigma_0^2 = 1$, $T_s = 0.1\,\text{ms}$).

Summing up, we can say that the tables system is — by using the MEDS (JM) — approximately four times (three times) faster than the corresponding direct system. The benefit of higher speed is confronted by the disadvantage of higher demand for storage elements, but this, however, is the only disadvantage of the tables system worth mentioning. Remember that the total demand for storage elements is proportional to the sampling frequency (see Figure 8.3). By choosing the sampling frequency f_s just above $f_{s,min}$, then the minimum number of storage elements is obtained without accepting appreciable losses in precision. However, a good compromise between the model's precision and complexity is obtained by choosing f_s within the range $20f_{max} \leq f_s \leq 30f_{max}$. When such a designed channel simulator is used as link between the transmitter and the receiver of a mobile communication system, then a sampling rate conversion by means of an interpolation (a decimation) filter is in general required in order to fit the sampling frequency of the channel simulator to the sampling frequency of the receiver's input (transmitter's output).

8.5 COMPARISON WITH THE FILTER METHOD

At this point, it is advisable to carry out a comparison with the filter method, which is also often used in the design of simulation models for mobile radio channels. Here, we restrict our investigations to the modelling of Rayleigh processes. For that purpose, we consider the discrete-time structure depicted in Figure 8.17.

Since white Gaussian noise is, strictly speaking, not realizable, we consider $\tilde{\nu}_1[k]$ and $\tilde{\nu}_2[k]$ as two realizable noise sequences whose statistical properties are sufficiently close to those of ideal white Gaussian random processes. In particular, we demand that these pseudo-random sequences $\tilde{\nu}_i[k]$ $(i = 1, 2)$ are uncorrelated, having a very long period, and fulfilling the properties $E\{\tilde{\nu}_i[k]\} = 0$ and $\text{Var}\,\{\tilde{\nu}_i[k]\} = 1$.

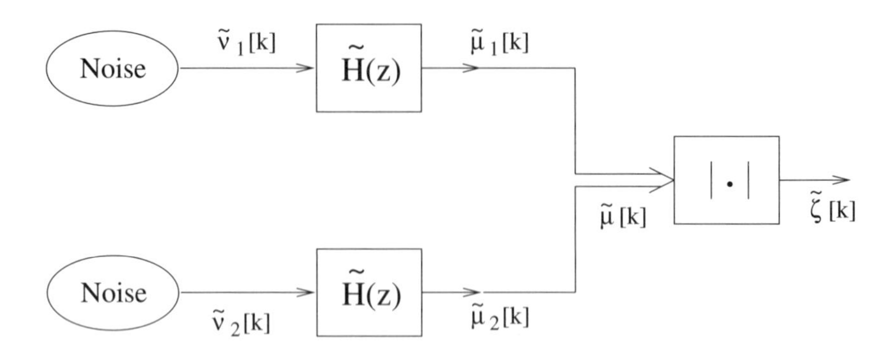

Figure 8.17: A simulation model for Rayleigh processes on basis of the filter method.

In Figure 8.17, $\tilde{H}(z)$ denotes the transfer function of a digital filter in the z-domain. In practice, recursive digital filters are widely in use for modelling of narrow-band random processes. The transfer function of such filters can be represented in the z-domain as follows

$$\tilde{H}(z) = A_0 \frac{\prod_{n=1}^{N_0/2} \left(z - \rho_{0n}\, e^{j\varphi_{0n}}\right)\left(z - \rho_{0n}\, e^{-j\varphi_{0n}}\right)}{\prod_{n=1}^{N_0/2} \left(z - \rho_{\infty n}\, e^{j\varphi_{\infty n}}\right)\left(z - \rho_{\infty n}\, e^{-j\varphi_{\infty n}}\right)}, \tag{8.69}$$

where N_0 denotes the order of the filter and A_0 is a constant which will be determined in such a way that the mean power at the output of the digital filter is equal to σ_0^2. As we already know, the principle of the filter method is to determine the coefficients of the transfer function of the filter in such a way that the deviations between the magnitude of the transfer function $|\tilde{H}(e^{j2\pi f T_s})|$ and the square root of the desired Doppler power spectral density $\sqrt{S_{\mu_i\mu_i}(f)}$ are minimal, or at least as small as possible, with respect to an appropriate error criterion. This problem is in general solved by applying numerical optimization procedures such as, e.g., the Fletcher-Powell algorithm [Fle63] or the Remez exchange procedure. An overview of commonly used optimization procedures can be found in [Fle87, Gro97, Ent76].

Particularly for the widely used Jakes power spectral density (3.8), a recursive digital filter of eighth order has been designed in [Bre86a], which very closely approximates the desired frequency response. In Table 8.2 the coefficients of the recursive digital filter adopted from [Hae88] are listed for a cut-off frequency f_c that has been normalized to the sampling frequency f_s according to $f_c = f_s/(110.5)$.

The resulting graph of the squared magnitude function $|\tilde{H}(e^{j2\pi f T_s})|^2$ and the desired Jakes power spectral density are both presented in Figure 8.18(a). The very good conformity between the corresponding autocorrelation functions is shown in Figure 8.18(b).

The cut-off frequency f_c is in case of the Jakes power spectral density identified with the maximum Doppler frequency f_{max}. This means that by changing of f_{max}

Table 8.2: Coefficients of the transfer function of the eighth order recursive filter [Hae88].

n	ρ_{0n}	φ_{0n}	$\rho_{\infty n}$	$\varphi_{\infty n}$
1	1.0	$5.730778 \cdot 10^{-2}$	0.991177	$4.542547 \cdot 10^{-2}$
2	1.0	$7.151706 \cdot 10^{-2}$	0.980664	$1.912862 \cdot 10^{-2}$
3	1.0	0.105841	0.998042	$5.507401 \cdot 10^{-2}$
4	1.0	0.264175	0.999887	$5.670618 \cdot 10^{-2}$

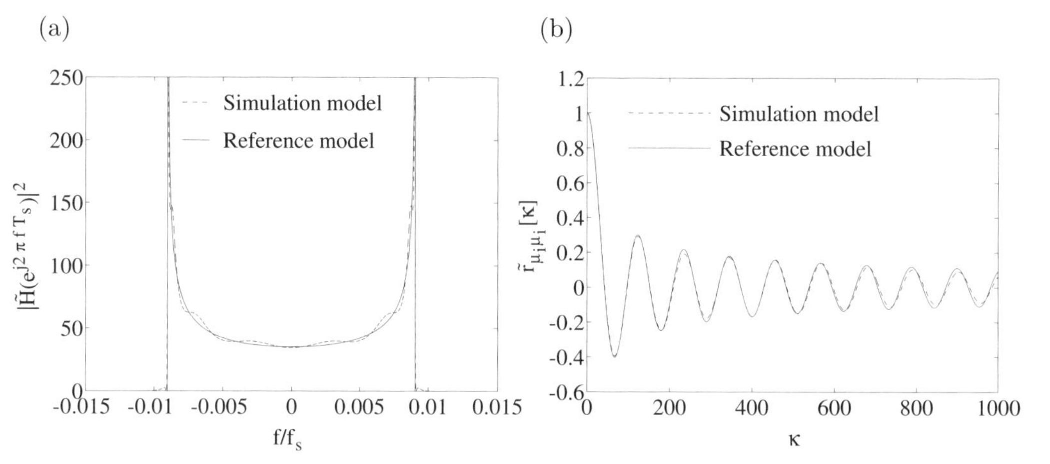

Figure 8.18: (a) Squared magnitude function $|\tilde{H}(e^{j2\pi f T_s})|^2$ of the eighth order recursive filter and (b) autocorrelation sequence $\tilde{r}_{\mu_i\mu_i}[\kappa]$ of the filtered random process $\tilde{\mu}_i[k]$ $(i = 1, 2)$ [Bre86a].

or f_c, all coefficients of the transfer function $\tilde{H}(z)$ have to be recalculated employing common lowpass-to-lowpass transformations [Opp75]. In this case it should be noted that due to these frequency transformations nonlinear frequency distortions occur, which in particular cannot be ignored when the relation f_c/f_s is small. In practice, this problem is solved by employing sampling rate conversion. Thereby, the digital filter operates with a small sampling rate that has to be converted afterwards by means of an interpolation filter to the mostly much higher sampling rate of the transmission system. We will not go into details of sampling rate conversion, since our aim here is to compare the computation speed of different channel simulators, which disregards anyway the conversion of the sampling rate for the reason of fairness. Otherwise, an interpolator would also be necessary for both the tables system and the direct system. Clearly, this is in principle always feasible, but with regard to a simple measurement of the computation speed, this will only lead to the fact that the computation speed becomes dependent, aside from other factors, on the chosen interpolation factor.

In [Pae00e], the structure shown in Figure 8.17 of the eighth order recursive digital filter described above has been implemented by using MATLAB on a workstation (HP 730) and the iteration time ΔT_{sim} has been measured according to the rule (8.68). The result of this measurement was $\Delta T_{sim} = 0.02\,\text{ms}$. It also turned out that approximately 70 per cent of the total computation time is required for the generation of the real-valued random sequences $\tilde{\nu}_1[k]$ and $\tilde{\nu}_2[k]$, whereas the filtering of these sequences only occupies the remaining 30 per cent of the computation time. From this result we conclude that a reduction of the filter order does not automatically lead to a significant reduction in iteration time ΔT_{sim}.

Relating now the iteration time obtained for the filter system of eighth order to the corresponding iteration time of the direct system and the tables system, it becomes apparent that by using the MEDS ($N_1 = 7$, $N_2 = 8$) the direct system is approximately 25 per cent slower than the filter system, whereas the speed of the tables system outperforms that of the filter system by approximately 300 per cent.

Appendix A

DERIVATION OF THE JAKES POWER SPECTRAL DENSITY AND THE CORRESPONDING AUTOCORRELATION FUNCTION

The derivation of the Jakes power spectral density is based on the following three assumptions:

(i) The propagation of the electromagnetic waves takes place in the two-dimensional (horizontal) plane, and the receiver is located in the centre of an isotropic scattering area.

(ii) The angles of arrival α of the waves arriving the receiving antenna are uniformly distributed in the interval $[-\pi, \pi)$.

(iii) The antenna radiation pattern of the receiving antenna is circular-symmetrical (omnidirectional antenna).

Due to the assumption that the angles of arrival α are random variables with the probability density function

$$p_\alpha(\alpha) = \begin{cases} \dfrac{1}{2\pi}, & \alpha \in [-\pi, \pi), \\ 0, & \text{else}, \end{cases} \tag{A.1}$$

it follows that the Doppler frequencies, defined by

$$f = f(\alpha) := f_{max} \cos(\alpha), \tag{A.2}$$

are also random variables. The probability density function of the Doppler frequencies f, denoted by $p_f(f)$, can easily be computed by using (2.38). Applying (2.38) to the present problem enables us to write the probability density function $p_f(f)$ in the following form

$$p_f(f) = \sum_{\nu=1}^{m} \frac{p_\alpha(\alpha_\nu)}{\left|\frac{d}{d\alpha} f(\alpha)\right|_{\alpha=\alpha_\nu}}, \tag{A.3}$$

where m is the number of solutions of the equation (A.2) within the interval $[-\pi, \pi)$. For $|f| > f_{max}$, the equation $f = f_{max}\cos(\alpha)$ has no real-valued solution and, consequently, $p_f(f) = 0$ for $|f| > f_{max}$. However, due to the ambiguity of the inverse function of the cosine function within the interval $[-\pi, \pi)$, two solutions exist for $|f| < f_{max}$, namely,

$$\alpha_1 = -\alpha_2 = \arccos\left(f/f_{max}\right), \tag{A.4}$$

so that $m = 2$. After elementary computations and by using (A.1)–(A.4), we find the following result for the probability density function $p_f(f)$ of the Doppler frequencies

$$p_f(f) = \begin{cases} \dfrac{1}{\pi f_{max}\sqrt{1-(f/f_{max})^2}}, & |f| < f_{max}, \\ 0, & |f| > f_{max}. \end{cases} \tag{A.5}$$

It can easily be seen that the probability density function $p_f(f)$ of the Doppler frequencies has to be directly proportional to the power spectral density $S_{\mu\mu}(f)$ of the scattered components $\mu(t) = \mu_1(t) + j\mu_2(t)$, received at the receiving antenna. Therefore, we imagine that $\mu(t)$ can be represented by a superposition of an infinite number of exponential functions according to

$$\mu(t) = \lim_{N\to\infty}\sum_{n=1}^{N} c_n\, e^{j(2\pi f_n t + \theta_n)}. \tag{A.6}$$

As a consequence of the idealized assumption of isotropic scattering propagation, all amplitudes $c_n = \sigma_0\sqrt{2/N}$ have the same size. The Doppler frequencies f_n in (A.6) are random variables whose probability density function is determined by (A.5). Likewise, the phases θ_n are random variables, but they are uniformly distributed in the interval $[0, 2\pi)$. One may note that the power spectral density $S_{\mu\mu}(f)$ of (A.6) is composed of an infinite number of discrete spectral lines and that within an infinitesimal frequency interval df the average power $S_{\mu\mu}(f)\,df$ can be observed. This power has to be proportional to the number of spectral lines contained in df. On the other hand, with (A.5) the number of spectral lines contained in the frequency interval df can also be represented by $p_f(f)\,df$. Hence, the following relation holds

$$S_{\mu\mu}(f)\,df \sim p_f(f)\,df, \tag{A.7}$$

and, thus,

$$S_{\mu\mu}(f) \sim p_f(f). \tag{A.8}$$

Consequently, due to $\int_{-\infty}^{\infty} S_{\mu\mu}(f)\,df = 2\sigma_0^2$ and $\int_{-\infty}^{\infty} p_f(f)\,df = 1$, it follows the relation

$$S_{\mu\mu}(f) = 2\sigma_0^2\, p_f(f). \tag{A.9}$$

Thus, by taking (A.5) into account, we find the power spectral density

$$S_{\mu\mu}(f) = \begin{cases} \dfrac{2\sigma_0^2}{\pi f_{max}\sqrt{1-(f/f_{max})^2}}, & |f| \le f_{max}, \\ 0, & |f| > f_{max}, \end{cases} \tag{A.10}$$

which in the literature is often called *Jakes power spectral density* or *Clarke power spectral density*. Strictly speaking, in the above equation, we should have used the less strict inequality $|f| < f_{max}$ instead of $|f| \leq f_{max}$. In other publications, however, the poles at $f = \pm f_{max}$ are commonly assigned to the range of the Jakes power spectral density. Without wanting to go into a detailed analysis of $S_{\mu\mu}(f)$ at $f = \pm f_{max}$, we will follow the conventional notation, particularly since this small modification does not have any effect on the subsequent computations anyway.

For the power spectral density of the real part and the imaginary part of $\mu(t) = \mu_1(t) + j\mu_2(t)$, the relation

$$S_{\mu_i\mu_i}(f) = \frac{S_{\mu\mu}(f)}{2} = \begin{cases} \dfrac{\sigma_0^2}{\pi f_{max}\sqrt{1-(f/f_{max})^2}}\,, & |f| \leq f_{max}\,, \\ 0\,, & |f| > f_{max}\,, \end{cases} \tag{A.11}$$

holds for $i = 1$ and $i = 2$, respectively.

Finally, we also compute the autocorrelation function $r_{\mu\mu}(\tau)$ of the scattered component $\mu(t) = \mu_1(t) + j\mu_2(t)$. At first, we choose the way over the inverse Fourier transform of the Jakes power spectral density (A.10) and obtain — by taking into account that $S_{\mu\mu}(f)$ is an even function — the expression

$$\begin{aligned} r_{\mu\mu}(\tau) &= \int_{-\infty}^{\infty} S_{\mu\mu}(f)\, e^{j2\pi f\tau}\, df \\ &= \frac{4\sigma_0^2}{\pi f_{max}} \int_0^{f_{max}} \frac{\cos(2\pi f\tau)}{\sqrt{1-(f/f_{max})^2}}\, df\,. \end{aligned} \tag{A.12}$$

The substitution of f by $f_{max}\cos(\alpha)$ first of all leads to

$$r_{\mu\mu}(\tau) = \sigma_0^2 \frac{4}{\pi} \int_0^{\pi/2} \cos(2\pi f_{max}\tau\cos\alpha)\, d\alpha\,, \tag{A.13}$$

from which, by using the integral representation of the zeroth-order Bessel function of the first kind [Gra81, eq. (3.715.19)]

$$J_0(z) = \frac{2}{\pi} \int_0^{\pi/2} \cos(z\cos\alpha)\, d\alpha\,, \tag{A.14}$$

the result

$$r_{\mu\mu}(\tau) = 2\sigma_0^2\, J_0(2\pi f_{max}\tau) \tag{A.15}$$

immediately follows.

An alternative to this computation method is the following one. Starting with the definition of the autocorrelation function

$$r_{\mu\mu}(\tau) := E\{\mu^*(t)\mu(t+\tau)\}\,, \tag{A.16}$$

introduced by (2.48) and using (A.6), we find

$$r_{\mu\mu}(\tau) = \lim_{N\to\infty}\lim_{M\to\infty}\sum_{n=1}^{N}\sum_{m=1}^{M} c_n c_m\; E\left\{e^{j[2\pi(f_m-f_n)t+2\pi f_m\tau+\theta_m-\theta_n]}\right\}. \tag{A.17}$$

The calculation of the expected value has to be performed with respect to the uniformly distributed phases as well as with respect to the Doppler frequencies distributed according to (A.5). Determining the expected value with respect to θ_m and θ_n results in $r_{\mu\mu}(\tau) = 0$ for $n \neq m$ and in

$$r_{\mu\mu}(\tau) = \lim_{N\to\infty}\sum_{n=1}^{N} c_n^2\; E\left\{e^{j2\pi f_n\tau}\right\} \tag{A.18}$$

for $n = m$. With the probability density function (A.5), we can — after a short intermediate computation similar to that of the first procedure — represent the expected value appearing the right-hand side of (A.18) by

$$\begin{aligned} E\left\{e^{j2\pi f_n\tau}\right\} &= \int_{-\infty}^{\infty} p_f(f)\; e^{j2\pi f\tau}\, df \\ &= J_0(2\pi f_{max}\tau)\,. \end{aligned} \tag{A.19}$$

Finally, we recall that the amplitudes c_n are determined, according to the assumptions made before, by $c_n = \sigma_0\sqrt{2/N}$. Thus, from (A.18) and under consideration of (A.19), it follows the expression

$$r_{\mu\mu}(\tau) = 2\sigma_0^2\; J_0(2\pi f_{max}\tau)\,, \tag{A.20}$$

which is identical to the result (A.15) obtained by computing the inverse Fourier transform of the Jakes power spectral density.

Appendix B

DERIVATION OF THE LEVEL-CROSSING RATE OF RICE PROCESSES WITH DIFFERENT SPECTRAL SHAPES OF THE UNDERLYING GAUSSIAN RANDOM PROCESSES

Let $\mu_1(t)$ and $\mu_2(t)$ be two uncorrelated zero-mean Gaussian random processes with identical variances but different spectral shapes, i.e., the corresponding autocorrelation functions are subject to the following conditions:

$$\text{(i)} \qquad r_{\mu_1\mu_1}(0) = r_{\mu_2\mu_2}(0) = \sigma_0^2 \,, \tag{B.1}$$

$$\text{(ii)} \qquad r_{\mu_1\mu_1}(\tau) \neq r_{\mu_2\mu_2}(\tau) \,, \quad \text{if } \tau > 0 \,, \tag{B.2}$$

$$\text{(iii)} \qquad \frac{d^n}{d\tau^n}\, r_{\mu_1\mu_1}(\tau) \neq \frac{d^n}{d\tau^n}\, r_{\mu_2\mu_2}(\tau) \,, \quad \text{if } \tau \geq 0 \,, \quad n = 1, 2, \ldots \tag{B.3}$$

For the purpose of further simplification of the problem, we assume that $f_\rho = 0$, i.e., the line-of-sight component m is supposed to be time invariant and, thus, is determined by (3.3).

Starting point for the computation of the level-crossing rate of the resulting Rice process is the joint probability density function of the stationary processes $\mu_{\rho_1}(t)$, $\mu_{\rho_2}(t)$, $\dot{\mu}_{\rho_1}(t)$, and $\dot{\mu}_{\rho_2}(t)$ [see (3.4)] at the same time t. Here, we have to take the following fact into account: if $\mu_{\rho_i}(t)$ is a real-valued (stationary) Gaussian random process with mean value $E\{\mu_{\rho_i}(t)\} = m_i \neq 0$ and variance $\text{Var}\{\mu_{\rho_i}(t)\} = \text{Var}\{\mu_i(t)\} = r_{\mu_i\mu_i}(0) = \sigma_0^2$, then its derivative with respect to time, denoted by $\dot{\mu}_{\rho_i}(t)$, is also a real-valued (stationary) Gaussian random process but with mean value $E\{\dot{\mu}_{\rho_i}(t)\} = \dot{m}_i = 0$ and variance $\text{Var}\{\dot{\mu}_{\rho_i}(t)\} = \text{Var}\{\dot{\mu}_i(t)\} = r_{\dot{\mu}_i\dot{\mu}_i}(0) = -\ddot{r}_{\mu_i\mu_i}(0) = \beta_i$ $(i = 1, 2)$. Due to (B.3), the inequality $\beta_1 \neq \beta_2$ holds for β_i. Furthermore, the processes $\mu_{\rho_i}(t)$ and $\dot{\mu}_{\rho_i}(t)$ are in pairs uncorrelated at the same time t. From this fact, it follows that the joint probability density function $p_{\mu_{\rho_1}\mu_{\rho_2}\dot{\mu}_{\rho_1}\dot{\mu}_{\rho_2}}(x_1, x_2, \dot{x}_1, \dot{x}_2)$ is given by the multivariate Gaussian distribution, which can, by using (2.20), be represented by

$$p_{\mu_{\rho_1}\mu_{\rho_2}\dot{\mu}_{\rho_1}\dot{\mu}_{\rho_2}}(x_1, x_2, \dot{x}_1, \dot{x}_2) = \frac{e^{-\frac{(x_1-m_1)^2}{2\sigma_0^2}}}{\sqrt{2\pi}\,\sigma_0} \cdot \frac{e^{-\frac{(x_2-m_2)^2}{2\sigma_0^2}}}{\sqrt{2\pi}\,\sigma_0} \cdot \frac{e^{-\frac{\dot{x}_1^2}{2\beta_1}}}{\sqrt{2\pi\beta_1}} \cdot \frac{e^{-\frac{\dot{x}_2^2}{2\beta_2}}}{\sqrt{2\pi\beta_2}} . \tag{B.4}$$

The transformation of the Cartesian coordinates (x_1, x_2) into polar coordinates (z, θ), by means of $z = \sqrt{x_1^2 + x_2^2}$ and $\theta = \arctan(x_2/x_1)$, leads to the following system of equations:

$$\begin{aligned} x_1 &= z\cos\theta\,, & \dot{x}_1 &= \dot{z}\cos\theta - \dot{\theta} z\sin\theta\,, \\ x_2 &= z\sin\theta\,, & \dot{x}_2 &= \dot{z}\sin\theta + \dot{\theta} z\cos\theta\,, \end{aligned} \tag{B.5}$$

for $z \geq 0$ and $|\theta| \leq \pi$. The application of the transformation rule (2.38) then results in the joint probability density function

$$p_{\xi\dot{\xi}\vartheta\dot{\vartheta}}(z, \dot{z}, \theta, \dot{\theta}) = |J|^{-1} p_{\mu_{\rho_1}\mu_{\rho_2}\dot{\mu}_{\rho_1}\dot{\mu}_{\rho_2}}(z\cos\theta, z\sin\theta, \dot{z}\cos\theta - \dot{\theta} z\sin\theta, \dot{z}\sin\theta + \dot{\theta} z\cos\theta), \tag{B.6}$$

where

$$J = J(z) = \begin{vmatrix} \frac{\partial x_1}{\partial z} & \frac{\partial x_1}{\partial \dot{z}} & \frac{\partial x_1}{\partial \theta} & \frac{\partial x_1}{\partial \dot{\theta}} \\ \frac{\partial x_2}{\partial z} & \frac{\partial x_2}{\partial \dot{z}} & \frac{\partial x_2}{\partial \theta} & \frac{\partial x_2}{\partial \dot{\theta}} \\ \frac{\partial \dot{x}_1}{\partial z} & \frac{\partial \dot{x}_1}{\partial \dot{z}} & \frac{\partial \dot{x}_1}{\partial \theta} & \frac{\partial \dot{x}_1}{\partial \dot{\theta}} \\ \frac{\partial \dot{x}_2}{\partial z} & \frac{\partial \dot{x}_2}{\partial \dot{z}} & \frac{\partial \dot{x}_2}{\partial \theta} & \frac{\partial \dot{x}_2}{\partial \dot{\theta}} \end{vmatrix}^{-1} = -\frac{1}{z^2} \tag{B.7}$$

denotes the Jacobian determinant [see (2.39)]. Inserting (B.5) and (B.7) into (B.6) results, after some algebraic calculations, in the following expression for the joint probability function $p_{\xi\dot{\xi}\vartheta\dot{\vartheta}}(z, \dot{z}, \theta, \dot{\theta})$

$$\begin{aligned} p_{\xi\dot{\xi}\vartheta\dot{\vartheta}}(z, \dot{z}, \theta, \dot{\theta}) &= \frac{z^2}{(2\pi\sigma_0)^2\sqrt{\beta_1\beta_2}} e^{-\frac{1}{2\sigma_0^2}[z^2+\rho^2-2z\rho\cos(\theta-\theta_\rho)]} \\ &\quad \cdot e^{-\frac{\dot{z}^2}{2}\left(\frac{\cos^2\theta}{\beta_1}+\frac{\sin^2\theta}{\beta_2}\right)-z^2\dot{\theta}^2\left(\frac{\cos^2\theta}{\beta_2}+\frac{\sin^2\theta}{\beta_1}\right)-z\dot{z}\dot{\theta}\left(\frac{\beta_1-\beta_2}{\beta_1\beta_2}\right)\cos\theta\sin\theta}, \end{aligned} \tag{B.8}$$

for $z \geq 0$, $|\dot{z}| < \infty$, $|\theta| \leq \pi$ and $|\dot{\theta}| < \infty$. Using (2.40), we can now compute the joint probability density function of the processes $\xi(t)$ and $\dot{\xi}(t)$ at the same time t by using the relation

$$p_{\xi\dot{\xi}}(z, \dot{z}) = \int_{-\infty}^{\infty} \int_{-\pi}^{\pi} p_{\xi\dot{\xi}\vartheta\dot{\vartheta}}\,(z, \dot{z}, \theta, \dot{\theta})\, d\theta\, d\dot{\theta}\,, \quad z \geq 0\,, \quad |\dot{z}| < \infty\,. \tag{B.9}$$

Inserting (B.8) into (B.9) finally results in

$$p_{\xi\dot{\xi}}(z,\dot{z}) = \frac{z}{(2\pi)^{3/2}\,\sigma_0^2}\, e^{-\frac{z^2+\rho^2}{2\sigma_0^2}} \int_{-\pi}^{\pi} e^{\frac{z\rho}{\sigma_0^2}\cos\,(\theta-\theta_\rho)} \cdot \frac{e^{-\frac{\dot{z}^2}{2(\beta_1\cos^2\theta+\beta_2\sin^2\theta)}}}{\sqrt{\beta_1\cos^2\theta+\beta_2\sin^2\theta}}\, d\theta\,. \tag{B.10}$$

Since the level-crossing rate $N_\xi(r)$ of Rice process $\xi(t)$ is generally defined by

$$N_\xi(r) := \int_0^\infty \dot{z}\; p_{\xi\dot{\xi}}(r,\dot{z}) d\dot{z}\,, \quad r \geq 0\,, \tag{B.11}$$

we obtain, by using the above expression (B.10), the result

$$N_\xi(r) = \frac{r\; e^{-\frac{r^2+\rho^2}{2\sigma_0^2}}}{(2\pi)^{3/2}\sigma_0^2} \cdot \int_{-\pi}^{\pi} e^{\frac{r\rho}{\sigma_0^2}\cos(\theta-\theta_\rho)} \sqrt{\beta_1\cos^2\theta+\beta_2\sin^2\theta}\; d\theta\,, \tag{B.12}$$

which holds for $\beta_1 \neq \beta_2$. Without restriction of generality, we may assume that $\beta_1 \geq \beta_2$ holds. On this condition, we can also express (B.12) by

$$N_\xi(r) = \sqrt{\frac{\beta_1}{2\pi}} \cdot \frac{r}{\sigma_0^2} e^{-\frac{r^2+\rho^2}{2\sigma_0^2}} \cdot \frac{1}{\pi}\int_0^\pi \cosh\left[\frac{r\rho}{\sigma_0^2}\cos(\theta-\theta_\rho)\right]\sqrt{1-k^2\sin^2\theta}\, d\theta\,,\; r \geq 0\,, \tag{B.13}$$

where $k = \sqrt{(\beta_1-\beta_2)/\beta_1}$.

It should be mentioned that for $\beta = \beta_1 = \beta_2 \neq 0$, i.e., $k = 0$, and by using the relation [Abr72, eq. (9.6.16)]

$$I_0(z) = \frac{1}{\pi}\int_0^\pi \cosh(z\cos\theta)\; d\theta\,, \tag{B.14}$$

the above expression for the level-crossing rate $N_\xi(r)$ can be reduced to the form (3.27), as it was to be expected.

At the end of Appendix B, we consider an approximation for the case that the relative deviation between β_1 and β_2 is very small. Thus, for a positive number ε with $\varepsilon/\beta_1 << 1$, it holds

$$\beta_1 = \beta_2 + \varepsilon\,. \tag{B.15}$$

Due to $k = \sqrt{(\beta_1-\beta_2)/\beta_1} = \sqrt{\varepsilon/\beta_1} << 1$, we may use the approximation

$$\begin{aligned}\sqrt{1-k^2\sin^2\theta} &\approx 1-\frac{k^2}{2}\sin^2\theta \\ &= 1-\frac{\varepsilon}{2\beta_1}\sin^2\theta\,,\end{aligned} \tag{B.16}$$

so that the relation (B.13) can be simplified for $\theta_\rho = 0$ by the following approximations

$$\begin{aligned}N_\xi(r)|_{\beta_1\approx\beta_2} &\approx \sqrt{\frac{\beta_1}{2\pi}}\cdot\frac{r}{\sigma_0^2}\,e^{-\frac{r^2+\rho^2}{2\sigma_0^2}}\left[I_0\left(\frac{r\rho}{\sigma_0^2}\right)-\frac{\varepsilon}{2\beta_1}I_1\left(\frac{r\rho}{\sigma_0^2}\right)/\left(\frac{r\rho}{\sigma_0^2}\right)\right] \\ &\approx \sqrt{\frac{\beta_1}{2\pi}}\cdot\frac{r}{\sigma_0^2}\,e^{-\frac{r^2+\rho^2}{2\sigma_0^2}}\,I_0\left(\frac{r\rho}{\sigma_0^2}\right) \\ &\approx \sqrt{\frac{\beta_1}{2\pi}}\cdot p_\xi(r)\,.\end{aligned} \tag{B.17}$$

For the derivation of this relation, we have made use of the integral representation of the first-order modified Bessel function of the first kind [Abr72, eq. (9.6.18)]

$$I_1(z) = \frac{z}{\pi}\int_0^\pi e^{\pm z\cos\theta}\sin^2\theta\,d\theta\,. \tag{B.18}$$

Hence, (B.17) shows that in case $\beta_1 \approx \beta_2$, the expression (3.27) approximately holds for the level-crossing rate of Rice processes $\xi(t)$, if the quantity β is substituted by β_1 in (3.27).

Appendix C

DERIVATION OF THE EXACT SOLUTION OF THE LEVEL-CROSSING RATE AND THE AVERAGE DURATION OF FADES OF DETERMINISTIC RICE PROCESSES

We start with the derivation of the exact solution of the level-crossing rate of deterministic Rice processes using a finite number of harmonic functions. The assumptions (4.61a) and (4.61b), which were made in Subsection 4.3.2 for the purpose of simplification, will be dropped here. After that follows the computation of the corresponding average duration of fades.

Let us consider two uncorrelated zero-mean deterministic Gaussian processes

$$\tilde{\mu}_i(t) = \sum_{n=1}^{N_i} c_{i,n} \cos(2\pi f_{i,n} t + \theta_{i,n}), \quad i = 1, 2, \tag{C.1}$$

with identical variances equal to $\mathrm{Var}\{\tilde{\mu}_i(t)\} = \tilde{\sigma}_{\mu_i}^2 = \sum_{n=1}^{N_i} c_{i,n}^2/2$, where the parameters $c_{i,n}$, $f_{i,n}$, and $\theta_{i,n}$ are nonzero real-valued constants. We demand that the discrete Doppler frequencies have to be different from each other for all $n = 1, 2, \ldots, N_i$ and $i = 1, 2$, so that in particular the sets $\{f_{1,n}\}_{n=1}^{N_1}$ and $\{f_{2,n}\}_{n=1}^{N_2}$ are disjoint, guaranteeing that the deterministic Gaussian processes $\tilde{\mu}_1(t)$ and $\tilde{\mu}_2(t)$ are uncorrelated. According to (4.34), the probability density function of $\tilde{\mu}_i(t)$ reads as follows

$$\tilde{p}_{\mu_i}(x) = 2 \int_0^\infty \left[\prod_{n=1}^{N_i} J_0(2\pi c_{i,n} \nu) \right] \cos(2\pi \nu x) d\nu, \quad i = 1, 2. \tag{C.2}$$

Since the differentiation with respect to time is a linear operation, it follows from (C.1) that

$$\dot{\tilde{\mu}}_i(t) = -2\pi \sum_{n=1}^{N_i} c_{i,n} f_{i,n} \sin(2\pi f_{i,n} t + \theta_{i,n}) \,, \quad i = 1, 2 \,, \tag{C.3}$$

also describes two uncorrelated zero-mean deterministic Gaussian processes, where the variance of this processes is equal to $\text{Var}\{\dot{\tilde{\mu}}_i(t)\} = \tilde{\beta}_i = 2\pi^2 \sum_{n=1}^{N_i} (c_{i,n} f_{i,n})^2$. For the corresponding probability density function $\tilde{p}_{\dot{\mu}_i}(\dot{x})$ of $\dot{\tilde{\mu}}_i(t)$, the expression

$$\tilde{p}_{\dot{\mu}_i}(\dot{x}) = 2 \int_0^{\infty} \left[\prod_{n=1}^{N_i} J_0 \left[(2\pi)^2 c_{i,n} f_{i,n} \nu \right] \right] \cos(2\pi\nu\dot{x}) \, d\nu \,, \quad i = 1, 2 \,, \tag{C.4}$$

holds.

In this connection it has to be taken into account that with (4.13), the cross-correlation function of $\tilde{\mu}_i(t)$ and $\dot{\tilde{\mu}}_i(t)$ can be expressed by

$$\tilde{r}_{\mu_i \dot{\mu}_i}(\tau) = \dot{\tilde{r}}_{\mu_i \mu_i}(\tau) = -\pi \sum_{n=1}^{N_i} c_{i,n}^2 f_{i,n} \sin(2\pi f_{i,n} \tau) \,, \tag{C.5}$$

and, thus, it becomes clear that $\tilde{\mu}_i(t)$ and $\dot{\tilde{\mu}}_i(t)$ are in general correlated. For the computation of the level-crossing rate, however, we are only interested in the behaviour of $\tilde{\mu}_i(t_1)$ and $\dot{\tilde{\mu}}_i(t_2)$ at the same time instant $t = t_1 = t_2$, which is equivalent to $\tau = t_2 - t_1 = 0$. Observe that from (C.5) it follows $\tilde{r}_{\mu_i \dot{\mu}_i}(\tau) = 0$ for $\tau = 0$, i.e., the deterministic Gaussian processes $\tilde{\mu}_i(t)$ and $\dot{\tilde{\mu}}_i(t)$ are uncorrelated at the same time t. Consequently, also the deterministic processes $\tilde{\mu}_1(t)$, $\tilde{\mu}_2(t)$, $\dot{\tilde{\mu}}_1(t)$, and $\dot{\tilde{\mu}}_2(t)$, are uncorrelated in pairs at the same time t. We know that if two random variables are uncorrelated, then they are not necessarily statistically independent. However, for Gaussian distributed random variables, uncorrelatedness is equivalent to independence [Pap91]. In the present case, the probability density functions $\tilde{p}_{\mu_i}(x_i)$ and $\tilde{p}_{\dot{\mu}_i}(\dot{x})$ [see (C.2) and (C.4), respectively] are both almost identical to the Gaussian distribution if $N_i \geq 7$. Therefore, we may assume that $\tilde{\mu}_1(t)$, $\tilde{\mu}_2(t)$, $\dot{\tilde{\mu}}_1(t)$, and $\dot{\tilde{\mu}}_2(t)$ are mutually statistically independent at the same time t. As a consequence, the joint probability density function of these processes can be expressed by the product of the individual probability density functions, i.e.,

$$\tilde{p}_{\mu_1 \mu_2 \dot{\mu}_1 \dot{\mu}_2}(x_1, x_2, \dot{x}_1, \dot{x}_2) = \tilde{p}_{\mu_1}(x_1) \cdot \tilde{p}_{\mu_2}(x_2) \cdot \tilde{p}_{\dot{\mu}_1}(\dot{x}_1) \cdot \tilde{p}_{\dot{\mu}_2}(\dot{x}_2) \,. \tag{C.6}$$

Considering the line-of-sight component (3.2), we assume — in order to simplify matter — that $f_\rho = 0$ holds, so that $m = m_1 + jm_2$ is a complex-valued constant, whose real and imaginary part is characterized by the discrete probability density function $p_{m_i}(x_i) = \delta(x_i - m_i)$, $i = 1, 2$. For the probability density functions of the complex deterministic processes $\tilde{\mu}_{\rho_i}(t) = \tilde{\mu}_i(t) + m_i$ and $\dot{\tilde{\mu}}_{\rho_i}(t) = \dot{\tilde{\mu}}_i(t) + \dot{m}_i = \dot{\tilde{\mu}}_i(t)$, the following relations hold for $i = 1, 2$:

$$\tilde{p}_{\mu_{\rho_i}}(x_i) = \tilde{p}_{\mu_i}(x_i) * p_{m_i}(x_i) = \tilde{p}_{\mu_i}(x_i - m_i) \,, \tag{C.7a}$$

$$\tilde{p}_{\dot{\mu}_{\rho_i}}(\dot{x}_i) = \tilde{p}_{\dot{\mu}_i}(\dot{x}_i) * p_{\dot{m}_i}(\dot{x}_i) = \tilde{p}_{\dot{\mu}_i}(\dot{x}_i) \,. \tag{C.7b}$$

Thus, for the joint probability density function of the deterministic processes $\tilde{\mu}_{\rho_1}(t)$, $\tilde{\mu}_{\rho_2}(t)$, $\dot{\tilde{\mu}}_{\rho_1}(t)$, and $\dot{\tilde{\mu}}_{\rho_2}(t)$, we may write

$$\tilde{p}_{\mu_{\rho_1}\mu_{\rho_2}\dot{\mu}_{\rho_1}\dot{\mu}_{\rho_2}}(x_1, x_2, \dot{x}_1, \dot{x}_2) = \tilde{p}_{\mu_1}(x_1 - m_1) \cdot \tilde{p}_{\mu_2}(x_2 - m_2) \cdot \tilde{p}_{\dot{\mu}_1}(\dot{x}_1) \cdot \tilde{p}_{\dot{\mu}_2}(\dot{x}_2) \,. \tag{C.8}$$

The transformation of the Cartesian coordinates $(x_1, x_2, \dot{x}_1, \dot{x}_2)$ to polar coordinates $(z, \dot{z}, \theta, \dot{\theta})$ [cf. Appendix B, eq. (B.5)] results in the joint probability density function of the processes $\tilde{\xi}(t)$, $\dot{\tilde{\xi}}(t)$, $\tilde{\vartheta}(t)$, and $\dot{\tilde{\vartheta}}(t)$ at the same time t according to

$$\begin{aligned}\tilde{p}_{\xi\dot{\xi}\vartheta\dot{\vartheta}}(z, \dot{z}, \theta, \dot{\theta}) = z^2 \cdot & \tilde{p}_{\mu_1}(z\cos\theta - \rho\cos\theta_\rho) \cdot \tilde{p}_{\mu_2}(z\sin\theta - \rho\sin\theta_\rho) \\ & \cdot \tilde{p}_{\dot{\mu}_1}(\dot{z}\cos\theta - \dot{\theta}z\sin\theta) \cdot \tilde{p}_{\dot{\mu}_2}(\dot{z}\sin\theta + \dot{\theta}z\cos\theta) \,, \end{aligned} \tag{C.9}$$

for $0 \leq z < \infty$, $|\dot{z}| < \infty$, $|\theta| \leq \pi$, and $|\dot{\theta}| < \infty$. From this expression, the joint probability density function $\tilde{p}_{\xi\dot{\xi}}(z, \dot{z})$ of the deterministic processes $\tilde{\xi}(t)$ and $\dot{\tilde{\xi}}(t)$ can be obtained after applying the relation (2.40). Hence,

$$\begin{aligned}\tilde{p}_{\xi\dot{\xi}}(z, \dot{z}) \quad = \quad & z^2 \int_{-\infty}^{\infty} \int_{-\pi}^{\pi} \tilde{p}_{\mu_1}(z\cos\theta - \rho\cos\theta_\rho) \cdot \tilde{p}_{\mu_2}(z\sin\theta - \rho\sin\theta_\rho) \\ & \cdot \tilde{p}_{\dot{\mu}_1}(\dot{z}\cos\theta - \dot{\theta}z\sin\theta) \cdot \tilde{p}_{\dot{\mu}_2}(\dot{z}\sin\theta + \dot{\theta}z\cos\theta) \, d\theta \, d\dot{\theta} \,, \end{aligned} \tag{C.10}$$

where $0 \leq z < \infty$ and $|\dot{z}| < \infty$. If we substitute the equation above into the definition of the level-crossing rate for deterministic Rice processes $\tilde{\xi}(t)$

$$\tilde{N}_\xi(r) := \int_0^\infty \dot{z}\, \tilde{p}_{\xi\dot{\xi}}(r, \dot{z}) \, d\dot{z} \,, \quad r \geq 0 \,, \tag{C.11}$$

then we obtain the expression

$$\begin{aligned}\tilde{N}_\xi(r) \quad = \quad & r^2 \int_{-\pi}^{\pi} \tilde{p}_{\mu_1}(r\cos\theta - \rho\cos\theta_\rho) \cdot \tilde{p}_{\mu_2}(r\sin\theta - \rho\sin\theta_\rho) \\ & \cdot \int_0^\infty \dot{z} \int_{-\infty}^{\infty} \tilde{p}_{\dot{\mu}_1}(\dot{z}\cos\theta - \dot{\theta}r\sin\theta) \cdot \tilde{p}_{\dot{\mu}_2}(\dot{z}\sin\theta + \dot{\theta}r\cos\theta) \, d\dot{\theta} \, d\dot{z} \, d\theta \,. \end{aligned} \tag{C.12}$$

It is mathematically convenient to express (C.12) as

$$\tilde{N}_\xi(r) = r^2 \int_{-\pi}^{\pi} w_1(r, \theta) \, w_2(r, \theta) \int_0^\infty \dot{z} \; f(r, \dot{z}, \theta) \, d\dot{z} \, d\theta \,, \tag{C.13}$$

where $w_1(r, \theta)$, $w_2(r, \theta)$, and $f(r, \dot{z}, \theta)$ are auxiliary functions defined by

$$w_1(r,\theta) \quad := \quad \tilde{p}_{\mu_1}(r\cos\theta - \rho\cos\theta_\rho)\,, \tag{C.14a}$$
$$w_2(r,\theta) \quad := \quad \tilde{p}_{\mu_2}(r\sin\theta - \rho\sin\theta_\rho)\,, \tag{C.14b}$$

and

$$\begin{aligned} f(r,\dot{z},\theta) \quad := \quad & 2\int_0^\infty \left[\prod_{n=1}^{N_1} J_0(4\pi^2 c_{1,n} f_{1,n}\nu_1)\right] \int_0^\infty \left[\prod_{m=1}^{N_2} J_0(4\pi^2 c_{2,m} f_{2,m}\nu_2)\right] \\ & \cdot \int_{-\infty}^{\infty} \left\{\cos\left[2\pi\dot{z}(\nu_1\cos\theta - \nu_2\sin\theta) - 2\pi\dot{\theta}r(\nu_1\sin\theta + \nu_2\cos\theta)\right]\right. \\ & \left. + \cos\left[2\pi\dot{z}(\nu_1\cos\theta + \nu_2\sin\theta) - 2\pi\dot{\theta}r(\nu_1\sin\theta - \nu_2\cos\theta)\right]\right\} d\dot{\theta}\, d\nu_1\, d\nu_2\,, \end{aligned} \tag{C.15}$$

respectively. The integration over $\dot{\theta}$ in (C.15) results in

$$\begin{aligned} & \int_{-\infty}^{\infty} \cos\left[2\pi\dot{z}(\nu_1\cos\theta \mp \nu_2\sin\theta) - 2\pi\dot{\theta}r(\nu_1\sin\theta \pm \nu_2\cos\theta)\right] d\dot{\theta} \\ & \qquad = \cos[2\pi\dot{z}(\nu_1\cos\theta \mp \nu_2\sin\theta)] \cdot \delta[r(\nu_1\sin\theta \pm \nu_2\cos\theta)]\,. \end{aligned} \tag{C.16}$$

Putting the relation

$$\delta[r(\nu_1\sin\theta \pm \nu_2\cos\theta)] = \frac{\delta(\tan\theta \pm \nu_1/\nu_2)}{|r\nu_1\cos\theta|} \tag{C.17}$$

into (C.16) and using the transformation of the variables $\varphi = \tan\theta$, then (C.13) can be represented by

$$\begin{aligned} \tilde{N}_\xi(r) \quad = \quad & 2r^2 \int_{-\infty}^{\infty} w_1(r,\arctan\varphi)\; w_2(r,\arctan\varphi) \\ & \cdot \int_0^\infty \dot{z}\, f(r,\dot{z},\arctan\varphi)\;\cos^2(\arctan\varphi)\; d\dot{z}\; d\varphi\,, \end{aligned} \tag{C.18}$$

where

$$\begin{aligned} f(r,\dot{z},\arctan\varphi) = {} & 2\int_0^\infty\!\!\int_0^\infty \frac{\left[\prod_{n=1}^{N_1} J_o(4\pi^2 c_{1,n} f_{1,n}\nu_1)\right]\left[\prod_{m=1}^{N_2} J_o(4\pi^2 c_{2,m} f_{2,m}\nu_2)\right]}{|r\nu_1\cos(\arctan\varphi)|} \\ & \cdot\left\{\cos\left[2\pi\dot{z}\nu_2\cos(\arctan\varphi)\left(\frac{\nu_1}{\nu_2} - \varphi\right)\right]\cdot\delta\left(\varphi + \frac{\nu_2}{\nu_1}\right)\right. \\ & \left. + \cos\left[2\pi\dot{z}\nu_2\cos(\arctan\varphi)\left(\frac{\nu_1}{\nu_2} + \varphi\right)\right]\cdot\delta\left(\varphi - \frac{\nu_2}{\nu_1}\right)\right\} d\nu_1\, d\nu_2\,. \end{aligned} \tag{C.19}$$

If we now substitute (C.19) into (C.18) and subsequently transform the Cartesian coordinates (ν_1, ν_2) to polar coordinates (z, θ) by means of $(\nu_1, \nu_2) \to (z\cos\theta, z\sin\theta)$, then we obtain

$$\begin{aligned} \tilde{N}_\xi(r) &= 2r \int_0^\infty \int_0^\pi w_1(r,\theta)\, [w_2(r,\theta) + w_2(r,-\theta)] \\ &\quad \cdot \int_0^\infty j_1(z,\theta)\, j_2(z,\theta)\, \dot{z}\, \cos(2\pi z\dot{z})\, dz\, d\theta\, d\dot{z}\,, \end{aligned} \tag{C.20}$$

where

$$j_1(z,\theta) = \prod_{n=1}^{N_1} J_0(4\pi^2 c_{1,n} f_{1,n} z\cos\theta)\,, \tag{C.21a}$$

$$j_2(z,\theta) = \prod_{n=1}^{N_2} J_0(4\pi^2 c_{2,n} f_{2,n} z\sin\theta)\,, \tag{C.21b}$$

and $w_1(r,\theta)$, $w_2(r,\theta)$ are the auxiliary functions introduced by (C.14a) and (C.14b), respectively. For the derivation of (C.20), we exploited the fact that $w_1(r,\theta)$ is an even function in θ, i.e., $w_1(r,\theta) = w_1(r,-\theta)$. Since $w_2(r,\theta)$ is neither even nor odd in θ if $\rho \neq 0$ (or $\theta_\rho \neq k\pi,\ k = 0, \pm1, \pm2, \ldots$), we may also write for the level-crossing rate of deterministic Rice processes [Pae99c]

$$\tilde{N}_\xi(r) = 2r \int_0^\infty \int_{-\pi}^\pi w_1(r,\theta)\, w_2(r,\theta) \int_0^\infty j_1(z,\theta)\, j_2(z,\theta)\, \dot{z}\cos(2\pi z\dot{z})\, dz\, d\theta\, d\dot{z}\,. \tag{C.22}$$

Further but only slight simplifications are possible for the level-crossing rate $\tilde{N}_\zeta(r)$ of deterministic Rayleigh processes $\tilde{\zeta}(t)$. Since $\rho = 0$ holds in this case, it follows that $w_2(r,\theta)$ is an even function in θ as well, so that from (C.22), the expression

$$\tilde{N}_\zeta(r) = 4r \int_0^\infty \int_0^\pi w_1(r,\theta)\, w_2(r,\theta) \int_0^\infty j_1(z,\theta)\, j_2(z,\theta)\, \dot{z}\cos(2\pi z\dot{z})\, dz\, d\theta\, d\dot{z} \tag{C.23}$$

can be obtained, where $w_1(r,\theta)$ and $w_2(r,\theta)$ have to be computed according to (C.14a) and (C.14b), respectively, by taking $\rho = 0$ into account. In (C.23), $j_1(z,\theta)$ and $j_2(z,\theta)$ again denote the functions (C.21a) and (C.21b), respectively.

By means of the exact solution of the level-crossing rate $\tilde{N}_\xi(r)$ of deterministic Rice processes $\tilde{\xi}(t)$ it now becomes obvious that apart from ρ and the number of harmonic functions N_i, $\tilde{N}_\xi(r)$ also depends on the quantities $c_{i,n}$ and $f_{i,n}$. In contrast to that, the Doppler phases $\theta_{i,n}$ have no influence on $\tilde{N}_\xi(r)$. Thus, for a given number of harmonic functions N_i, the deviations between the level-crossing rate of the simulation model and that of the reference model are essentially determined by the method applied for the computation of the model parameters $c_{i,n}$ and $f_{i,n}$. For the purpose of illustration and verification of the obtained results, the normalized level-crossing rates, computed according to (C.22) and (C.23), are depicted in Figure C.1 together with the pertinent simulation results. The methods (MEDS, MEA, MCM) applied for the determination of the model parameters $c_{i,n}$ and $f_{i,n}$ are described in detail in Chapter 5. For the MCM it should in addition be noted that the results shown in Figure C.1 are only valid for a certain realization of the set of discrete Doppler frequencies $\{f_{i,n}\}_{n=1}^{N_i}$. Another realization for $\{f_{i,n}\}_{n=1}^{N_i}$ may give better or worse results for $\tilde{N}_\xi(r)$. The reason for this is in the nature of the MCM, according to which the discrete Doppler frequencies $f_{i,n}$ are random variables, so that the deviations between $\tilde{N}_\xi(r)$ and $N_\xi(r)$ can only be described statistically. Further details on this subject are described in Subsection 5.1.4.

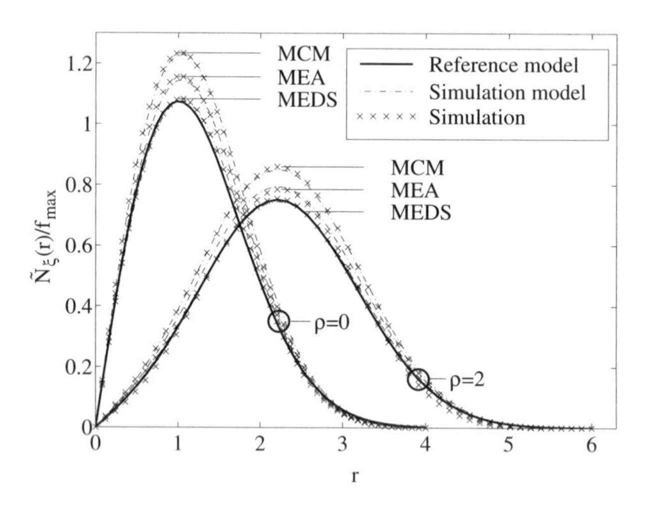

Figure C.1: Normalized level-crossing rate of deterministic Rice and Rayleigh processes both realized with $N_1 = 7$ and $N_2 = 8$ (Jakes PSD, $f_{max} = 91\,\text{Hz}$, $\sigma_0^2 = 1$, $\theta_\rho = \pi/4$).

Next, we want to show that the level-crossing rate of deterministic Rice processes indeed converges to the level-crossing rate of the reference model as $N_i \to \infty$, i.e.,

$$\tilde{N}_\xi(r) = N_\xi(r)\,, \quad N_i \to \infty\,. \tag{C.24}$$

Therefore, we merely assume that the autocorrelation function $\tilde{r}_{\mu_i\mu_i}(\tau)$ of $\tilde{\mu}_i(t)$ fulfils the following two conditions:

$$\text{(i)} \qquad \tilde{r}_{\mu_i\mu_i}(0) = r_{\mu_i\mu_i}(0) \quad \Longleftrightarrow \quad \tilde{\sigma}_{\mu_i}^2 = \tilde{\sigma}_0^2 = \sigma_0^2\,, \tag{C.25a}$$

$$\text{(ii)} \qquad \ddot{\tilde{r}}_{\mu_i\mu_i}(0) = \ddot{r}_{\mu_i\mu_i}(0) \quad \Longleftrightarrow \quad \tilde{\beta}_i = \tilde{\beta} = \beta\,. \tag{C.25b}$$

The first condition (i) imposes the so-called *power constraint* on the simulation model. If the power constraint is fulfilled, then the mean power of the deterministic process $\tilde{\mu}_i(t)$ is identical to the variance of the stochastic process $\mu_i(t)$. By analogy to the power constraint (C.25a), we will in the following denote (C.25b) as *curvature constraint.* The curvature constraint imposes on the simulation model that the curvature of the autocorrelation function $\tilde{r}_{\mu_i\mu_i}(\tau)$ of $\tilde{\mu}_i(t)$ is identical to the curvature of the autocorrelation function $r_{\mu_i\mu_i}(\tau)$ of $\mu_i(t)$ at $\tau = 0$. It should be mentioned that the power constraint is a necessary condition and the curvature constraint is a sufficient condition for the validity of the relation (C.24).

In order to prove (C.24), we once again consider the time t as a uniformly distributed random variable, and we remember that as $N_i \to \infty$, it follows from the central limit theorem (2.16) that the probability density function of (C.1) converges to a Gaussian distribution with the mean 0 and the variance $\tilde{\sigma}_0^2$, i.e.,

$$\lim_{N_i\to\infty} \tilde{p}_{\mu_i}(x_i) = \frac{1}{\sqrt{2\pi}\tilde{\sigma}_0} e^{-\frac{x_i^2}{2\tilde{\sigma}_0^2}}, \quad i = 1, 2, \tag{C.26}$$

where

$$\tilde{\sigma}_0^2 = \lim_{N_i\to\infty} \tilde{r}_{\mu_i\mu_i}(0) = \lim_{N_i\to\infty} \sum_{n=1}^{N_i} \frac{c_{i,n}^2}{2}. \tag{C.27}$$

If we now substitute the result (C.26) into (C.14a) and (C.14b), then it follows

$$w_1(r,\theta) = \frac{1}{\sqrt{2\pi}\tilde{\sigma}_0} e^{-\frac{(r\cos\theta - \rho\cos\theta_\rho)^2}{2\tilde{\sigma}_0^2}}, \quad \text{as } N_1 \to \infty, \tag{C.28a}$$

$$w_2(r,\theta) = \frac{1}{\sqrt{2\pi}\tilde{\sigma}_0} e^{-\frac{(r\sin\theta - \rho\sin\theta_\rho)^2}{2\tilde{\sigma}_0^2}}, \quad \text{as } N_2 \to \infty. \tag{C.28b}$$

Applying the Fourier transform on the right-hand side of (C.2) and (C.26), we realize that (4.38) can be expressed more generally by

$$\lim_{N_i\to\infty} \prod_{n=1}^{N_i} J_0(2\pi c_{i,n}\nu) = e^{-2(\pi\tilde{\sigma}_0\nu)^2}, \tag{C.29}$$

where $\tilde{\sigma}_0^2$ is given by (C.27). Furthermore, by replacing the quantities $c_{i,n}$ with $2\pi c_{i,n} f_{i,n}$ in (C.29), the relation

$$\lim_{N_i\to\infty} \prod_{n=1}^{N_i} J_0(4\pi^2 f_{i,n} c_{i,n}\nu) = e^{-2\tilde{\beta}_i(\pi\nu)^2} \tag{C.30}$$

can easily be derived, where $\tilde{\beta}_i$ denotes the quantity introduced by (4.22). Thus, it becomes clear that in the limit $N_i \to \infty$, the functions $j_1(z,\theta)$ [see (C.21a)] and $j_2(z,\theta)$ [see (C.21b)] converge to

$$j_1(z,\theta) = e^{-2\tilde{\beta}_1(\pi z\cos\theta)^2}, \quad \text{as } N_1 \to \infty, \tag{C.31a}$$

$$j_2(z,\theta) = e^{-2\tilde{\beta}_2(\pi z\sin\theta)^2}, \quad \text{as } N_2 \to \infty, \tag{C.31b}$$

respectively. If we now substitute the obtained results (C.28a), (C.28b), (C.31a), and (C.31b) into $\tilde{N}_\xi(r)$ according to (C.22), then — on the condition that $\tilde{\beta} = \tilde{\beta}_1 = \tilde{\beta}_2$ holds — it follows the expression

$$\lim_{N_i \to \infty} \tilde{N}_\xi(r) = \frac{r}{\pi \tilde{\sigma}_0^2} e^{-\frac{r^2+\rho^2}{2\tilde{\sigma}_0^2}} \int_0^\infty \int_{-\pi}^{\pi} \dot{z}\, e^{\frac{r\rho}{\tilde{\sigma}_0^2}\cos(\theta-\theta_\rho)} \int_0^\infty e^{-2\tilde{\beta}(\pi z)^2} \cos(2\pi z \dot{z})\, dz\, d\theta\, d\dot{z}\,. \tag{C.32}$$

Using the integral [Gra81, eq. (3.896.4)]

$$\int_0^\infty e^{-ux^2} \cos(bx)\, dx = \frac{1}{2}\sqrt{\frac{\pi}{u}}\, e^{-\frac{b^2}{4u}}\,, \qquad \mathrm{Re}\,\{u\} > 0\,, \tag{C.33}$$

(C.32) can be simplified to

$$\lim_{N_i \to \infty} \tilde{N}_\xi(r) = \frac{r}{\sqrt{2\pi\tilde{\beta}\tilde{\sigma}_0^2}} e^{-\frac{r^2+\rho^2}{2\tilde{\sigma}_0^2}} \cdot \frac{1}{2\pi}\int_{-\pi}^{\pi} e^{\frac{r\rho}{\tilde{\sigma}_0^2}\cos(\theta-\theta_\rho)} d\theta \cdot \int_0^\infty \dot{z}\, e^{-\frac{\dot{z}^2}{2\tilde{\beta}}}\, d\dot{z}\,. \tag{C.34}$$

The remaining two integrals over θ and $\dot{z}$ can be solved without great expense by using the integral representation of the zeroth-order modified Bessel function of the first kind [Abr72, eq. (9.6.16)]

$$I_0(z) = \frac{1}{\pi}\int_0^\pi e^{\pm z\cos\theta}\, d\theta \tag{C.35}$$

and the integral [Gra81, eq. (3.461.3)]

$$\int_0^\infty x^{2n+1}\, e^{-px^2}\, dx = \frac{n!}{2p^{n+1}}\,, \qquad p > 0\,. \tag{C.36}$$

Finally, we obtain

$$\lim_{N_i \to \infty} \tilde{N}_\xi(r) = \sqrt{\frac{\tilde{\beta}}{2\pi}} \cdot \frac{r}{\tilde{\sigma}_0^2}\, e^{-\frac{r^2+\rho^2}{2\tilde{\sigma}_0^2}}\, I_0\left(\frac{r\rho}{\tilde{\sigma}_0^2}\right). \tag{C.37}$$

Taking the power constraint (C.25a) and the curvature constraint (C.25b) into account, the right-hand side of the above equation can now directly be identified with (2.62), which proves the validity of (C.24).

For completeness, we will also give the exact solution for the average duration of fades $\tilde{T}_{\xi_-}(r)$ of deterministic Rice processes $\tilde{\xi}(t)$. Since we need an expression for the cumulative distribution function $\tilde{F}_{\xi_-}(r)$ of $\tilde{\xi}(t)$, we will first of all derive this by substituting (4.50) into

$$\tilde{F}_{\xi_-}(r) = \int_0^r \tilde{p}_\xi(z)\, dz\,. \tag{C.38}$$

The integration over z can be carried out by using the indefinite integral [Gra81, eq. (5.56.2)]

$$\int z\, J_0(z)\, dz = z\, J_1(z)\,, \tag{C.39}$$

so that after some algebraic manipulations, the following result is obtained

$$\tilde{F}_{\xi_-}(r) = 2r \int_0^\infty J_1(2\pi r y) \int_0^\pi h_1(y,\theta)\, h_2(y,\theta)\, \cos[2\pi\rho y \cos(\theta - \theta_\rho)]\, d\theta\, dy\,, \tag{C.40}$$

where

$$h_1(y,\theta) = \prod_{n=1}^{N_1} J_0(2\pi c_{1,n} y \cos\theta)\,, \tag{C.41a}$$

$$h_2(y,\theta) = \prod_{n=1}^{N_2} J_0(2\pi c_{2,n} y \sin\theta)\,. \tag{C.41b}$$

With the cumulative distribution function (C.40) presented above and the solution for the level-crossing rate (C.22) found before, the average duration of fades $\tilde{T}_{\xi_-}(r)$ of deterministic Rice processes $\tilde{\xi}(t)$ can now be analysed analytically using

$$\tilde{T}_{\xi_-}(r) = \frac{\tilde{F}_{\xi_-}(r)}{\tilde{N}_\xi(r)}\,. \tag{C.42}$$

In order to illustrate the obtained results, we consider Figure C.2, which shows the normalized average duration of fades of deterministic Rice and Rayleigh processes, according to the theoretical results (C.42), in comparison with the corresponding simulation results.

Subsequently, we want to prove that in the conditions (C.25a) and (C.25b), the average duration of fades $\tilde{T}_{\xi_-}(r)$ of deterministic Rice processes $\tilde{\xi}(t)$ converges to the average duration of fades $T_{\xi_-}(r)$ of stochastic Rice processes $\xi(t)$ as the number of harmonic functions N_i tends to infinity, i.e.,

$$\tilde{T}_{\xi_-}(r) = T_{\xi_-}(r)\,, \quad \text{as } N_i \to \infty\,. \tag{C.43}$$

Due to (C.24) and the general relation (C.42), it is sufficient here to show that

$$\tilde{F}_{\xi_-}(r) = F_{\xi_-}(r)\,, \quad \text{as } N_i \to \infty\,, \tag{C.44a}$$

or, equivalently, that

$$\tilde{p}_\xi(r) = p_\xi(r)\,, \quad \text{as } N_i \to \infty\,, \tag{C.44b}$$

holds. Due to (C.29), we therefore first realize that the functions $h_1(y,\theta)$ [see (C.41a)] and $h_2(y,\theta)$ [see (C.41b)] tend to

$$h_1(y,\theta) = e^{-2(\pi\tilde{\sigma}_0 y \cos\theta)^2}\,, \quad \text{as } N_1 \to \infty\,, \tag{C.45a}$$

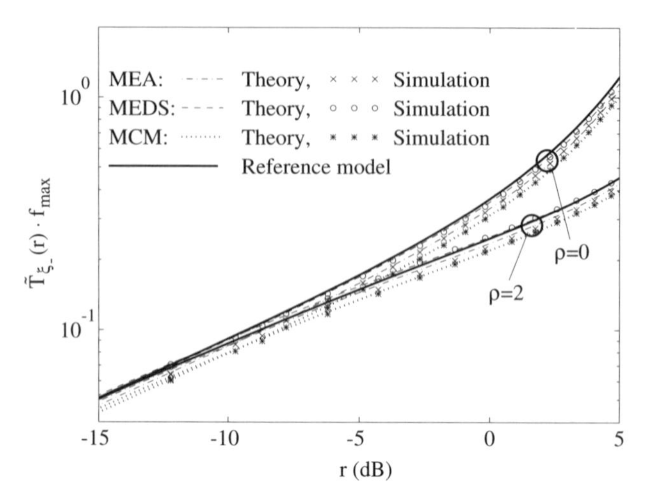

Figure C.2: Normalized average duration of fades of deterministic Rice and Rayleigh processes both realized with $N_1 = 7$ and $N_2 = 8$ (Jakes PSD, $f_{max} = 91$ Hz, $\sigma_0^2 = 1,\ \theta_\rho = \pi/4$).

and

$$h_2(y, \theta) = e^{-2(\pi\tilde{\sigma}_0 y sin\theta)^2}, \quad \text{as } N_2 \to \infty, \tag{C.45b}$$

respectively. With this result it follows from (4.50)

$$\lim_{N_i \to \infty} \tilde{p}_\xi(z) = (2\pi)^2 z \int_0^\infty e^{-2(\pi\tilde{\sigma}_0 y)^2} J_0(2\pi z y)\, \frac{1}{\pi} \int_0^\pi \cos[2\pi\rho y \cos(\theta - \theta_\rho)] d\theta\, y\, dy\,. \tag{C.46}$$

The integral representation of the zeroth-order Bessel function [Abr72, eq. (9.1.18)]

$$J_0(z) = \frac{1}{\pi} \int_0^\pi \cos(z \cos\theta)\, d\theta \tag{C.47}$$

enables us to write the expression (C.46) in the form

$$\lim_{N_i \to \infty} \tilde{p}_\xi(z) = (2\pi)^2 z \int_0^\infty e^{-2(\pi\tilde{\sigma}_0 y)^2} J_0(2\pi z y)\, J_0(2\pi\rho y)\, y\, dy\,. \tag{C.48}$$

The remaining integral can be solved by using [Gra81, eq. (6.633.2)]

$$\int_0^\infty e^{-(ax)^2} J_0(\alpha x)\, J_0(\beta x)\, x\, dx = \frac{1}{2a^2}\, e^{-\frac{\alpha^2+\beta^2}{4a^2}}\, I_0\left(\frac{\alpha\beta}{2a^2}\right). \tag{C.49}$$

Thus, we finally obtain

$$\tilde{p}_\xi(z) = \frac{z}{\tilde{\sigma}_0^2}\, e^{-\frac{z^2+\rho^2}{2\tilde{\sigma}_0^2}}\, I_0\left(\frac{z\rho}{\tilde{\sigma}_0^2}\right), \quad N_i \to \infty\,. \tag{C.50}$$

With the power constraint (C.25a), i.e., $\tilde{\sigma}_0^2 = \sigma_0^2$, the Rice distribution (2.26) follows from the right-hand side of (C.50) proving the validity of (C.44b) and, consequently, also the validity of (C.43).

Appendix D

ANALYSIS OF THE RELATIVE MODEL ERROR BY USING THE MONTE CARLO METHOD IN CONNECTION WITH THE JAKES POWER SPECTRAL DENSITY

We consider the relative model error

$$\frac{\Delta\beta_i}{\beta} = \frac{\tilde{\beta}_i - \beta}{\beta}, \quad i = 1, 2, \tag{D.1}$$

where the quantities β and $\tilde{\beta}_i$ are, especially for the Jakes power spectral density, given by

$$\beta = 2(\pi\sigma_0 f_{max})^2 \tag{D.2}$$

and

$$\tilde{\beta}_i = \frac{2\beta}{f_{max}^2 N_i} \sum_{n=1}^{N_i} f_{i,n}^2 , \tag{D.3}$$

respectively. If we use the Monte Carlo method for the computation of the model parameter, then the discrete Doppler frequencies $f_{i,n}$ are random variables, where each random variable $f_{i,n}$ is characterized by the probability density function

$$p_{f_{i,n}}(f_{i,n}) = \begin{cases} \dfrac{2}{\pi f_{max}\sqrt{1-(f_{i,n}/f_{max})^2}}, & 0 < f \leq f_{max} , \\ 0 , & \text{else} . \end{cases} \tag{D.4}$$

With the Chebyshev inequality (2.15), the relation

$$P\left(\left| \frac{\Delta\beta_i}{\beta} - E\left\{\frac{\Delta\beta_i}{\beta}\right\} \right| \geq \varepsilon \right) \leq \frac{\text{Var}\ \{\Delta\beta_i/\beta\}}{\varepsilon^2} \tag{D.5}$$

holds for all $\varepsilon > 0$. Using (D.4), we find

$$E\left\{f_{i,n}^2\right\} = \frac{f_{max}^2}{2} \tag{D.6}$$

and

$$\begin{aligned}\operatorname{Var}\left\{f_{i,n}^2\right\} &= E\left\{f_{i,n}^4\right\} - \left(E\left\{f_{i,n}^2\right\}\right)^2 \\ &= \frac{3}{8} f_{max}^4 - \frac{f_{max}^4}{4} \\ &= \frac{f_{max}^4}{8} \,.\end{aligned} \tag{D.7}$$

Hence, for the mean value and the variance of the relative model error $\Delta\beta_i/\beta$ [cf. (D.1)], we obtain the following expression in connection with (D.3)

$$E\left\{\frac{\Delta\beta_i}{\beta}\right\} = 0 \tag{D.8}$$

and

$$\begin{aligned}\operatorname{Var}\left\{\frac{\Delta\beta_i}{\beta}\right\} &= \operatorname{Var}\left\{\frac{\beta_i}{\beta}\right\} \\ &= \left(\frac{2}{f_{max}^2 N_i}\right)^2 \operatorname{Var}\left\{\sum_{n=1}^{N_i} f_{i,n}^2\right\} \\ &= \left(\frac{2}{f_{max}^2 N_i}\right)^2 \sum_{n=1}^{N_i} \operatorname{Var}\left\{f_{i,n}^2\right\} \\ &= \frac{1}{2N_i}\,,\end{aligned} \tag{D.9}$$

respectively. Thus, with the Chebyshev inequality (D.5) the relation

$$P\left(\left|\frac{\Delta\beta_i}{\beta}\right| \geq \varepsilon\right) \leq \frac{1}{2N_i\varepsilon^2} \tag{D.10}$$

follows. For example, let $\varepsilon = 0.02$ and $N_i = 2500$ (!), then the above inequality can be interpreted as follows: the probability that the absolute value of the relative model error $|\Delta\beta_i/\beta|$ is greater than or equal to 2 per cent is smaller than or equal to 50 per cent.

Appendix E

SPECIFICATION OF FURTHER $\mathcal{L}$-PATH CHANNEL MODELS ACCORDING TO COST 207

In addition to the 4-path and 6-path channel models presented in Table 7.3, further $\mathcal{L}$-path channel models have been specified by COST 207 [COS89]. They are quoted in this appendix for completeness.

Table E.1: Rural Area.

Path no. ℓ	Propagation delay τ'_ℓ	Path power (lin.)	Path power (dB)	Category of the Doppler PSD	Delay spread $B^{(2)}_{\tau'\tau'}$
Rural Area: 6-path channel model (alternative)					
0	0 μs	1	0	"Rice"	0.1 μs
1	0.1 μs	0.4	-4	"Jakes"	
2	0.2 μs	0.16	-8	"Jakes"	
3	0.3 μs	0.06	-12	"Jakes"	
4	0.4 μs	0.03	-16	"Jakes"	
5	0.5 μs	0.01	-20	"Jakes"	

Table E.2: Typical Urban.

Path no. ℓ	Propagation delay τ'_ℓ	Path power (lin.)	Path power (dB)	Category of the Doppler PSD	Delay spread $B^{(2)}_{\tau'\tau'}$
(i) Typical Urban: 12-path channel model					
0	0.0 μs	0.4	-4	"Jakes"	1.0 μs
1	0.2 μs	0.5	-3	"Jakes"	
2	0.4 μs	1	0	"Jakes"	
3	0.6 μs	0.63	-2	"Gauss I"	
4	0.8 μs	0.5	-3	"Gauss I"	
5	1.2 μs	0.32	-5	"Gauss I"	
6	1.4 μs	0.2	-7	"Gauss I"	
7	1.8 μs	0.32	-5	"Gauss I"	
8	2.4 μs	0.25	-6	"Gauss II"	
9	3.0 μs	0.13	-9	"Gauss II"	
10	3.2 μs	0.08	-11	"Gauss II"	
11	5.0 μs	0.1	-10	"Gauss II"	
(ii) Typical Urban: 12-path channel model (alternative)					
0	0.0 μs	0.4	-4	"Jakes"	1.0 μs
1	0.1 μs	0.5	-3	"Jakes"	
2	0.3 μs	1	0	"Jakes"	
3	0.5 μs	0.55	-2.6	"Jakes"	
4	0.8 μs	0.5	-3	"Gauss I"	
5	1.1 μs	0.32	-5	"Gauss I"	
6	1.3 μs	0.2	-7	"Gauss I"	
7	1.7 μs	0.32	-5	"Gauss I"	
8	2.3 μs	0.22	-6.5	"Gauss II"	
9	3.1 μs	0.14	-8.6	"Gauss II"	
10	3.2 μs	0.08	-11	"Gauss II"	
11	5.0 μs	0.1	-10	"Gauss II"	
(iii) Typical Urban: 6-path channel model (alternative)					
0	0.0 μs	0.5	-3	"Jakes"	1.0 μs
1	0.2 μs	1	0	"Jakes"	
2	0.5 μs	0.63	-2	"Jakes"	
3	1.6 μs	0.25	-6	"Gauss I"	
4	2.3 μs	0.16	-8	"Gauss II"	
5	5.0 μs	0.1	-10	"Gauss II"	

Table E.3: Bad Urban.

Path no. ℓ	Propagation delay τ'_ℓ	Path power (lin.)	Path power (dB)	Category of the Doppler PSD	Delay spread $B^{(2)}_{\tau'\tau'}$
(i) Bad Urban: 12-path channel model					
0	0.0 μs	0.2	-7	"Jakes"	2.5 μs
1	0.2 μs	0.5	-3	"Jakes"	
2	0.4 μs	0.79	-1	"Jakes"	
3	0.8 μs	1	0	"Gauss I"	
4	1.6 μs	0.63	-2	"Gauss I"	
5	2.2 μs	0.25	-6	"Gauss II"	
6	3.2 μs	0.2	-7	"Gauss II"	
7	5.0 μs	0.79	-1	"Gauss II"	
8	6.0 μs	0.63	-2	"Gauss II"	
9	7.2 μs	0.2	-7	"Gauss II"	
10	8.2 μs	0.1	-10	"Gauss II"	
11	10.0 μs	0.03	-15	"Gauss II"	
(ii) Bad Urban: 12-path channel model (alternative)					
0	0.0 μs	0.17	-7.7	"Jakes"	2.5 μs
1	0.1 μs	0.46	-3.4	"Jakes"	
2	0.3 μs	0.74	-1.3	"Jakes"	
3	0.7 μs	1	0	"Gauss I"	
4	1.6 μs	0.59	-2.3	"Gauss I"	
5	2.2 μs	0.28	-5.6	"Gauss II"	
6	3.1 μs	0.18	-7.4	"Gauss II"	
7	5.0 μs	0.72	-1.4	"Gauss II"	
8	6.0 μs	0.69	-1.6	"Gauss II"	
9	7.2 μs	0.21	-6.7	"Gauss II"	
10	8.1 μs	0.1	-9.8	"Gauss II"	
11	10.0 μs	0.03	-15.1	"Gauss II"	
(iii) Bad Urban: 6-path channel model (alternative)					
0	0.0 μs	0.56	-2.5	"Jakes"	2.5 μs
1	0.3 μs	1	0	"Jakes"	
2	1.0 μs	0.5	-3	"Gauss I"	
3	1.6 μs	0.32	-5	"Gauss I"	
4	5.0 μs	0.63	-2	"Gauss II"	
5	6.6 μs	0.4	-4	"Gauss II"	

Table E.4: Hilly Terrain.

Path no. ℓ	**Propagation delay** τ'_ℓ	**Path power** (lin.)	(dB)	**Category of the Doppler PSD**	**Delay spread** $B^{(2)}_{\tau'\tau'}$
(i) Hilly Terrain: 12-path channel model					
0	0.0 μs	0.1	-10	"Jakes"	5.0 μs
1	0.2 μs	0.16	-8	"Jakes"	
2	0.4 μs	0.25	-6	"Jakes"	
3	0.6 μs	0.4	-4	"Gauss I"	
4	0.8 μs	1	0	"Gauss I"	
5	2.0 μs	1	0	"Gauss I"	
6	2.4 μs	0.4	-4	"Gauss II"	
7	15.0 μs	0.16	-8	"Gauss II"	
8	15.2 μs	0.13	-9	"Gauss II"	
9	15.8 μs	0.1	-10	"Gauss II"	
10	17.2 μs	0.06	-12	"Gauss II"	
11	20.0 μs	0.04	-14	"Gauss II"	
(ii) Hilly Terrain: 12-path channel model (alternative)					
0	0.0 μs	0.1	-10	"Jakes"	5.0 μs
1	0.1 μs	0.16	-8	"Jakes"	
2	0.3 μs	0.25	-6	"Jakes"	
3	0.5 μs	0.4	-4	"Jakes"	
4	0.7 μs	1	0	"Gauss I"	
5	1.0 μs	1	0	"Gauss I"	
6	1.3 μs	0.4	-4	"Gauss I"	
7	15.0 μs	0.16	-8	"Gauss II"	
8	15.2 μs	0.13	-9	"Gauss II"	
9	15.7 μs	0.1	-10	"Gauss II"	
10	17.2 μs	0.06	-12	"Gauss II"	
11	20.0 μs	0.04	-14	"Gauss II"	
(iii) Hilly Terrain: 6-path channel model (alternative)					
0	0.0 μs	1	0	"Jakes"	5.0 μs
1	0.1 μs	0.71	-1.5	"Jakes"	
2	0.3 μs	0.35	-4.5	"Jakes"	
3	0.5 μs	0.18	-7.5	"Jakes"	
4	15 μs	0.16	-8.0	"Gauss II"	
5	17.2 μs	0.02	-17.7	"Gauss II"	

MATLAB-PROGRAMS

In the following, a selection of MATLAB-programs (*m-files*) is presented, which save the user from programming effort during the realization of the methods used to design the model parameters of deterministic processes, and which will help him find his way in the topic of simulation and analysis of mobile radio channel models. MATLAB stands for **mat**rix **lab**oratory, an interpreter language developed by The Math Works, Inc., for the numerical computation and visualization of matrices. The *m-files* presented below require the *Signal Processing Toolbox* and the *Optimization Toolbox*.

References on necessary subroutines (*functions*) and a description of the input and output parameters of the individual programs can be found in the program header block of each program.

At first, the *m-files* for the computation of the model parameters are presented by making use of the methods described in Chapter 5. Here, these methods are subdivided depending on the type of power spectral density (Jakes/Gauss) of the deterministic Gaussian processes to be realized. Subsequently, functions for the time-domain simulation of various frequency-nonselective (Chapter 6) and frequency-selective (Chapter 7) mobile radio channels are presented. Finally, further tools are provided, with the help of which the designed channel simulators can be analysed with respect to their statistical properties such as the probability density function, the cumulative distribution function, the level-crossing rate, and the average duration of fades.

```
%-------------------------------------------------------------------
% parameter_Jakes.m ------------------------------------------------
%
% Program for the computation of the discrete Doppler frequencies,
% Doppler coefficients and Doppler phases by using the Jakes power
% spectral density.
%
% Used m-files: LPNM_opt_Jakes.m, fun_Jakes.m,
%                          grad_Jakes.m, acf_mue.m
%-------------------------------------------------------------------
% [f_i_n,c_i_n,theta_i_n]=parameter_Jakes(METHOD,N_i,sigma_0_2,...
```

```
%                                                f_max,PHASE,PLOT)
%-------------------------------------------------------------------
% Explanation of the input parameters:
%
% METHOD:
% |----------------------------------------------|------------------|
% | Methods for the computation of the discrete  |      Input       |
% | Doppler frequencies and Doppler coefficients |                  |
% |----------------------------------------------|------------------|
% |----------------------------------------------|------------------|
% | Method of equal distances (MED)              |     'ed_j'       |
% |----------------------------------------------|------------------|
% | Mean square error method  (MSEM)             |     'ms_j'       |
% |----------------------------------------------|------------------|
% | Method of equal areas (MEA)                  |     'ea_j'       |
% |----------------------------------------------|------------------|
% | Monte Carlo method (MCM)                     |     'mc_j'       |
% |----------------------------------------------|------------------|
% | Lp-norm method (LPNM)                        |     'lp_j'       |
% |----------------------------------------------|------------------|
% | Method of exact Doppler spread (MEDS)        |     'es_j'       |
% |----------------------------------------------|------------------|
% | Jakes method (JM)                            |     'jm_j'       |
% |----------------------------------------------|------------------|
%
% N_i: number of harmonic functions
% sigma_0_2: average power of the real deterministic Gaussian
%            process mu_i(t)
% f_max: maximum Doppler frequency
%
% PHASE:
% |----------------------------------------------|------------------|
% | Methods for the computation of the Doppler   |      Input       |
% | phases                                       |                  |
% |----------------------------------------------|------------------|
% |----------------------------------------------|------------------|
% | Random Doppler phases                        |     'rand'       |
% |----------------------------------------------|------------------|
% | Permuted Doppler phases                      |     'perm'       |
% |----------------------------------------------|------------------|
%
% PLOT: plot of the ACF and the PSD of mu_i(t), if PLOT==1

function [f_i_n,c_i_n,theta_i_n]=parameter_Jakes(METHOD,N_i,...
                                 sigma_0_2,f_max,PHASE,PLOT)

if nargin<6,
```

```
   error('Not enough input parameters')
end

sigma_0=sqrt(sigma_0_2);

% Method of equal distances (MED)
if     METHOD=='ed_j',
       n=(1:N_i)';
       f_i_n=f_max/(2*N_i)*(2*n-1);
       c_i_n=2*sigma_0/sqrt(pi)*(asin(n/N_i)-asin((n-1)/N_i)).^0.5;
       K=1;

% Mean square error method (MSEM)
elseif METHOD=='ms_j',
       n=(1:N_i)';
       f_i_n=f_max/(2*N_i)*(2*n-1);
       Tp=1/(2*f_max/N_i);
       t=linspace(0,Tp,5E3);
       Jo=besselj(0,2*pi*f_max*t);
       c_i_n=zeros(size(f_i_n));
       for k=1:length(f_i_n),
           c_i_n(k)=2*sigma_0*...
                    sqrt(1/Tp*( trapz( t,Jo.*...
                    cos(2*pi*f_i_n(k)*t )) ));
       end
       K=1;

% Method of equal areas (MEA)
elseif METHOD=='ea_j'
       n=(1:N_i)';
       f_i_n=f_max*sin(pi*n/(2*N_i));
       c_i_n=sigma_0*sqrt(2/N_i)*ones(size(n));
       K=1;

% Monte Carlo method (MCM)
elseif METHOD=='mc_j'
       n=rand(N_i,1);
       f_i_n=f_max*sin(pi*n/2);
       c_i_n=sigma_0*sqrt(2/N_i)*ones(size(n));

       K=1;

% Lp-norm method (LPNM)
elseif METHOD=='lp_j',
       if   exist('fminu')~=2
            disp([' =====> This method requires ',...
                  'the Optimization Toolbox !!'])
```

```
            return
       else
            N=1E3;
            p=2;    % Norm
            s_o=1;
            [f_i_n,c_i_n]=LPNM_opt_Jakes(N,f_max,sigma_0,p,N_i,s_o);
            K=1;
       end

% Method of exact Doppler spread (MEDS)
elseif METHOD=='es_j',
       n=(1:N_i)';
       f_i_n=f_max*sin(pi/(2*N_i)*(n-1/2));
       c_i_n=sigma_0*sqrt(2/(N_i))*ones(size(f_i_n));
       K=1;

% Jakes method (JM)
elseif METHOD=='jm_j',
       n=1:N_i-1;
       f_i_n=f_max*[[cos(pi*n/(2*(N_i-1/2))),1]',...
                   [cos(pi*n/(2*(N_i-1/2))),1]'];
       c_i_n=2*sigma_0/sqrt(N_i-1/2)*[[sin(pi*n/(N_i-1)),1/2]',...
                                      [cos(pi*n/(N_i-1)),1/2]'];
       K=1;
       theta_i_n=zeros(size(f_i_n));
       PHASE='none';

else
       error('Method is unknown')
end

% Computation of the Doppler phases:
if     PHASE=='rand',
       theta_i_n=rand(N_i,1)*2*pi;

elseif PHASE=='perm',
       n=(1:N_i)';
       Z=rand(size(n));
       [dummy,I]=sort(Z);
       theta_i_n=2*pi*n(I)/(N_i+1);
end;

if PLOT==1,
   if  METHOD=='jm_j'
       subplot(2,3,1)
       stem([-f_i_n(N_i:-1:1,1);f_i_n(:,1)],...
             1/4*[c_i_n(N_i:-1:1,1);c_i_n(:,1)].^2)
```

```
        title('i=1')
        xlabel('f (Hz)')
        ylabel('PSD')
        subplot(2,3,2)
        stem([-f_i_n(N_i:-1:1,2);f_i_n(:,2)],...
              1/4*[c_i_n(N_i:-1:1,2);c_i_n(:,2)].^2)
        title('i=2')
        xlabel('f (Hz)')
        ylabel('PSD')
        tau_max=N_i/(K*f_max);
        tau=linspace(0,tau_max,500);
        r_mm=sigma_0^2*besselj(0,2*pi*f_max*tau);

        r_mm_tilde1=acf_mue(f_i_n(:,1),c_i_n(:,1),tau);
        subplot(2,3,4)
        plot(tau,r_mm,'r-',tau,r_mm_tilde1,'g--')
        title('i=1')
        xlabel('tau (s)')
        ylabel('ACF')
        r_mm_tilde2=acf_mue(f_i_n(:,2),c_i_n(:,2),tau);
        subplot(2,3,5)
        plot(tau,r_mm,'r-',tau,r_mm_tilde2,'g--')
        title('i=2')
        xlabel('tau (s)')
        ylabel('ACF')
        subplot(2,3,3)
        stem([-f_i_n(N_i:-1:1,1);f_i_n(:,1)],...
              1/4*[c_i_n(N_i:-1:1,1);c_i_n(:,1)].^2+...
              1/4*[c_i_n(N_i:-1:1,2);c_i_n(:,2)].^2)
        title('i=1,2')
        xlabel('f (Hz)')
        ylabel('PSD')
        subplot(2,3,6)
        plot(tau,2*r_mm,'r-',tau,r_mm_tilde1+r_mm_tilde2,'g--')
        title('i=1,2')
        xlabel('tau (s)')
        ylabel('ACF')
    else
        subplot(1,2,1)
        stem([-f_i_n(N_i:-1:1);f_i_n],...
              1/4*[c_i_n(N_i:-1:1);c_i_n].^2)
        xlabel('f/Hz')
        ylabel('LDS')
        tau_max=N_i/(K*f_max);
        tau=linspace(0,tau_max,500);
        r_mm=sigma_0^2*besselj(0,2*pi*f_max*tau);
        r_mm_tilde=acf_mue(f_i_n,c_i_n,tau);
```

```
        subplot(1,2,2)
        plot(tau,r_mm,'r-',tau,r_mm_tilde,'g--')
        xlabel('tau (s)')
        ylabel('ACF')
    end
end

%--------------------------------------------------------------------
% LPNM_opt_Jakes.m --------------------------------------------------
%
% Program for the computation of the discrete Doppler frequencies
% employing the Jakes PSD by using a numerical optimization method.
%
% Used m-files: parameter_Jakes.m, fun_Jakes.m,
%                              grad_Jakes.m, acf_mue.m
%--------------------------------------------------------------------
% [f_i_n,c_i_n]=LPNM_opt_Jakes(N,f_max,sigma_0_2,p,N_i,PLOT)
%--------------------------------------------------------------------
% Explanation of the input parameters:
%
% N: length of vector tau
% f_max: maximum Doppler frequency
% sigma_0_2: average power of the real Gaussian process mu_i(t)
% p: parameter of the Lp-norm (here: p=2,4,6,...)
% N_i: number of harmonic functions
% PLOT: display of the intermediate optimization results, if PLOT==1

function [f_i_n,c_i_n]=LPNM_opt_Jakes(N,f_max,sigma_0_2,p,N_i,PLOT)

tau=linspace(0,N_i/(2*f_max),N);
Jo=sigma_0_2*besselj(0,2*pi*f_max*tau);
c_i_n=sqrt(sigma_0_2)*sqrt(2/N_i)*ones(N_i,1);

save data Jo tau N_i c_i_n p PLOT

% Initial values:
[f_i_n,dummy1,dummy2]=parameter_Jakes('es_j',N_i,...
                      sqrt(sigma_0_2),f_max,'none',0);
o=foptions;
o(1)=1;
o(1)=0;
o(2)=1e-9;
o(14)=N_i/10*200;
o(9)=0;

xo=f_i_n;
```

```
x=fminu('fun_Jakes',xo,o,'grad_Jakes');

load x

f_i_n=x;

%-----------------------------------------------------------------------
% fun_Jakes.m ----------------------------------------------------------
%
% Computation of the error function according to Eq.(5.61) for the
% optimization of the discrete Doppler frequencies (Jakes PSD).
%
% Used m-file: acf_mue.m
%-----------------------------------------------------------------------
% F=fun_Jakes(x)
%-----------------------------------------------------------------------
% Explanation of the input parameters:
%
% x: parameter vector to be optimized

function F=fun_Jakes(x)

load data

f_i_n=x;
r=acf_mue(f_i_n,c_i_n,tau);
F=norm(abs(Jo-r),p);
if PLOT==1,
   subplot(1,2,1)
   stem(f_i_n,c_i_n)
   xlabel('f_i_n')
   ylabel('c_i_n')
   title(['N_i = ',num2str(N_i)])
   subplot(1,2,2)
   plot(tau,Jo,tau,r)
   xlabel('tau (s)')
   ylabel('ACF')
   title(['Error-norm=',num2str(F)])
   pause(0)
end

save x x

%-----------------------------------------------------------------------
% grad_Jakes.m ---------------------------------------------------------
%
% Computation of the analytical gradient of the error function for
% the optimization of the discrete Doppler frequencies (Jakes PSD).
```

```
%
% Used m-file: acf_mue.m
%-------------------------------------------------------------------
% G=grad_Jakes(x)
%-------------------------------------------------------------------
% Explanation of the input parameters:
%
% x: parameter vector to be optimized

function G=grad_Jakes(x)

load data

f_i_n=x;
r=acf_mue(f_i_n,c_i_n,tau);
D=Jo-r;
F=norm(D,p);
G=[];
for k=1:N_i,
    g=F^(1-p)*D.^(p-1)*(2*pi*c_i_n(k)^2*tau.*...
      sin(2*pi*f_i_n(k)*tau)).';
    G=[G;g];
end

%-------------------------------------------------------------------
% acf_mue.m --------------------------------------------------------
%
% Computation of the ACF of deterministic Gaussian processes mu_i(t)
%
%-------------------------------------------------------------------
% r_mm=acf_mue(f,c,tau)
%-------------------------------------------------------------------
% Explanation of the input parameters:
%
% f: discrete Doppler frequencies
% c: Doppler coefficients
% tau: time separation variable

function r_mm=acf_mue(f,c,tau)

r_mm=0;
for n=1:length(c),
    r_mm=r_mm+0.5*c(n)^2*cos(2*pi*f(n)*tau);
end

%-------------------------------------------------------------------
% parameter_Gauss.m ------------------------------------------------
%
```

```
% Program for the computation of the discrete Doppler frequencies,
% Doppler coefficients, and Doppler phases by using the Gaussian
% power spectral density.
%
% Used m-files: LPNM_opt_Gauss.m, fun_Gauss.m,
%                         grad_Gauss.m, acf_mue.m
%-----------------------------------------------------------------
% [f_i_n,c_i_n,theta_i_n]=parameter_Gauss(METHOD,N_i,sigma_0_2,...
%                                        f_max,f_c,PHASE,PLOT)
%-----------------------------------------------------------------
% Explanation of the input parameters:
%
% METHOD:
% |----------------------------------------------|------------------|
% | Methods for the computation of the discrete  |      Input       |
% | Doppler frequencies and Doppler coefficients |                  |
% |----------------------------------------------|------------------|
% |----------------------------------------------|------------------|
% | Method of equal distances (MED)              |      'ed_g'      |
% |----------------------------------------------|------------------|
% | Mean square error method  (MSEM)             |      'ms_g'      |
% |----------------------------------------------|------------------|
% | Method of equal areas (MEA)                  |      'ea_g'      |
% |----------------------------------------------|------------------|
% | Monte Carlo method (MCM)                     |      'mc_g'      |
% |----------------------------------------------|------------------|
% | Lp-norm method (LPNM)                        |      'lp_g'      |
% |----------------------------------------------|------------------|
% | Method of exact Doppler spread (MEDS)        |      'es_g'      |
% |----------------------------------------------|------------------|
%
% N_i: number of harmonic functions
% sigma_0_2: average power of the real deterministic Gaussian
%            process mu_i(t)
% f_max: maximum Doppler frequency
% f_c: 3-dB-cutoff frequency
%
% PHASE:
% |----------------------------------------------|------------------|
% | Methods for the computation of the Doppler   |      Input       |
% | phases                                       |                  |
% |----------------------------------------------|------------------|
% |----------------------------------------------|------------------|
% | Random Doppler phases                        |      'rand'      |
% |----------------------------------------------|------------------|
% | Permuted Doppler phases                      |      'perm'      |
% |----------------------------------------------|------------------|
```

```
%
% PLOT: plot of the ACF and the PSD of mu_i(t), if PLOT==1

function [f_i_n,c_i_n,theta_i_n]=parameter_Gauss(METHOD,N_i,...
                                 sigma_0_2,f_max,f_c,PHASE,PLOT)

if nargin<7,
   error('Not enough input parameters')
end

sigma_0=sqrt(sigma_0_2);
kappa_c=f_max/f_c;

% Method of equal distances (MED)
if      METHOD=='ed_g',
        n=(1:N_i)';
        f_i_n=kappa_c*f_c/(2*N_i)*(2*n-1);
        c_i_n=sigma_0*sqrt(2)*sqrt(erf(n*kappa_c*...
              sqrt(log(2))/N_i)-erf((n-1)*kappa_c*...
              sqrt(log(2))/N_i) );
        K=1;

% Mean square error method (MSEM)
elseif METHOD=='ms_g',
        n=(1:N_i)';
        f_i_n=kappa_c*f_c/(2*N_i)*(2*n-1);
        tau_max=N_i/(2*kappa_c*f_c);
        N=1E3;
        tau=linspace(0,tau_max,N);
        f1=exp(-(pi*f_c*tau).^2/log(2));
        c_i_n=zeros(size(f_i_n));
        for k=1:length(c_i_n),
            c_i_n(k)=2*sigma_0*sqrt(trapz(tau,f1.*...
                     cos(2*pi*f_i_n(k)*tau))/tau_max);
        end
        K=1;

% Method of equal areas (MEA)
elseif METHOD=='ea_g'
        n=(1:N_i)';
        c_i_n=sigma_0*sqrt(2/N_i)*ones(size(n));
        f_i_n=f_c/sqrt(log(2))*erfinv(n/N_i);
        f_i_n(N_i)=f_c/sqrt(log(2))*erfinv(0.9999999);
        K=1;

% Monte Carlo method (MCM)
elseif METHOD=='mc_g'
```

```
        n=rand(N_i,1);
        f_i_n=f_c/sqrt(log(2))*erfinv(n);
        c_i_n=sigma_0*sqrt(2/N_i)*ones(size(n));
        K=1;

% Lp-norm method (LPNM)
elseif METHOD=='lp_g',

        if    exist('fminu')~=2
              disp([' =====> This method requires ',...
                     'the Optimization Toolbox !!'])
              return
        else
              N=1e3;
              p=2;
              [f_i_n,c_i_n]=LPNM_opt_Gauss(N,f_max,f_c,...
                                sigma_0_2,p,N_i,PLOT);
              K=2;
        end

% Method of exact Doppler spread (MEDS)
elseif METHOD=='es_g',
        n=(1:N_i)';
        c_i_n=sigma_0*sqrt(2/N_i)*ones(size(n));
        f_i_n=f_c/sqrt(log(2))*erfinv((2*n-1)/(2*N_i));
        K=1;
else
        error([setstr(10),'Method is unknown'])
end

% Computation of the Doppler phases:
if      PHASE=='rand',
        theta_i_n=rand(N_i,1)*2*pi;
elseif PHASE=='perm',
        n=(1:N_i)';
        Z=rand(size(n));
        [dummy,I]=sort(Z);
        theta_i_n=2*pi*n(I)/(N_i+1);
end

if PLOT==1,
   subplot(1,2,1)
   stem([-f_i_n(N_i:-1:1);f_i_n],...
        1/4*[c_i_n(N_i:-1:1);c_i_n].^2)
   xlabel('f (Hz)')
   ylabel('PSD')
   tau_max=N_i/(K*kappa_c*f_c);
```

```
   tau=linspace(0,tau_max,500);
   r_mm=sigma_0_2*exp(-(pi*f_c/sqrt(log(2))*tau).^2);
   r_mm_tilde=acf_mue(f_i_n,c_i_n,tau);
   subplot(1,2,2)
   plot(tau,r_mm,'r-',tau,r_mm_tilde,'g--')
   xlabel('tau (s)')
   ylabel('ACF')
end

%-----------------------------------------------------------------------
% LPNM_opt_Gauss.m -----------------------------------------------------
%
% Program for the computation of the discrete Doppler frequencies
% employing the Gaussian PSD by using a numerical optimization
% method.
%
% Used m-files: parameter_Gauss.m, fun_Gauss.m,
%               grad_Gauss.m, acf_mue.m
%-----------------------------------------------------------------------
% [f_i_n,c_i_n]=LPNM_opt_Gauss(N,f_max,f_c,sigma_0_2,p,N_i,PLOT)
%-----------------------------------------------------------------------
% Explanation of the input parameters:
%
% N: length of vector tau
% f_max: maximum Doppler frequency
% f_c: 3-dB-cutoff frequency
% sigma_0_2: average power of the real Gaussian process mu_i(t)
% p: parameter of the Lp-norm (here: p=2,4,6,...)
% N_i: number of harmonic functions
% PLOT: display of the intermediate optimization results, if PLOT==1

function [f_i_n,c_i_n]=LPNM_opt_Gauss(N,f_max,f_c,sigma_0_2,...
                                     p,N_i,PLOT)

kappa_c=f_max/f_c;

F_list=[];
save F_list F_list

tau_max=N_i/(2*kappa_c*f_c);
tau=linspace(0,tau_max,N);
r_mm=sigma_0_2*exp(-(pi*f_c/sqrt(log(2))*tau).^2);

[f_i_n,c_i_n,dummy]=parameter_Gauss('es_g',N_i,sigma_0_2,f_max,...
                                    f_c,'none',PLOT);

save data r_mm tau N_i c_i_n p PLOT
```

```
o=foptions;
o(1)=1;
o(1)=0;
o(2)=1e-9;
o(14)=N_i/10*200;
o(9)=0;

xo=f_i_n;

x=fminu('fun_Gauss',xo,o,'grad_Gauss');

load x

f_i_n=sort(abs(x));

%--------------------------------------------------------------------
% fun_Gauss.m ------------------------------------------------------
%
% Computation of the error function according to Eq.(5.61) for the
% optimization of the discrete Doppler frequencies (Gaussian PSD).
%
% Used m-file: acf_mue.m
%--------------------------------------------------------------------
% F=fun_Gauss(x)
%--------------------------------------------------------------------
% Explanation of the input parameters:
%
% x: parameter vector to be optimized

function F=fun_Gauss(x)

load data

f_i_n=x;

r=acf_mue(f_i_n,c_i_n,tau);
F=norm(abs(r_mm-r),p);
if PLOT==1,
   subplot(1,2,1)
   stem(f_i_n,c_i_n)
   xlabel('f_i_n')
   ylabel('c_i_n')

   title(['N_i = ',num2str(N_i)])
   subplot(1,2,2)
   plot(tau,r_mm,tau,r)
```

```
   xlabel('tau (s)')
   ylabel('ACF')
   title(['Error-norm=',num2str(F)])
   pause(0)

end

save x x

%----------------------------------------------------------------------
% grad_Gauss.m --------------------------------------------------------
%
% Computation of the analytical gradient of the error function for
% the optimization of the discrete Doppler frequencies
% (Gaussian PSD).
%
% Used m-file: acf_mue.
%----------------------------------------------------------------------
% G=grad_Gauss(x)
%----------------------------------------------------------------------
% Explanation of the input parameters:
%
% x: parameter vector to be optimized

function G=grad_Gauss(x)

load data

f_i_n=x;
r=acf_mue(f_i_n,c_i_n,tau);
D=r_mm-r;
F=norm(D,p);
G=[];
for k=1:N_i,
    g=F^(1-p)*D.^(p-1)*(2*pi*c_i_n(k)^2*tau.*...
      sin(2*pi*f_i_n(k)*tau)).';
    G=[G;g];
end

%----------------------------------------------------------------------
% Mu_i_t.m ------------------------------------------------------------
%
% Program for the simulation of real deterministic Gaussian processes
% mu_i(t) [see Fig. 4.2(b)].
%----------------------------------------------------------------------
% mu_i_t=Mu_i_t(c,f,th,T_s,T_sim,PLOT)
%----------------------------------------------------------------------
% Explanation of the input parameters:
```

```
%
% f:  discrete Doppler frequencies
% c:  Doppler coefficients
% th: Doppler phases
% T_s: sampling interval
% T_sim: duration of the simulation
% PLOT: plot of the deterministic Gaussian process mu_i(t),
%       if PLOT==1

function mu_i_t=Mu_i_t(c,f,th,T_s,T_sim,PLOT)

if nargin==5,
   PLOT=0;
end

N=ceil(T_sim/T_s);
t=(0:N-1)*T_s;
mu_i_t=0;
for k=1:length(f),
    mu_i_t=mu_i_t+c(k)*cos(2*pi*f(k)*t+th(k));
end

if PLOT==1,
   plot(t,mu_i_t)
   xlabel('t (s)')
   ylabel('mu_i(t)')
end

%----------------------------------------------------------------------
% Rice_proc.m ---------------------------------------------------------
%
% Program for the simulation of deterministic Rice processes xi(t)
% (see Fig. 4.3).
%
% Used m-file: Mu_i_t.m
%----------------------------------------------------------------------
% xi_t=Rice_proc(f1,c1,th1,f2,c2,th2,rho,f_rho,theta_rho,...
%                                                    T_s,T_sim,PLOT)
%----------------------------------------------------------------------
% Explanation of the input parameters:
%
% f1, c1, th1: discrete Doppler frequencies, Doppler coefficients,
%              and Doppler phases of mu_1(t)
% f2, c2, th2: discrete Doppler frequencies, Doppler coefficients,
%              and Doppler phases of mu_2(t)
% rho: amplitude of the LOS component m(t)
% f_rho: Doppler frequency of the LOS component m(t)
```

```
% theta_rho: phase of the LOS component m(t)
% T_s: sampling interval
% T_sim: duration of the simulation
% PLOT: plot of the deterministic Rice process xi(t), if PLOT==1

function xi_t=Rice_proc(f1,c1,th1,f2,c2,th2,rho,f_rho,theta_rho,...
                        T_s,T_sim,PLOT)

if nargin==10,
   PLOT=0;
end

N=ceil(T_sim/T_s);
t=(0:N-1)*T_s;
arg=2*pi*f_rho*t+theta_rho;

xi_t=abs(Mu_i_t(c1,f1,th1,T_s,T_sim)+rho*cos(arg)+...
        j*(Mu_i_t(c2,f2,th2,T_s,T_sim)+rho*sin(arg)) );

if PLOT==1,
   plot(t,20*log10(xi_t))
   xlabel('t (s)')
   ylabel('20 log xi(t)')
end

%--------------------------------------------------------------------
% Suzuki_Type_I.m ---------------------------------------------------
%
% Program for the simulation of deterministic extended Suzuki
% processes of Type I (see Fig. 6.9).
%
% Used m-files: parameter_Jakes.m, parameter_Gauss.m, Mu_i_t.m
%--------------------------------------------------------------------
% eta_t=Suzuki_Type_I(N_1,N_2,N_3,sigma_0_2,kappa_0,f_max,sigma_3,...
%                     m_3,rho,f_rho,theta_rho,f_c,T_s,T_sim,PLOT)
%--------------------------------------------------------------------
% Explanation of the input parameters:
%
% N_1, N_2, N_3: number of harmonic functions of the real deter-
%                ministic Gaussian processes nu_1(t), nu_2(t),
%                and nu_3(t), respectively
% sigma_0_2: average power of the real deterministic Gaussian
%            processes mu_1(t) and mu_2(t)
% kappa_0: frequency ratio f_min/f_max (0<=kappa_0<=1)
% f_max: maximum Doppler frequency
% sigma_3: square root of the average power of the real deterministic
%          Gaussian process nu_3(t)
```

```
% m_3: average value of the third real deterministic Gaussian
%      process mu_3(t)
% rho: amplitude of the LOS component m(t)
% f_rho: Doppler frequency of the LOS component m(t)
% theta_rho: phase of the LOS component m(t)
% f_c: 3-dB-cut-off frequency
% T_s: sampling interval
% T_sim: duration of the simulation
% PLOT: plot of the deterministic extended Suzuki process eta(t) of
%       Type I, if PLOT==1

function eta_t=Suzuki_Type_I(N_1,N_2,N_3,sigma_0_2,kappa_0,f_max,...
               sigma_3,m_3,rho,f_rho,theta_rho,f_c,T_s,T_sim,PLOT)

if nargin==14,
   PLOT=0;
end

[f1,c1,th1]=parameter_Jakes('es_j',N_1,sigma_0_2,f_max,'rand',0);
c1=c1/sqrt(2);

N_2_s=ceil(N_2/(2/pi*asin(kappa_0)));
[f2,c2,th2]=parameter_Jakes('es_j',N_2_s,sigma_0_2,f_max,'rand',0);
f2 =f2(1:N_2);
c2 =c2(1:N_2)/sqrt(2);
th2=th2(1:N_2);

[f3,c3,th3]=parameter_Gauss('es_g',N_3,1,f_max,f_c,'rand',0);
gaMma=(2*pi*f_c/sqrt(2*log(2)))^2;
f3(N_3)=sqrt(gaMma*N_3/(2*pi)^2-sum(f3(1:N_3-1).^2));

N=ceil(T_sim/T_s);
t=(0:N-1)*T_s;

arg=2*pi*f_rho*t+theta_rho;

xi_t=abs(Mu_i_t(c1,f1,th1,T_s,T_sim)+...
         Mu_i_t(c2,f2,th2,T_s,T_sim)+rho*cos(arg)+...
         j*(Mu_i_t(c1,f1,th1-pi/2,T_s,T_sim)-...
         Mu_i_t(c2,f2,th2-pi/2,T_s,T_sim)+rho*sin(arg)));
lambda_t=exp(Mu_i_t(c3,f3,th3,T_s,T_sim)*sigma_3+m_3);

eta_t=xi_t.*lambda_t;

if PLOT==1,
   plot(t,20*log10(eta_t),'b-')
   xlabel('t (s)')
```

```
   ylabel('20 log eta(t)')
end

%------------------------------------------------------------------------
% Suzuki_Type_II.m ---------------------------------------------------------
%
% Program for the simulation of deterministic extended Suzuki
% processes of Type II (see Fig. 6.23).
%
% Used m-files: parameter_Jakes.m, parameter_Gauss.m, Mu_i_t.m
%------------------------------------------------------------------------
% eta_t=Suzuki_Type_II(N_1,N_3,sigma_0_2,kappa_0,theta_0,f_max,...
%                  sigma_3,m_3,rho,theta_rho,f_c,T_s,T_sim,PLOT)
%------------------------------------------------------------------------
% Explanation of the input parameters:
%
% N_1, N_3: number of harmonic functions of the real deterministic
%           Gaussian processes nu_0(t) and nu_3(t), respectively
% sigma_0_2: average power of the real deterministic Gaussian
%            process mu_0(t)  (for kappa_0=1)
% kappa_0: frequency ratio f_min/f_max (0<=kappa_0<=1)
% theta_0: phase shift between mu_1_n(t) and mu_2_n(t)
% f_max: maximum Doppler frequency
% sigma_3: square root of the average power of the real deterministic
%          Gaussian process nu_3(t)
% m_3: average value of the real deterministic Gaussian
%      process mu_3(t)
% rho: amplitude of the LOS component m(t)
% theta_rho: phase of the LOS component m(t)
% f_c: 3-dB-cut-off frequency
% T_s: sampling interval
% T_sim: duration of the simulation
% PLOT: plot of the deterministic extended Suzuki process eta(t) of
%       Type II, if PLOT==1

function eta_t=Suzuki_Type_II(N_1,N_3,sigma_0_2,kappa_0,theta_0,...
                            f_max,sigma_3,m_3,rho,theta_rho,f_c,...
                            T_s,T_sim,PLOT)
if nargin==13,
   PLOT=0;
end

N_1_s=ceil(N_1/(2/pi*asin(kappa_0)));
[f1,c1,th1]=parameter_Jakes('es_j',N_1_s,sigma_0_2,f_max,'rand',0);
f1 =f1(1:N_1);
c1 =c1(1:N_1);
th1=th1(1:N_1);
```

```
[f3,c3,th3]=parameter_Gauss('es_g',N_3,1,f_max,f_c,'rand',0);
gaMma=(2*pi*f_c/sqrt(2*log(2)))^2;
f3(N_3)=sqrt(gaMma*N_3/(2*pi)^2-sum(f3(1:N_3-1).^2));

N=ceil(T_sim/T_s);
t=(0:N-1)*T_s;

xi_t=abs(Mu_i_t(c1,f1,th1,T_s,T_sim)+rho*cos(theta_rho)+...
         j*(Mu_i_t(c1,f1,th1-theta_0,T_s,T_sim)+...
            rho*sin(theta_rho) ) );

lambda_t=exp(Mu_i_t(c3,f3,th3,T_s,T_sim)*sigma_3+m_3);

eta_t=xi_t.*lambda_t;


if PLOT==1,
   plot(t,20*log10(eta_t),'b-')
   xlabel('t (s)')
   ylabel('20 log eta(t)')
end

%----------------------------------------------------------------------
% gen_Rice_proc.m ---------------------------------------------------
%
% Program for the simulation of deterministic generalized Rice
% processes (see Fig. 6.29).
%
% Used m-files: parameter_Jakes.m, Mu_i_t.m
%----------------------------------------------------------------------
% xi_t=gen_Rice_proc(N_1,N_2,sigma_1_2,sigma_2_2,kappa_0,...
%                    theta_0,rho,theta_rho,f_max,...
%                    T_s,T_sim,PLOT)
%----------------------------------------------------------------------
% Explanation of the input parameters:
%
% N_1, N_2: number of harmonic functions of the real deterministic
%           Gaussian processes nu_1(t) and nu_2(t), respectively
% sigma_1_2: average power of the real deterministic Gaussian
%            process nu_1(t)
% sigma_2_2: average power of the real deterministic Gaussian
%            process nu_2(t)
% kappa_0: frequency ratio f_min/f_max (0<=kappa_0<=1)
% theta_0: phase shift between mu_1_n(t) and mu_2_n(t)
% rho: amplitude of the LOS component m(t)
% theta_rho: phase of the LOS component m(t)
```

```
% f_max: maximum Doppler frequency
% T_s: sampling interval
% T_sim: duration of the simulation
% PLOT: plot of the deterministic generalized Suzuki process xi(t),
%       if PLOT==1

function xi_t=gen_Rice_proc(N_1,N_2,sigma_1_2,sigma_2_2,kappa_0,...
                            theta_0,rho,theta_rho,f_max,T_s,...
                            T_sim,PLOT)

if nargin==11,
   PLOT=0;
end

[f1,c1,th1]=parameter_Jakes('es_j',N_1,sigma_1_2,f_max,'rand',0);
c1=c1/sqrt(2);

N_2_s=ceil(N_2/(2/pi*asin(kappa_0)));
[f2,c2,th2]=parameter_Jakes('es_j',N_2_s,sigma_2_2,f_max,'rand',0);
f2 =f2(1:N_2);
c2 =c2(1:N_2)/sqrt(2);
th2=th2(1:N_2);

N=ceil(T_sim/T_s);
t=(0:N-1)*T_s;

xi_t=abs(Mu_i_t(c1,f1,th1,T_s,T_sim)+...
         Mu_i_t(c2,f2,th2,T_s,T_sim)+rho*cos(theta_rho)+...
         j*(Mu_i_t(c1,f1,th1-theta_0,T_s,T_sim)+...
         Mu_i_t(c2,f2,th2+theta_0,T_s,T_sim)+...
         rho*sin(theta_rho)));

if PLOT==1,
   plot(t,20*log10(xi_t),'b-')
   xlabel('t (s)')
   ylabel('20 log xi(t)')
end

%----------------------------------------------------------------------
% det_mod_Loo.m -------------------------------------------------------
%
% Program for the simulation of modified Loo processes.
%
% Used m-files: parameter_Jakes.m, parameter_Gauss.m, Mu_i_t.m
%----------------------------------------------------------------------
% rho_t=det_mod_Loo(N_1,N_2,N_3,sigma_1_2,kappa_1,sigma_2_2,...
%                   kappa_2,f_max,sigma_3,m_3,f_rho,...
```

```
%                   theta_rho,f_c,T_s,T_sim,PLOT)
%------------------------------------------------------------------
% Explanation of the input parameters:
%
% N_1, N_2, N_3: number of harmonic functions of the real determi-
%                nistic Gaussian processes nu_1(t), nu_2(t), and
%                nu_3(t), respectively
% sigma_1_2: average power of the real deterministic Gaussian
%            process nu_1(t)
% kappa_1: frequency ratio f_min/f_max (0<=kappa_0<=1) of nu_1(t)
% sigma_2_2: average power of the real deterministic Gaussian
%            process nu_2(t)
% kappa_2: frequency ratio f_min/f_max (0<=kappa_0<=1) of nu_2(t)
% f_max: maximum Doppler frequency
% sigma_3: square root of the average power of the real deterministic
%          Gaussian process nu_3(t)
% m_3: average value of the third real deterministic Gaussian
%      process mu_3(t)
% f_rho: Doppler frequency of the LOS component m(t)
% theta_rho: phase of the LOS component m(t)
% f_c: 3-dB-cut-off frequency
% T_s: sampling interval
% T_sim: duration of the simulation
% PLOT: plot of the time-domain signal rho(t), if PLOT==1

function rho_t=det_mod_Loo(N_1,N_2,N_3,sigma_1_2,kappa_1,...
               sigma_2_2,kappa_2,f_max,sigma_3,m_3,f_rho,...
               theta_rho,f_c,T_s,T_sim,PLOT)

if nargin==15,
   PLOT=0;
end

sigma_1=sqrt(sigma_1_2);
sigma_2=sqrt(sigma_2_2);

N_1_s=ceil(N_1/(2/pi*asin(kappa_1)));
[f1,c1,th1]=parameter_Jakes('es_j',N_1_s,sigma_1_2,f_max,'rand',0);
f1 =f1(1:N_1);
c1 =c1(1:N_1)/sqrt(2);
th1=th1(1:N_1);

N_2_s=ceil(N_2/(2/pi*asin(kappa_2)));
[f2,c2,th2]=parameter_Jakes('es_j',N_2_s,sigma_2_2,f_max,'rand',0);
f2 =f2(1:N_2);
c2 =c2(1:N_2)/sqrt(2);
th2=th2(1:N_2);
```

```
[f3,c3,th3]=parameter_Gauss('es_g',N_3,1,f_max,f_c,'rand',0);
gaMma=(2*pi*f_c/sqrt(2*log(2)))^2;
f3(N_3)=sqrt(gaMma*N_3/(2*pi)^2-sum(f3(1:N_3-1).^2));

N=ceil(T_sim/T_s);
t=(0:N-1)*T_s;

arg=2*pi*f_rho*t+theta_rho;

RHO_t=exp(Mu_i_t(c3,f3,th3,T_s,T_sim)*sigma_3+m_3);

rho_t=abs(Mu_i_t(c1,f1,th1,T_s,T_sim)+...
          Mu_i_t(c2,f2,th2,T_s,T_sim)+RHO_t.*cos(arg)+...
          j*(Mu_i_t(c1,f1,th1-pi/2,T_s,T_sim)-...
          Mu_i_t(c2,f2,th2-pi/2,T_s,T_sim)+RHO_t.*sin(arg)));

if PLOT==1,
   plot(t,20*log10(rho_t),'b-',t,20*log10(RHO_t),'y--')
   xlabel('t (s)')
   ylabel('20 log rho(t)')
end

%----------------------------------------------------------------------
% F_S_K.m -------------------------------------------------------------
%
% Program for the simulation of deterministic frequency-selective
% mobile radio channels.
%
%----------------------------------------------------------------------
% [y_t,T,t_0]=F_S_K(x_t,f_max,m_s,T,t_0,q_l,...
%                   C1,F1,TH1,C2,F2,TH2,F01,F02,RHO,F_RHO,PLOT)
%----------------------------------------------------------------------
% Explanation of the input parameters:
%
% x_t: time-domain input signal of the channel simulator (sampled
%      with T_s=0.2E-6 s)
% f_max: maximum Doppler frequency
% m_s: sampling rate ratio
% T: contents of the delay elements of the time variant FIR filter
% t_0: offset in time
% q_l: q_l=tau_l/T_s+1
%----------------------------------------------------------------------
% The following matrices are generated in F_S_K_p.m:
% F1, F2: discrete Doppler frequencies
% C1, C2: Doppler coefficients
% TH1, TH2: Doppler phases
```

```
% F01, F02: frequency shift value of the Doppler PSD according to
%           Gauss I and Gauss II, respectively
% RHO: amplitude of the direct component
% F_RHO: Doppler frequency of the direct component
%----------------------------------------------------------------------
% PLOT: plot of the output signal of the channel, if PLOT==1

function [y_t,T,t_0]=F_S_K(x_t,f_max,m_s,T,t_0,q_l,...
                    C1,F1,TH1,C2,F2,TH2,F01,F02,RHO,F_RHO,PLOT)

T_s=0.2E-6;

% Initialization:
mu_l=zeros(size(q_l));
y_t=zeros(size(x_t));

for n=0:length(x_t)-1,
    if rem(n/m_s,m_s)-fix(rem(n/m_s,m_s))==0,
       mu_l=sum((C1.*cos(2*pi*F1*f_max*(n*T_s+t_0)+TH1)).').*...
            exp(-j*2*pi*F01*f_max*(n*T_s+t_0))+j*...
            (sum((C2.*cos(2*pi*F2*f_max*(n*T_s+t_0)+TH2)).').*...
            exp(-j*2*pi*F02*f_max*(n*T_s+t_0)))+...
            RHO.*exp(j*2*pi*F_RHO*f_max*(n*T_s+t_0));
    end
    T(1)=x_t(n+1);
    y_t(n+1)=sum(mu_l.*T(q_l));
    T(2:length(T))=T(1:length(T)-1);
end

t_0=length(x_t)*T_s+t_0;

if PLOT==1,
   plot((0:length(y_t)-1)*T_s,20*log10(abs(y_t)),'g-')
end

%----------------------------------------------------------------------
% F_S_K_p.m -----------------------------------------------------------
%
% Program for the generation of the matrices used in F_S_K.m.
%
% Used m-file: pCOST207.m
%----------------------------------------------------------------------
% [C1,F1,TH1,C2,F2,TH2,F01,F02,RHO,F_RHO,q_l,T]=
%                       F_S_K_p(N_1,AREA,f_max)
%----------------------------------------------------------------------
% Explanation of the input parameters:
%
```

```
% N_1: minimum number of discrete Doppler frequencies
% AREA: according to COST 207, 4 types of channels are specified:
%           1) Rural Area:    'ra'
%           2) Typical Urban: 'tu'
%           3) Bad Urban:     'bu'
%           4) Hilly Terrain: 'ht'
% f_max: maximum Doppler frequency

function [C1,F1,TH1,C2,F2,TH2,F01,F02,RHO,F_RHO,q_l,T]=...
         F_S_K_p(N_1,AREA,f_max)

% The greatest common divisor of the discrete propagation delays
% defines the sampling interval T_s:
T_s=0.2E-6;

if     all(lower(AREA)=='ra'),
       a_l=[1,0.63,0.1,0.01];
       tau_l=[0,0.2,0.4,0.6]*1E-6;
       DOPP_KAT=['RI';'JA';'JA';'JA'];
elseif all(lower(AREA)=='tu'),
       a_l=[0.5,1,0.63,0.25,0.16,0.1];
       tau_l=[0,0.2,0.6,1.6,2.4,5]*1E-6;
       DOPP_KAT=['JA';'JA';'G1';'G1';'G2';'G2'];
elseif all(lower(AREA)=='bu'),
       a_l=[0.5,1,0.5,0.32,0.63,0.4];
       tau_l=[0,0.4,1.0,1.6,5.0,6.6]*1E-6;
       DOPP_KAT=['JA';'JA';'G1';'G1';'G2';'G2'];
elseif all(lower(AREA)=='ht'),
       a_l=[1,0.63,0.4,0.2,0.25,0.06];
       tau_l=[0,0.2,0.4,0.6,15,17.2]*1E-6;
       DOPP_KAT=['JA';'JA';'JA';'JA';'G2';'G2'];
end

% Generate the parameters and assign them to the matrices:
num_of_taps=length(DOPP_KAT);
F1=zeros(num_of_taps,N_1+2*num_of_taps-1);
F2=F1;C1=F1;C2=F1;TH1=F1;TH2=F1;
F01=zeros(1,num_of_taps);F02=F01;
RHO=zeros(1,num_of_taps);F_RHO=RHO;
NN1=N_1+2*(num_of_taps-1):-2:N_1;
for k=1:num_of_taps,
    [f1,f2,c1,c2,th1,th2,rho,f_rho,f01,f02]=...
    pCOST207(DOPP_KAT(k,:),NN1(k));
    F1(k,1:NN1(k))=f1;
    C1(k,1:NN1(k))=c1*sqrt(a_l(k));
    TH1(k,1:NN1(k))=th1;
    F2(k,1:NN1(k)+1)=f2;
```

```
    C2(k,1:NN1(k)+1)=c2*sqrt(a_l(k));
    TH2(k,1:NN1(k)+1)=th2;
    F01(k)=f01;F02(k)=f02;
    RHO(k)=rho;F_RHO(k)=f_rho;
end

% Determine indices of the delay elements of the FIR filter:
q_l=tau_l/T_s+1;

% Initialization of the delay elements of the FIR filter:
T=zeros(1,max(q_l));

%-------------------------------------------------------------------
% pCOST207.m -------------------------------------------------------
%
% Program for the derivation of the channel parameters of the
% Doppler PSDs defined by COST 207.
%
%-------------------------------------------------------------------
%[f1,f2,c1,c2,th1,th2,rho,f_rho,f01,f02]=pCOST207(D_S_T,N_i)
%-------------------------------------------------------------------
% Explanation of the input parameters:
%
% D_S_T: type of the Doppler PSD:
%            Jakes:    D_S_T='JA'
%            Rice:     D_S_T='RI'
%            Gauss  I: D_S_T='G1'
%            Gauss II: D_S_T='G2'
% N_i: number of harmonic functions

function [f1,f2,c1,c2,th1,th2,rho,f_rho,f01,f02]=pCOST207(D_S_T,N_i)

if     all(lower(D_S_T)=='ri'), % RICE
       n=(1:N_i);
       f1=sin(pi/(2*N_i)*(n-1/2));
       c1=0.41*sqrt(1/N_i)*ones(1,N_i);
       th1=rand(1,N_i)*2*pi;
       n=(1:N_i+1);
       f2=sin(pi/(2*(N_i+1))*(n-1/2));
       c2=0.41*sqrt(1/(N_i+1))*ones(1,N_i+1);
       th2=rand(1,N_i+1)*2*pi;
       f01=0;f02=0;
       rho=0.91;f_rho=0.7;
elseif all(lower(D_S_T)=='ja'), % JAKES
       n=(1:N_i);
       f1=sin(pi/(2*N_i)*(n-1/2));
       c1=sqrt(1/N_i)*ones(1,N_i);
```

```
        th1=rand(1,N_i)*2*pi;
        n=(1:N_i+1);
        f2=sin(pi/(2*(N_i+1))*(n-1/2));
        c2=sqrt(1/(N_i+1))*ones(1,N_i+1);
        th2=rand(1,N_i+1)*2*pi;
        f01=0;f02=0;
        rho=0;f_rho=0;
elseif all(lower(D_S_T)=='g1'), % GAUSS I
        n=(1:N_i);
        sgm_0_2=5/6;
        c1=sqrt(sgm_0_2*2/N_i)*ones(1,N_i);
        f1=sqrt(2)*0.05*erfinv((2*n-1)/(2*N_i));
        th1=rand(1,N_i)*2*pi;
        sgm_0_2=1/6;
        c2=[sqrt(sgm_0_2*2/N_i)*ones(1,N_i),0]/j;
        f2=[sqrt(2)*0.1*erfinv((2*n-1)/(2*N_i)),0];
        th2=[rand(1,N_i)*2*pi,0];
        f01=0.8;f02=-0.4;
        rho=0;f_rho=0;
elseif all(lower(D_S_T)=='g2'), % GAUSS II
        n=(1:N_i);
        sgm_0_2=10^0.5/(sqrt(10)+0.15);
        c1=sqrt(sgm_0_2*2/N_i)*ones(1,N_i);
        f1=sqrt(2)*0.1*erfinv((2*n-1)/(2*N_i));
        th1=rand(1,N_i)*2*pi;
        sgm_0_2=0.15/(sqrt(10)+0.15);
        c2=[sqrt(sgm_0_2*2/N_i)*ones(1,N_i),0]/j;
        f2=[sqrt(2)*0.15*erfinv((2*n-1)/(2*N_i)),0];
        th2=[rand(1,N_i)*2*pi,0];
        f01=-0.7;f02=0.4;
        rho=0;f_rho=0;
end

%-----------------------------------------------------------------
% cdf_sim.m ------------------------------------------------------
%
% Program for the computation of cumulative distribution
% functions F(r).
%
%-----------------------------------------------------------------
% F_r=cdf_sim(xi_t,r,PLOT)
%-----------------------------------------------------------------
% Explanation of the input parameters:
%
% xi_t: deterministic process or time-domain signal to be analysed
%       with respect to the cumulative distribution function F(r).
% r: level vector
```

```
% PLOT: plot of the resulting cumulative distribution function F(r),
%       if PLOT==1

function F_r=cdf_sim(xi_t,r,PLOT)

if nargin==2,
   PLOT=0;
end

F_r=zeros(size(r));

for l=1:length(r),
    F_r(l)=length(find(xi_t<=r(l)));
end

F_r=F_r/length(xi_t);

if PLOT==1,
   plot(r,F_r,'rx')
   xlabel('r')
   ylabel('F(r)')
end

%-----------------------------------------------------------------------
% pdf_sim.m ------------------------------------------------------------
%
% Program for the computation of probability density functions p(z).
%
%-----------------------------------------------------------------------
% p_z=pdf_sim(xi_t,z,PLOT)
%-----------------------------------------------------------------------
% Explanation of the input parameters:
%
% xi_t: deterministic process or time-domain signal to be analysed
%       with respect to the probability density function p(z).
% z: equidistant level vector
% PLOT: plot of the resulting probability density function p(z),
%       if PLOT==1

function p_z=pdf_sim(xi_t,z,PLOT)

if nargin==2,
   PLOT=0;
end

p_z=hist(xi_t,z)/length(xi_t)/abs(z(2)-z(1));
```

```
if PLOT==1,
   plot(z,p_z,'mx')
   xlabel('z')
   ylabel('p(z)')
end

%----------------------------------------------------------------------
% lcr_sim.m -----------------------------------------------------------
%
% Program for the computation of the level-crossing rate N(r).
%
%----------------------------------------------------------------------
% N_r=lcr_sim(xi_t,r,T_sim,PLOT)
%----------------------------------------------------------------------
% Explanation of the input parameters:
%
% xi_t: deterministic process or time-domain signal to be analysed
%       with respect to the level-crossing rate N(r)
% r: level vector
% T_sim: duration of the simulation
% PLOT: plot of the resulting level-crossing rate N(r), if PLOT==1

function N_r=lcr_sim(xi_t,r,T_sim,PLOT)

if nargin==3,
   PLOT=0;
end

N_r=zeros(size(r));

for k=1:length(r),
    N_r(k)=sum(xi_t(2:length(xi_t)) < r(k) & ...
               xi_t(1:length(xi_t)-1) >= r(k) );
end

N_r=N_r/T_sim;

if PLOT==1,
   plot(r,N_r,'yx')
   xlabel('r')
   ylabel('N(r)')
end

%----------------------------------------------------------------------
% adf_sim.m -----------------------------------------------------------
%
% Program for the computation of the average duration of fades T_(r).
%
```

```
% Used m-files: cdf_sim.m, lcr_sim.m
%----------------------------------------------------------------------
% adf=adf_sim(xi_t,r,T_sim,PLOT)
%----------------------------------------------------------------------
% Explanation of the input parameters:
%
% xi_t: deterministic process or time-domain signal to be analysed
%       with respect to the average duration of fades T_(r)
% r: equidistant level vector
% T_sim: duration of the simulation
% PLOT: plot of the resulting average duration of fades T_(r),
%       if PLOT==1

function adf=adf_sim(xi_t,r,T_sim,PLOT)

cdf=cdf_sim(xi_t,r);
lcr=lcr_sim(xi_t,r,T_sim);

adf=cdf./lcr;

if PLOT==1,
   plot(r,adf,'yx')
   xlabel('r')
   ylabel('T_(r)')
end
```

ABBREVIATIONS

ACF	autocorrelation function
ATDMA	Advanced Time Division Multiple Access
BMFT	Bundesministerium für Forschung und Technologie
BU	Bad Urban
CEPT	Conference of European Posts and Telecommunications Administrations
COST	European Cooperation in the Field of Scientific and Technical Research
DCS	Digital Cellular System
DECT	Digital European Cordless Telephone
DGUS	deterministic Gaussian uncorrelated scattering
ETSI	European Telecommunications Standards Institute
FIR	finite impulse response
FPLMTS	Future Public Land Mobile Telecommunications System
GSM	Global System for Mobile Communications (former: Groupe Spécial Mobile)
GWSSUS	Gaussian wide-sense stationary uncorrelated scattering
HT	Hilly Terrain
IMT 2000	International Mobile Telecommunications 2000
INMARSAT	International Maritime Satellite Organisation
ISI	intersymbol interference
JM	Jakes method
LEO	low earth orbit
LOS	line-of-sight
LPNM	L_p-norm method
MBS	Mobile Broadband System
MCM	Monte Carlo method
MEA	method of equal areas
MED	method of equal distances

MEDS	method of exact Doppler spread
MEO	medium earth orbit
MMEA	modified method of equal areas
MSEM	mean-square-error method
PCN	Personal Communications Network
PSD	power spectral density
RA	Rural Area
RACE	Research on Advanced Communications in Europe
TU	Typical Urban
UMTS	Universal Mobile Telecommunications System
WGN	white Gaussian noise
WSSUS	wide-sense stationary uncorrelated scattering

SYMBOLS

SET THEORY

$\mathbb{C}$	set of complex numbers
$\mathbb{N}$	set of natural numbers
$\mathbb{R}$	set of real numbers
$\mathbb{Z}$	set of integer numbers
$\in$	is an element of
$\notin$	is not an element of
$\forall$	for all
$\subset$	subset
$\cup$	union
$\cap$	intersection
$A \setminus B$	difference of set A and set B
$\emptyset$	empty set or null set
$[a, b]$	set of real numbers within the closed interval from a to b, i.e., $[a, b] = \{x \in \mathbb{R} \mid a \leq x \leq b\}$
$[a, b)$	set of real numbers within the right-hand side open interval from a to b, i.e., $[a, b) = \{x \in \mathbb{R} \mid a \leq x < b\}$
$(a, b]$	set of real numbers within the left-hand side open interval from a to b, i.e., $(a, b] = \{x \in \mathbb{R} \mid a < x \leq b\}$
$\{x_n\}_{n=1}^{N}$	set of elements $x_1, x_2, \ldots, x_N$

OPERATORS AND MISCELLANEOUS SYMBOLS

$\arg\{x\}$	argument of $x = x_1 + jx_2$
$\mathrm{Cov}\,\{x_1, x_2\}$	covariance between x_1 and x_2
e^x	exponential function
$E\{x\}$	(statistical) mean value or expected value of x

$F\{x(t)\}$	Fourier transform of $x(t)$
$\exp\{x\}$	exponential function
$F^{-1}\{X(f)\}$	inverse Fourier transform of $X(f)$
$\gcd\{x_n\}_{n=1}^N$	greatest common divisor (also known as highest common factor) of $x_1, x_2, \ldots, x_N$
$\text{Im}\{x\}$	imaginary part of $x = x_1 + jx_2$
$\text{lcm}\,\{x_n\}_{n=1}^N$	least common multiple of $x_1, x_2, \ldots, x_N$
lim	limit
$\ln x$	natural logarithm of x
$\log_a x$	logarithm of x to base a
$\max\{x_n\}_{n=1}^N$	largest element of the set $\{x_1, x_2, \ldots, x_N\}$
$\min\{x_n\}_{n=1}^N$	smallest element of the set $\{x_1, x_2, \ldots, x_N\}$
mod	modulo operation
$n!$	factorial function
$P(\mu \leq x)$	probability that the event μ is less than or equal to x
$\text{Re}\{x\}$	real part of $x = x_1 + jx_2$
$\text{round}\{x\}$	nearest integer to x
$\text{sgn}\,(x)$	sign of the number x: 1 if $x > 0$, -1 if $x < 0$
$\text{Var}\,\{x\}$	variance of x
$x_1(t) * x_2(t)$	convolution of $x_1(t)$ and $x_2(t)$
x^*	complex conjugate of the complex number $x = x_1 + jx_2$
$\lvert x \rvert$	absolute value of x
$\sqrt{x}$	principal value of the square root of x, i.e., $\sqrt{x} \geq 0$ for $x \geq 0$
$\prod_{n=1}^N$	multiple product
$\sum_{n=1}^N$	multiple sum
$\int_a^b x(t)dt$	integral of the function $x(t)$ over the interval $[a, b]$
$\dot{x}(t)$	derivative of the function $x(t)$ with respect to time t
$\breve{x}(t)$	Hilbert transform of $x(t)$
$x \to a$	x tends to a or x approaches a
$\lceil x \rceil$	ceiling function, the smallest integer greater than or equal to x
$\lfloor x \rfloor$	floor function, the greatest integer less than or equal to x
$\approx$	approximately equal
$\sim$	distributed according to (statistics) or asymptotically equal (analysis)
$\leq$	less than or equal to
$\ll$	much less than

$=$	equal
$\neq$	unequal
$\circ\!\!-\!\!\bullet$	Fourier transform

MATRICES AND VECTORS

$(a_{m,n})$	matrix with $a_{m,n}$ as the entry of the mth row and the nth column
$\boldsymbol{A}^T$	transpose matrix of the matrix $\boldsymbol{A}$
$\boldsymbol{A}^{-1}$	inverse matrix of the matrix $\boldsymbol{A}$
$\boldsymbol{C}_{\mu_\rho}$	covariance matrix of the vector process $\boldsymbol{\mu}_\rho(t) = (\mu_{\rho_1}(t), \mu_{\rho_2}(t), \dot{\mu}_{\rho_1}(t), \dot{\mu}_{\rho_2}(t))^T$
$\det \boldsymbol{A}$	determinant of the matrix $\boldsymbol{A}$
J	Jacobian determinant
$\boldsymbol{m}$	column vector of m_1, m_2, $\dot{m}_1$, and $\dot{m}_2$, i.e., $\boldsymbol{m} = (m_1, m_2, \dot{m}_1, \dot{m}_2)^T$
$\boldsymbol{R}_\mu$	autocorrelation matrix of the vector process $\boldsymbol{\mu}(t) = (\mu_1(t), \mu_2(t), \dot{\mu}_1(t), \dot{\mu}_2(t))^T$
$\mathrm{tr}(\boldsymbol{A})$	trace of the matrix $\boldsymbol{A} = (a_{m,n}) \in \mathbb{R}^{N\times N}$, i.e., $\mathrm{tr}(\boldsymbol{A}) = \sum_{n=1}^{N} a_{n,n}$
$\boldsymbol{x}$	column vector of x_1, x_2, $\dot{x}_1$, and $\dot{x}_2$, i.e., $\boldsymbol{x} = (x_1, x_2, \dot{x}_1, \dot{x}_2)^T$
$\boldsymbol{\Omega}$	parameter vector

SPECIAL FUNCTIONS

$\mathrm{erf}\,(\cdot)$	error function
$\mathrm{erfc}\,(\cdot)$	complementary error function
$E(\cdot,\cdot)$	elliptic integral of the second kind
$\boldsymbol{E}(\cdot)$	complete elliptic integral of the second kind
$F(\cdot,\cdot;\cdot;\cdot)$	hypergeometric function
$H_0(\cdot)$	Struve's function of order zero
$I_\nu(\cdot)$	modified Bessel function of the first kind of order ν
$J_\nu(\cdot)$	Bessel function of the first kind of order ν
$Q_m(\cdot,\cdot)$	generalized Marcum's Q-function
$\mathrm{rect}\,(\cdot)$	rectangular function
$\mathrm{sinc}\,(\cdot)$	sinc function
$\delta(\cdot)$	delta function
$\Gamma(\cdot)$	gamma function

STOCHASTIC PROCESSES

B_C	coherence bandwidth
$B^{(1)}_{\mu_i\mu_i}$	average Doppler shift of $\mu_i(t)$
$B^{(2)}_{\mu_i\mu_i}$	Doppler spread of $\mu_i(t)$
$B^{(1)}_{\tau'\tau'}$	average delay
$B^{(2)}_{\tau'\tau'}$	delay spread
c_0	speed of light
c_R	Rice factor
$E_2(\mathbf{\Omega})$	mean-square-error norm
f	Doppler frequency
f_0	carrier frequency
f_c	cut-off frequency
f_{max}	maximum Doppler frequency
f_{min}	lower cut-off frequency of the left-hand side restricted Jakes power spectral density
f_s	sampling rate
f_{sym}	symbol rate
f_ρ	Doppler frequency of the line-of-sight component $m(t)$
$F_{\zeta_-}(r)$	cumulative distribution function of Rayleigh processes $\zeta(t)$
$F_{\eta_-}(r)$	cumulative distribution function of Suzuki processes $\eta(t)$
$F_{\eta_+}(r)$	complementary cumulative distribution function of Suzuki processes $\eta(t)$
$F_\vartheta(\varphi)$	cumulative distribution function of the phase $\vartheta(t)$ of $\mu(t) = \mu_1(t) + j\mu_2(t)$
$F_{\mu_i}(r)$	cumulative distribution function of Gaussian random processes $\mu_i(t)$
$F_{\xi_-}(r)$	cumulative distribution function of Rice processes $\xi(t)$
$F_{\xi_+}(r)$	complementary cumulative distribution function of Rice processes $\xi(t)$
$F_{\varrho_-}(r)$	cumulative distribution function of Loo processes $\varrho(t)$
$F_{\varrho_+}(r)$	complementary cumulative distribution function of Loo processes $\varrho(t)$
$h(\tau')$	time-invariant impulse response
$h(\tau', t)$	time-variant impulse response
$H(f)$	transfer function of linear time-invariant systems
$H(f', t)$	time-variant transfer function
$\check{H}(f)$	Hilbert transformer
$\mathcal{L}$	number of discrete paths

$m(t)$	(time-variant) line-of-sight component
m_s'	sampling rate ratio, i.e., $m_s' = f_s'/f_s = T_s/T_s'$
m_{μ_i}	mean value of $\mu_i(t)$
$N_\zeta(r)$	level-crossing rate of Rayleigh processes $\zeta(t)$
$N_\eta(r)$	level-crossing rate of Suzuki processes $\eta(t)$
$N_\xi(r)$	level-crossing rate of Rice processes $\xi(t)$
$N_\varrho(r)$	level-crossing rate of Loo processes $\varrho(t)$
$p_{0_-}(\tau_-;r)$	probability density function of the fading intervals τ_- of Rayleigh processes $\zeta(t)$
$p_{1_-}(\tau_-;r)$	approximate solution for $p_{0_-}(\tau_-;r)$
$p_\zeta(z)$	Rayleigh distribution
$p_\eta(z)$	Suzuki distribution
$p_\lambda(z)$	lognormal distribution
$p_\vartheta(\theta)$	probability density function of the phase $\vartheta(t)$
$p_{\mu_i}(x)$	Gaussian distribution
$p_{\mu_{\rho_1}\mu_{\rho_2}\dot{\mu}_{\rho_1}\dot{\mu}_{\rho_2}}$	joint probability density function of $\mu_{\rho_1}(t)$, $\mu_{\rho_2}(t)$, $\dot{\mu}_{\rho_1}(t)$, and $\dot{\mu}_{\rho_2}(t)$
$p_\xi(z)$	Rice distribution
$p_\varrho(z)$	probability density function of Loo processes $\varrho(t)$
$p_\omega(z)$	Nakagami distribution
$p_{\xi\dot{\xi}}(z,\theta)$	joint probability density function of $\xi(t)$ and $\dot{\xi}(t)$
$p_{\xi\dot{\xi}\vartheta\dot{\vartheta}}(z,\dot{z},\theta,\dot{\theta})$	joint probability density function of $\xi(t)$, $\dot{\xi}(t)$, $\vartheta(t)$, and $\dot{\vartheta}(t)$
$Q_m(\cdot,\cdot)$	generalized Marcum's Q-function
r	amplitude level
$r_{hh}(\cdot,\cdot;\cdot,\cdot)$	autocorrelation function of $h(\tau',t)$
$r_{HH}(v',\tau)$	time-frequency correlation function of WSSUS models
$r_{HH}(\cdot,\cdot;\cdot,\cdot)$	autocorrelation function of $H(f',t)$
$r_{ss}(\cdot,\cdot;\cdot,\cdot)$	autocorrelation function of $s(\tau',f)$
$r_{TT}(\cdot,\cdot;\cdot,\cdot)$	autocorrelation function of $T(f',f)$
$r_{xx}(t_1,t_2)$	autocorrelation function of $x(t)$, i.e., $r_{xx}(t_1,t_2) = E\{x^*(t_1)x(t_2)\}$
$r_{yy}(t_1,t_2)$	autocorrelation function of $y(t)$, i.e., $r_{yy}(t_1,t_2) = E\{y^*(t_1)y(t_2)\}$
$r_{\mu\mu}(\tau)$	autocorrelation function of $\mu(t) = \mu_1(t) + j\mu_2(t)$
$r_{\mu_i\mu_i}(\tau)$	autocorrelation function of $\mu_i(t)$
$\hat{r}_{\mu_i\mu_i}(\tau)$	autocorrelation function of $\hat{\mu}_i(t)$
$r_{\mu_1\mu_2}(\tau)$	cross-correlation function of $\mu_1(t)$ and $\mu_2(t)$
$r_{\tau'\tau'}(v')$	frequency correlation function
$s(\tau',f)$	Doppler-variant impulse response

$S(\tau', f)$	scattering function of WSSUS models
$S_{hh}(\tau', \tau)$	delay cross-power spectral density of WSSUS models
$S_{TT}(\upsilon', f)$	Doppler cross-power spectral density of WSSUS models
$S_{\tau'\tau'}(\tau')$	delay power spectral density
$S_{\mu\mu}(f)$	power spectral density of $\mu(t) = \mu_1(t) + j\mu_2(t)$
$S_{\mu_i\mu_i}(f)$	power spectral density of $\mu_i(t)$
$S_{\mu_1\mu_2}(f)$	cross-power spectral density of $\mu_1(t)$ and $\mu_2(t)$
t	time variable
$T(f', f)$	Doppler-variant transfer function
T_C	coherence time
T_s	sampling interval
T_{sym}	symbol interval
$T_{\zeta_-}(r)$	average duration of fades of Rayleigh processes $\zeta(t)$
$T_{\eta_-}(r)$	average duration of fades of Suzuki processes $\eta(t)$
$T_{\xi_-}(r)$	average duration of fades of Rice processes $\xi(t)$
$T_{\varrho_-}(r)$	average duration of fades of Loo processes $\varrho(t)$
u_n	random variable, uniformly distributed in the interval $(0, 1]$
v	speed of the mobile unit
$W_i(\cdot)$	weighting function
$x(t)$	input signal
$y(t)$	output signal
β	negative curvature of the autocorrelation function $r_{\mu_i\mu_i}(\tau)$ at the origin, i.e., $\beta = \beta_i = -\ddot{r}_{\mu_i\mu_i}(0) \quad (i = 1, 2)$
γ	negative curvature of the autocorrelation function $r_{\nu_3\nu_3}(\tau)$ at the origin, i.e., $\gamma = -\ddot{r}_{\nu_3\nu_3}(0)$
$\zeta(t)$	Rayleigh process
$\eta(t)$	Suzuki process
θ_ρ	phase of the line-of-sight component $m(t)$
$\vartheta(t)$	phase of $\mu_\rho(t)$, i.e., $\vartheta(t) = \arg\{\mu_\rho(t)\}$
κ_0	frequency ratio f_{min} over f_{max}
κ_c	frequency ratio f_{max} over f_c
$\lambda(t)$	lognormal process
$\mu(t)$	zero-mean complex Gaussian random process
$\mu_i(t)$	real Gaussian random process (stochastic reference model)
$\hat{\mu}_i(t)$	real stochastic process (stochastic simulation model)
$\mu_\rho(t)$	complex Gaussian random process with mean $m(t)$

$\nu_i(t)$	white Gaussian noise
$\xi(t)$	Rice process
ρ	amplitude of the line-of-sight component $m(t)$
$\varrho(t)$	Loo process
σ_0^2	mean power of $\mu_i(t)$
τ	time difference between t_2 and t_1, i.e., $\tau = t_2 - t_1$
τ_-	fading interval
τ_+	connecting interval
$\tau_q(r)$	length of the time interval that comprises q % of all fading intervals of the process $\zeta(t)$ at level r
τ'	continuous propagation delay
τ'_ℓ	discrete propagation delay of the ℓth path
τ'_{max}	maximum propagation delay
$\Delta\tau'_\ell$	propagation delay difference between τ'_ℓ and $\tau'_{\ell-1}$, i.e., $\Delta\tau'_\ell = \tau'_\ell - \tau'_{\ell-1}$
ϕ_0	symbol for the cross-correlation function $r_{\mu_1\mu_2}(\tau)$ at $\tau = 0$
ψ_0	symbol for the autocorrelation function $r_{\mu_i\mu_i}(\tau)$ at $\tau = 0$
$\Psi_{\mu_i}(\nu)$	characteristic function of $\mu_i(t)$

DISCRETE-TIME DETERMINISTIC PROCESSES

$a_{i,n}[k]$	address of the table $\mathrm{Tab}_{i,n}$ at the discrete time k
$\bar{B}^{(1)}_{\mu_i\mu_i}$	average Doppler shift of $\bar{\mu}_i[k]$
$\bar{B}^{(2)}_{\mu_i\mu_i}$	Doppler spread of $\bar{\mu}_i[k]$
$c_{i,n}$	Doppler coefficient of the nth component of $\bar{\mu}_i[k]$
$\bar{f}_{i,n}$	quantised Doppler frequency of the nth component of $\bar{\mu}_i[k]$
f_s	sampling frequency
$f_{s,min}$	minimum sampling frequency
$\bar{F}_{\zeta_-}(r)$	cumulative distribution function of discrete deterministic Rayleigh processes $\bar{\zeta}[k]$
$\bar{F}_{\vartheta}(\varphi)$	cumulative distribution function of the phase $\bar{\vartheta}[k]$ of $\bar{\mu}[k] = \bar{\mu}_1[k] + j\bar{\mu}_2[k]$
$\bar{F}_{\mu_i}(r)$	cumulative distribution function of discrete deterministic Gaussian processes $\bar{\mu}_i[k]$
k	discrete time variable ($t = kT_s$)
K	number of simulated samples of a discrete deterministic process
L	period of $\bar{\zeta}[k]$

$\hat{L}$	upper limit of the period of $\bar{\zeta}[k]$
L_i	period of $\bar{\mu}_i[k]$
$\hat{L}_i$	upper limit of the period of $\bar{\mu}_i[k]$
$L_{i,n}$	period of the nth component of $\bar{\mu}_i[k]$
$\bar{m}_{\mu_i}$	mean value of the sequence $\bar{\mu}_i[k]$
$\boldsymbol{M}_i$	channel matrix; contains the complete information for the reconstruction of $\bar{\mu}_i[k]$
$\bar{N}_\zeta(r)$	level-crossing rate of discrete deterministic Rayleigh processes $\bar{\zeta}[k]$
$\bar{p}_\zeta(z)$	probability density function of discrete deterministic Rayleigh processes $\bar{\zeta}[k]$
$\bar{p}_\vartheta(\theta)$	probability density function of the phase $\bar{\vartheta}[k]$ of $\bar{\mu}[k] = \bar{\mu}_1[k] + j\bar{\mu}_2[k]$
$\bar{p}_{\mu_i}(x)$	probability density function of discrete deterministic Gaussian processes $\bar{\mu}_i[k]$
$\text{Reg}_{i,n}$	register; contains one period of the harmonic elementary sequence $\bar{\mu}_{i,n}[k]$
$\bar{r}_{\mu_i\mu_i}[\kappa]$	autocorrelation sequence of $\bar{\mu}_i[k]$
$\bar{r}_{\mu_1\mu_2}[\kappa]$	cross-correlation sequence of $\bar{\mu}_1[k]$ and $\bar{\mu}_2[k]$
$\boldsymbol{S}_i$	selection matrix
$\bar{S}_{\mu_i\mu_i}(f)$	power spectral density of $\bar{\mu}_i[k]$
$\bar{S}_{\mu_1\mu_2}(f)$	cross-power spectral density of $\bar{\mu}_1[k]$ and $\bar{\mu}_2[k]$
$\text{Tab}_{i,n}$	table; contains one period of the harmonic elementary sequence $\bar{\mu}_{i,n}[k]$
T_s	sampling interval
T_{sim}	simulation time
ΔT_{sim}	iteration time
$\bar{T}_{\zeta_-}(r)$	average duration of fades of discrete deterministic Rayleigh processes $\bar{\zeta}[k]$
$\bar{\beta}_i$	negative curvature of the autocorrelation sequence $\bar{r}_{\mu_i\mu_i}[\kappa]$ at the origin, i.e., $\bar{\beta}_i = -\ddot{\bar{r}}_{\mu_i\mu_i}[0] \quad (i = 1, 2)$
$\Delta\bar{\beta}_i$	model error of $\ddot{\bar{r}}_{\mu_i\mu_i}[0]$, i.e., $\Delta\bar{\beta}_i = \bar{\beta}_i - \beta$
$\Delta_{n,m}^{(i,j)}$	auxiliary function for the determination of the minimum sampling frequency $f_{A,min}$
$\varepsilon_{\bar{f}_{i,n}}$	relative error of the quantized Doppler frequencies $\bar{f}_{i,n}$
$\bar{\zeta}[k]$	discrete-time deterministic Rayleigh process
$\bar{\theta}_{i,n}$	quantized Doppler phase of the nth component of $\bar{\mu}_i[k]$
$\bar{\vartheta}[k]$	phase of $\bar{\mu}[k] = \bar{\mu}_1[k] + j\bar{\mu}_2[k]$, i.e., $\bar{\vartheta}[k] = \arg\{\bar{\mu}[k]\}$
κ	time difference between the instants k_2 and k_1, i.e., $\kappa = k_2 - k_1$
$\bar{\mu}[k]$	complex discrete-time deterministic Gaussian process

$\bar{\mu}_i[k]$	real discrete-time deterministic Gaussian process
$\bar{\mu}_{i,n}[k]$	nth harmonic elementary function of $\bar{\mu}_i[k]$
$\bar{\sigma}^2_{\mu_i}$	mean power of $\bar{\mu}_i[k]$

CONTINUOUS-TIME DETERMINISTIC PROCESSES

$\tilde{a}_\ell$	delay coefficient of the ℓth path
$\tilde{B}_C$	coherence bandwidth of DGUS models
$\tilde{B}^{(1)}_{\mu_i\mu_i}$	average Doppler shift of $\tilde{\mu}_i(t)$
$\tilde{B}^{(2)}_{\mu_i\mu_i}$	Doppler spread of $\tilde{\mu}_i(t)$
$\tilde{B}^{(1)}_{\tau'\tau'}$	average delay of DGUS models
$\tilde{B}^{(2)}_{\tau'\tau'}$	delay spread of DGUS models
$c_{i,n}$	Doppler coefficient of the nth component of $\tilde{\mu}_i(t)$
$c_{i,n,\ell}$	Doppler coefficient of the nth component of $\tilde{\mu}_{i,\ell}(t)$
$E_{p_{\mu_i}}$	mean-square error of $\tilde{p}_{\mu_i}(x)$
$E_{r_{\mu_i\mu_i}}$	mean-square error of $\tilde{r}_{\mu_i\mu_i}(\tau)$
$f_{i,n}$	discrete Doppler frequency of the nth component of $\tilde{\mu}_i(t)$
$f_{i,n,\ell}$	discrete Doppler frequency of the nth component of $\tilde{\mu}_{i,\ell}(t)$
F_i	greatest common divisor $f_{i,1}, f_{i,2}, \ldots, f_{i,N_i}$, i.e., $F_i = \gcd\{f_{i,n}\}_{n=1}^{N_i}$
$\tilde{F}_{\zeta_-}(r)$	cumulative distribution function of deterministic Rayleigh processes $\tilde{\zeta}(t)$
$\tilde{F}_{\eta_-}(r)$	cumulative distribution function of deterministic Suzuki processes $\tilde{\eta}(t)$
$\tilde{F}_{\vartheta}(\varphi)$	cumulative distribution function of the phase $\tilde{\vartheta}(t)$ of $\tilde{\mu}(t) = \tilde{\mu}_1(t) + j\tilde{\mu}_2(t)$
$\tilde{F}_{\mu_i}(r)$	cumulative distribution function of deterministic Gaussian processes $\tilde{\mu}_i(t)$
$\tilde{F}_{\xi_-}(r)$	cumulative distribution function of deterministic Rice processes $\tilde{\xi}(t)$
$\tilde{F}_{\varrho_-}(r)$	cumulative distribution function of deterministic Loo processes $\tilde{\varrho}(t)$
$\tilde{h}(\tau')$	time-invariant impulse response of DGUS models
$\tilde{h}(\tau', t)$	time-variant impulse response of DGUS models
$\tilde{H}(f', t)$	time-variant transfer function of DGUS models
$\tilde{m}_{\mu_i}$	mean value of $\tilde{\mu}_i(t)$
N	smallest number of N_1 and N_2, i.e., $N = \min\{N_1, N_2\}$
N_s	number of sampling values
N_i	number of harmonic functions of $\tilde{\mu}_i(t)$

$N_{i,\ell}$	number of harmonic functions of $\tilde{\mu}_{i,\ell}(t)$
N_i'	virtual number of harmonic functions of $\tilde{\mu}_i(t)$
$\tilde{N}_\zeta(r)$	level-crossing rate of deterministic Rayleigh processes $\tilde{\zeta}(t)$
$\tilde{N}_\eta(r)$	level-crossing rate of deterministic Suzuki processes $\tilde{\eta}(t)$
$\tilde{N}_\xi(r)$	level-crossing rate of deterministic Rice processes $\tilde{\xi}(t)$
$\tilde{N}_\varrho(r)$	level-crossing rate of deterministic Loo processes $\tilde{\varrho}(t)$
$\tilde{p}_{0_-}(\tau_-;r)$	probability density function of the fading intervals τ_- of $\tilde{\zeta}(t)$
$\tilde{p}_{0_{-+}}(\tau_-,\tau_+;r)$	joint probability density function of fading and connecting intervals of $\tilde{\zeta}(t)$
$\tilde{p}_{1_-}(\tau_-;r)$	approximate solution for $\tilde{p}_{0_-}(\tau_-;r)$
$\tilde{p}_\zeta(z)$	probability density function of deterministic Rayleigh processes $\tilde{\zeta}(t)$
$\tilde{p}_\eta(z)$	probability density function of deterministic Suzuki processes $\tilde{\eta}(t)$
$\tilde{p}_\vartheta(\theta)$	probability density function of the phase $\tilde{\vartheta}(t)$ of $\tilde{\mu}(t)=\tilde{\mu}_1(t)+j\tilde{\mu}_2(t)$
$\tilde{p}_{\mu_i}(x)$	probability density function of deterministic Gaussian processes $\tilde{\mu}_i(t)$
$\tilde{p}_\xi(z)$	probability density function of deterministic Rice processes $\tilde{\xi}(t)$
$\tilde{p}_\varrho(z)$	probability density function of deterministic Loo processes $\tilde{\varrho}(t)$
$\tilde{p}_{\xi\dot{\xi}}(z,\theta)$	joint probability density function of $\tilde{\xi}(t)$ and $\dot{\tilde{\xi}}(t)$
$\tilde{r}_{hh}(\cdot,\cdot;\cdot,\cdot)$	autocorrelation function $\tilde{h}(\tau',t)$
$\tilde{r}_{HH}(\upsilon',\tau)$	time-frequency correlation function of DGUS models
$\tilde{r}_{ss}(\cdot,\cdot;\cdot,\cdot)$	autocorrelation function of $\tilde{s}(\tau',f)$
$\tilde{r}_{TT}(\cdot,\cdot;\cdot,\cdot)$	autocorrelation function of $\tilde{T}(f',f)$
$\tilde{r}_{\mu\mu}(\tau)$	autocorrelation function of $\tilde{\mu}(t)=\tilde{\mu}_1(t)+j\tilde{\mu}_2(t)$
$\tilde{r}_{\mu_i\mu_i}(\tau)$	autocorrelation function of $\tilde{\mu}_i(t)$
$\tilde{r}_{\mu_{i,\ell}\mu_{i,\ell}}(\tau)$	autocorrelation function of $\tilde{\mu}_{i,\ell}(t)$
$\tilde{r}_{\mu_\ell\mu_\ell}(\tau)$	autocorrelation function of $\tilde{\mu}_\ell(t)$
$\tilde{r}_{\mu_1\mu_2}(\tau)$	cross-correlation function of $\tilde{\mu}_1(t)$ and $\tilde{\mu}_2(t)$
$\tilde{r}_{\tau'\tau'}(\upsilon')$	frequency correlation function of DGUS models
$\tilde{s}(\tau',f)$	Doppler-variant impulse response of DGUS models
$\tilde{S}(\tau',f)$	scattering function of DGUS models
$\tilde{S}_{hh}(\tau',\tau)$	delay cross-power spectral density of DGUS models
$\tilde{S}_{TT}(\upsilon',f)$	Doppler cross-power spectral density of DGUS models
$\tilde{S}_{\mu_i\mu_i}(f)$	power spectral density of $\tilde{\mu}_i(t)$
$\tilde{S}_{\mu_\ell\mu_\ell}(f)$	power spectral density of $\tilde{\mu}_\ell(t)$
$\tilde{S}_{\mu_1\mu_2}(f)$	cross-power spectral density of $\tilde{\mu}_1(t)$ and $\tilde{\mu}_2(t)$
$\tilde{S}_{\tau'\tau'}(\tau')$	delay power spectral density of DGUS models

$\tilde{T}(f', f)$	Doppler-variant transfer function of DGUS models
$\tilde{T}_C$	coherence time of DGUS models
T_i	period of $\tilde{\mu}_i(t)$
T_s, T_s'	sampling intervals
T_{sim}	simulation time
$\tilde{T}_{\zeta_-}(r)$	average duration of fades of deterministic Rayleigh processes $\tilde{\zeta}(t)$
$\tilde{T}_{\eta_-}(r)$	average duration of fades of deterministic Suzuki processes $\tilde{\eta}(t)$
$\tilde{T}_{\xi_-}(r)$	average duration of fades of deterministic Rice processes $\tilde{\xi}(t)$
$\tilde{T}_{\varrho_-}(r)$	average duration of fades of deterministic Loo processes $\tilde{\varrho}(t)$
$\tilde{\beta}_i$	negative curvature of the autocorrelation function $\tilde{r}_{\mu_i\mu_i}(\tau)$ at the origin, i.e., $\tilde{\beta}_i = -\ddot{\tilde{r}}_{\mu_i\mu_i}(0) \quad (i = 1, 2)$
$\Delta\beta_i$	model error of the simulation model, i.e., $\Delta\beta_i = \tilde{\beta}_i - \beta$
$\tilde{\gamma}$	negative curvature of the autocorrelation function of $\tilde{r}_{\nu_3\nu_3}(\tau)$ at the origin, i.e., $\tilde{\gamma} = -\ddot{\tilde{r}}_{\nu_3\nu_3}(0)$
ε_{N_ξ}	relative error of the level-crossing rate $\tilde{N}_\xi(r)$
$\varepsilon_{T_{\xi_-}}$	relative error of the average duration of fades $\tilde{T}_{\xi_-}(r)$
$\tilde{\zeta}(t)$	continuous-time deterministic Rayleigh process
$\tilde{\eta}(t)$	continuous-time deterministic Suzuki process
θ_0	phase difference between $\tilde{\mu}_{1,n}(t)$ and $\tilde{\mu}_{2,n}(t)$
$\theta_{i,n}$	Doppler phase of the nth component of $\tilde{\mu}_i(t)$
$\theta_{i,n,\ell}$	Doppler phase der nth component of $\tilde{\mu}_{i,\ell}(t)$
$\vec{\theta}_i$	Doppler phase vector
$\vec{\Theta}_i$	standard phase vector
$\tilde{\vartheta}(t)$	phase of $\tilde{\mu}_\rho(t)$, i.e., $\tilde{\vartheta}(t) = \arg\{\tilde{\mu}_\rho(t)\}$
$\tilde{\lambda}(t)$	continuous-time deterministic lognormal process
$\tilde{\mu}(t)$	zero-mean complex continuous-time deterministic Gaussian process
$\tilde{\mu}_i(t)$	zero-mean real continuous-time deterministic Gaussian process
$\tilde{\mu}_{i,\ell}(t)$	real deterministic Gaussian process of the ℓth path of DGUS models
$\tilde{\mu}_{i,n}(t)$	nth harmonic elementary function of $\tilde{\mu}_i(t)$
$\tilde{\mu}_\ell(t)$	complex deterministic Gaussian process of the ℓth path of DGUS models
$\tilde{\mu}_\rho(t)$	complex deterministic Gaussian process with mean value $m(t)$
$\tilde{\xi}(t)$	continuous-time deterministic Rice process
$\tilde{\varrho}(t)$	continuous-time deterministic Loo process
$\tilde{\sigma}_\mu^2$	mean power of $\tilde{\mu}(t)$

$\tilde{\sigma}^2_{\mu_i}$	mean power of $\tilde{\mu}_i(t)$
$\tilde{\tau}'_\ell$	discrete propagation delay of the ℓth path
$\Delta\tilde{\tau}'_\ell$	propagation delay difference between $\tilde{\tau}'_\ell$ and $\tilde{\tau}'_{\ell-1}$, i.e., $\Delta\tilde{\tau}'_\ell = \tilde{\tau}'_\ell - \tilde{\tau}'_{\ell-1}$
$\tilde{\tau}_q(r)$	length of the time interval that comprises q % of all fading intervals of the process $\tilde{\zeta}(t)$ at level r
$\tilde{\phi}_0$	symbol for the cross-correlation function $\tilde{r}_{\mu_1\mu_2}(\tau)$ at $\tau = 0$
$\tilde{\psi}_0$	symbol for the autocorrelation function $\tilde{r}_{\mu_i\mu_i}(\tau)$ at $\tau = 0$
$\tilde{\Xi}_\ell(f)$	Fourier transform of $\tilde{\mu}_\ell(t)$
$\tilde{\Psi}_{\mu_i}(\nu)$	characteristic function of $\tilde{\mu}_i(t)$
$\Omega_{i,n}$	normalized discrete Doppler frequency, i.e., $\Omega_{i,n} = 2\pi f_{i,n} T_s$

Bibliography

[Abr72] M. Abramowitz and I. A. Stegun, *Handbook of Mathematical Functions with Formulas, Graphs, and Mathematical Tables.* Washington: National Bureau of Standards, 1972.

[Aka97] Y. Akaiwa, *Introduction to Digital Mobile Communication.* New York: John Wiley & Sons, 1997.

[Akk86] A. S. Akki and F. Haber, "A statistical model of mobile-to-mobile land communication channel," *IEEE Trans. Veh. Technol.*, vol. 35, no. 1, pp. 2–7, Feb. 1986.

[Akk94] A. S. Akki, "Statistical properties of mobile-to-mobile land communication channels," *IEEE Trans. Veh. Technol.*, vol. 43, no. 4, pp. 826–831, Nov. 1994.

[Ald82] M. Aldinger, "Die Simulation des Mobilfunk-Kanals auf einem Digitalrechner," *FREQUENZ*, vol. 36, no. 4/5, pp. 145–152, 1982.

[And92] J. B. Andersen and P. Eggers, "A heuristic model of power delay profiles in landmobile communications," in *Proc. URSI Int. Symp. Electromagnetic Theory*, Sydney, Australia, Aug. 1992, pp. 55–57.

[And95] J. B. Andersen, T. S. Rappaport, and S. Yoshida, "Propagation measurements and models for wireless communications channels," *IEEE Commun. Mag.*, vol. 33, no. 1, pp. 42–49, Jan. 1995.

[Arr73] G. Arredondo, W. Chriss, and E. Walker, "A multipath fading simulator for mobile radio," *IEEE Trans. Veh. Technol.*, vol. 22, no. 4, pp. 241–244, May 1973.

[Aul79] T. Aulin, "A modified model for the fading signal at the mobile radio channel," *IEEE Trans. Veh. Technol.*, vol. 28, no. 3, pp. 182–203, Aug. 1979.

[Baj82] A. S. Bajwa and J. D. Parsons, "Small-area characterisation of UHF urban and suburban mobile radio propagation," *Inst. Elec. Eng. Proc.*, vol. 129, no. 2, pp. 102–109, April 1982.

[Bei97] F. Beichelt, *Stochastische Prozesse für Ingenieure.* Stuttgart: Teubner, 1997.

[Bel63] P. A. Bello, "Characterization of randomly time-variant linear channels," *IEEE Trans. Comm. Syst.*, vol. 11, no. 4, pp. 360–393, Dec. 1963.

[Bel73] P. A. Bello, "Aeronautical channel characterization," *IEEE Trans. Commun.*, vol. 21, pp. 548–563, May 1973.

[Ben48] W. R. Bennett, "Distribution of the sum of randomly phased components," *Quart. Appl. Math.*, vol. 5, pp. 385–393, May 1948.

[Ber86] D. Berthoumieux and J. M. Pertoldi, "Hardware propagation simulator of the frequency-selective fading channel at 900 MHz," in *Proc. 2nd Nordic Seminar on Land Mobile Radio Communications*, Stockholm, Sweden, 1986, pp. 214–217.

[Bla72] D. M. Black and D. O. Reudink, "Some characteristics of mobile radio propagation at 836 MHz in the Philadelphia area," *IEEE Trans. Veh. Technol.*, vol. 21, pp. 45–51, Feb. 1972.

[Bla84] R. E. Blahut, *Theory and Practice of Error Control Codes.* Reading, Massachusetts: Addison-Wesley, 1984.

[Boe98] J. F. Böhme, *Stochastische Signale.* Stuttgart: Teubner, 2nd ed., 1998.

[Bra91] W. R. Braun and U. Dersch, "A physical mobile radio channel model," *IEEE Trans. Veh. Technol.*, vol. 40, no. 2, pp. 472–482, May 1991.

[Bre70] H. Brehm, *Ein- und zweidimensionale Verteilungsdichten von Nulldurchgangsabständen stochastischer Signale.* Ph.D. dissertation, University Frankfurt/Main, Frankfurt, Germany, June 1970.

[Bre78] H. Brehm, *Sphärisch invariante stochastische Prozesse.* Habilitation thesis, University Frankfurt/Main, Frankfurt, Germany, 1978.

[Bre86a] H. Brehm, W. Stammler, and M. Werner, "Design of a highly flexible digital simulator for narrowband fading channels," in *Signal Processing III: Theories and Applications*, Amsterdam, The Netherlands: Elsevier Science Publishers (North-Holland), EURASIP, Sep. 1986, pp. 1113–1116.

[Bre86b] H. Brehm and M. Werner, "Generalized Rayleigh fading in a mobile radio channel," in *Proc. 2nd Nordic Seminar on Digital Land Mobile Radio Communication*, Stockholm, Oct. 1986, pp. 210–214.

[Bre89] H. Brehm, "Pegelkreuzungen bei verallgemeinerten Gauß-Prozessen," *Archiv Elektr. Übertr.*, vol. 43, no. 5, pp. 271–277, 1989.

[Bro91] I. N. Bronstein and K. A. Semendjajew, *Taschenbuch der Mathematik.* Frankfurt/Main: Harri Deutsch, 25th ed., 1991.

[But83] J. S. Butterworth and E. E. Matt, "The characterization of propagation effects for land mobile satellite services," in *Inter. Conf. on Satellite Systems for Mobile Communications and Navigations*, June 1983, pp. 51–54.

[Cap80] E. Caples, K. Massad, and T. Minor, "A UHF channel simulator for digital mobile radio," *IEEE Trans. Veh. Technol.*, vol. 29, no. 2, pp. 281–289, May 1980.

[Cas88] E. F. Casas and C. Leung, "A simple digital fading simulator for mobile radio," in *Proc. IEEE Vehicular Technology Conference*, Sep. 1988, pp. 212–217.

[Cas90] E. F. Casas and C. Leung, "A simple digital fading simulator for mobile radio," *IEEE Trans. Veh. Technol.*, vol. 39, no. 3, pp. 205–212, Aug. 1990.

[Cha79] U. Charash, "Reception through Nakagami fading multipath channels with random delays," *IEEE Trans. Commun.*, vol. 27, no. 4, pp. 657–670, April 1979.

[Cla68] R. H. Clarke, "A statistical theory of mobile-radio reception," *Bell Syst. Tech. Journal*, vol. 47, pp. 957–1000, July/Aug. 1968.

[Cor94] G. E. Corazza and F. Vatalaro, "A statistical model for land mobile satellite channels and its application to nongeostationary orbit systems systems," *IEEE Trans. Veh. Technol.*, vol. 43, no. 3, pp. 738–742, Aug. 1994.

[COS86] COST 207 WG1, "Proposal on channel transfer functions to be used in GSM tests late 1986," *COST 207 TD (86)51 Rev. 3*, Sep. 1986.

[COS89] COST 207, "Digital land mobile radio communications," Office for Official Publications of the European Communities, Final Report, Luxembourg, 1989.

[Cox72] D. C. Cox, "Delay Doppler characteristics of multipath propagation at 910 MHz in a suburban mobile radio environment," *IEEE Trans. Antennas Propagat.*, vol. 20, no. 5, pp. 625–635, Sep. 1972.

[Cox73] D. C. Cox, "910 MHz urban mobile radio propagation: Multipath characteristics in New York City," *IEEE Trans. Veh. Technol.*, vol. 22, no. 4, pp. 104–110, Nov. 1973.

[Cre95] P. M. Crespo and J. Jiménez, "Computer simulation of radio channels using a harmonic decomposition technique," *IEEE Trans. Veh. Technol.*, vol. 44, no. 3, pp. 414–419, Aug. 1995.

[Cyg88] D. Cygan, M. Dippold, and J. Finkenzeller, "Kanalmodelle für die satellitengestützte Kommunikation landmobiler Teilnehmer," *Archiv Elek. Übertr.*, vol. 42, no. 6, pp. 329–339, Nov./Dec. 1988.

[Dav58] W. B. Davenport and W. L. Root, *An Introduction to the Theory of Random Signals and Noise*. New York: McGraw-Hill, 1958.

[Dav70] W. B. Davenport, *Probability and Random Processes*. New York: McGraw-Hill, 1970.

[Dav87] F. Davarian, "Channel simulation to facilitate mobile-satellite communications research," *IEEE Trans. Commun.*, vol. 35, no. 1, pp. 47–56, Jan. 1987.

[Den93] P. Dent, G. E. Bottomley, and T. Croft, "Jakes fading model revisited," *Electronics Letters*, vol. 29, no. 13, pp. 1162–1163, June 1993.

[Der93] U. Dersch and R. J. Rüegg, "Simulations of the time and frequency selective outdoor mobile radio channel," *IEEE Trans. Veh. Technol.*, vol. 42, no. 3, pp. 338–344, Aug. 1993.

[Dup86] J. Dupraz, *Probability, Signals, Noise.* London: North Oxford Academic Publishers, 1986.

[Ehr82] L. Ehrman, L. B. Bates, J. F. Eschle, and J. M. Kates, "Real-time software simulation of the HF radio channel," *IEEE Trans. Commun.*, vol. 30, no. 8, pp. 1809–1817, Aug. 1982.

[Ent76] W. Entenmann, *Optimierungsverfahren.* Heidelberg: Hüthig-Verlag, 1976.

[Ert98] R. B. Ertel, P. Cardieri, K. W. Sowerby, T. S. Rappaport, and J. H. Reed, "Overview of spatial channel models for antenna array communication systems," *IEEE Personal Commun.*, pp. 10–22, Feb. 1998.

[Fec93a] S. A. Fechtel, "A novel approach to modeling and efficient simulation of frequency-selective fading radio channels," *IEEE J. Select. Areas Commun.*, vol. 11, no. 3, pp. 422–431, April 1993.

[Fec93b] S.A. Fechtel, *Verfahren und Algorithmen der robusten Synchronisation für die Datenübertragung über dispersive Schwundkanäle.* Ph.D. dissertation, Rheinisch-Westfälische Technische Hochschule Aachen, Aachen, Germany, 1993.

[Feh95] K. Feher, *Wireless Digital Communications: Modulation and Spread Spectrum Applications.* Upper Saddle River, New Jersey: Prentice-Hall, 1995.

[Fel94] T. Felhauer, *Optimale erwartungstreue Algorithmen zur hochauflösenden Kanalschätzung mit Bandspreizsignalformen.* Düsseldorf: VDI-Verlag, Fortschritt-Berichte, series 10, no. 278, 1994.

[Fet96] A. Fettweis, *Elemente nachrichtentechnischer Systeme.* Stuttgart: Teubner, 2nd ed., 1996.

[Fle63] R. Fletcher and M. J. D. Powell, "A rapidly convergent descent method for minimization," *Computer Journal*, vol. 6, no. 2, pp. 163–168, 1963.

[Fle87] R. Fletcher, *Practical Methods of Optimization.* New York: John Wiley & Sons, 2nd ed., 1987.

[Fle90] B. H. Fleury, *Charakterisierung von Mobil- und Richtfunkkanälen mit schwach stationären Fluktuationen und unkorrelierter Streuung (WSSUS).* Ph.D. dissertation, Swiss Federal Institute of Technology Zurich, Zurich, Switzerland, 1990.

[Fle96] B. H. Fleury and P. E. Leuthold, "Radiowave propagation in mobile communications: An overview of European research," *IEEE Commun. Mag.*, vol. 34, no. 2, pp. 70–81, Feb. 1996.

[Fli91] N. Fliege, *Systemtheorie.* Stuttgart: Teubner, 1991.

[Fon97] F. P. Fontan, J. Pereda, M. J. Sedes, M. A. V. Castro, S. Buonomo, and P. Baptista, "Complex-envelope three-state Markov chain simulator for the LMS channel," *Int. J. Sat. Commun.*, pp. 1–15, Jan. 1997.

[For72] G. D. Forney, "Maximum-likelihood sequence estimation for digital sequences in the presence of intersymbol interference," *IEEE Trans. Inform. Theory*, vol. 18, pp. 363–378, May 1972.

[Gan72] M. J. Gans, "A power-spectral theory of propagation in the mobile-radio environment," *IEEE Trans. Veh. Technol.*, vol. 21, no. 1, pp. 27–38, Feb. 1972.

[Gel82] H. J. Gelbrich, K. Löw, and R. W. Lorenz, "Funkkanalsimulation und Bitfehler-Strukturmessungen an einem digitalen Kanal," *FREQUENZ*, vol. 36, no. 4/5, pp. 130–138, 1982.

[Gib96] J. D. Gibson, Ed., *The Mobile Communications Handbook.* CRC Press in Cooperation with IEEE Press, 1996.

[Goe92a] M. Göller and K.D. Masur, "Ergebnisse von Funkkanalmessungen im 900 MHz Bereich auf Neubaustrecken der Deutschen Bundesbahn," *Nachrichtentechnik–Elektronik*, vol. 42, no. 4, pp. 146–149, 1992.

[Goe92b] M. Göller and K. D. Masur, "Ergebnisse von Funkkanalmessungen im 900 MHz Bereich auf Neubaustrecken der Deutschen Bundesbahn," *Nachrichtentechnik–Elektronik*, vol. 42, no. 5, pp. 206–210, 1992.

[Gol89] J. Goldhirsh and W. J. Vogel, "Mobile satellite system fade statistics for shadowing and multipath from roadside trees at UHF on L-band," *IEEE Trans. Antennas Propagat.*, vol. 37, no. 4, pp. 489–498, April 1989.

[Gol92] J. Goldhirsh and W. J. Vogel, "Propagation effects for land mobile satellite systems: Overview of experimental and modeling results," *NASA Reference Publication 1274*, Feb. 1992.

[Gol96] G. H. Golub and C. F. van Loan, *Matrix Computations.* Baltimore: The Johns Hopkins University Press, 3rd ed., 1996.

[Gra81] I. S. Gradstein and I. M. Ryshik, *Tables of Series, Products, and Integrals.* Frankfurt: Harri Deutsch, 5th ed., vol. I and II, 1981.

[Gre78] L. J. Greenstein, "A multipath fading channel model for terrestrial digital radio systems," *IEEE Trans. Commun.*, vol. 26, no. 8, pp. 1247–1250, 1978.

[Gro97] C. Großmann and J. Terno, *Numerik der Optimierung.* Stuttgart: Teubner, 2nd ed., 1997.

[Hae88] R. Häb, *Kohärenter Empfang bei Datenübertragung über nichtfrequenzselektive Schwundkanäle.* Ph.D. dissertation, Rheinisch-Westfälische Technische Hochschule Aachen, Aachen, Germany, 1988.

[Hae97] E. Hänsler, *Statistische Signale.* Berlin: Springer, 1997.

[Hag82] J. Hagenauer and W. Papke, "Der gespeicherte Kanal — Erfahrungen mit einem Simulationsverfahren für Fading-Kanäle," *FREQUENZ*, vol. 36, no. 4/5, pp. 122–129, 1982.

[Han77] F. Hansen and F. I. Meno, "Mobile fading — Rayleigh and lognormal superimposed," *IEEE Trans. Veh. Technol.*, vol. 26, no. 4, pp. 332–335, Nov. 1977.

[Hed99] R. Heddergott and P. Truffer, "Comparison of high resolution channel parameter measurements with ray tracing simulations in a multipath environment," in *Proc. 3rd European Personal Mobile Communications Conference*, EPMCC'99, Paris, France, March 1999, pp. 167–172.

[Hes80] G. C. Hess, "Land-mobile satellite excess path loss measurements," *IEEE Trans. Veh. Technol.*, vol. 29, no. 2, pp. 290–297, May 1990.

[Hes93] G. C. Hess, *Land-Mobile Radio System Engineering.* Boston, MA: Artech House, 1993.

[Hoe90] P. Höher, *Kohärenter Empfang trelliscodierter PSK-Signale auf frequenzselektiven Mobilfunkkanälen — Entzerrung, Decodierung und Kanalparameterschätzung.* Düsseldorf: VDI-Verlag, Fortschritt-Berichte, series 10, no. 147, 1990.

[Hoe92] P. Höher, "A statistical discrete-time model for the WSSUS multipath channel," *IEEE Trans. Veh. Technol.*, vol. 41, no. 4, pp. 461–468, Nov. 1992.

[Hor85] R. A. Horn and C. R. Johnson, *Matrix Analysis.* New York: Cambridge University Press, 1985.

[Huc83] R. W. Huck, J. S. Butterworth, and E. E. Matt, "Propagation measurements for land mobile satellite services," in *Proc. IEEE 33rd Vehicular Technology Conference*, Toronto, Canada, 1983, pp. 265–268.

[Jah95] A. Jahn, "Propagation data and channel model for LMS systems," *ESA Purchase Order 141742*, Final Report, DLR, Institute for Communications Technology, Jan. 1995.

[Jak93] W. C. Jakes, Ed., *Microwave Mobile Communications*. Piscataway, NJ: IEEE Press, 1993.

[Joh94] G. E. Johnson, "Constructions of particular random processes," *Proc. of the IEEE*, vol. 82, no. 2, pp. 270–285, Feb. 1994.

[Jun97] P. Jung, *Analyse und Entwurf digitaler Mobilfunksysteme*. Stuttgart: Teubner, 1997.

[Kad91] G. Kadel and R. W. Lorenz, "Breitbandige Ausbreitungsmessungen zur Charakterisierung des Funkkanals beim GSM-System," *FREQUENZ*, vol. 45, no. 7/8, pp. 158–163, 1991.

[Kad92] G. Kadel and R. W. Lorenz, "Wideband propagation measurements of the mobile radio channel," in *Proc. ISAP-92*, Sappore, Japan, 1992, pp. 81–84.

[Kai80] T. Kailath, *Linear Systems*. Englewood Cliffs, New Jersey: Prentice-Hall, 1980.

[Kam96] K. D. Kammeyer, *Nachrichtenübertragung*. Stuttgart: Teubner, 2nd ed., 1996.

[Kam98] K. D. Kammeyer and K. Kroschel, *Digitale Signalverarbeitung*. Stuttgart: Teubner, 4th ed., 1998.

[Kat95] R. Kattenbach and H. Früchting, "Calculation of system and correlation functions for WSSUS channels from wideband measurements," *FREQUENZ*, vol. 49, no. 3/4, pp. 42–47, 1995.

[Kit82] L. Kittel, "Analoge und diskrete Kanalmodelle für die Signalübertragung beim beweglichen Funk," *FREQUENZ*, vol. 36, no. 4/5, pp. 153–160, 1982.

[Kra88] A. Krantzik and D. Wolf, "Simulation and analysis of Suzuki fading processes," in *Proc. of the 1988 IEEE Int. Conf. on Acoustics, Speech, and Signal Processing*, New York, USA, 1988, pp. 2184–2187.

[Kra90a] A. Krantzik and D. Wolf, "Distribution of the fading-intervals of modified Suzuki processes," in *Signal Processing V: Theories and Applications*, L. Torres, E. Masgrau, and M. A. Lagunas, Eds., Amsterdam, The Netherlands: Elsevier Science Publishers, B.V, 1990, pp. 361–364.

[Kra90b] A. Krantzik and D. Wolf, "Statistische Eigenschaften von Fadingprozessen zur Beschreibung eines Landmobilfunkkanals," *FREQUENZ*, vol. 44, no. 6, pp. 174–182, June 1990.

[Kub00] E. Kubista, F. P. Fontan, M. A. V. Castro, S. Buonomo, B. R. Arbesser-Rastburg, and J. P. V. P. Baptista, "Ka-band propagation measurements and statistics for land mobile satellite applications," *IEEE Trans. Veh. Technol.*, vol. 49, no. 3, pp. 973–983, May 2000.

[Kuc82] H. P. Kuchenbecker, "Statistische Eigenschaften von Schwund- und Verbindungsdauer beim Mobilfunk-Kanal," *FREQUENZ*, vol. 36, no. 4/5, pp. 138–144, 1982.

[Lam97] U. Lambrette, S. Fechtel, and H. Meyer, "A frequency domain variable data rate frequency hopping channel model for the mobile radio channel," in *Proc. IEEE 47th Veh. Technol. Conf., VTC'97*, Phoenix, Arizona, USA, May 1997.

[Lau94] D. I. Laurenson and G. J. R. Povey, "Channel modelling for a predictive rake receiver system," in *Proc. 5th IEEE Int. Symp. Personal, Indoor and Mobile Radio Commun., PIMRC'94*, The Hague, The Netherlands, Sep. 1994, pp. 715–719.

[Lee82] W. C. Y. Lee, *Mobile Communications Engineering*. New York: McGraw-Hill, 1982.

[Lee93] W. C. Y. Lee, *Mobile Communications Design Fundamentals*. New York: John Wiley & Sons, 2nd ed., 1993.

[Lee95] W. C. Y. Lee, *Mobile Cellular Telecommunications: Analog and Digital Systems*. New York: McGraw-Hill, 2nd ed., 1995.

[Lib99] J. C. Liberty and T. S. Rappaport, *Smart Antennas for Wireless Communications: IS-95 and Third Generation CDMA Applications*. Upper Saddle River, New Jersey: Prentice-Hall, 1999.

[Lon62] M. S. Longuet-Higgins, "The distribution of intervals between zeros of a stationary random function," *Phil. Trans. Royal. Soc.*, vol. A 254, pp. 557–599, 1962.

[Loo85] C. Loo, "A statistical model for a land mobile satellite link," *IEEE Trans. Veh. Technol.*, vol. 34, no. 3, pp. 122–127, Aug. 1985.

[Loo87] C. Loo, "Measurements and models of a land mobile satellite channel and their applications to MSK signals," *IEEE Trans. Veh. Technol.*, vol. 35, no. 3, pp. 114–121, Aug. 1987.

[Loo90] C. Loo, "Digital transmission through a land mobile satellite channel," *IEEE Trans. Commun.*, vol. 38, no. 5, pp. 693–697, May 1990.

[Loo91] C. Loo and N. Secord, "Computer models for fading channels with applications to digital transmission," *IEEE Trans. Veh. Technol.*, vol. 40, no. 4, pp. 700–707, Nov. 1991.

[Loo96] C. Loo, "Statistical models for land mobile and fixed satellite communications at Ka band," in *Proc. IEEE 46th Veh. Technol. Conf., VTC'96*, Atlanta, Georgia, USA, April/May 1996, pp. 1023–1027.

[Loo98] C. Loo and J. S. Butterworth, "Land mobile satellite channel measurements and modeling," *Proc. of the IEEE*, vol. 86, no. 7, pp. 1442–1463, July 1998.

[Lor79] R. W. Lorenz, "Theoretische Verteilungsfunktionen von Mehrwegeschwundprozessen im beweglichen Funk und die Bestimmung ihrer Parameter aus Messungen," *Technischer Bericht des Forschungsinstituts der DBP beim FTZ, FI 455 TBr 66*, March 1979.

[Lor85] R. W. Lorenz, "Zeit- und Frequenzabhängigkeit der Übertragunsfunktion eines Funkkanals bei Mehrwegeausbreitung mit besonderer Berücksichtigung des Mobilfunkkanals," *Der Fernmelde-Ingenieur*, Verlag für Wissenschaft und Leben Georg Heidecker, vol. 39, no. 4, April 1985.

[Lor86] R. W. Lorenz, "Modell und Simulation des Mobilfunkkanals zur Analyse von Signalverzerrungen durch frequenzselektiven Schwund," *FREQUENZ*, vol. 40, no. 9/10, pp. 241–248, 1986.

[Lue90] H. D. Lüke, *Signalübertragung — Grundlagen der digitalen und analogen Nachrichtensysteme.* Berlin: Springer, 1990.

[Lut85] E. Lutz and E. Plöchinger, "Generating Rice processes with given spectral properties," *IEEE Trans. Veh. Technol.*, vol. 34, no. 4, pp. 178–181, Nov. 1985.

[Lut91] E. Lutz, D. Cygan, M. Dippold, F. Dolainsky, and W. Papke, "The land mobile satellite communication channel — Recording, statistics, and channel model," *IEEE Trans. Veh. Technol.*, vol. 40, no. 2, pp. 375–386, May 1991.

[Lut96] E. Lutz, "A Markov model for correlated land mobile satellite channels," *Int. J. Sat. Commun.*, vol. 13, pp. 333–339, 1996.

[Man95] Mannesmann Mobilfunk, *Geschäftsbericht.* Düsseldorf, Germany, 1995.

[Mar92] U. Martin, "Ein System zur Messung der Eigenschaften von Mobilfunkkanälen und ein Verfahren zur Nachverarbeitung der Meßdaten," *FREQUENZ*, vol. 46, no. 7/8, pp. 178–188, 1992.

[Mar94a] U. Martin, *Ausbreitung in Mobilfunkkanälen: Beiträge zum Entwurf von Meßgeräten und zur Echoschätzung.* Ph.D. dissertation, University Erlangen–Nuremberg, Erlangen, Germany, 1994.

[Mar94b] U. Martin, "Modeling the mobile radio channel by echo estimation," *FREQUENZ*, vol. 48, no. 9/10, pp. 198–212, Sep./Oct. 1994.

[Mar99] U. Martin, J. Fuhl, I. Gaspard, M. Haardt, A. Kuchar, C. Math, A. F. Molisch, and R. Thomä, "Model scenarios for direction-selective adaptive antennas in cellular mobile communication systems — Scanning the literature," *Wireless Personal Communications, Special Issue on Space Division Multiple Access*, Kluwer Academic Publishers, pp. 109–129, Oct. 1999.

[McF56] J. A. McFadden, "The axis-crossing intervals of random functions I," *Trans. Inst. Rad. Eng.*, vol. 2, pp. 146–150, 1956.

[McF58] J. A. McFadden, "The axis-crossing intervals of random functions II," *Trans. Inst. Rad. Eng.*, vol. 4, pp. 14–24, 1958.

[Meh94] A. Mehrotra, *Cellular Radio Performance Engineering.* Boston, MA: Artech House, 1994.

[Mey95] H. Mey, "Chancen der europäischen Industrie im Bereich mobiler Endgeräte," in *Proc. 2. ITG-Fachtagung Mobile Kommunikation'95*, Neu-Ulm, Germany, Sep. 1995, p. 11.

[Mid60] D. Middleton, *An Introduction to Statistical Communication Theory.* New York: McGraw-Hill, 1960.

[Mil95] M. J. Miller, B. Vucetic, and L. Berry, Eds., *Satellite Communications: Mobile and Fixed Services.* Boston, MA: Kluwer Academic Publishers, 3rd ed., 1995.

[Mun82] T. Munakata and D. Wolf, "A novel approach to the level-crossing problem of random processes," in *Proc. of the 1982 IEEE Int. Symp. on Inf. Theory*, Les Arcs, France, 1982, vol. IEEE-Cat. 82 CH 1767-3 IT, pp. 149–150.

[Mun83] T. Munakata and D. Wolf, "On the distribution of the level-crossing time-intervals of random processes," in *Proc. of the 7th Int. Conf. on Noise in Physical Systems*, Montpelier, USA, M. Savelli, G. Lecoy, and J. P. Nougier, Eds., North-Holland Publ. Co., Amsterdam, The Netherlands, 1983, pp. 49–52.

[Mun86] T. Munakata, *Mehr-Zustände-Modelle zur Beschreibung des Pegelkreuzungsverhaltens stationärer stochastischer Prozesse.* Ph.D. dissertation, University Frankfurt/Main, Frankfurt, Germany, March 1986.

[Nak60] M. Nakagami, "The *m*-distribution: A general formula of intensity distribution of rapid fading," in *Statistical Methods in Radio Wave Propagation*, W. G. Hoffman, Ed., Oxford, UK: Pergamon Press, 1960.

[Neu87] A. Neul, J. Hagenauer, W. Papke, F. Dolainsky, and F. Edbauer, "Aeronautical channel characterization based on measurement flights," *IEEE Conf. GLOBECOM'87*, Tokyo, Japan, pp. 1654–1659, Nov. 1987.

[Neu89] A. Neul, *Modulation und Codierung im aeronautischen Satellitenkanal.* Ph.D. dissertation, University of the Federal Armed Forces Munich, Munich, Germany, Sep. 1989.

[Nie78] D. Nielson, "Microwave propagation measurements for mobile digital radio application," *IEEE Trans. Veh. Technol.*, vol. 27, no. 3, pp. 117–131, Aug. 1978.

[Nie92] M. J. J. Nielen, "UMTS: A third generation mobile system," in *Proc. 3rd IEEE Int. Symp. Personal, Indoor and Mobile Radio Commun., PIMRC'92*, Boston, Massachusetts, USA, 1992, pp. 17–21.

[Nyl68] H. W. Nylund, "Characteristics of small-area signal fading on mobile circuits in the 150 MHz band," *IEEE Trans. Veh. Technol.*, vol. 17, pp. 24–30, Oct. 1968.

[Oht80] K. Ohtani and H. Omori, "Distribution of burst error lengths in Rayleigh fading radio channels," *Electronics Letters*, vol. 16, no. 23, pp. 889–891, 1980.

[Oht81] K. Ohtani, K. Daikoku, and H. Omori, "Burst error performance encountered in digital land mobile radio channel," *IEEE Trans. Veh. Technol.*, vol. 23, no. 1, pp. 156–160, 1981.

[Oku68] Y. Okumura, E. Ohmori, T. Kawano, and K. Fukuda, "Field strength and its variability in VHF and UHF land mobile radio services," *Rev. Elec. Commun. Lab.*, vol. 16, pp. 825–873, Sep./Oct. 1968.

[Olm99] J. J. Olmos, A. Gelonch, F. J. Casadevall, and G. Femenias, "Design and implementation of a wide-band real-time mobile channel emulator," *IEEE Trans. Veh. Technol.*, vol. 48, no. 3, pp. 746–764, May 1999.

[Opp75] A. V. Oppenheim and R. W. Schafer, *Digital Signal Processing.* Englewood Cliffs, New Jersey: Prentice-Hall, 1975.

[Pad95] J. E. Padgett, C. G. Günther, and T. Hattori, "Overview of wireless personal communications," *IEEE Communication Magazine*, vol. 33, no. 1, pp. 28–41, Jan. 1995.

[Pae94a] M. Pätzold, U. Killat, Y. Shi, and F. Laue, "A discrete simulation model for the WSSUS multipath channel derived from a specified scattering function," in *Proc. 8. Aachener Kolloquium über Mobile Kommunikationssysteme*, RWTH Aachen, Aachen, Germany, March 1994, pp. 347–351.

[Pae94b] M. Pätzold, U. Killat, and F. Laue, "A deterministic model for a shadowed Rayleigh land mobile radio channel," in *Proc. 5th IEEE Int. Symp. Personal, Indoor and Mobile Radio Commun., PIMRC'94*, The Hague, The Netherlands, Sep. 1994, pp. 1202–1210.

[Pae95a] M. Pätzold, U. Killat, and F. Laue, "Ein erweitertes Suzukimodell für den Satellitenmobilfunkkanal," in *Proc. 40. Internationales Wissenschaftliches Kolloquium*, Technical University Ilmenau, Ilmenau, Germany, Sep. 1995, vol. I, pp. 321–328.

[Pae95b] M. Pätzold, U. Killat, and F. Laue, "A new deterministic simulation model for WSSUS multipath fading channels," in *Proc. 2. ITG-Fachtagung Mobile Kommunikation '95*, Neu-Ulm, Germany, Sep. 1995, pp. 301–312.

[Pae96a] M. Pätzold, U. Killat, Y. Shi, and F. Laue, "A deterministic method for the derivation of a discrete WSSUS multipath fading channel model," *European Trans. Telecommun. and Related Technologies (ETT)*, vol. 7, no. 2, pp. 165–175, March/April 1996.

[Pae96b] M. Pätzold, U. Killat, F. Laue, and Y. Li, "An efficient deterministic simulation model for land mobile satellite channels," in *Proc. IEEE 46th Veh. Technol. Conf., VTC'96*, Atlanta, Georgia, USA, April/May 1996, pp. 1028–1032.

[Pae96c] M. Pätzold, U. Killat, F. Laue, and Y. Li, "A new and optimal method for the derivation of deterministic simulation models for mobile radio channels," in *Proc. IEEE 46th Veh. Technol. Conf., VTC'96*, Atlanta, Georgia, USA, April/May 1996, pp. 1423–1427.

[Pae96d] M. Pätzold, U. Killat, and F. Laue, "A deterministic digital simulation model for Suzuki processes with application to a shadowed Rayleigh land mobile radio channel," *IEEE Trans. Veh. Technol.*, vol. 45, no. 2, pp. 318–331, May 1996.

[Pae96e] M. Pätzold, U. Killat, F. Laue, and Y. Li, "On the problems of Monte Carlo method based simulation models for mobile radio channels," in *Proc. IEEE 4th Int. Symp. on Spread Spectrum Techniques & Applications, ISSSTA'96*, Mayence, Germany, Sep. 1996, pp. 1214–1220.

[Pae97a] M. Pätzold, U. Killat, Y. Li, and F. Laue, "Modeling, analysis, and simulation of nonfrequency-selective mobile radio channels with asymmetrical Doppler power spectral density shapes," *IEEE Trans. Veh. Technol.*, vol. 46, no. 2, pp. 494–507, May 1997.

[Pae97b] M. Pätzold, F. Laue, and U. Killat, "A frequency hopping Rayleigh fading channel simulator with given correlation properties," in *Proc. IEEE Int. Workshop on Intelligent Signal Processing and Communication Systems, ISPACS'97*, Kuala Lumpur, Malaysia, Nov. 1997, pp. S8.1.1–S8.1.6.

[Pae97c] M. Pätzold and F. Laue, "Generalized Rice processes and generalized Suzuki processes for modeling of frequency-nonselective mobile radio channels," unpublished, 1997.

[Pae97d] M. Pätzold and F. Laue, "Fundamental design methods for fading channel models with Gaussian Doppler power spectrum," unpublished, 1997.

[Pae98a] M. Pätzold, *Stochastische und deterministische Modelle zur Modellierung von nichtfrequenzselektiven Mobilfunkkanälen.* Habilitation thesis, Technical University Hamburg-Harburg, Hamburg, Germany, 1998.

[Pae98b] M. Pätzold, U. Killat, F. Laue, and Y. Li, "On the statistical properties of deterministic simulation models for mobile fading channels," *IEEE Trans. Veh. Technol.*, vol. 47, no. 1, pp. 254–269, Feb. 1998.

[Pae98c] M. Pätzold, Y. Li, and F. Laue, "A study of a land mobile satellite channel model with asymmetrical Doppler power spectrum and lognormally distributed line-of-sight component," *IEEE Trans. Veh. Technol.*, vol. 47, no. 1, pp. 297–310, Feb. 1998.

[Pae98d] M. Pätzold, U. Killat, and F. Laue, "An extended Suzuki model for land mobile satellite channels and its statistical properties," *IEEE Trans. Veh. Technol.*, vol. 47, no. 2, pp. 617–630, May 1998.

[Pae98e] M. Pätzold and F. Laue, "Statistical properties of Jakes' fading channel simulator," in *Proc. IEEE 48th Veh. Technol. Conf., VTC'98*, Ottawa, Ontario, Canada, May 1998, pp. 712–718.

[Pae98f] M. Pätzold and R. García, "A new procedure for the design of fast simulation models for Rayleigh fading channels," in *Proc. IEEE Int. Symp. on Wireless Communications*, ISWC'98, Montreal, Quebec, Canada, May 1998, p. 28.

[Pae99a] M. Pätzold, A. Szczepanski, and F. Laue, "Flexible stationäre und nichtstationäre Kanalmodelle für den Satellitenmobilfunkkanal und deren Anpassung an die statistischen Eigenschaften von gemessenen Kanälen," in *Proc. ITG-Diskussionssitzung Meßverfahren im Mobilfunk*, Günzburg, Germany, March 1999, pp. 59–60.

[Pae99b] M. Pätzold and R. García, "Design and performance of fast channel simulators for Rayleigh fading channels," in *Proc. 3rd European Personal Mobile Communications Conference*, EPMCC'99, Paris, France, March 1999, pp. 280–285.

[Pae99c] M. Pätzold and F. Laue, "Level-crossing rate and average duration of fades of deterministic simulation models for Rice fading channels," *IEEE Trans. Veh. Technol.*, vol. 48, no. 4, pp. 1121–1129, July 1999.

[Pae99d] M. Pätzold, *Mobilfunkkanäle — Modellierung, Analyse und Simulation.* Wiesbaden: Vieweg, 1999.

[Pae00a] M. Pätzold, "Perfect channel modeling and simulation of measured wide-band mobile radio channels," in *Proc. 1st International Conference on 3G Mobile Communication Technologies*, IEE 3G2000, London, UK, March 2000, pp. 288–293.

[Pae00b] M. Pätzold, A. Szczepanski, S. Buonomo, and F. Laue, "Modeling and simulation on nonstationary land mobile satellite channels by using extended Suzuki and handover processes," in *Proc. IEEE 51st Veh. Technol. Conf., VTC2000-Spring*, Tokyo, Japan, May 2000, pp. 1787–1792.

[Pae00c] M. Pätzold and A. Szczepanski, "Methods for modeling of specified and measured multipath power delay profiles," in *Proc. IEEE 51st Veh. Technol. Conf., VTC2000-Spring*, Tokyo, Japan, May 2000, pp. 1828–1834.

[Pae00d] M. Pätzold and F. Laue, "The performance of deterministic Rayleigh fading channel simulators with respect to the bit error probability," in *Proc. IEEE 51st Veh. Technol. Conf., VTC2000-Spring*, Tokyo, Japan, May 2000, pp. 1998–2003.

[Pae00e] M. Pätzold, R. García, and F. Laue, "Design of high-speed simulation models for mobile fading channels by using table look-up techniques," *IEEE Trans. Veh. Technol.*, vol. 49, no. 4, pp. 1178–1190, July 2000.

[Pae00f] M. Pätzold, T. Jargstorff, and F. Laue, "A procedure for the design of deterministic spatial channel models," in *Proc. 3rd Int. Symp. on Wireless Personal Multimedia Communications, WPMC'00*, Bangkok, Thailand, Nov. 2000, pp. 452–459.

[Pae00g] M. Pätzold, A. Szczepanski, and N. Youssef, "Methods for modeling of specified and measured multipath power delay profiles," *IEEE Trans. Veh. Technol.*, submitted for publication, 2000.

[Pap77] A. Papoulis, *Signal Analysis.* New York: McGraw-Hill, 1977.

[Pap91] A. Papoulis, *Probability, Random Variables, and Stochastic Processes.* New York: McGraw-Hill, 3rd ed., 1991.

[Par82] J. D. Parsons and A. S. Bajwa, "Wideband characterisation of fading mobile radio channels," *Inst. Elec. Eng. Proc.*, vol. 129, no. 2, pp. 95–101, April 1982.

[Par89] J. D. Parsons and J. G. Gardiner, *Mobile Communication Systems.* Glasgow: Blackie & Son, 1989.

[Par92] J. D. Parsons, *The Mobile Radio Propagation Channel.* London: Pentech Press, 1992.

[Ped00] K. I. Pedersen, P. E. Mogensen, and B. H. Fleury, "A stochastic model of the temporal and azimuthal dispersion seen at the base station in outdoor propagation environments," *IEEE Trans. Veh. Technol.*, vol. 49, no. 2, March 2000.

[Pee93] P. Z. Peebles, *Probability, Random Variables, and Random Signal Principles.* New York: McGraw-Hill, 3rd ed., 1993.

[Poo98] H. V. Poor and G. W. Wornell, Eds., *Wireless Communications: Signal Processing Perspectives.* Upper Saddle River, New Jersey: Prentice-Hall, 1998.

[Pro95] J. Proakis, *Digital Communications.* New York: McGraw-Hill, 3rd ed., 1995.

[Qu99] S. Qu and T. Yeap, "A three-dimensional scattering model for fading channels in land mobile environment," *IEEE Trans. Veh. Technol.*, vol. 48, no. 3, pp. 765–781, May 1999.

[Rab75] L. R. Rabiner and B. Gold, *Theory and Applications of Digital Signal Processing.* Englewood Cliffs, New Jersey: Prentice-Hall, 1975.

[Rai65] A. J. Rainal, "Axis-crossing intervals of Rayleigh processes," *Bell Syst. Tech. J.*, vol. 44, pp. 1219–1224, 1965.

[Rap96] T. S. Rappaport, *Wireless Communications: Principles and Practice.* Upper Saddle River, New Jersey: Prentice-Hall, 1996.

[Red95] S. M. Redl, M. K. Weber, and M. W. Oliphant, *An Introduction to GSM.* Boston, MA: Artech House, 1995.

[Reu72] D. O. Reudink, "Comparison of radio transmission at X-Band frequencies in suburban and urban areas," *IEEE Trans. Ant. Prop.*, vol. 20, pp. 470–473, July 1972.

[Ric44] S. O. Rice, "Mathematical analysis of random noise," *Bell Syst. Tech. J.*, vol. 23, pp. 282–332, July 1944.

[Ric45] S. O. Rice, "Mathematical analysis of random noise," *Bell Syst. Tech. J.*, vol. 24, pp. 46–156, Jan. 1945.

[Ric48] S. O. Rice, "Statistical properties of a sine wave plus random noise," *Bell Syst. Tech. J.*, vol. 27, pp. 109–157, Jan. 1948.

[Ric58] S. O. Rice, "Distribution of the duration of fades in radio transmission: Gaussian noise model," *Bell Syst. Tech. J.*, vol. 37, pp. 581–635, May 1958.

[Sad98] J. S. Sadowsky and V. Kafedziski, "On the correlation and scattering functions of the WSSUS channel for mobile communications," *IEEE Trans. Veh. Technol.*, vol. 47, no. 1, pp. 270–282, Feb. 1998.

[Sch89] H. W. Schüßler, J. Thielecke, K. Preuss, W. Edler, and M. Gerken, "A digital frequency-selective fading simulator," *FREQUENZ*, vol. 43, no. 2, pp. 47–55, 1989.

[Sch90] R. Schwarze, *Ein Systemvorschlag zur Verkehrsinformationsübertragung mittels Rundfunksatelliten.* Ph.D. dissertation, Universität-Gesamthochschule-Paderborn, Paderborn, Germany, 1990.

[Sch91] H. W. Schüßler, *Netzwerke, Signale und Systeme, Bd. 2: Theorie kontinuierlicher und diskreter Systeme.* Berlin: Springer, 1991.

[Schu89] H. Schulze, "Stochastische Modelle und digitale Simulation von Mobilfunkkanälen," in *U.R.S.I/ITG Conf. in Kleinheubach 1988,* Germany (FR), Proc. Kleinheubacher Reports of the German PTT, Darmstadt, Germany, 1989, vol. 32, pp. 473–483.

[Sha88] K. S. Shanmugan and A. Breipohl, *Random Signals: Detection, Estimation, and Data Analysis.* New York: John Wiley & Sons, 1988.

[She77] N. H. Shepherd, "Radio wave loss deviation and shadow loss at 900 MHz," *IEEE Trans. Veh. Technol.*, vol. 26, no. 4, Nov. 1977.

[Ste87] S. Stein, "Fading channel issues in system engineering," *IEEE J. Select. Areas Commun.*, vol. 5, no. 2, pp. 68–89, Feb. 1987.

[Ste94] H. Steffan, "Adaptive generative radio channel models," in *Proc. 5th IEEE Int. Symp. Personal, Indoor and Mobil Radio Commun., PIMRC'94*, The Hague, The Netherlands, Sep. 1994, pp. 268–273.

[Stee94] R. Steele, Ed., *Mobile Radio Communications.* Piscataway NJ: IEEE Press, 1994.

[Stu96] G. L. Stüber, *Principles of Mobile Communications.* Boston, MA: Kluwer Academic Publishers, 1996.

[Suz77] H. Suzuki, "A statistical model for urban radio propagation," *IEEE Trans. Commun.*, vol. 25, no. 7, pp. 673–680, July 1977.

[Tez87] R. Tetzlaff, J. Wehhofer, and D. Wolf, "Simulation and analysis of Rayleigh fading processes," in *Proc. of the 9th Int. Conf. on Noise in Physical Systems*, Montreal, Canada, 1987, pp. 113–116.

[The92] C. W. Therrien, *Discrete Random Signals and Statistical Signal Processing.* Englewood Cliffs, New Jersey: Prentice-Hall, 1992.

[Tho99] R. S. Thomä and U. Martin, "Richtungsaufgelöste Messung von Mobilfunkkanälen," in *Proc. ITG-Diskussionssitzung Meßverfahren im Mobilfunk*, Günzburg, March 1999, pp. 34–36.

[Unb90] R. Unbehauen, *Systemtheorie.* München: R. Oldenbourg Verlag, 5th ed., 1990.

[Vog88] W. J. Vogel and J. Goldhirsh, "Fade measurements at L-band and UHF in mountainous terrain for land mobile satellite systems," *IEEE Trans. Antennas Propagat.*, vol. 36, no. 1, pp. 104–113, Jan. 1988.

[Vog90] W. J. Vogel and J. Goldhirsh, "Mobile satellite system propagation measurements at L-band using MARECS-B2," *IEEE Trans. Antennas Propagat.*, vol. 38, no. 2, pp. 259–264, Feb. 1990.

[Vog95] W. J. Vogel and J. Goldhirsh, "Multipath fading at L band for low elevation angle, land mobile satellite scenarios," *IEEE J. Select. Areas Commun.*, vol. 13, no. 2, pp. 197–204, Feb. 1995.

[Vuc90] B. Vucetic and J. Du, "Channel modeling and simulation in satellite mobile communication systems," in *Proc. Int. Conf. Satel. Mobile Commun.*, Adelaide, Australia, Aug. 1990, pp. 1–6.

[Vuc92] B. Vucetic and J. Du, "Channel modeling and simulation in satellite mobile communication systems," *IEEE J. Select. Areas in Commun.*, vol. 10, no. 8, pp. 1209–1218, Oct. 1992.

[Wer91] M. Werner, *Modellierung und Bewertung von Mobilfunkkanälen.* Ph.D. dissertation, Technical Faculty of the University Erlangen–Nuremberg, Erlangen, Germany, 1991.

[Wit97] M. Wittmann, J. Marti, and T. Kürner, "Impact of the power delay profile shape on the bit error rate in mobile radio systems," *IEEE Trans. Veh. Technol.*, vol. 46, no. 2, pp. 329–339, May 1997.

[Wol83a] D. Wolf, T. Munakata, and J. Wehhofer, "Die Verteilungsdichte der Pegelunterschreitungszeitintervalle bei Rayleigh-Fadingkanälen," *NTG-Fachberichte 84*, pp. 23–32, 1983.

[Wol83b] D. Wolf, T. Munakata, and J. Wehhofer, "Statistical properties of Rice fading processes," in *Signal Processing II: Theories and Applications, Proc. EUSIPCO'83 Second European Signal Processing Conference*, H. W. Schüßler, Erlangen, Ed., Elsevier Science Publishers B.V. (North-Holland), 1983, pp. 17–20.

[Xie00] Y. Xie and Y. Fang, "A general statistical channel model for mobile satellite systems," *IEEE Trans. Veh. Technol.*, vol. 49, no. 3, pp. 744–752, May 2000.

[Yac99] M. D. Yacoub, J. E. V. Bautista, and L. G. de Rezende Guedes, "On higher order statistics of the Nakagami-m distribution," *IEEE Trans. Veh. Technol.*, vol. 48, no. 3, pp. 790–794, May 1999.

[Yip95] K.-W. Yip and T.-S. Ng, "Efficient simulation of digital transmission over WSSUS channels," *IEEE Trans. Commun.*, vol. 43, no. 12, pp. 2907–2913, Dec. 1995.

[You52] W. R. Young, "Comparison of mobile radio transmission at 150, 450, 900, and 3700 MHz," *Bell Syst. Tech. J.*, vol. 31, pp. 1068–1085, Nov. 1952.

[You96] N. Youssef, T. Munakata, and M. Takeda, "Fade statistics in Nakagami fading environments," in *Proc. IEEE 4th Int. Symp. on Spread Spectrum Techniques & Applications, ISSSTA'96*, Mayence, Germany, Sep. 1996, pp. 1244–1247.

[Zol93] E. Zollinger, *Eigenschaften von Funkübertragungsstrecken in Gebäuden.* Ph.D. dissertation, Swiss Federal Institute of Technology Zurich, Zurich, Switzerland, 1993.

[Zur92] R. Zurmühl and S. Falk, *Matrizen und ihre Anwendungen — 1. Grundlagen.* Berlin: Springer, 6th ed., 1992.

INDEX

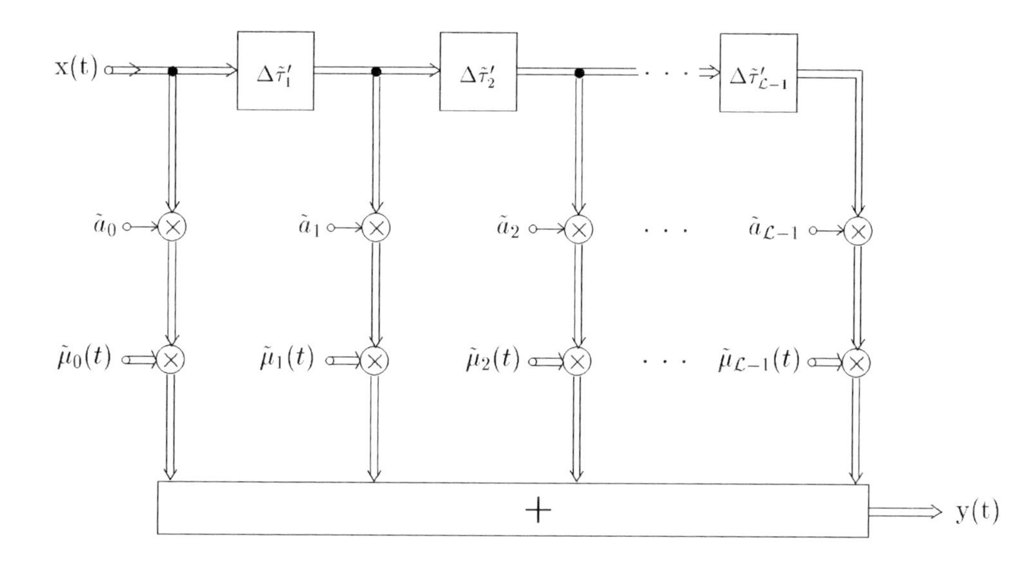

Figure 7.11: Deterministic simulation model for a frequency-selective mobile radio channel in the equivalent complex baseband.

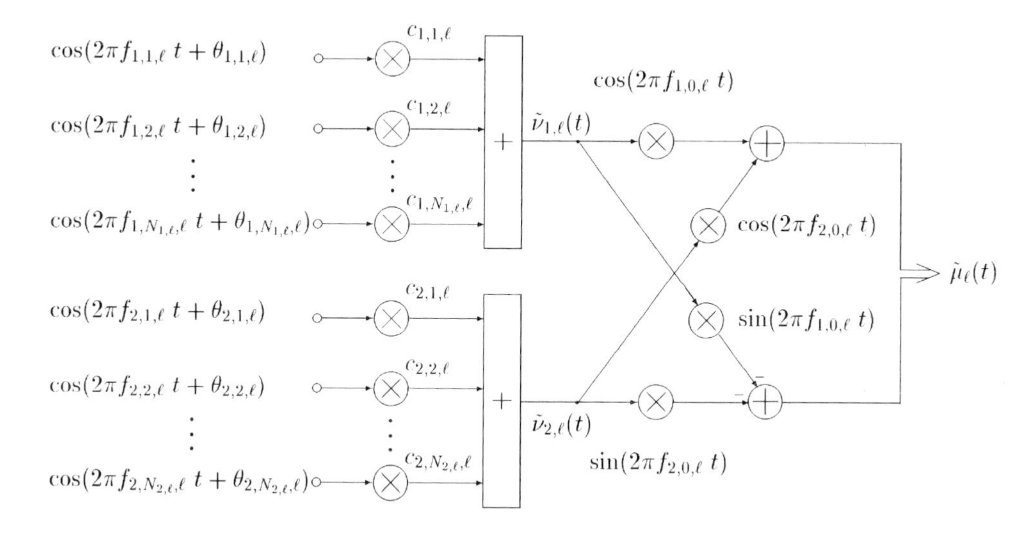

Figure 7.16: Simulation model for complex deterministic Gaussian processes $\tilde{\mu}_\ell(t)$ by using the frequency-shifted Gaussian power spectral densities according to COST 207 [see Table 7.2].

Errata

Mobile Fading Channels

Edited by
Matthias Pätzold

0 471 49549 2

Please note that figures 7.10, 7.11 and 7.16 were printed incorrectly. The correct versions are printed below.

The publishers very much regret this error and hope that this clarification will help to avoid any confusion.

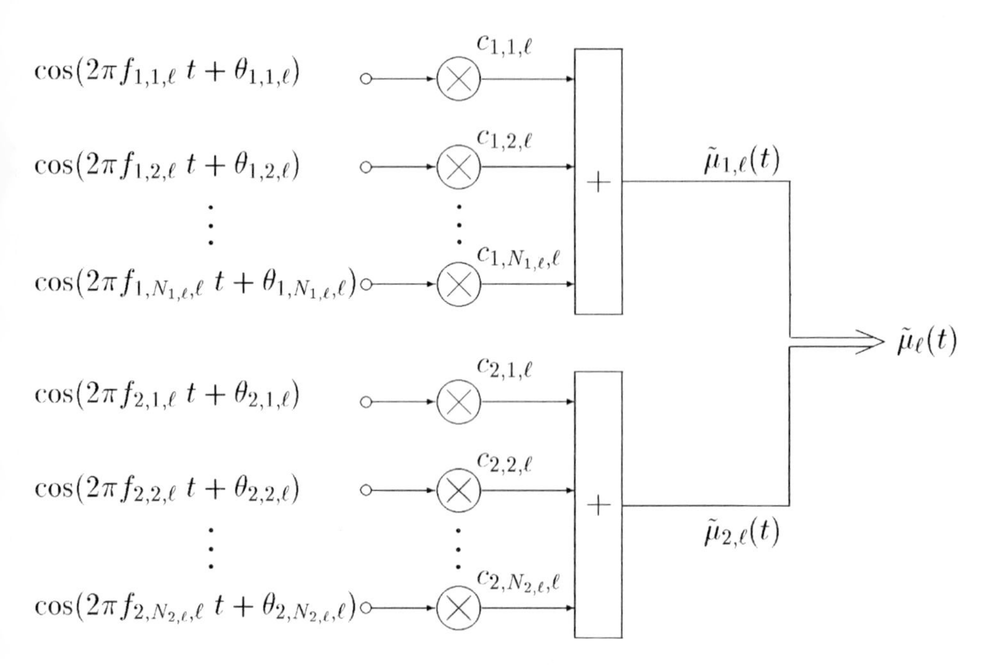

Figure 7.10: Simulation model for complex deterministic Gaussian processes $\tilde{\mu}_e(t)$

DATE DUE
